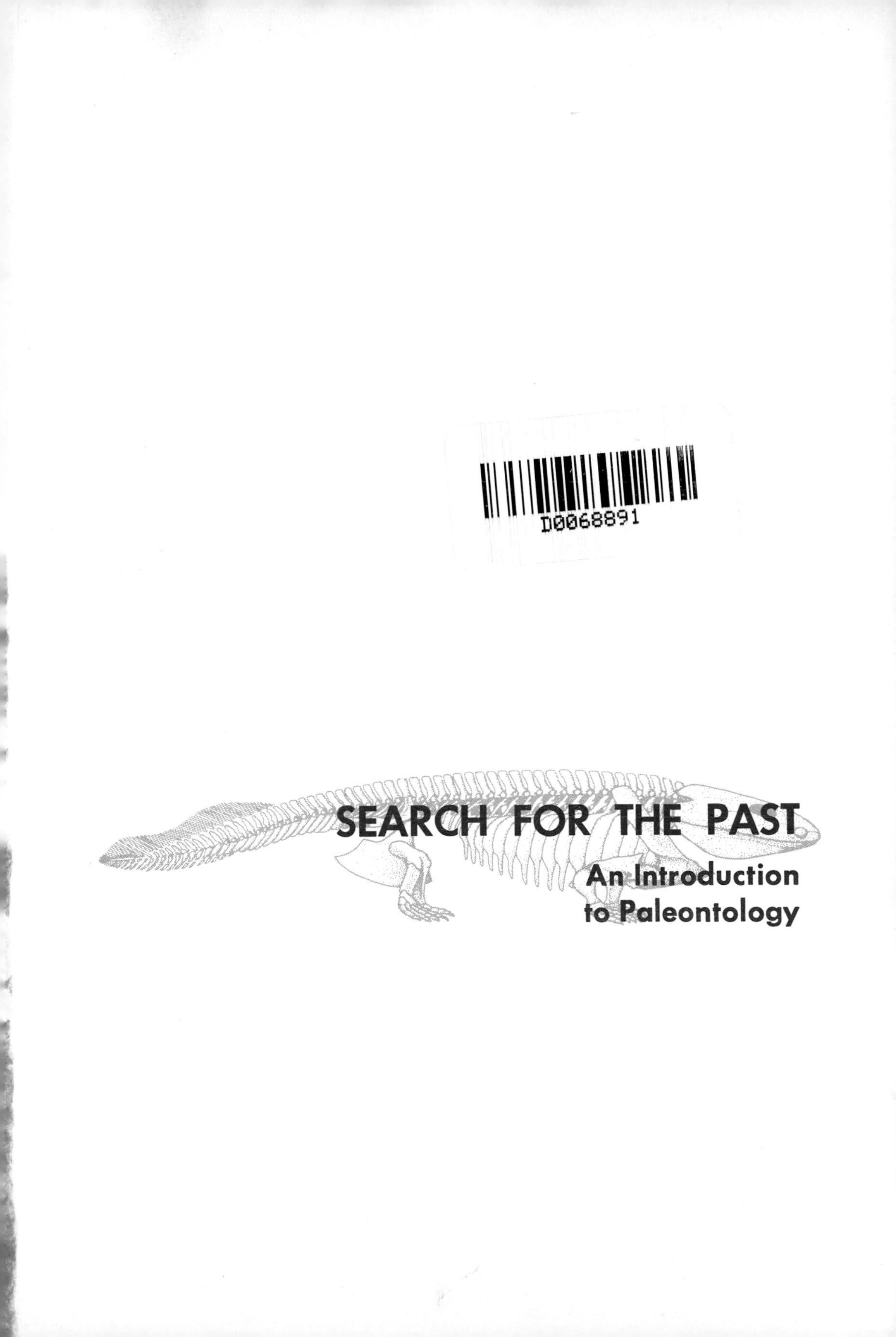

SEARCH FOR THE PAST

An Introduction to Paleontology

James R. Beerbower
Department of Geology
McMaster University

SEARCH FOR

PRENTICE-HALL, INC., Englewood Cliffs, N.J.

THE PAST

An Introduction to Paleontology

SECOND EDITION

SEARCH FOR THE PAST
An Introduction to Paleontology, 2nd ed.
by James R. Beerbower

PRENTICE-HALL INTERNATIONAL, INC., London
PRENTICE-HALL OF AUSTRALIA, PTY. LTD., Sydney
PRENTICE-HALL OF CANADA, LTD., Toronto
PRENTICE-HALL OF INDIA PRIVATE LTD., New Delhi
PRENTICE-HALL OF JAPAN, INC., Tokyo

Library of Congress Catalog Card Number 68-18060
Printed in the United States of America

Current printing (last digit):

10 9 8 7 6 5

Preface

The first edition of this book was the somewhat defiant assertion of a young man and of what seemed the culmination of a revolution in paleontology. The revolutionary ardor cools, however, and one begins to wonder if it really was tiger's milk we drank as graduate students. The result: A new edition, polished, perfected, updated, consolidated—and middle-aged. I can only hope in this strait to assert that the revolution has not yet taken place, and that I will follow the noble example from *Alice in Wonderland:*

> "You are old, Father William," the young man said
> "And your hair has become very white;
> "And yet you incessantly stand on your head—
> Do you think, at your age, it is right?"

At this moment, most of the vitality of paleontology seems to lie in the geochemical (shell mineralogy and organic chemistry) and geologic (sedimentary petrology and stratigraphic model building) aspects. We teeter on the knife edge of the computer revolution in taxonomy. Theoretical morphology—as opposed to descriptive efforts—is reviving. Paleoecology, I fear, is marking time, asserting again and again that ancient organisms grew, lived, and died just as do modern ones. The disciples of this latter field clearly need to define new questions and to seek new techniques from geochemistry and from population studies (paleomicroecology). I have been unable to resist the temptation to add several sections to the text. The later chapters—on the systematics, morphology, ecology, and evolution of the major invertebrate taxa—follow with few exceptions the Treatise on Invertebrate Paleontology. The material and presentation are somewhat more advanced than in the first edition; this, hopefully, anticipates an improvement in the caliber and preparedness of undergraduate students.

I have very much appreciated the interest and enthusiasm raised by the first edition; I hope I do not disappoint with this revision. If there is to be another edition it will depend on you to make this one obsolete.

Acknowledgments

I would hope first that this book might be a worthy tribute to the late James Dyson who through my twelve years at Lafayette College served as elder brother, teacher, and inspiration, and who as much as anyone made the first edition possible. I owe a special debt also to Everett Clair Olson who as master to my apprenticeship showed me what was possible.

I must again acknowledge the assistance of Ralph Johnson and Marvin Weller with the first edition. Many of the changes—hopefully, improvements—in this version arose from their comments and suggestions on the earlier manuscript. I also very much appreciate the useful suggestions and criticisms of the first edition supplied by many of my colleagues. Robert Sloan, Archie MacAlpin, and G. Z. Foldvary have been particularly helpful in this regard.

A. R. Palmer, Keith Young, and Ernest Lundelius read the manuscript in its entirety and contributed generously from their knowledge and insight to its improvement. David Raup, Everett Olson, Lee McAlester, A. R. Palmer, James Brower, Robert Bader, Nicholas Hotton, Karl Waage, Keith Young, John Wilson, Ernest Lundelius, David Kitts, Francis Stehli, Porter Kier, William Oliver, Gerd Westermann, Robert Denison, Zeddie Bowen, and Laing Ferguson have read one or more sections of the manuscript and have lent me the benefit of their specialized knowledge in the various fields of paleontology. This revision would have been impossible without their help; I trust that it justifies their efforts and does not fall too far below the value of their suggestions.

Gay Parsons Walker prepared the new illustrations; she assimilated my crude sketches and suggestions and created from them something beyond my power to conceive. My wife, Honesty, has contributed in a variety of ways as patient audience, critic, and enthusiast. She is also chiefly responsible for the index, and she struggled with my scratching to distill a clean manuscript. Judy Harriss typed most of one draft; Dianna Edwards also contributed in sundry ways to manuscript preparation. Dan Serebrakian of Prentice-Hall by some miracle made a book of it.

To all, my thanks.

JAMES R. BEERBOWER

Table of contents

6

Evolution 133

7

Patterns of evolution 162

8

Fossils and stratigraphy: Adventures in space and time 187

13

Segments: The Annelida 298

14

Jointed limbs: The Arthropoda 307

15

Some that crawled: Chitons, snails, and clams 333

16

To correlate: The Cephalopoda 360

17

Arms, stems, and spines: The Echinodermata 377

18

Brothers under the skin: Protochordata 421

SEARCH FOR THE PAST

An Introduction to Paleontology

1 Fossils and science

Paleontology is the study of fossils. Fossils are the traces of ancient life. There is a book in thirteen words—but only a trace of a book, and pretty well petrified at that. Explanation of this definition requires ten thousand times thirteen words. You must visualize Aristotle turning a fossil shell in his hands, and a contemporary scientist with calipers, computer, and a mass spectrometer, examining the same shell. And from that you must visualize the living beast on the sea floor, five million years ancient.

Paleontology comprises fossils and men and ideas about fossils. A mammoth in the deep freeze of the Siberian tundra and the shells of clams in the shales of western New York are parts of this story; a paleontologist who studies the microscopic fossils in well cuttings is another part; the assignment of a relative age to coal beds in West Virginia and the interpretation of the origin and evolution of complex terrestrial communities in the Permian of Texas also build "paleontology."

But this may be clearer if you will examine with me a slab of rock chiseled from a layer of shale (Figure 1-1). The first glance shows, simply,

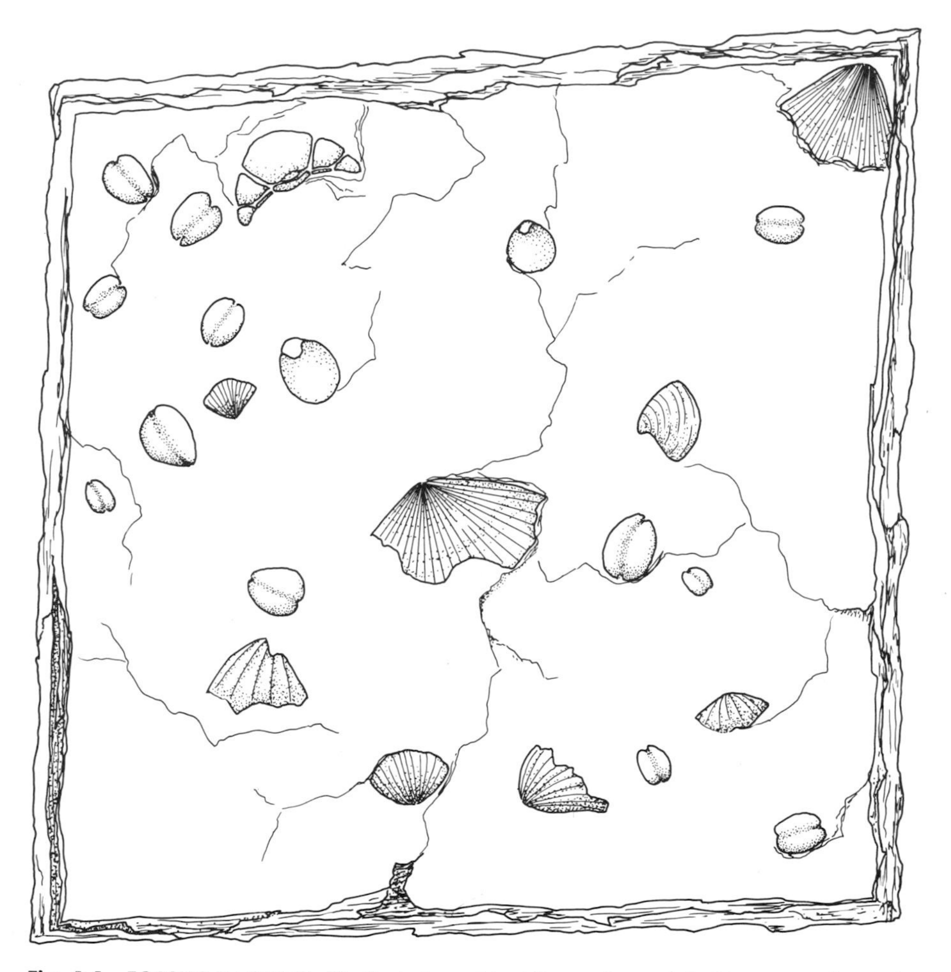

Fig. 1-1. FOSSILS IN SHALE. Vertical view of bedding surface; slab 10 cm on side. Ludlowville Shale; Middle Devonian, Cazenovia Cr., Erie Co., N.Y.

odd pebbles, but the oddity of their shape and of their presence in an otherwise fine-grained rock demands closer examination. At the upper border, toward the left, is a vaguely ovoid shape with a striking pattern of bumps, ridges, and grooves.* Let's enlarge this one as Figure 1-2A. A little study indicates that the bumps are nonrandom—that there is a symmetry (Figure 1-2B). Continued examination reveals further detailed patterns, and we can draw a picture of our "pebble" as Figure 1-2C. Since other blocks of the shale contain similar objects, let's call this kind of "pebble" a *trilobite.* Two things should be apparent: 1) the symmetry and order in form are quite different from an inorganic crystal, a sedi-

* If you have some knowledge of fossils, you will immediately tag it as "trilobite," but this jumps ahead of the story—one can describe its form without this name.

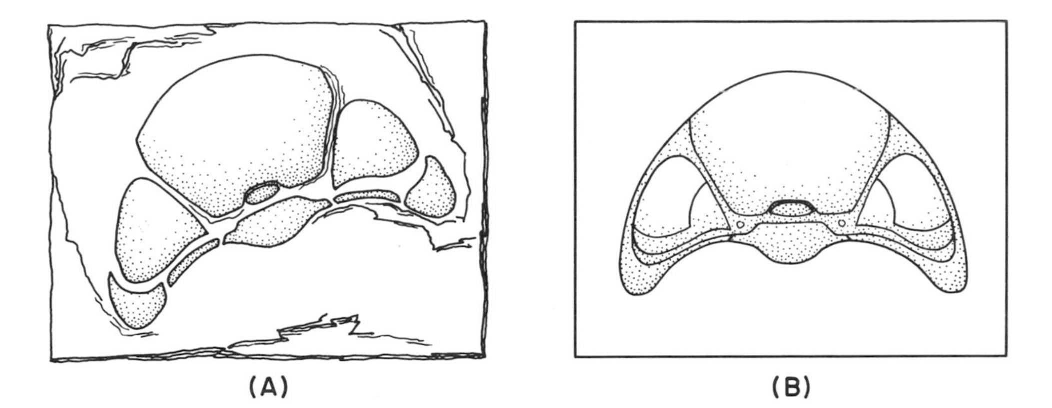

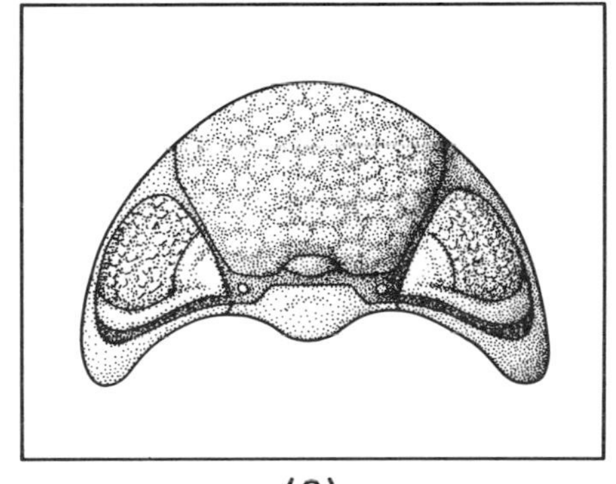

Fig. 1-2. OBSERVATION OF FOSSILS. **(A)** View of single fossil from shale block of Fig. 1-1. No detail or orientation. **(B)** Same fossil, removed from shale and oriented. **(C)** Same fossil with details shaded.

mentary structure or an aggregate of grains formed during conversion of sediment to rock and 2) the pattern is strikingly similar to that in the head of some modern animals. We can now assert that we're "doing paleontology"; we're looking at a "fossil." At the same time we've moved two (subjective) steps away from the jumble of shadows in Figure 1-1, the first step to a description (Figure 1-2C) and the second to a connection with life. In so doing, we've tagged this particular "pebble" with some labels that may have scientific as well as aesthetic valuc.

What other labels can we add? The fossil has a position within the block, on the bedding surface so many millimeters from the upper left (northwest corner). The block in turn has a position on a rock exposure; the exposure has a position in a series of rock layers—375 cm. below a distinctive limestone within a previously defined formation and has a geographic location—Cazenovia Creek, Erie County, New York. The fossil can be tagged with the characteristics of the rock within which it rests: a gray, calcareous shale. It also may bear the label, "Fossil *B*, one millimeter to the left, Fossil *C*, five centimeters to the right."

But this was once an animal. How did it grow? How did it live? Why do we find this kind of animal at this level in a sequence of layers? Clearly, any fossil demands an explanation, as well as a description and a statement of its relationships in form and in space. These three actions, description, relation, and explanation, are the skeleton of paleontology (Weller, 1965).

But the skeleton without the flesh of life is dead—and deadly to our interest. Ink scribbled in a notebook, measurement, chemical analyses, all the paraphernalia of precise observation and all the hours devoted to comparison with pictures and descriptions in reference books or with identified specimens won't recreate the living animal. The paleontologist applies his imagination and reason to reconstruct the living past from the traces of the once-living. From observations, imagination suggests certain hypotheses to account for their nature, and reason judges, from a set of given rules, which hypothesis is most probable.

The characteristics of fossils

Fossils resemble, to a greater or lesser degree, modern organisms, or rather parts of those organisms. Fossils are found only in sedimentary rocks (some of the rare exceptions are discussed below), and these rocks resemble modern sediments in which similar organisms are sometimes buried. Different kinds of fossils are associated in the rocks in the same way in which different organisms are now associated in life or in burial.

Many fossils resemble modern organisms in composition and form, some even in microscopic detail. The shell of a clam living in a tidal channel is very like the shells cast up by waves along the beach and like the shells buried by sand high on the beach. In turn, similar shells occur in the partly consolidated rocks from a hill beyond the shore, and still others—though perhaps less similar—in a dense sandstone on a distant mountain (Figure 1-3). Or a section of a fossil beneath a microscope may show the fine details of shell structure.

Other structures found in rocks, however, are less obviously fossils, and some that originated by inorganic processes (crystallization, compaction, etc.) suggest animal or plant structures. There is no sure way of identifying these anomalous objects; but as a rule of thumb, inorganic structures are no more complicated than crystal growths, vary greatly in general form or detail (or both), and may cut across the sedimentary beds or show other evidence of later formation.

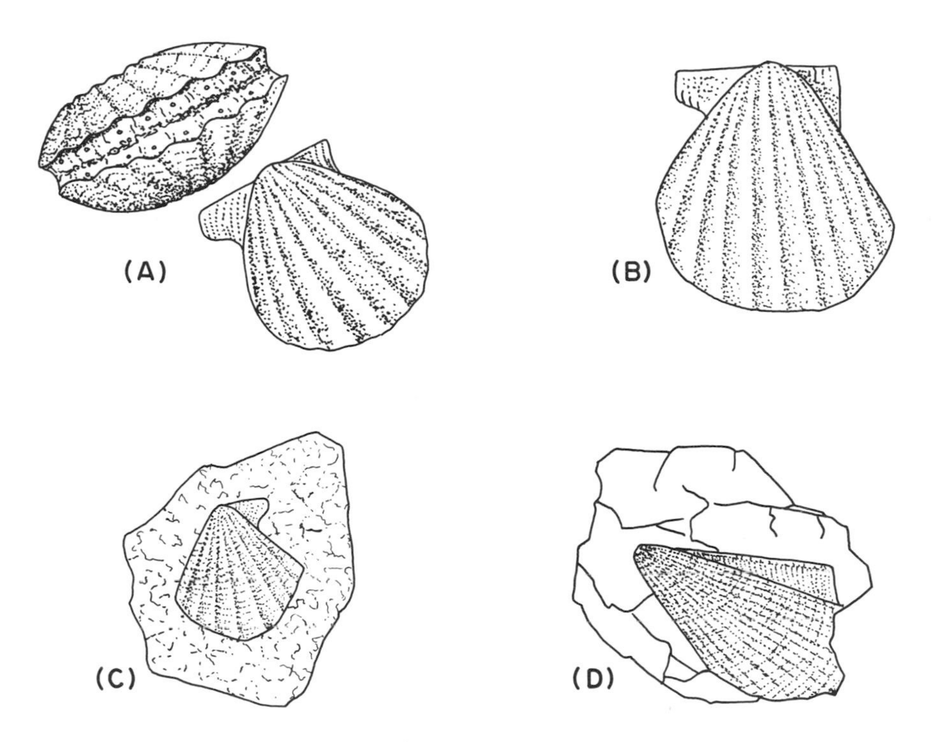

Fig. 1-3. SIMILARITY OF RECENT ANIMALS AND FOSSILS. **(A)** Two recent clams, lower one lying on one side with one valve of shell visible; the other, edge-on, with both valves and some of the viscera shown. **(B)** Shell of a similar clam collected from ocean bottom sediments. **(C)** Somewhat similar shell from late Mesozoic rocks. **(D)** Shell from middle Paleozoic rocks. Note that the shell has "ribs" and an "ear" like **(A)**, **(B)**, and **(C)**, although its shape is considerably different.

If similarity of fossil to recent form extends to microscopic features, you might expect equal similarity in chemical composition. Decay, oxidation, reduction, solution, and the activities of other organisms, however, destroy or alter the normal composition before, during, or after burial. The organic compounds are least stable, and most are preserved only in special circumstances (the frozen mammoths of Siberia) or as traces in shell or bone. The inorganic compounds (calcium carbonate, silica, etc.) from which some animals construct their skeletons resist destruction or alteration but will yield to intense chemical activity. Because of this attrition only a small proportion of the animals living at any particular time were preserved as fossils, and most of these in fragmental and altered form.

The trilobite discussed in the first part of this chapter illustrates these characteristics: It resembles the skeleton of modern animals in form and in composition. It also illustrates the association with rock type and other fossils. One would expect to find fossils in sedimentary rocks, because these are deposited on the earth's surface, but the association goes further than that. For just as trilobites of this kind are found only in certain sandstones, shales, and limestones, so any one kind of fossil is limited to certain rock types. Rock characteristics reflect the environment at the place of deposition and the life environment of the fossil—or at least the environment into which it was washed for burial. The structure of the fossil also tells something of the place in which the organism lived, so that interpretation of environment based on rock association becomes a check against interpretation based on fossils. The two usually come out quite close, and those that fail to jibe—like the armored terrestrial dinosaur in marine limestones—can be explained by transportation after death.

The association of different kinds of fossils in the same rocks is also similar to that anticipated from a knowledge of modern animal and plant associations. Again anomalies exist, but these usually have a logical explanation. Coral reef associations, freshwater lake associations, shallow marine associations—all persist in the fossil record.

But granite gneiss associations? Absurd, of course, but some igneous and metamorphic rocks do contain fossils. Lava flows cover trees and animals, and, even if the flesh and skeleton are burnt away, the form may be preserved as a mold in the cooling lava. Likewise, metamorphism of sedimentary rock tends to destroy fossils by recrystallization and by distortion, but some metamorphic rocks do retain traces of fossils (Neuman, 1965).

Fossils and the living world

The question I'd now ask rhetorically is whether these observations are consistent with the original definition of fossils as the traces of organic activity. You should answer, "Yes, obviously. They resemble modern organisms in form, composition, and occurrence." Particularly, certain forms, structures, and chemical compounds exist at present only as parts of organisms or as results of organic activity. Therefore their occurrence in rocks implies the existence of organisms in the past. If this interpretation is correct—and it could be disproved only by finding all of these features developing in association without organic intervention—a further set of principles must be sought. These principles should extend the hypothesis and permit its application to detailed problems of paleontology. If someone decided that fossils were formed under the influence of the stars (as suggested by some late medieval philosophers), he, as a "paleoastrologer," would seek rules to relate the character of fossils to the movement of the planets. He might assume certain fossil shapes were the results of specific conjunctions: that the heart urchin was associated with

Venus, perhaps. Reason would then have rules by which it might constrain imagination. Just so, paleontologists need a set of rules.

Uniformitarianism

If fossils derive from living things, one reasonably assumes that life processes were the same in the past as they are at present.* The observations summarized above imply this concept, although a different set of processes might conceivably give rise to these same structures and association. The test is to see whether this first principle of "biological uniformitarianism" serves to predict in a simple manner the results of further observations. Many of the following chapters will be concerned directly or indirectly with this testing.

But biological uniformitarianism forms only a part of a broader concept of uniformitarianism. Presumably rivers flowed to the sea in the Cambrian as they do now—flowed, cut valleys, and built deltas. Rain fell, rocks weathered, and waves broke along the shore. Geologists assume that the same physical and biological processes occurred in the past as at present, and they arrive at the familiar dictum: The present is the key to the past. Although only an assumption, it agrees with observation and serves usefully in the interpretation of earth history.

Rivers, however, flow at different rates and carry different amounts and kinds of sediment. Rocks weather rapidly or slowly and in different ways as proportions of moisture and temperature vary. Biological processes vary in their rates and in their interactions. Probably the rates and combinations were "about" the same, but "about" covers a wide range of variation. Further, man's knowledge of earth (and biological) processes is limited by his short period of observation. If you lived so rapidly that a minute spanned your life, could you say that the hour hand of the clock moved? Or that the alarm mechanism had significance if you did not hear it ring? In the same way, an event that occurred but once in a billion years might be inexplicable, or a process that acted slowly over that same period might be unrecognized.

Perhaps one might better define uniformitarianism as the existence of a single set of physical and chemical processes throughout geologic time. These "eternal" processes are represented, albeit imperfectly, by the physical and chemical "laws" derived by modern science, e.g., the second law of thermodynamics. These laws Simpson (1963) calls *universal* or *immanent* since the restrictions in their definition free them of place and time. Geology is concerned with the operation (*configuration*) of these laws in a single system, the earth, during a particular time interval, and so, to follow

* Since fossils are recognized by their "biologic" attributes, this statement approaches a tautology.

Gould (1965), uniformitarianism is merely a special statement of inductive logic. The problem is to determine rates and combinations; uniformity of rates forms the simplest hypothesis and can be accepted in the initial formulation of a theory because of simplicity and economy in testing. Uniformity of configuration (as opposed to uniformitarianism of process), however, is only a working principle and not an invariable rule or a scientific law.

Time, the perjurer

Universal processes acting through billions of years have shaped the earth and its life. The paleontologist concerns himself with the history of that shaping. Yet as the chisel of a sculptor obliterates in succeeding blows its earlier marks, so these processes obliterate the traces of their own action, leaving for the paleontologist only a small sample of the past. Properly, he desires a complete reconstruction: a jungle from which he may hear the roar of lions and a coral reef through which undulate varicolored fish. Further, as a biologist, he wishes to know the details of the living system in some exact, quantitative fashion. How does he bridge the gap between a fossil and the living world which it represents?

If he had a time machine, he might move back through successive moments to take measurements, stop along some Devonian shore to count the trilobites, note what they eat, observe their growth, and analyze the water in which they live. These measurements, although only a few of the many possible, would serve to describe a living system. If he can make some of the same measurements on the fossils, they will serve in the same fashion. Then, if the relation between the measures and the dynamics of the living system are understood, he has a reconstruction.

Obviously, most of the desirable measurements cannot be made on fossils. Of these trilobites, perhaps one in a hundred thousand is fossilized, and when fossilized, only the skeleton of a complex animal is left. Likewise lost are the soft-bodied creatures that lived beside the trilobite, on some of which it fed. Little evidence is left in the rocks of currents or temperature or salinity of the water. But some things may be undisturbed, like the proportion of one kind of trilobite to another and the type of bottom sediment. The paleontologist searches for such meaningful measurements.

Please note, though, that these measurements are made either on a very small system, e.g., a single trilobite, or on a very small sample of a large and complex system, e.g., a Devonian mud-bottom marine community. Even in the simple description on pp. 2 and 3, we selected certain features from the total bundle of sensory impressions—look again at 1-2A-C. This particular specimen was only one of many exposed in this outcrop. Was it picked at random or was it chosen for some special feature of size or appearance? Obviously, a "selected" specimen represents the others in the outcrop only in a very special and distorted fashion. But even if the sample is taken at random, it is still far from a random sample of the biological system (Figure 1-4). For example, clams and

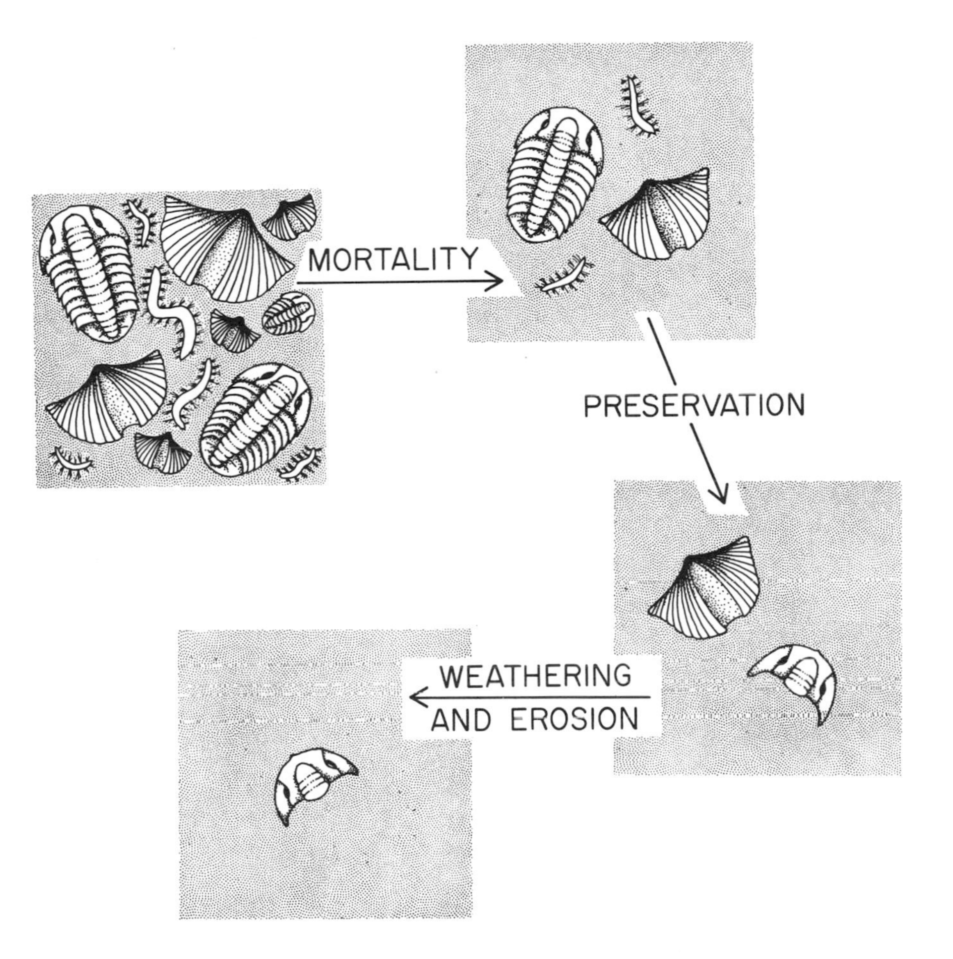

Fig. 1-4. THE FOSSIL SAMPLE AS A RESULT OF SELECTIVE MORTALITY AND PRESERVATION.

snails appear to be more abundant in limestone concretions taken from the same bed than in the shale itself. Is this an aspect of their biology? Or is it more probably the differential preservation of skeletons composed originally of a comparatively unstable mineral, aragonite? Did the animals live in patches, as they now occur, or were they heaped together by wave and current?

Because of these difficulties, a paleontologist cannot claim the same degree of certainty for his conclusions as does a biologist, nor can he reach any conclusions at all about some problems. With this reservation, the range and validity of fossil interpretations vary with the number of different kinds of observations, with the "fairness" of the sample, with its size, and with the availability of information on modern biological systems. An increasing number of independent observations permits joint estimates of the same characteristic of the biological system, and, in turn, a larger, unbiased sample and more accurate knowledge of the interrelation of measures in biological systems increase the accuracy of those estimates.

Answers in search of questions

The study of ancient biological systems begins with the brute fact of fossils, with their form and composition, with their orientation, with the associations of rock and fossil, and with the distribution and variation of these properties in space. The answers are, in effect, given and fixed; the key problem is to ask the proper kinds of questions. This problem is still far from a complete solution, but one can define the general classes of questions that may be asked. In essence, these come down to "who lived," as revealed in form and variation; "how they lived," as implied in form, composition, orientation, and association; "where they lived," as indicated by geographic distribution; "when they lived," as related to position in the rock sequence; and finally, simply, and inexorably, "whither and why" this pattern in space and time. "Who," "how," "where," and "when" are chiefly of interest to geologists; "why" and "whither" are biological problems.

A paleontologist answers such questions by explanations—for example, as he explains the flattened body of a trilobite in terms of life on a sea floor. Or he may explain it as a consequence of hereditary material determining growth under specific external conditions. Or as a sample from a pool of hereditary material selected through evolutionary sequence. The spectrum of explanation runs from universal process—support and protection on the sea bottom—to configurational history—the changes of hereditary material through many generations. The universal and the configurational are inextricably linked; in this case by a process, natural selection, operating in a series of unique environments.

Explanations derive from two kinds of statements about the natural world parallel to the universal and configurational modes. The flattened body of a trilobite has certain invariable functional properties, e.g., its hydrodynamic properties in a current. These are universal statements; they hold true for a fish as well as for a trilobite, for the Cambrian as well as for the Pleistocene; for an insect in a wind storm as well as for a snail in tidal swash. In general, such statements are certain, and explanations derived from them equally certain. The explanations, however, are likely to be imprecise, e.g., a trilobite was stable in a current, but the critical velocity can not be specified unless the density of the body can be determined.

Opposed to these are other statements that are certain if and only if certain conditions hold, i.e., they are "normally" but not invariably true, and therefore are called *normic* by some philosophers of science. Flattened bodies normally occur in marine animals that live and feed on the bottom—but not all flattened animals are bottom inhabitants nor are all bottom inhabitants flattened. Specific conditions about locomotion, method of feeding, protection, etc., are necessary to make the statement certain. Normic statements can obviously yield rather precise explanations, but as precision increases certainty may decrease because the number of auxiliary conditions increases.

The meaning of the past

Paleontologists thus produce interpretations that are rarely both precise and certain. What, then, is the value of paleontology, beyond satisfying the "monkey curiosity" of the paleontologists?

It is, first, a Chinese box, concealing different meanings for successive observers. For over 150 years, since Fuchsel, Smith, Cuvier, and Brongniart, it has assisted stratigraphers in determining relative age of rock layers. The pioneers found, empirically, that fossils displayed a definite succession of types within a sedimentary rock sequence and that this succession remained constant wherever found. Subsequently, fossils were discovered to characterize certain ancient environments and therefore could be used to determine past geographic relationships and the physical evolution of the earth's surface and climate.

The value of paleontology to its other related discipline, biology, is less fundamental but exists none the less. Since life has a historical context represented by paleontology, no understanding of a modern biological system is complete until its historical development is known. Further, paleontologists can discern certain general patterns of evolution from this historical record. Finally, the relation of evolutionary changes to changes within the biological system and its environment suggest problems in evolutionary dynamics and solutions to these problems.

Beyond this, however, paleontology is an adventure, an exploration into the jungles of the past. Because it is our own past and our own history, it forms for each one of us an immediate, personal mystery. In solving that mystery we see ourselves in the perspective of time.

REFERENCES

Gould, S. J. 1965. "Is Uniformitarianism Necessary?," *American Jour. of Science*, vol. 263, pp. 223-228.

Kitts, David. 1963. "The Theory of Geology," in C. C. Albritton, *The Fabric of Geology*. Reading, Mass.: Addison-Wesley. Pp. 49-68.

Neuman, R. B. 1965. "Collecting in Metamorphic Rocks," in: B. Kummel and D. Raup, *Handbook of Paleontological Techniques*. San Francisco: W. H. Freeman. Pp. 159-163.

Newell, N. D. 1959. "Adequacy of the Fossil Record," *Jour. of Paleontology*, vol. 33, pp. 488-499.

Rudwick, M. J. S. 1964. "The Inference of Function from Structure in Fossils," *British Jour. Philos. of Science*, vol. 15, pp. 27-40. A consideration of the explanatory power of paleontology.

Schindewolf, O. H. 1950. *Grundfragen der Paleontologie*. Stuttgart: Schweizerbart. A stimulating, difficult book on some major problems of paleontology.

Shafer, W. 1962. *Aktuo-Paleontologie*. Frankfurt am Main: Verlag Waldemar Kramer. A study of biological traces in modern sediments.

Simpson, G. G. 1963. "Historical Science," in C. C. Albritton, *The Fabric of Geology*. Reading, Mass.: Addison-Wesley. Pp. 24-48.

Weller, J. M. 1960. "The Development of Paleontology," *Jour. of Paleontology*, vol. 34, pp. 1001-1019. A historical review.

———. 1965. "The Status of Paleontology," *Jour. of Paleontology*, vol. 39, pp. 741-772.

2

The shape of yesterday

A reptile crosses the moist sand of a beach, and a trackway imprisons the shape of his foot. That track is a sample of the living animal and relates to the observer at least a small part of its history. A skeleton or a shell is, in the same sense, a sample of the life of which it formed a part. The interpretation of the sample depends upon the observation of fossil shape and structure, upon a knowledge of the relation of form in modern animals and plants to their lives, and finally upon the connection of this knowledge to the fossil sample.

Examining an organism, different people see different things; a painter sees form and color; a biologist may see an organization of parts or a relation of form to the environment; another biologist sees the plant or animal as a consequence of development and growth, and still another as a bundle of chemical processes. Obviously, the full analysis of form is complex. It involves the observation of form and its development during the life of the organism. It requires detection of the adjustment of form to function. It includes a comparison of form, development, and function among many individuals. It implies an explanation of the origins, i.e., evolution, of form in its various aspects. Further, the paleontologist in his study of a fossil must comprehend the effects of decay, burial, diagenesis, deformation, etc., upon his sample of the living past.

What is form?

A student of form (a morphologist) is really involved in the study of organization. Organization is another splendid abstract term, but one can move from this to a clear definition in practice. Consider again the trilobite head illustrated and discussed in Chapter 1 (pp. 2 and 3, Figure 1-2C). One can divide this head into a series of elements laid out as in Figure 2-1A on a Cartesian grid: a large semicirculate plate (*cph*), a large sub-

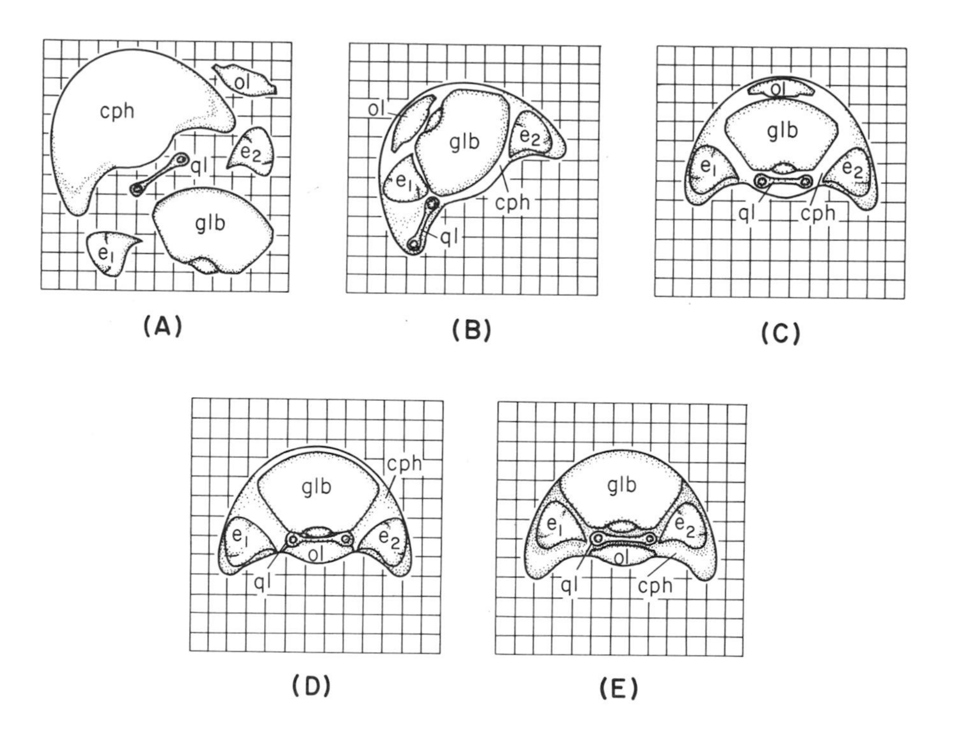

Fig. 2-1. ORGANIZATION OF TRILOBITE HEAD. **(A)** Components of head. **(B)** Hierarchic ordering. **(C)** Symmetrical ordering. **(D)** Polarity ordering. **(E)** Distance ordering.

circular lump (*glb*), two small subcircular lumps (e_1, e_2), a short, dumbbell shaped rod (*gl*), and a short straight rod (*ol*). What kind of geometric constraints will build these blobs back into a trilobite head?

First, try a command that sets up a simple *hierarchy,* i.e., some elements become members of or lie inside the boundaries of other elements. Say that *glb*, e_1, e_2, *gl*, and *ol* are elements of *cph*. This command yields 2-1B. Next, one can arrange these elements on the grid symmetrically so that unpaired elements are perpendicular to and bisected by the *y*-axis, and the paired elements are disposed at equal distances on either side of the *y*-axis and at equal distances above the *x*-axis. If one wished formal rigor, this ordering could all be formulated in terms of analytical geometry; in any case, the result is in Figure 2-1C; the command is a *symmetry* statement. Next, order that *glb* be in front of *gl* and that element in turn in front of *ol*. This ordered arrangement along the *y*-axis can be conveniently called a *polarity* command (Figure 2-1D). Finally, one may instruct that the various elements be spaced at specific distances—a *spacing* command —as illustrated in Figure 2-1E. This last order results in something very like Figure 1-2C, albeit simplified. Raup and Michelson (1965) in somewhat similar fashion have written, for the distribution of points in space, a series of conditions which, when programmed into a computer, generate coiled "shells" like those actually constructed by snails (Figure 2-2).

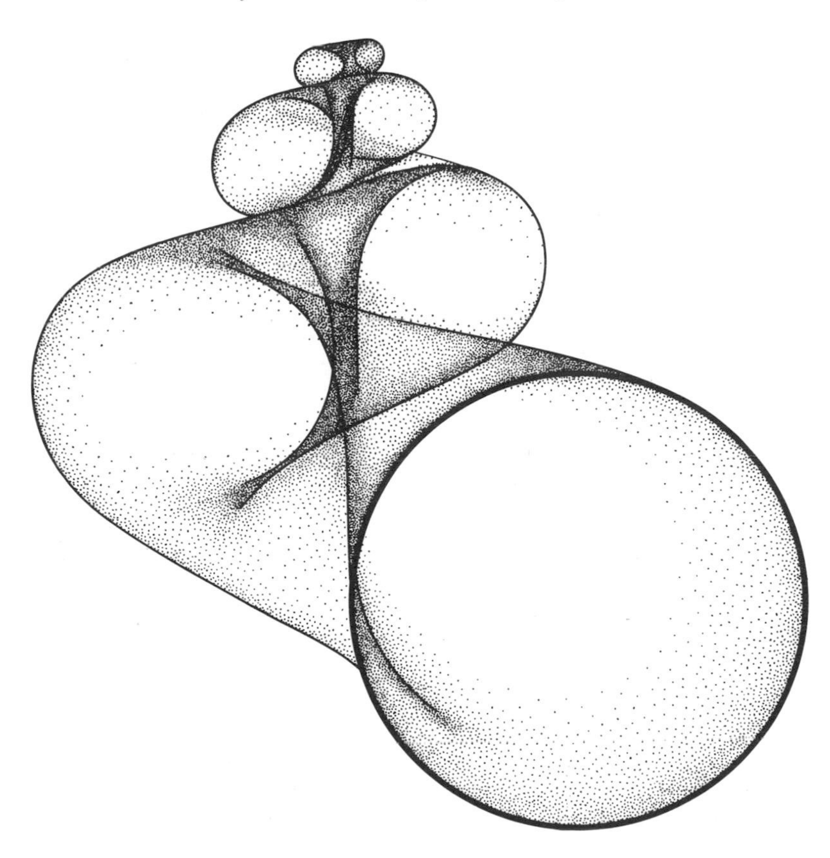

Fig. 2-2. SNAIL SHELL DRAWN BY A COMPUTER. *(After Raup and Michelson.)*

Symmetry and other geometries

The orders or conditions imposed upon the arrangement of organic parts are the basic elements of morphology; it is reasonably obvious when

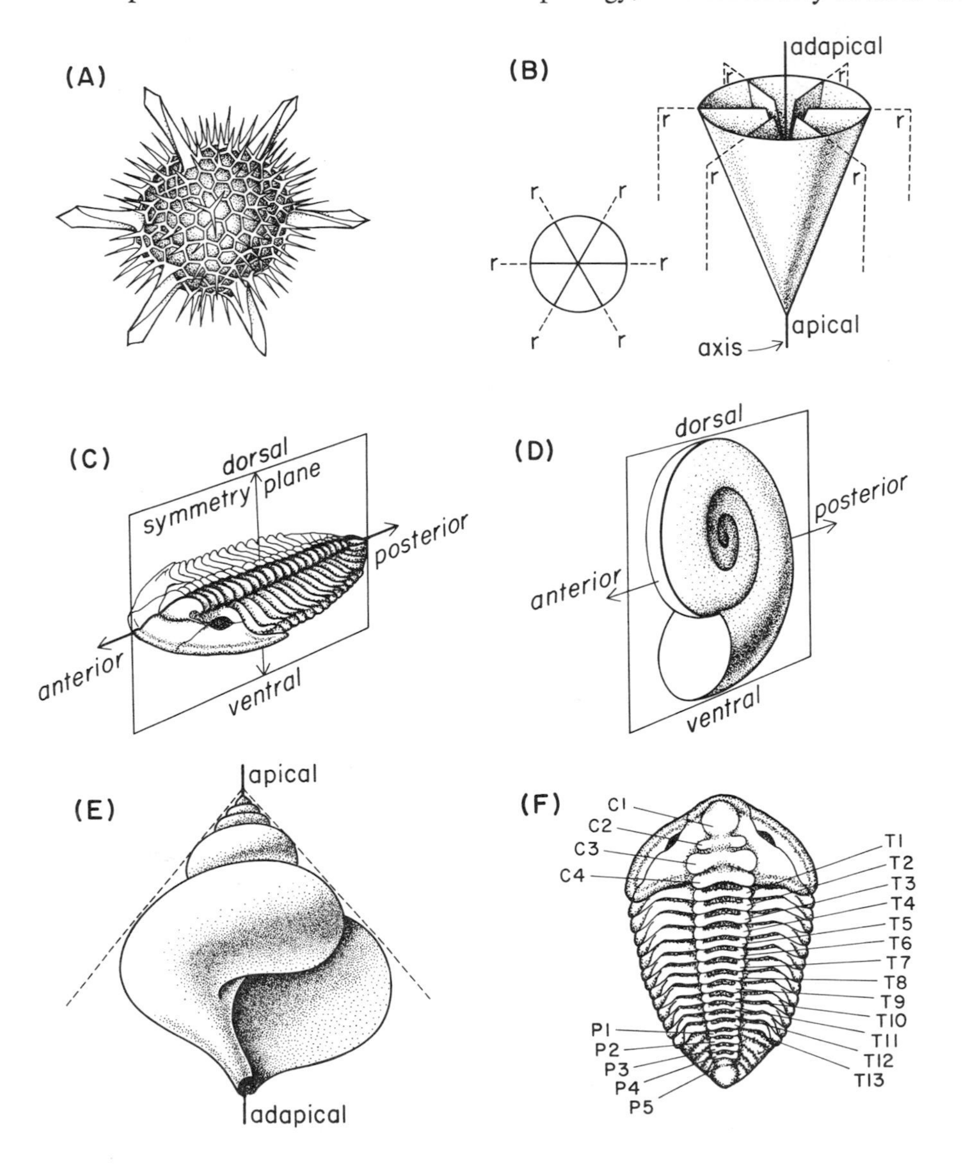

Fig. 2-3. SYMMETRY AND POLARITY. **(A)** Spherical symmetry. **(B)** Polarity and radial symmetry: three symmetry planes (*r*) intersecting in symmetry axis with apical and adapical poles. **(C)** Polarity and bilateral symmetry: single symmetry plane with anterior and posterior poles and dorsal and ventral directions. **(D)** Coiled form with bilateral symmetry (planospiral). **(E)** Form coiled on surface of cone (conispiral). **(F)** Repetition of similar elements perpendicular to symmetry (metamerism).

one looks at plants and animals that a relatively small number of kinds of orders are adequate for their description; the most general of these position parts about an abstract point, line, or plane. The parts of the protozoan skeleton in Figure 2-3A can be ordered in fewest terms if one defines a single element (spine) and repeats it at regular intervals over the surface of a sphere—thus *spherical symmetry.* A coral skeleton (Figure 2-3B) requires that the elements be disposed radially about an *axial line* (*radial symmetry*) and, further, that their radial ends lie on the surface of a cone. The apex of the cone provides a direction and thus polarity for the axial line. These directions, apical (toward) and abapical (away from) correspond to functional directions, abapical defining the position of the mouth, apical the location of the sea floor.

The conical "polarity" of the coral is quite different than the front-to-back (*anterior* to *posterior*) polarity of the trilobite. In trilobites and in the many other *bilaterally symmetrical* animals, the elements are disposed on either side of a single plane of symmetry. This plane bisects the organism; the bisection generates an axis, and the elements on one side of the plane form a mirror image of those on the other (Figure 2-3C). The axis possesses purely functional polarity—anterior toward the mouth and posterior away from it. In the ideal case, the symmetry plane is functionally perpendicular to the surface on which the animal lives and thus a second set of directions is defined as *ventral* (down) and *dorsal* (up). Even in animals that do not have this ideal posture these directions are referred back to the presumed original condition: in ourselves, functional "up" is morphologic "anterior," "front" is "ventral," "down" is "posterior," and "back" is "dorsal."

Not all organic geometry can be resolved into symmetry and polarity.

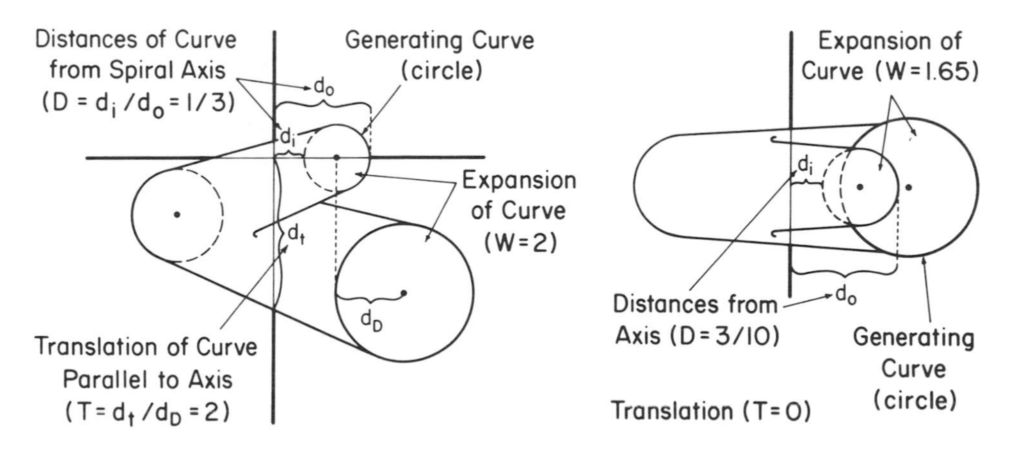

Fig. 2-4. PARAMETERS FOR GENERATION OF SPIRAL FORM. **(A)** Conispiral. **(B)** Planospiral. (*After Raup.*)

For example, in animals such as snails that show spiral growth, the form may be expressed in four components (Raup, 1966), 1) the shape of the shell in cross-section, the *generating curve,* 2) the expansion of that curve during its revolution about the axis of the spiral, 3) the distance of the generating curve from the spiral axis, and 4) shift or translation of the curve parallel to the axis. Thus in Figure 2-4A, the generating curve is a circle, the curve doubles its dimensions in a complete revolution, the distance of the inner margin of the curve from the axis is one-third the distance of the outer margin, and the curve moves parallel to the axis twice as far as it is displaced from it. In Figure 2-4B, the generating curve is again a circle, but it expands by a factor of 1.65 in each revolution; the distance of the inner margin from the axis is three-tenths the distance of the outer margin, and translation parallel to the axis is zero. The first set of commands produces a spiral on a conical surface (*conispiral*); the second a spiral in a plane (*planospiral*). Quite different is the serial repetition of similar (though not necessarily identical) units along a polarity axis (Figure 2-3F); this ordering, called *metamerism,* occurs in trilobites, annelid worms, and along our own backbones.

Each of the major animal groups, corals, vertebrates, snails, etc., is typically characterized by a particular set of geometric commands; the individual kinds of animals within a group play variations on these themes by varying the number, specific form, and spacing of the elements. If one varies the relative size of the *e* elements in the trilobite of Figure 2-1E and their position relative to *glb,* the result is a series of quite distinct kinds of trilobites, as in Figure 2-5. Morphologists subsume these varia-

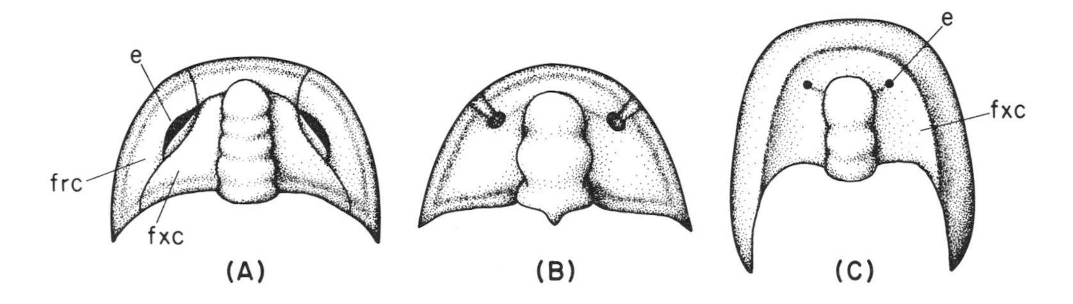

Fig. 2-5. VARIATION IN FORM BY CHANGES IN PROPORTIONS, SPACING, AND HIERARCHY OF ELEMENTS. The element "e" is reduced in size and transposed from *frc* to *fxc*.

tions under the term *topography* and compare the *topographic* relationships in various organisms. Changes in spacing may also place similar elements in different hierarchic assemblages; thus in Figure 2-5A the *e* elements lie with the boundaries of a larger element, *frc*; in Figure 2-5C the *e* elements are placed rather within *fxc.*

How to call one's parts

In the discussion of form to this point I have talked about abstract "elements": e_1, e_2, *glb*, etc. Obviously e_1 is an eye—why bother with this abstract code? The simplest answer is that e_1 is obvious but that *glb* or *ocl* have no counterparts in everyday terminology. Theoretically, different "code" terms could be applied to every plant or animal studied; the description would begin with a definition of terms to be used and proceed to consider their size and geometric arrangement. Thus a second trilobite might be said to comprise elements *gz*, f_1, f_2, *mx*, and *prx*. The element *gz* would be a large subcircular lump, f_1 and f_2 small subcircular lumps, *mx* a dumbbell-shaped rod, etc. Such a system, however, would rapidly become unwieldy as the number of trilobites under study increased. It seems much more sensible to develop "names" for trilobite parts in general as has been done for the parts of our own body. Therefore, $gz = glb$ and one applies a name, *glabella.*

A difficulty lies in determining that *gz* really does approximate *glb*. The problem takes a variety of forms and involves the function of the element, its growth, and its evolutionary history as well as its geometry. At this point I will simply suggest that elements that have a similar position in the geometric matrix should receive the same name. Thus, the large subcircular lump nearest the anterior pole of any trilobite head,* bisected by the symmetry plane, and medial to a pair of subcircular lumps (eyes) will be called the "anterior lobe of the glabella." Each major group of organisms with its characteristic geometry will therefore be described by a distinctive set of names. Paleontologists however may use the same term, septum, for example, in several different groups, but this use does not necessarily mean the same part—the nasal septum of a man resembles the septum of a coral in having a plate-like form, but occupies quite a different position in the geometry of structure.

Restoration of the ruins

The fossils shown in Figure 2-6A and B display a geometric pattern not considered in the preceding discussion. Are these the forms of the living animals or do they result from the stresses imposed during rock formation and deformation? In all features other than the peculiar ellipsoid coiling, Figure 2-6A is nearly identical with other fossils typified by Figure 2-6C, but this elongated spiral occurs in all of the similar forms from the same outcrop and is associated with warping of the plane of bilateral symmetry. Other, noncoiled fossils from the same place also show elongation

* Side-stepping the question of deciding that a particular structure in the rock is a "head" and belongs to a "trilobite."

or warping. If one measures the orientation of elongation for the various fossils in the outcrop a marked parallelism emerges.

In contrast, Figure 2-6B exemplifies in its elongation numerous similar fossils from many localities; bilateral symmetry is preserved; orientation varies randomly in an outcrop; and the fossils associated with Figure 2-6B show no elongation. What can one conclude? Clearly the specimen in Figure 2-6A has been deformed and may yield clues to stress vectors. The specimen in Figure 2-6B, however, grew this way, and its elongation possesses biological significance.

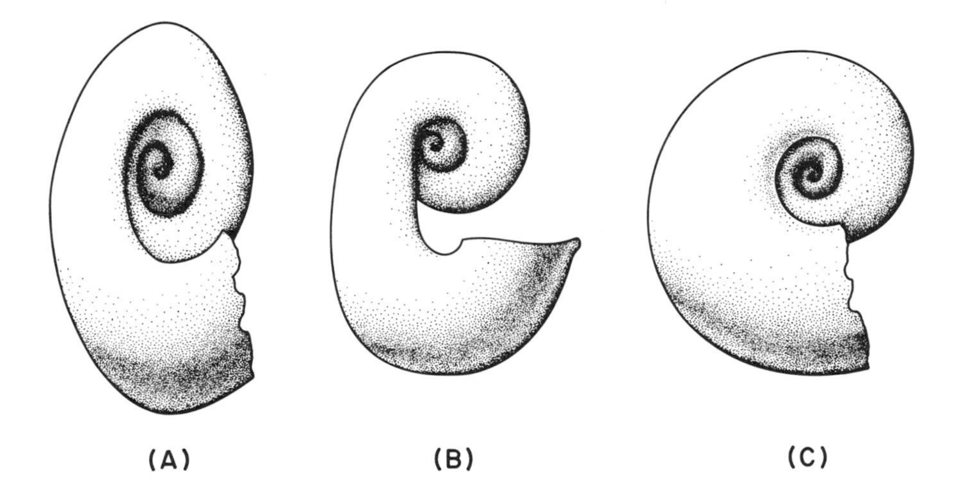

Fig. 2-6. VARIATION IN FORM. **(A)** Distortion from simple spiral curve caused by geologic deformation. **(B)** Distortion from simple spiral caused by variation in growth. **(C)** Example of **(A)** in undistorted form.

Another disturbing factor is the separation of fossil parts, leaf from tree, jaw from skull, tail from trunk. Mesozoic mammals are known almost entirely from isolated teeth and jaw fragments; total skeletal form is unknown. And how does one associate into a single organic structure the various separated fragments of a trilobite? No complete or final answer seems possible, but some reconstruction is reasonable. A knowledge of complete, articulated trilobites requires that one place a series of narrow segments behind the head and that these, in general, become smaller toward the back. The segments will work—move one over another—only if the bumps and hollows are locked together in a specific way. Different kinds of trilobites have different tails—how can one associate a specific type of tail with a specific head if they are not articulated? If tail *A* always occurs in the same layers with head *Q* but only rarely with head *R* then the association of *Q* and *A* is probable.

What if the heads *R* and *Q* are very similar? Could one reconstruct the tail of *R* as being similar? The answer would have to be generally yes —but frequently, no. The classical exception is a group of extinct mammals very similar to horses in the pattern of teeth and skull. The pattern was always (?) associated with hoofed feet. The discovery of articulated skeletons demonstrated, however, that this particular animal had clawed feet like those of bears and that the association is actually normic, as might be expected, and not universal (see p. 11).

And what of parts—nerves, muscles, blood vessels—that are seldom if ever preserved? Once again, the details of preserved parts reveal some clues. Figure 2-7 displays the undersurface of the skull of a Permian

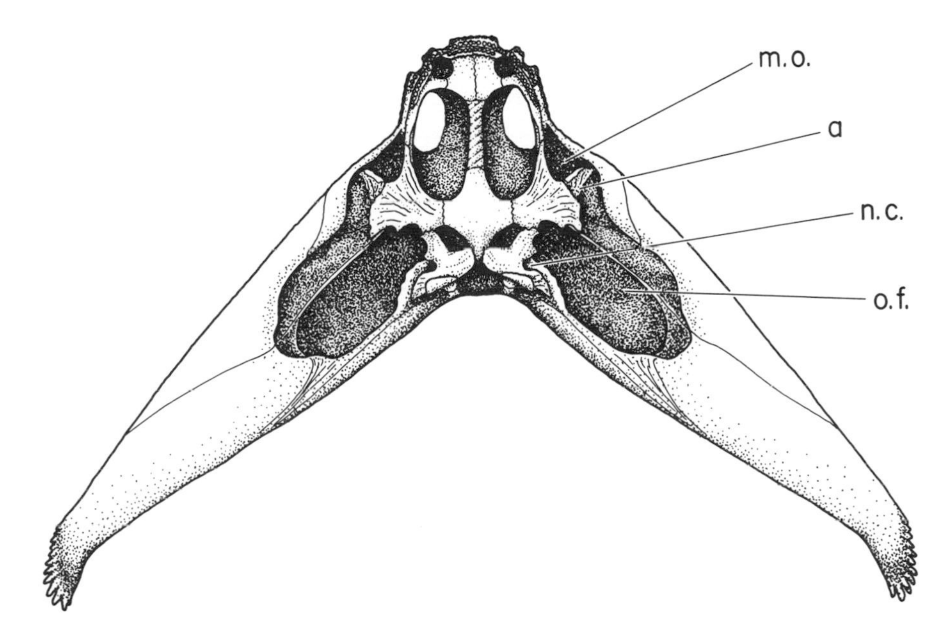

Fig. 2-7. RECONSTRUCTION OF STRUCTURE. *m.o.* = muscle origin; *a* = jaw articulation; *n.c.* = nerve canal; *o.f.* = branchial pouch.

amphibian. The small canal labeled *n.c.* probably provided egress for a large nerve from the brain to the roof of the mouth. At least, a similar canal with similar relations to the various bones serves this function in recent amphibians and reptiles. The canal, however, opens to the side rather than to the back as in living forms and thus indicates a modification in position of the nerve. This shift may be related to the expansion of the throat region to the sides of the head—indicated by the structures labeled *o.f.* The *o.f.* structures probably mark the position of large throat pouches. On the other hand, I cannot determine from the fossil material whether these pouches enclosed gills. I suspect they did, but this is simply guesswork. In contrast to this uncertainty the jaw articulation—labeled *a*—

demonstrates in its facets that stress was directed inward as well as to the back by the muscles closing the jaw. In turn, I reason that a large mass of muscle lay along the inside of the jaw and originated in the area labeled *m.o.* This muscle would pull the jaw up, back, and in against the articular facet. Note that I've reconstructed this nonpreservable organ, a muscle, without reference to muscle location in modern animals. I can then test my reconstruction by observing that modern reptiles and amphibians may have a very powerful internal jaw muscle.

Molecular paleontology

Beyond the order of gross form is the architecture of fine structure—down to the molecular and submolecular levels. An obvious paleontological classification is in terms of preservability: structures likely to be preserved *vs.* those unlikely to be preserved. The boundaries between the classes roughly approximate the boundary between inorganic and organic realms in chemistry. Structures composed principally of inorganic compounds, i.e., minerals like calcite, apatite, etc., are far more likely to be preserved; they are largely restricted to skeletons and to feeding structures where hardness and rigidity are advantageous. Organic skeletal material —chitin and similar compounds—are unlikely to be preserved intact; they may persist in degraded form as in the carbonized skeleton of graptolites or in the carbonized cellulose of leaves and bark. Coal, and presumably oil and gas as well, represents massed, degraded remnants of complicated organic compounds. Surprisingly, however, complex organic compounds have been recorded in shales and, more significantly from a paleontologic viewpoint, in fossil shell and bone. This aspect of molecular paleontology has reached sufficient status to deserve a separate section in my discussion.

Organic compounds of fossils

Research on the organic components of fossils has followed two lines: the study of their stability relations and thermal degradation and, more recently, work on the biochemistry of skeletal material in different kinds of animals. The former work has established a geologic stability scale related apparently to thermal stability. In 1964, Hare and Abelson published some preliminary studies of amino acids in recent mollusc shells. They reported that closely related species differed consistently in a few amino acid ratios but that distantly related species showed greater than tenfold differences. All the cephalopods studied, and those snails and clams regarded as belonging to groups that have evolved little since the origin of molluscs, had ratios marked by a rough equality of glycine and alanine and a predominance (40-50% of total) of these amino acids. The gastropod shells, however, had a distinctly higher percentage of aspartic acid than those of the clams and cephalopods. Shells of groups that evolved later have in contrast much more glycine than alanine.

Grégoire (1950) has used electron microscopy to study the gross organization of the organic components in shells. He found a trabecular organic mat between the layers of aragonite in Cenozoic and Mesozoic mollusc shells. The structure of the mat resembles that in recent molluscs and apparently consists of a similar protein (conchiolin). Differences in physical behavior of the fossil mat from that in recent shells suggest, however, loss of some of the less stable amino acids. Finally, the structure differs between fossil snails and fossil cephalopods in very much the same way that it differs between recent snails and recent cephalopods.

And where does this all lead? The possibilities are intriguing: biochemical "fingerprinting" of major evolutionary lineages and even observation of primary hereditary material in populations long extinct. The problems are equally challenging: chromatographic methodology permits identification of minor traces of organic compounds, but it also identifies minor contaminants; very little is known about the comparative biochemistry of modern skeletal material—much less its relation to hereditary constitution. At present, "molecular paleontologists" are speleologists peering into the entry of a dark and tortuous cavern (for recent work, see Degens, 1967).

Skeletal mineralogy and structure

The inorganic chemistry of fossil material is somewhat better known than that of the organic—probably for this reason its possibilities seem more limited. It is pursued on several levels: isotopic, elemental, mineralogic, and microstructural, grading upward to gross morphology. Grégoire in the work summarized in the preceding section also reported that the arrangement of mineral grains was similar in Mesozoic and recent shells and that the mineral composition, aragonite, was similar. Stehli described (1956) a similar preservation of microstructure and mineralogy in Pennsylvanian fossils, but such finds are rare even in the late Mesozoic.

Studies on the submolecular level are a mixed bag. Organisms serve as entropy pumps, and their inorganic products are in equilibrium only in a narrow biologic system, not with the organism as a whole or its external chemical environment. Given the isotopic or elemental composition of skeletal material, one must know the scope of the system and its nature —whether open or closed. Furthermore, the system will change with the death of the organism; it may change again and again between burial in the sediment and appearance in the chemical retort. Some of the positive aspects of isotopic and elemental shell chemistry in environmental interpretation are reported on pp. 94 and 95. Figure 2-8 summarizes work on recent sea urchins with the percentage of magnesium carbonate plotted against water temperature. You will note that variations in composition are a consequence of the kind of sea urchin studied as well as of environmental

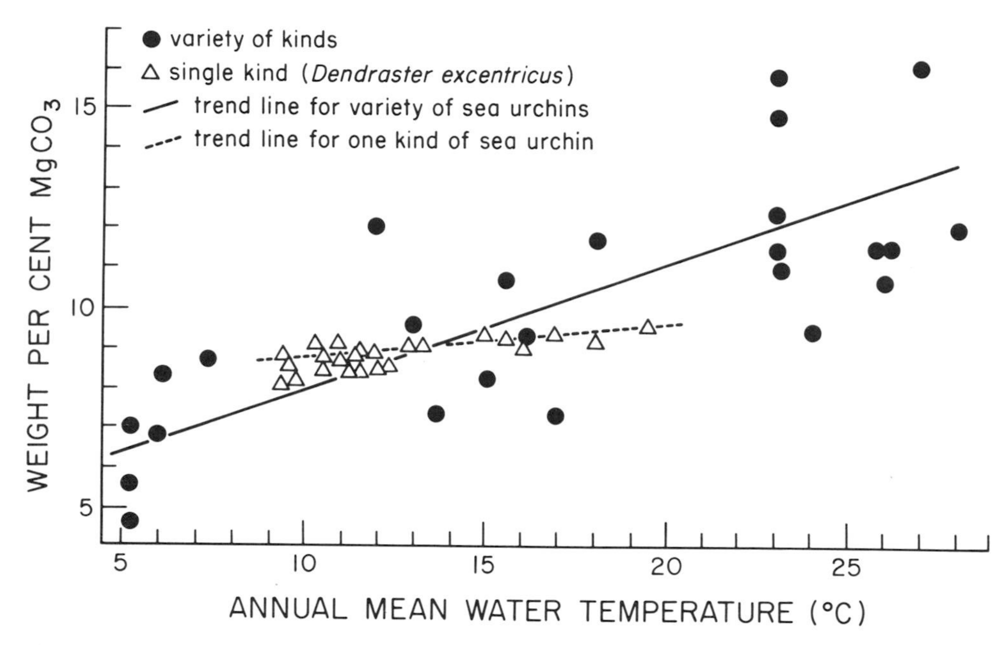

Fig. 2-8. RELATION OF MAGNESIUM CONTENT TO WATER TEMPERATURE IN SEA URCHINS. *(After Pilkey and Hower.)*

variation. Turekian and Armstrong found very great variation in the percentages of strontium, barium, manganese, and iron in Cretaceous molluscs from South Dakota. Some shells which had unaltered mineralogy (aragonite) and microarchitecture contained up to 3 times as much strontium, 10 times as much barium, and 100 times as much manganese as modern shells. They postulate three end-members in fossil mollusc shell chemistry: 1) unaltered, primary aragonite 2) completely altered, secondary calcite 3) primary aragonite with ions adsorbed on the shell material.

These examples demonstrate that study of shell chemistry does not provide any "black box" from which one may readily read out water temperature, water composition or evolutionary relationship. This chemistry does, however, reflect the biological, environmental, and diagenetic processes that formed it; some consistent, useful results have been achieved, and further advances should be possible with an increase of information on 1) the chemistry of shell formation, 2) comparative shell chemistry, and 3) the diagenetic alteration of shells. The application of new techniques, e.g., mapping of elemental distribution with an electron probe, also offers possibilities for improved interpretation.

The consequences of growth

Since an organism consists of units with a definite spatial arrangement, its topography may be mapped—the axes and planes of symmetry as the grid; the major units as the dominating hills and valleys; and the different cells as contours of the living landscape. This topography is obviously more strictly organized than any real land surface—the very word organism implies this—and it must develop as does a landscape, in some orderly fashion but with more efficient control.

How does this development concern a student of fossils? I would answer, in a general way, that he cannot grasp the significance of form (and its variation and transformation in time) without understanding its development. More immediately, many fossils such as the clams represent an accumulation of growth stages, and many fossils are of immature organisms. The method of study depends on their manner of growth.

Mechanics of growth

The development of all organisms consists of cell proliferation, growth, and differentiation. These activities generate the units of the body. For example, in a coral this is seen most simply in the colonial types. In these a layer of tissue connects the individual animals that compose the colony. Acceleration of cell proliferation and of cell growth forms "buds" in this layer. The cells divide, grow, and redivide. Shortly, these processes localize within certain portions of the bud so that the structures of the mature coral appear; for example, one localization occurs within the body wall of the bud at a position adjacent to a mesentery (the sheet of tissue that partitions the coral's interior) of nearby adults. Here the cells divide more rapidly, produce, as a consequence, a ridge projecting into the body cavity, and ultimately grow to form mesenteries. At the same time, the simple, uniform cells of the bud change and differentiate into a variety of types characteristic of the adult.

In division, the new cells possess the same kind of hereditary material as the parent; consequently, each one has identical heredity. If they grow and divide at different rates or if they develop into different kinds of cells, they do so in spite of this identity. The control of mesentery position by the location of adjacent mesenteries suggests that the environment of the cell triggers a particular growth rate or a particular kind of differentiation.

Analysis of sexual reproduction and consequent development suggests other factors and controls. The eggs and sperm, formed in the inner cell layer of the adult, have only half of the hereditary material of the parent cell. After maturation, they are released into the water, where fertilization occurs. Since both egg and sperm cells bear only half the normal complement of hereditary material, the fusion of the two brings the amount of hereditary material up to the original level.

This initial cell of the new individual now divides in the same manner as the cells in asexual reproduction. The replication continues until a hollow ball is formed (Fig. 2-9). The ball then folds into itself to form

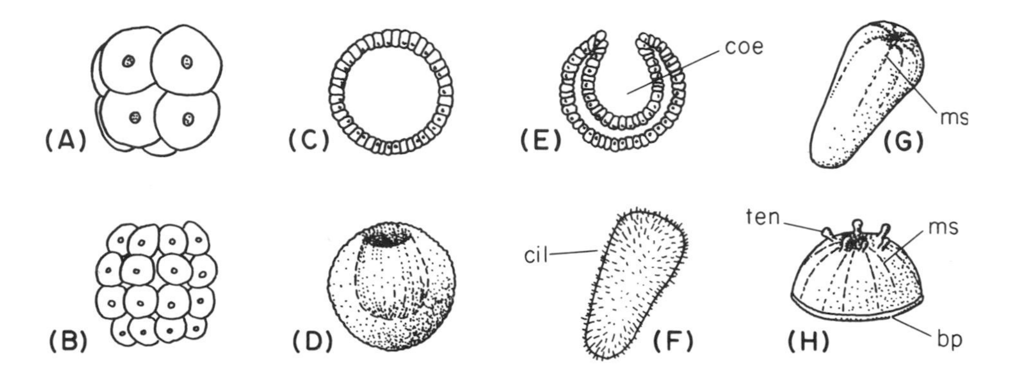

Fig. 2-9. DEVELOPMENT OF CORAL. **(A)** Early, eight cells. **(B)** Many celled. **(C)** Many celled, hollow ball stage. Section cut across embryo. **(D)** After infolding of one side of ball to form embryonic coelenteron. **(E)** Cross section showing embryonic coelenteron. **(F)** Slightly later stage somewhat elongated and ciliated (*cil*). **(G)** Formulation of initial mesenteries (*ms*). **(H)** Embryo after formation of flat basal plate (*bp*). First tentacles (*ten*) present.

an elongate cup with a wall of two cell layers. Mesenteries appear in pairs within the cup. The animal then settles to the bottom, forms a *basal plate* of aragonite, develops internal skeletal plates, the initial *septa,* beneath the intermesenteric spaces. The development of tentacles on the edges of the cup and of additional mesenteries and septa elaborate the structure further.

Here again the development is orderly, though more complicated than asexual budding. But this order, characteristic of both asexual and sexual reproduction and development, is mutable. If buds produced asexually by a single colony are separated but are allowed to develop in the same environment, development and form will be similar in each bud. On the other hand, if a series of fertilized eggs are placed in the same environment for development, the process and the resultant form will differ in details. Since the asexual buds have identical heredity and fertilized eggs differ in heredity, biologists conclude that the hereditary material controls, in some fashion, the pattern of development.

But not completely, for asexually produced buds placed in different environments will develop in different ways. This is most obvious when a decrease in available food reduces gross size, but other factors such as temperature or chemistry of the water are also effective. Apparently, heredity determines the potentialities of growth and development, but environment regulates the specific manner in which an individual develops. Further, each cell has the other cells of the organism as a part of its environment, and its position in the developing organism conditions its physiology and structure. In this fashion, the chemical activity of the cells in the mesenteries of the colonial corals induces the formation of mesenteries in the bud. Studies of other animal groups and of plants show the same results and demonstrate that the generalization is valid throughout the organic world.

Building skeletons

Paleontologists are most immediately concerned with the formation of the fossilizable parts, the skeleton. In general, the microelements forming the skeleton are built within and/or upon an organic matrix. They are in some sense in chemical equilibrium with the matrix; their form is determined jointly by the inherent growth pattern of the molecules produced and by the physical structure of the matrix. They are partly or completely isolated from the external physical and chemical environment. Gross mineralogy, gross chemical composition, and basic microarchitecture are controlled almost entirely by cellular activity—as determined by heredity and the intraorganic environment. Isotopic chemistry, trace element composition, and the elaboration of microstructure into macrostructure are influenced if not controlled by extraorganic environment. Thus in a brachiopod, Figure 2-10, one observes formation of mantle lobes; the cells on the external surface of the lobes differentiate to form the skeletal matrix; mineral "seeds" grow on or within this matrix to form the shell; a thin sheet of organic material separates the "seeds" from the surrounding water; further matrix differentiates at the edges of the lobe as it grows; further mineral seeds are then added on both the internal and marginal matrix to thicken and expand the shell.

Variations played upon this basic type of skeleton accretion determine variations in growth pattern. In oysters the first shell stages are aragonite. Subsequently, calcite is added marginally. Variation in the rate of marginal accretion produces *growth bands* separated by *growth lines* which represent intervals of nondeposition. Shell material added internally, in subplanar sheets, is also calcitic except where the muscles attach to the shell. There, possibly because of variation in magnesium ion concentration in the muscle, aragonite deposition continues. The oyster shell, therefore, records on its margins and in cross section its entire period

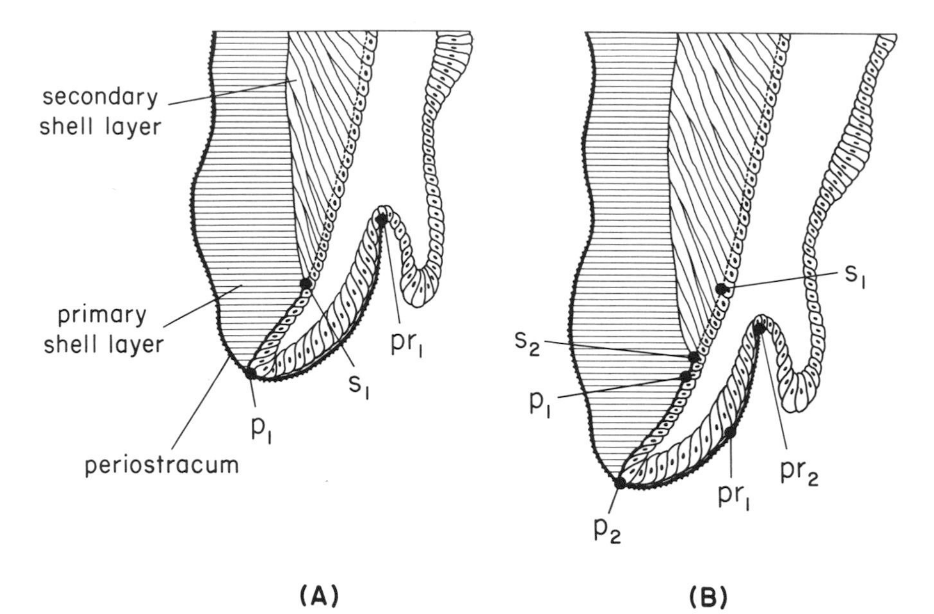

Fig. 2-10. GROWTH OF BRACHIOPOD SHELL. **(A)** Surficial organic layer (periostracum) secreted at pr_1; prisms of outer (primary) shell layer nucleated on cell surface at p_1; rods of inner shell layer (secondary) nucleated within cell at s_1. **(B)** Later stage in growth with new nucleation sites, pr_2, p_2, and s_2. *(After Williams.)*

of growth. Brachiopods, bryozoans, molluscs, echinoderms, and corals all show this type of growth record. In contrast, the skeletal matrix of the arthropods, including trilobites, encloses the body completely, like armor, without free margins for accretionary growth. As the animal grows, a new, unhardened matrix develops along the inner surface of the external armor. The armor then splits apart; the new matrix layer expands and, in turn, hardens into a new suit of armor; the remnants of the old skeleton are left behind as a "molt" to record the early stage of growth.

Echinoderms and vertebrates display a third type of growth. The skeletal elements form as isolated plates or rods within their matrix. They may then grow peripherally on any or all surfaces; new elements may be inserted between earlier ones; portions of older elements may be resorbed.* An echinoderm plate typically preserves, as growth bands, its own accretion, but in vertebrate bone the growth record is partly or largely eliminated.

Studies of growth and development in fossils

From the preceding discussion you would deduce that the sampling of growth in fossils is limited and controlled by the growth pattern of the

* Only very rarely in echinoderms

particular group. The absence of a fixed time scale complicates interpretation. Thus one may measure and plot the spacing of the "growth lines" in a shell, but a particular growth band might represent a day, month, or year. Another band might represent quite a different period. The best one can do is plot the increments serially. If, however, one examines serial increments in a number of shells, as in Figure 2-11, a periodicity of sea-

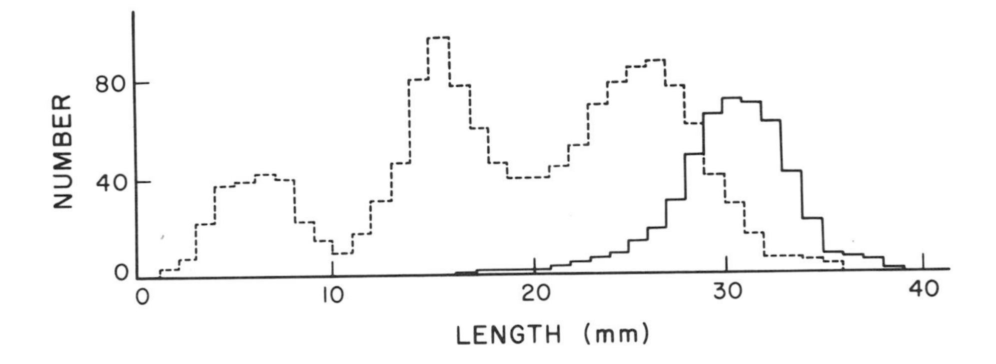

Fig. 2-11. SEASONAL SPACING OF GROWTH INCREMENTS. Solid line, histogram of shell length in sample of clams. Dotted line, histogram of number of growth lines per 1 mm distance from initial point of shells in same sample. High frequency in particular interval, e.g. 4-8 mm, indicates period of slow growth. *(After Craig and Hallam.)*

sonal or annual scope may appear. Wells (1963) has found such a cycle of increments in Devonian corals suggesting that the Devonian year contained 405 days. Periodicity may also be indicated by alternate bands of fine and coarse grains in thin sections of the shell.

The molt stages of ostracods record growth in quite a different way. Although theoretically one might find all the molts of a single animal, the odds are much higher that the skeletons collected represent many animals. They record therefore not only differences incurred during growth but also differences between individuals. Figure 2-12 provides a diagrammatic representation of this phenomenon and also illustrates discontinuities introduced by the episodic nature of skeletal formation. Note here that the ordinal scale is set by height of shell rather than by time so that growth is relative, not absolute—presumably, size correlates with time, although the correlation is imperfect and nonlinear. Relative growth, however, is significant in itself.

Expression, qualitative or quantitative, of growth depends primarily on the mode of growth. For structures that originate, disappear, or are extensively remodeled during development, a simple description of the structure, its position, and its time of appearance relative to the other structures suffices. A trilobite growth series considered in this way consists of a de-

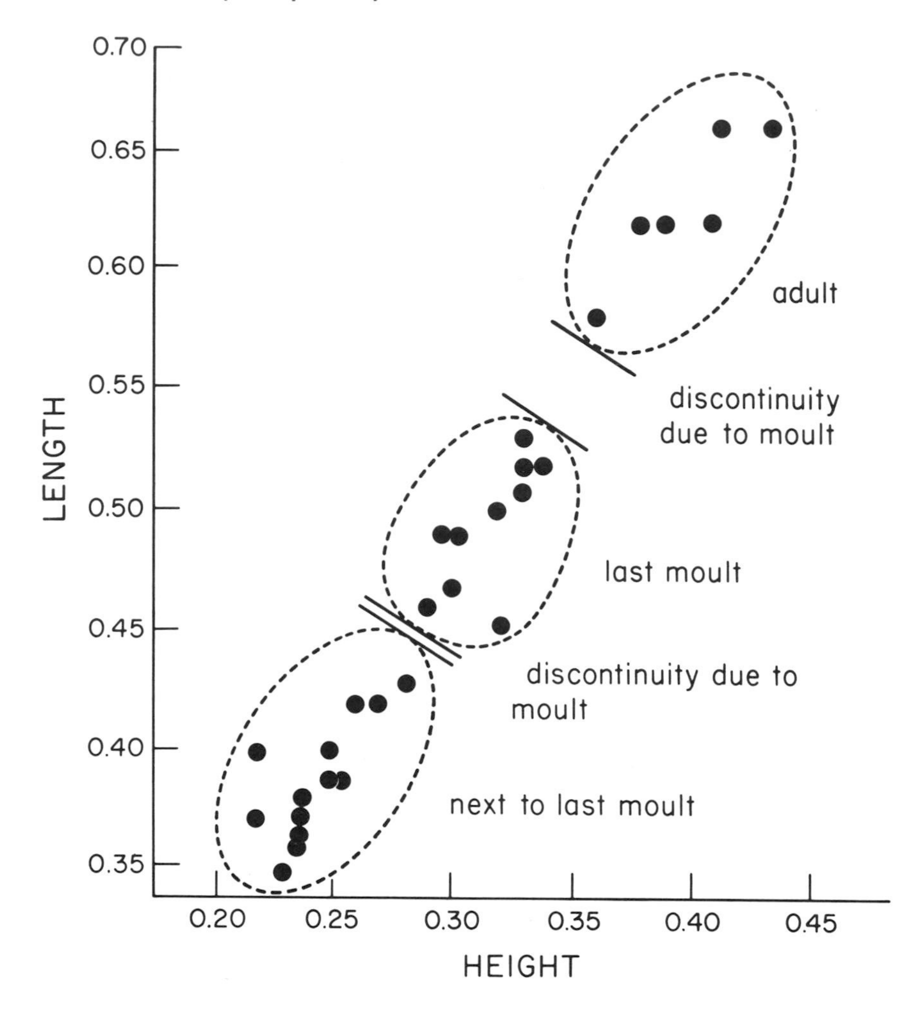

Fig. 2-12. MOULT STAGES IN GROWTH. Clustering of points along growth line (length *vs.* height) caused by periodic moults. (*After Reyment.*)

scription of the number and position of the paired spines on the head shield, relative to size, number of body segments, development of the eyelobe, and so on (Figure 2-13). However, much of the modification in form during development results from changes in proportions of structures already present. The oversized feet of adolescent boys and dogs illustrate this sort of *relative growth.* Relative growth may be measured by modification of a coordinate system where the positions of certain points are plotted on a grid and the change in relative position distorts the shape of the grid (Figure 2-14). More simply, two dimensions can be measured on individuals in the growth series. In this fashion, length and width were measured on a series of 22 brachiopods; each pair of measurements from

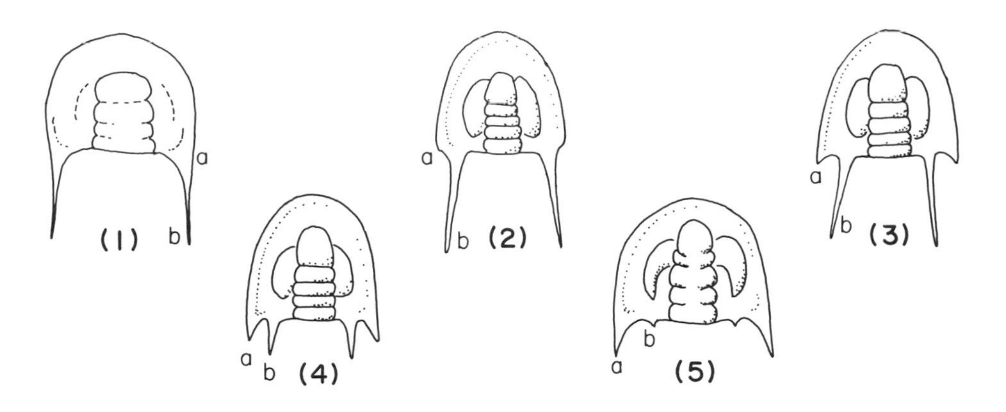

Fig. 2-13. CHANGES IN STRUCTURE IN TRILOBITE MOULTS. Note changes of head in proportions and spines (*a* and *b*). **(1)** Very early, 1 mm length. **(2)** Length 2 mm. **(3)** Length 3 mm. **(4)** Length 5 mm. **(5)** Length 13 mm.

an individual was then plotted on a graph (Figure 2-15). The points form a cloud elongated in one direction. A line fitted to the points expresses the relative growth of width with respect to length. The line can be expressed mathematically in the form:

$$y = bx^a$$

This line does not express the change in a single brachiopod but is an average of growth in several individuals of the same kind (see further p. 80).

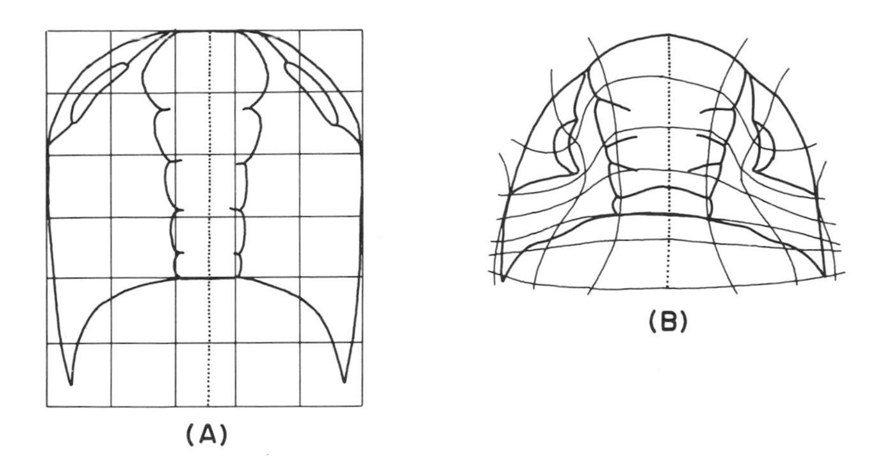

Fig. 2-14. RELATIVE GROWTH. Deformation of grid shows modification of form in development of adult trilobite **(B)** from larval **(A)**. Close spacing of lines in **(B)** indicates relatively slow growth in that region and direction; wide spacing indicates relatively rapid growth. (Coordinates imposed on figures in Piveteau, ed. *Traité de Paléontologie,* © Masson et Cie, Paris. *Used with permission.*)

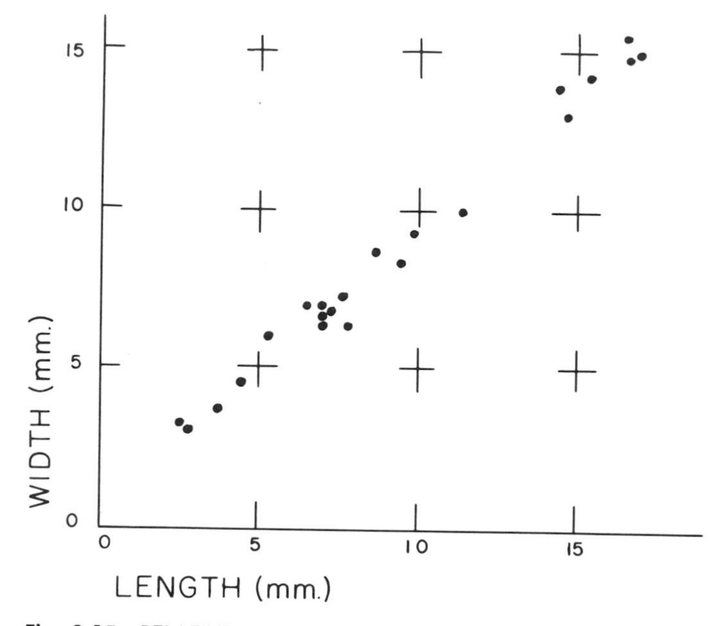

Fig. 2-15. RELATIVE GROWTH. Plot of length and width measurements on a series of fossil brachiopods. Each dot represents a pair of measurements made on one specimen.

So far I have described growth and development in terms of the resulting adult morphology. The distinction is useful for literary purposes, but becomes misleading if applied too literally; the organism at any given stage in its life history is a functioning system with form adapted to function. In the early stages, a primary function is growth, but this depends on the other functions—a dead larva is rarely capable of growing into a live adult. An animal, consequently, is a four dimensional system—the immediate three-dimensioned form and the time dimension of development.

Form and function

The relationship between the morphology of an organism and its mode of life is obvious. The fishlike body of a dolphin, the legs of a horse, and the carapace of a beetle are widely known examples. These relationships are called *adaptation,* i.e., that aspect of an organism which relates the requirements of the organism for life to the environment in which it lives. In a formal sense, adaptation of form or behavior is an "operation" which connects organic function with environment.

As obvious as the existence of adaptation is its importance to paleontologic interpretation. Adaptive aspects of skeletal form are preserved in fossilization. Since adaptation relates organic requirements, which are not directly preserved, to environments, which also arc not directly preserved, it yields the most important clues to ancient biological activities and to ancient environments. Before I consider the mode of interpretation, however, let me review the basic organic requirements.

The requirements for life

What do organisms need? *First,* nutrients. Their acquisition involves search or passive collecting, and, by implication, the means of search or collection: the moving tentacles of the coral, a tree's far-spreading roots, the hunting spiral of the hawk, or the movement of the elk to mountain pastures. It involves also the seizure of the prey or the cropping of the grass. *Second,* an ability to initiate and utilize the release of energy. *Third,* ability to adjust to, to control, or to resist the stresses of the environment.

But what does one animal, an aquatic salamander for example, do to survive and what arc its adaptations in form? Food is necessary—perhaps small worms seized in the toothed jaws. The prey is passed down the esophagus to the stomach, where it is engulfed by digestive juices and thence to the intestine for completion of digestion and for absorption into the blood.

The blood is pumped out through ramifying passages by a muscular chamber of the heart, and finally the food becomes available to the individual cells. The molecules of organic tissue are broken for the energy that lies in the breaking, and some of the molecules or parts of them are recombined to build the tissues of the amphibian. These transformations, however, require free oxygen that can only be acquired outside the animal, so the blood, in its path through the body, passes through the delicate filaments of the gills. Oxygen diffuses through the thin gill walls into the blood and is carried with the food molecules to the cells.

The energy-yielding reactions and the molecular syntheses are controlled within the cell by enzymes. The production of enzymes in turn is controlled by the hereditary materials acting within the cellular system. And so the tissues are built or repaired.

Each cell, however, poisons itself with the by-products of these reactions. The wastes of cellular activity, of course, diffuse back into the bathing blood, but the concentration there must be reduced by the filtering action of the kidneys and by diffusion of the gaseous wastes back into the surrounding water.

And the requirements? Nutrients, energy, and a stable environment for each cell. But how did the worm arrive within the salamander's jaws? The prey must have been perceived: an image on the retina, an image coordinated to previous feeding experience and present need in the central nervous system and then related to pursuit. On command of impulses from the brain and spinal ganglia, some muscles contract as others relax, and the animal lunges, jaws agape, at the passing worm. Similarly, it apprehends the threat of an approaching predator and flees beneath a log, or it seeks cooler, oxygenated water in a deep pool. In general then, organisms through their behavior adjust to, control, or resist the modification of their environment beyond the limits of tolerance.

But ultimately the limits are transgressed—the water becomes too cold, worms too scarce, the strike of the feeding heron too quick. Survival becomes that of the race rather than of the individual, and reproduction the ultimate adaptation. In salamanders the sexes are separate, and the egg and sperm are produced in the specialized tissues of ovary and testes. The female deposits eggs in the water; the male releases sperm over them. The fertilized eggs then initiate a new set of life histories.

One can construct a general list of functions from this analysis. First, food getting; second, the digestion of the food; third, respiration; fourth, circulation of nutrients; fifth, the metabolism of food molecules; sixth, excretion of the wastes from cellular activities. These are the necessary functions of any life (except that green plants substitute sunlight, carbon dioxide, and water for organic molecules, but all organisms known at present also perceive features of the external world, relate these features however crudely to the requirements of survival, and behave on the basis of these relations. Further, all organisms have some protective mechanisms that contribute (usually directly) to reproduction, and thus to perpetuation of the species.

Function from form

Adaptation occurs in all organisms but is expressed in different fashion in each. A squid as a vigorous predator has highly developed eyes; a clam as a sluggish bottom-dweller needs only the slightest ability to perceive "light or not light." Each is adapted to a particular environment and has narrower limits of environmental tolerance than organisms as a whole. From this viewpoint, form assumes more than abstract significance. Bilateral symmetry is not only a repetition of parts ordered by the mechanisms of growth and development; it is an index to the general mode of life of the organism.

Fine! But how does one make a functional interpretation of a fossil? What can one learn about the fossil amphibian of Figure 2-16? First, recognition as an amphibian restricts possible interpretations of functional requirements. In general, living amphibians are entirely or partially aquatic; they are either predators or herbivores but are less likely to be the latter. In the water they swim and/or crawl along the bottom. They require oxygen; they obtain it through the skin in part but may also em-

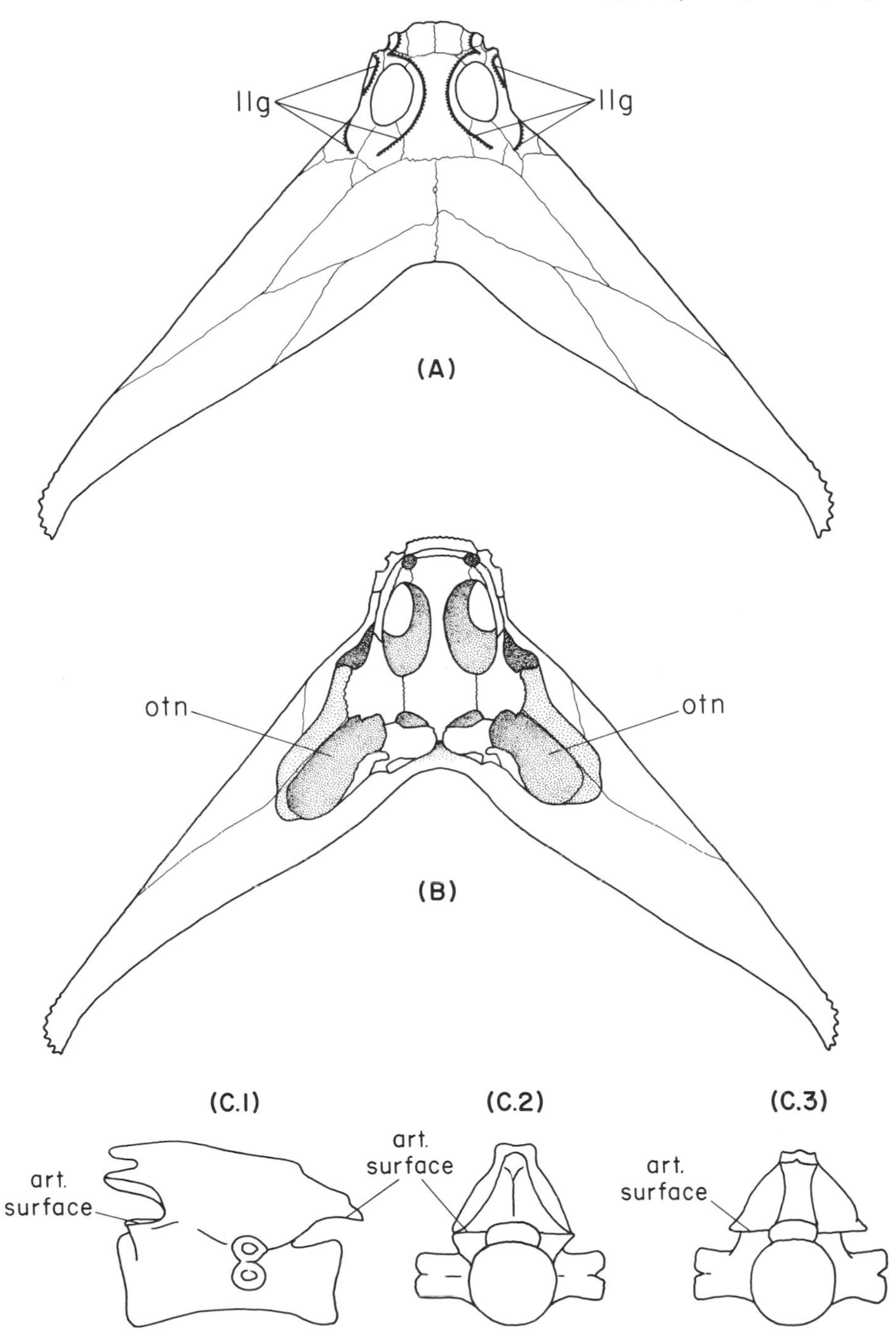

Fig. 2-16. FOSSIL AMPHIBIAN. **(A)** Dorsal view of skull, about 20 cm between horn tips. Structure *llg,* lateral line grooves. **(B)** Ventral view of skull. Structure *otn,* otic notch. **(C)** Trunk vertebra. Structure *art. surface,* articular surface. **(C.1)** Lateral view. **(C.2)** Anterior view. **(C.3)** Posterior view.

ploy lungs, external gills, and/or the membrane lining the throat. It is reasonable, though not absolutely necessary, to accept the fossil amphibian as displaying a functional system within these broad limits.

Let me next examine some of the alternatives within these limits. Was the fossil amphibian entirely aquatic? The grooves, *llg,* over the surface of the skull resemble similar grooves in modern amphibians. Empirically, such grooves are highly correlated with the aquatic mode, and in turn a fossil amphibian with strongly developed lateral line grooves should by inference be aquatic. Does one have any deductive justification for this inference? The lateral line grooves in modern fish and amphibians house canals that sense slight pressure fluctuations in a fluid medium. This establishes a function (pressure or wave detection) in the fossil connecting through a structure (lateral line groove) to a specific environmental feature (aquatic medium).*

Another approach to interpretation is through mechanical analysis. The vertebrae articulated so that body motion was entirely side to side. Propulsive efficiency in swimming is proportional, in part, to the surface area perpendicular to direction of body motion. Thus this fossil amphibian with a very small lateral surface area cannot have had effective body-tail propulsion in swimming. This relationship is strictly required by the mechanical relationships between vertebrae and by the dynamic relationships between propulsive mode and propeller form; they are deductively independent of any analogy with modern amphibians.

A fourth mode of functional analysis depends upon inference from environment as determined by rock characteristics or associated fossils. Specimens of this particular amphibian are found exclusively in lake deposits; general characteristics of the deposits suggest shallow, warm ponds with rather low oxygen tension. The latter feature would require modification of the structures related to respiration—lungs, skin, external gills, and throat. The lung structure is indeterminable of course. The elaborate pitting of the bones of the skull may reflect elaborate vasculization of the skin. The juxtaposition of the skull and shoulder above the throat allows no space for external gills. The large cavities (*otn*) on the undersurface of the skull are most reasonably related to large throat pouches which are adaptive for respiration with or without internal gills, and, in fact, are inexplicable without knowledge of the environment in which the animal functioned.

Let me now review the principles for functional interpretation of form. First, one expects that the requirements of an ancient organism will resemble to some extent those of living relatives. Second, the function of a particular fossil structure may be signified by the function of similar struc-

* This is derived from a "normic statement," since the connection between grooves, canals, and life in the water is conditional and not necessary. If one accepts the statement, one can make universal statements about the effectiveness of a system in detecting wave trains from specific directions and about its maximum resolution in perception.

tures in living organisms, related or not. Third, strict mechanics or statics of form implies function; conversely any ascribed function must be consistent with mechanical and/or static analysis. This analysis might include the orientation of the fossil in the rock. Fourth, definition of specific environmental factors from the associated rock and fossils may permit definition of specific functional requirements. Conversely, any function ascribed to form must be workable within an available environmental framework.

Because the fossil yields a very partial information about the living animal, interpretations of function are equally partial. Complex skeletons adapted to a large number of functions provide much grist for the mill of interpretation. Thus the habits and habitat of many fossil vertebrates are known in detail, whereas they are far less well known for fossil corals or clams.

Another difficulty arises from lack of information on living animals. Zoologists have described the form of most animal groups in at least moderate detail, but the function of various structures is less often inferred, much less demonstrated. Those groups of fossils that lack close living relatives pose special problems, as do those with structures without analogues among living animals.

Finally, specific functional interpretations are only likely, not certain, and attain a high degree of likelihood only as they are supported by other interpretations based on independent evidence.

Gravings in the rocks

A fossil not only signifies an organism living in an environment; it also represents a coincidence of events (or nonevents) over the thousands or millions of years since the death of that organism. Reconstruction of biological form from the fossil was discussed in a preceding section; it seems logical to conclude by looking at form as a consequence of fossilization. The variables are, first, the intrinsic properties of organisms on which fossilization operates and, second, the physical and chemical processes of fossilization.

The selection of immortals

Organic components decay; inorganic ones are fossilized. Is it that simple? Chave has reported (1964, 1962) a group of experimental studies that bear on the question. When he abraded a variety of recent marine skeletons in a tumbling barrel he found striking differences in resistance. In one experiment, the shells of a snail (*Nerita*) were worn but intact after

183 hours; over half the calcareous algae and bryozoans were reduced to small fragments in 30 minutes; the "half-life" of sea urchin skeletons and of small clams was about an hour; large clams of the same species had "half-lives" of about 100 hours. Varying the experimental conditions changes relative as well as absolute durability. Chave found no relation between durability and mineral composition; he did find variation with size, with microstructure, and with disposition of the organic matrix. Clearly, the chance of preservation in a wave traction environment is not simple.

Another interesting set of experiments (Chave, *et al.* 1962) concerns chemical destruction. Figure 2-17 reports that relative solubility of arag-

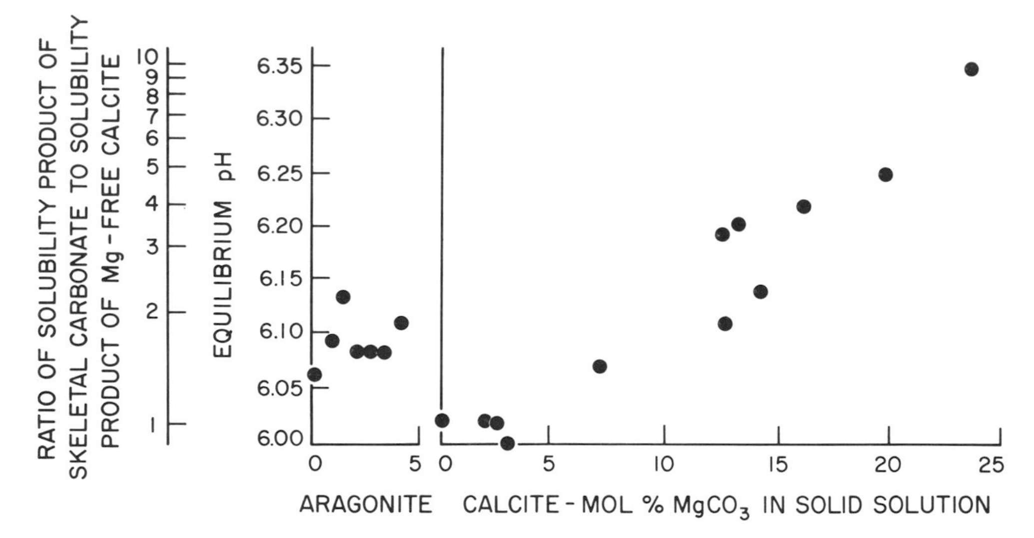

Fig. 2-17. VARIATION IN SOLUBILITY OF CALCITE WITH ARAGONITE AND MAGNESIUM CONTENT. (*After Chave.*)

onite, high-magnesium calcite, and low-magnesium calcite. The relatively high solubility of aragonite and of high-magnesium calcite is striking; the imperfect preservation of Paleozoic molluscs compared to the excellent preservation of brachiopods may reflect the difference in shell mineralogy, because many recent mollusc shells are aragonitic and brachiopod shells are calcitic. Empirically, phosphatic skeletons—linguloid brachiopods, conodonts, vertebrate bone—are more resistant than calcitic. Some skeletons, such as those of arthropods and graptolites, are hard organic material; under abnormally low pH conditions they apparently resist solution better than calcite, but under normal conditions of accumulation and burial they are selectively removed. If organic skeletons are impregnated with mineral grains as in the calcified trilobites, they possess a rather high probability of fossilization. The opaline silica laid down by some protozoans, sponges, and plants likewise resists solution although the small size of the skeletal elements exposes a large surface area to chemical action.

Fossilization

The Burgess shale of British Columbia contains abundant fossils of soft-bodied creatures; the brown coals of the Geisel valley in Germany preserve the microstructure of skin and muscle cells. On the other hand, many sedimentary rocks contain few or no fossils. Obviously the "en-

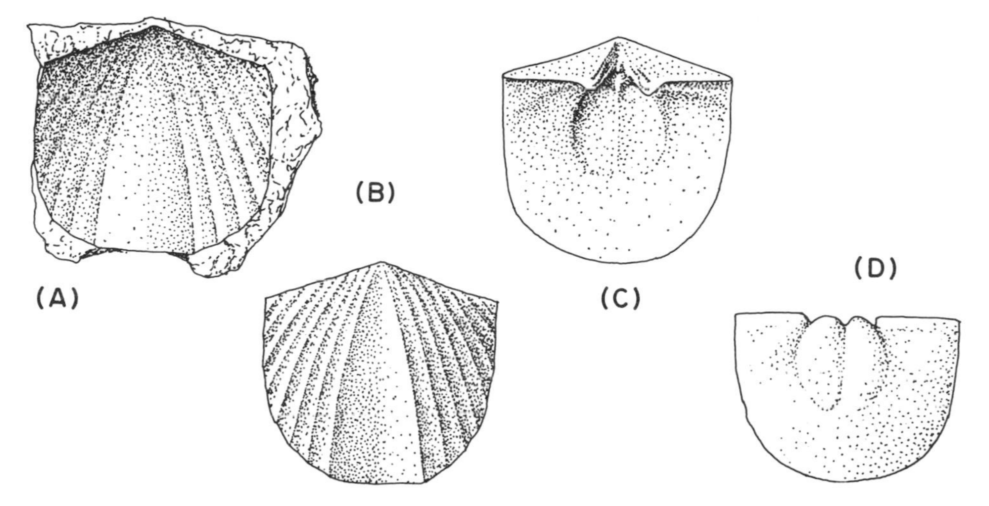

Fig. 2-18. FOSSIL MOLDS AND CASTS. Types of fossil preservation shown by a brachiopod shell. **(A)** External mold—the impression of the shell in the rock. **(B)** Cast—external surface—formed by replacement of original shell material. **(C)** Cast—internal surface. **(D)** Internal mold—impression formed by the inner surface of the shell on the mud or sand that filled the interior of the shell.

vironments of fossilization" are extremely complex; they are also a poorly known system. In general, the more rapidly an organism is buried and the tighter the seal of its sedimentary tomb, the better the chances of preservation. The logic is simple enough, for the skeletal tissue need only equilibrate in a very small closed physical and chemical system. In contrast, if an organism lies exposed on the bottom after death it must approach equilibrium with a much more extensive system including an extremely active biological component. Similarly, if burial does not close the system, e.g., if fluids percolate through the sediment or if burrowers rework the sediment, the system in which the organic and skeletal tissues must equilibrate

TABLE 2-1

THE NATURE OF FOSSILIZATION

Preservation without alteration	
Organic compounds	
Soft parts	Frozen; mummified. Such finds are rare and limited largely to Pleistocene deposits.
Skeletal parts	Organic constituents of bone or shell. Chitin in arthropods, graptolites, and some other invertebrates. Cartilage in some vertebrates.
Inorganic compounds	
Calcium carbonate:	
Calcite	Fairly stable, found in many invertebrate phyla.
Aragonite	Moderately stable, rare in rocks older than Mesozoic. Corals and molluscs.
Tricalcium phosphate	Brachiopods, arthropods, vertebrates. Quite stable.
Silica (opaline)	Moderately stable. Rare in rocks older than Cenozoic. Sponges and some protozoans.
Altered in fossilization	
Organic compounds	
Soft parts	Films of carbon. Rare. Found in fine shales deposited in anerobic environments.
Skeletal parts	Carbonized. Particularly the chitinous skeletons of arthropods and graptolites.
Inorganic compounds	
Permineralized	Deposition of minerals in interstices of skeleton. Commonly $CaCO_3$. Less frequently SiO_2, glauconite, iron compounds, etc.
Recrystallized	Less stable inorganic compounds alter in physical form to more stable state without change in chemical composition, e.g., aragonite to calcite. May be very common mode of preservation, but difficult to distinguish from replacement.
Replacement	Removal of original skeleton material by solution and deposition of new compounds, carbonates, silica, iron compounds, etc., in its place. Very common—intergrades with permineralization and recrystallization.
Preservation as molds or casts	
Organic compounds	
Soft parts	Imprints in fine-grained laminated shales and lithographic limestones.
Skeletal parts	Imprints or casts.
Inorganic compounds	
Molds	External and internal molds formed by sediment around or within skeletal parts (Figure 2-4).
Casts	Filling of mold after skeletal parts are dissolved (Fig. 2-4). Intergrades with replacement.

TABLE 2-1 (continued)

Evidences of animal activities	
Tracks	Mode of locomotion. Preserved as molds and casts.
Burrows	Animal habitat and behavior. Mode of burrowing. Preserved as molds and casts.
Coprolites	Fossilized excrement. Diet. Structure of gut. May be preserved in any of ways in Table 2-1.
Borings and tooth marks	Evidence of predation.

is very large and potentially that much more destructive.* Finally, if the system is reopened—for example, by metamorphism or by weathering—the fossil must reequilibrate in another environment. Of course, the sediment may buffer the system and increase the probability of preservation.

If the preservational environment becomes too severe, the fossil will be lost completely; less extreme departures will yield partial modification—physical or chemical. Table 2-1 indicates the general modes of modification. The amount of biological information obtained from a fossil will tend to decrease from the top to the bottom of the table: first, with loss of soft anatomy, second, with modification of original shell composition, third, with destruction of microstructure of the skeleton, and fourth, with blurring of macrostructure. The cellular preservation of soft tissue in Geisenthal stands at one extreme; the deformed and recrystallized brachiopods from the Paleozoic metamorphic rocks of New England at the other.

Some recent studies have dealt with details of chemical alteration in preservation. Krinsley and Bieri (1959) found enrichment of aluminum in snail shells from ocean bottom cores varied from 10^2 to 10^3 times, magnesium from 3 to 10^1, manganese up to 10^2, but the same shells showed no apparent increase in copper or strontium. Turekian and Armstrong (1961), in analyses of fresh-appearing translucent aragonitic shell material from Cretaceous cephalopods, noted significant additions of iron, manganese, and barium. These increments were greatest in samples with high calcite content (4-5%). Strontium, used in some studies as an index of water temperature, varied in a single shell by a factor of two and independently of the calcite ratio. They also noted differences from modern cephalopods in shell microstructure: more diffuse growth bands and organic laminae and larger aragonite crystals. But in spite of these and other studies, paleontologists still know little of why and how fossils are fossilized.

* Even this generalization has exceptions. If skeletal tissue is entombed with large amounts of organic tissue, decay of the latter may lead to destruction of the former in a closed system.

REFERENCES

Beerbower, J. R. 1963. "Morphology, Paleoecology, and Phylogeny of the Permo-Pennsylvanian Amphibian, *Diploceraspis*," *Bull. Mus. Comp. Zoology*, vol. 130, pp. 31-108.

Chave, K. E. 1964. "Skeletal Durability and Preservation," in: J. Imbrie, and N. D. Newell, Eds. *Approaches to Paleoecology*, New York: John Wiley and Sons. Pp. 377-387.

———, *et al.* 1962. "Observations on the Solubility of Skeletal Carbonates in Aqueous Solutions," *Science*, vol. 137, pp. 33-34.

Clark, W. E. Le Gros, and P. B. Medawar. 1945. *Essays on Growth and Form*, Oxford: Clarendon Press.

Craig, E. Y. and A. Hallam. 1963. "Size-frequency and Growth-ring Analyses of *Mytilus edulis* and *Cardium edule*, and Their Palaeoecological Signficance," *Paleontology*, vol. 6, pp. 731-750.

Degens, E. T. *et al.* 1967. "Paleobiochemistry of Molluscan Shell Proteins," *Comp. Biochem. Physiol.*, vol. 20, pp. 553-579.

Grégoire, C. 1959. "A Study on the Remains of Organic Components in Fossil Mother-of-pearl," *Institut royal nat. Belgique, Bull.*, Tome 35, no. 13.

Hare, P. E. and P. H. Abelson. 1964. "Comparative Biochemistry of the Amino Acids in Molluscan Shell Structures," *Geol. Soc. of America, Spec. Paper*, No. 82, p. 84.

Krinsley, D. and R. Bieri. 1959. "Changes in the Chemical Composition of Pteropod Shells after Deposition on the Sea Floor," *Jour. Paleontology*, vol. 33, pp. 682-684.

Lowenstam, H. A. 1963. "Biologic Problems Relating to the Composition and Diagenesis of Sediments," in T. W. Donelly (Ed.) *The Earth Sciences: Problems and Progress in Current Research*. Chicago: University of Chicago Press, pp. 137-195.

McAlester, A. L. 1962. "Mode of Preservation in Early Paleozoic Pelecypods and its Morphologic and Ecologic Significance," *Jour. Paleontology*, vol. 36, pp. 69-73.

Pilkey, O. H. and J. Hower. 1960. "The Effect of Environment on the Concentration of Skeletal Magnesium and Strontium in Dendraster," *Jour. Geology*, vol. 68, pp. 203-214.

Raup, D. M. 1960. "Ontogenetic Variation in the Crystallography of Echinoid Calcite," *Jour. Paleontology*, vol. 34, p. 1041-1050.

——— and Michelson, A. 1965. "Theoretical Morphology of the Coiled Shell," *Science*, vol. 147, pp. 1294-1295.

Rudwick, M. J. S. 1964. "The Inference of Function from Structure in Fossils," *British Jour. Philos. of Sci.*, vol. 15, pp. 27-40.

Seilacher, A. 1967. "Fossil Behavior," *Sci. American*, vol. 217 (August), pp. 72-80.

Stehli, F. G. 1956. "Shell Mineralogy in Paleozoic Invertebrates," *Science*, vol. 123, pp. 1031-1032.

Trueman, E. R. 1964. "Adaptive Morphology in Paleoecologic Interpretation," in J. Imbrie and N. D. Newell (Eds.) *Approaches to Paleoecology*, New York: John Wiley and Sons. Pp. 45-74.

Turekian, K. K. and R. L. Armstrong. 1961. "Chemical and Mineralogical Composition of Fossil Molluscan Shells from the Fox Hills Formation, South Dakota," *Geol. Soc. of America, Bull.*, vol. 72, pp. 1817-1828.

Wells, J. W. 1963. "Coral Growth and Geochronometry," *Nature*, vol. 197, pp. 948-950.

Williams, A. and A. J. Rowell. 1965. "Brachiopod Anatomy," in: R. C. Moore (Ed.) *Treatise on Invertebrate Paleontology*, Part H. New York: Geological Society of America, pp. 6-57.

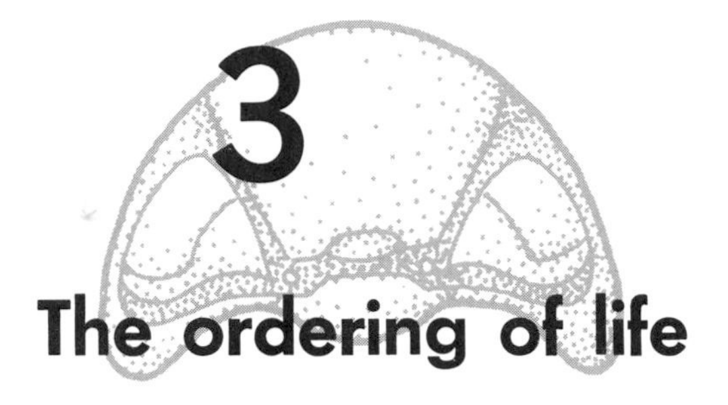

3 The ordering of life

A ledge of limestone may contain thousands of fossils; a single ten-centimeter-square surface of a shale (Figure 1-1) may contain dozens. No two fossils are exactly alike, just as no two humans—even "identical" twins—are exactly alike. But some of them are quite similar. Casual inspection demonstrates that the variation in form among the fossils is not—as suggested in Figure 3-1A—continuous and random but is, rather, discontinuous and nonrandom as in Figure 3-1B. One can divide the sample into a series of discrete groups (*1*, *2*, *3*, and *4*) the members of which overlap in a considerable number of properties and which have few properties in common with the members of other groups. A set of overlapping properties defines a *structural plan* for the group.

This orderly pattern of discontinuous variation is obviously related to a similar pattern observed among contemporary organisms: On the veldt, one can see lions, zebras, gnus, and rhinos, but one finds no "gnu-bras" or "zeb-ions."

Further study of the fossils shown in Figure 3-1B suggests that some of the groups are more like each other than they are like the remainder; more formally, they overlap in some properties (Groups *1* and *4; 2* and *3*).

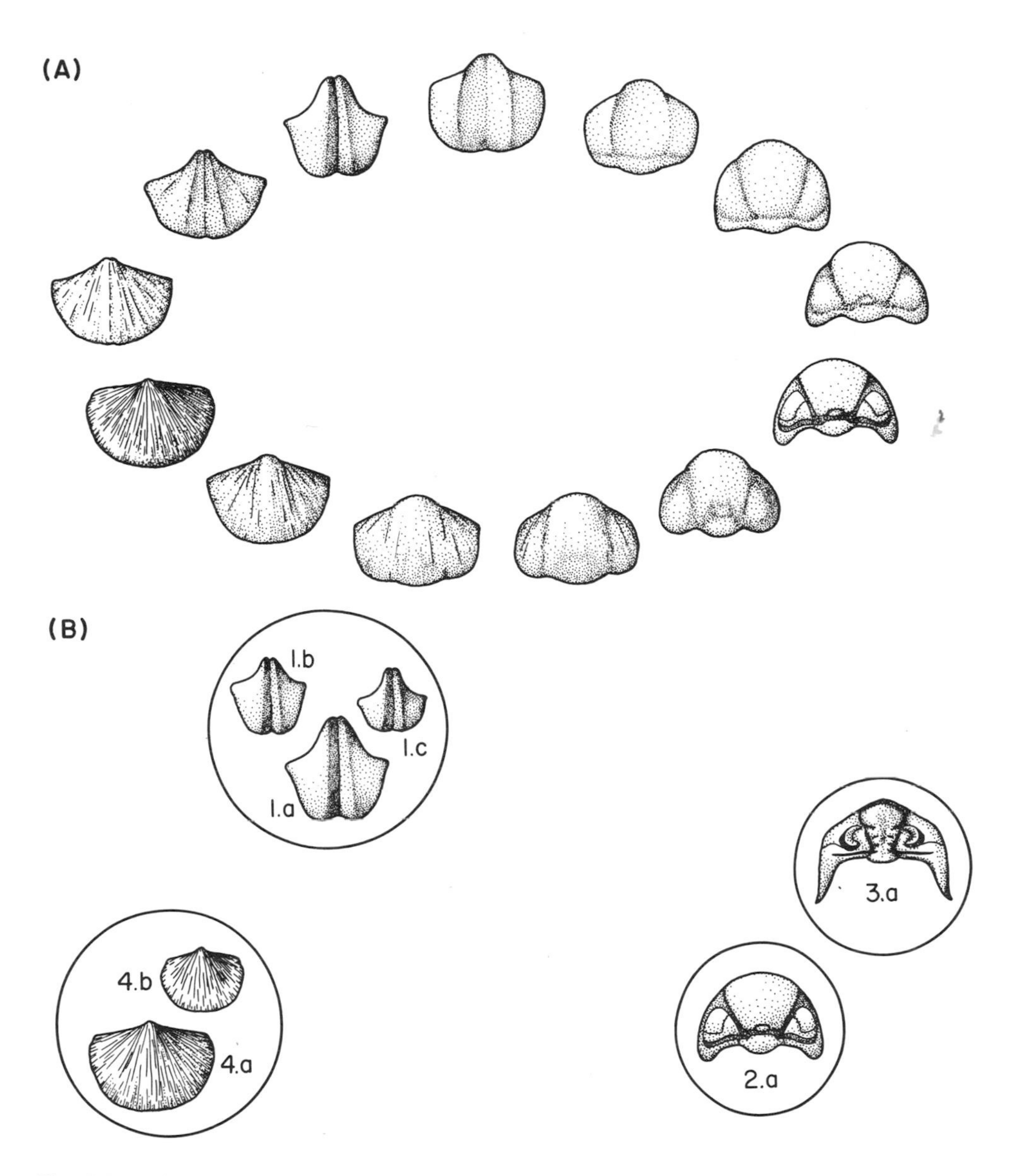

Fig. 3-1. CONTINUOUS AND DISCONTINUOUS VARIATION IN FOSSIL SAMPLES. **(A)** Hypothetical sample with continuous variation. **(B)** Observed sample with discontinuities defining discrete groups **1, 2, 3,** and **4.**

This recognition has a biological parallel also; the concept "cat" includes lions, tigers, cheetahs, and leopards, as well as house cats. Pending further definition, I'll term the elemental groups *species* and the more inclusive groupings "*supraspecific.*"

A ledge of the same limestone layer on the far side of the valley will share many species and will differ in only a few. But a sandstone lens within the limestone layer may contain a quite different assortment of fossils—one much like that found in sandstones of similar age many miles away. This pattern of variation also accords with biological experience, for a sand bar supports quite different groups of animals than an adjacent muddy bank.

If one collects from the layers immediately above and below the limestone, he finds, typically, that adjacent layers contain many species in common. As he climbs farther up (or down) the hillside, this similarity decreases; he discovers, ultimately, that many of the species first defined in the limestone occur only through one portion of the rock sequence and are totally absent above and below these limits. Obviously, the species were restricted in time as well as in environment.* On the other hand, inspection of species from different time units reveals supraspecific groups that consist of species successive in time—or even of an intergrading sequence between species.

These patterns of variation within and between fossil samples must correspond to and map the structure of biologic systems—as they exist at any moment and as they change through time. This cryptic map, when properly interpreted as much as any pirate map, can lead one to treasure, the history of life.

Causes of order

Since this chapter and the five that follow deal with "maps" in terms of biological systems, a brief overview is appropriate here. A system has dynamics, i.e., a flow of energy and information, and a structure, i.e., an ordered set of channels through which the flux is accomplished. The smallest complete system, the individual organism, was dealt with in the preceding chapter. It is convenient to think of the individual as a consequence of two overlapping subsystems, one of growth and development, or *ontogenetic,* and the other of activity, behavior, and function, or *physiological.* The structure for these systems is the form and composition of the individual; the form and composition of an individual fossil therefore maps—albeit incompletely—these systems.

A second system appears through reproduction. A large amount of information—the hereditary material—is transmitted from parent to off-

* Only a series of superposed layers demonstrate temporal restriction directly. The temporal limits of a species on one side of a mountain range may be quite different from those on the other side. A variety of indirect evidence however supports the conclusion that the temporal range of a species wherever it occurs is also limited.

spring within the physiological framework of reproduction and, for sexual organisms, within the breeding structure of the population. This *genetic system* will be considered in Chapter 4; here it suffices to note that the discrete "species" within a fossil sample correspond in some fashion to genetic systems and to observe that for sexually reproducing organisms membership in a morphologic "species" suggests membership in an interbreeding population. Conversely, distinct "species" are likely to represent distinct, noninterbreeding populations.

A third sort of biological system involves the change of genetic systems with time and so is that of *evolution* or, perhaps better, is phylogenetic.* The transfer is, again, of information in a genetic system but is modified by changes in information (mutation) and by the elimination of some information (natural selection). The lineage of populations connecting ancestor and descendents forms the structural correlative. Some, but not all, supraspecific groups of fossils reflect phylogeny, because the contemporaneous "species" belonging to a supraspecific group may have a common ancestry, i.e., ancestry in a single genetic system, and because a group of noncontemporaneous species may either belong to a single population lineage or possess common ancestry. Chapters 6 and 7 treat the major features of phylogenetic systems.

The interchange of energy (and of considerable material information as well) among contemporaneous organisms and with the environment constitutes a fourth, the ecological, system. Among fossils, the presence or absence of species in a sample and the association of species among samples must map this system. Both its processes and its structures are complex and heterogeneous, e.g., they involve predation, competition, and parasitism among organisms and the distribution of organisms within an environment. They can be conveniently considered in terms of a system centered: *a*) on single individuals (in part the "physiological system" discussed in Chapter 2); *b*) on single interbreeding populations (in Chapter 4), and, *c*) on complexes involving several such populations (Chapter 5). In the same way that genetic systems extend in time as phylogenetic ones, so ecologic systems might also extend as "evolving ecosystems"; this question will be argued out in Chapter 7.

All four (or five) systems are elements of an integrated whole: the "physiological" complex of an individual organism is explained, in part, by the individual's "ontogeny;" ontogeny, in turn, by the hereditary material supplied during reproduction; the supply of hereditary material is set by the previous evolution of the interbreeding population. But evolution involves the survival and reproduction of individuals in the context of an ecological system, a survival which is determined by the physiological complex.

* Phylogeny expresses the ancestor-descendent relations between noncontemporaneous populations and the common ancestry of distinct contemporaneous populations.

Fossils, form, and classification

An individual fossil has form and composition; these provide the elemental data for a map of the past. Mapping organic history requires classification of these elements; to some extent all other operations stem from or refer back to classification. A classification "map" symbolizes one or more relationships (similarities and/or differences) between objects through a set of conventions; one can in turn compare and classify maps purely on the basis of these conventions; one could even construct a "metaclassification" of various classification maps. I think a reasonable starting place for discussion then is the framework of convention in paleontologic classification.

Building a classification

Here is a tray of fossils! Classify them!

Given that command, how would one proceed? A classification implies a formal grouping of objects; one must decide on that form. A classification symbolizes a relationship between objects; one must choose the relationship(s) to be mapped and decide how to translate that relationship to the formal symbolism. A classification requires that the relationships between real, obdurate objects be inferred; one must choose rules of inference. Obviously if there is to be equivalency among classifications, the form must be parallel, the relationships similar, and the rules of inference for each classification correspond to some accepted standard. In general, paleontologists have agreed to use the same system of classification as the biologists.

Formal structure

In the example that opened this chapter, comparison of form in a set of fossils divided them into several distinct groups or *taxa* (sg. *taxon*). This particular classification included only two kinds of taxa, supraspecific and specific—these were the *categories* of the classification. Each individual belonged to one and only one specific taxon or species; each species belonged to one and only one supraspecific taxon; the arrangement was *hierarchic.* The categories in this hierarchy formed two distinct levels or ranks, the "lower," specific taxa nested within "higher," supraspecific taxa. This scheme, with necessary elaboration, approximates the biological method of classification.

Biologists and paleontologists employ seven basic categories, in rank from top to bottom: *Kingdom, Phylum, Class, Order, Family, Genus* (pl.

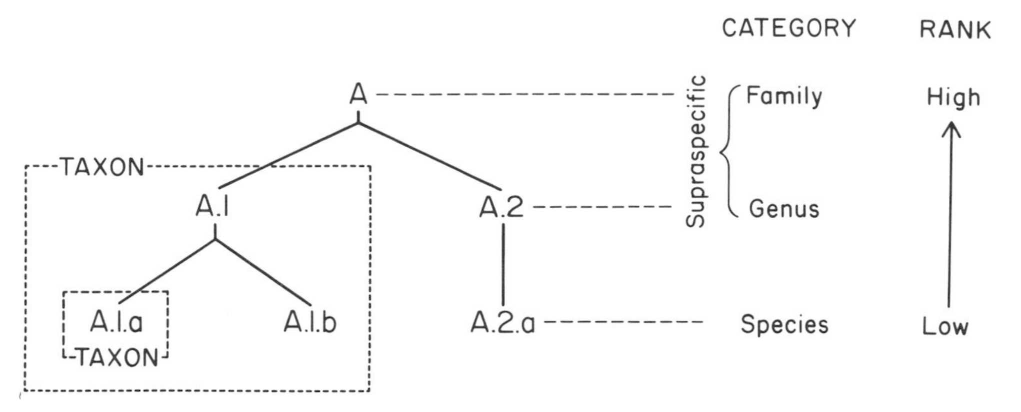

Fig. 3-2. HIERARCHIC CLASSIFICATION. The lowest-ranking category, "species," includes taxa **A.1.a**, **A.1.b**, and **A.2.a**. The higher-ranking category, "genus," includes taxa **A.1** and **A.2**, each of which comprises at least one taxon of lower rank. A lower-ranking taxon belongs to only one taxon of the next higher rank.

Genera), *Species* (pl. Species); additional categories can be formed as necessary by adding prefixes such as "super-," "infra-," and "sub-." Thus the individuals of Group 1 in Figure 3-1B are members of a very inclusive phylum, Phylum Brachiopoda, of a less inclusive Order Spiriferida, of a still narrower genus *Ambocoelia* and of the very exclusive species, *Ambocoelia umbonata.* In the hierarchy, each taxon, except the highest, must belong to one and only one taxon in the next highest ranking category; conversely every taxon except in the lowest ranking category must include at least one taxon from the next lowest ranking.

A system of naming accompanies this classification scheme. Each taxon receives a formal name, e.g., Hominidae. For taxa in categories above the genus up to superfamily, the name is formed by adding a distinctive suffix to the name of one of the component genera—thus "Hominidae" derives from the genus *Homo* and the suffix "-idae" indicating a family.

Each species is given a name consisting of two parts and written so: *Diploceraspis burkei.* (In print the species name is italicized; in typewritten or longhand manuscript it is underlined.) The first term is the name of the genus in which the species is placed; the second, the specific modifier, tells which one of the species of *Diploceraspis* is being considered. The generic name applies to one genus and to no other in the animal kingdom; the specific adjective is used only once within any genus, but may be reused in other genera. The whole procedure is like writing: Jones, John; Jones, Robert; Clark, John; Clark, Robert, and so on. A full citation of a species should also include the name of the individual who first described it and the year the description was published, e.g., *Diploceraspis burkei* Romer 1952. This full statement permits easy reference to the original description. Rules for formation, modification, and suppression of taxonomic names are complex (see Mayr, Linsley, and Usinger, 1952), but promote stability, so that *Diplocераspis* means just one taxon to all paleontologists, and allow flexibility also in accord with new knowledge gathered about the group.

Relationships in a classification

What does the classification of Figure 3-1C express? From the previous discussion, it expresses the similarities and differences in anatomy among the fossils—the *phenetic relationship* to use Sokal and Sneath's term (1963). To anatomy one may add geographic and stratigraphic information and for modern organisms biochemical, physiological, behavioral, and ecologic data. The classification then conveys or summarizes a very large amount of information, and some biologists, particularly the group called numerical taxonomists, believe phenetic classifications to be the most satisfactory. From the logical point of view, however, even more information is imported if genetic and phylogenetic relationships are included; most biologic classification is in theory phylogenetic, but, since phylogeny is inferred and not observed, phylogenetic classifications must derive from interpretation of phenetic data.

Members of a biological species (and of subspecific taxa) possess a genetic relationship (membership in a breeding population) for sexual organisms and a phylogenetic one (descendents of a common ancestral population or individual) for both sexual and asexual ones. Individuals from different contemporaneous species and the species themselves can possess only a phyletic relationship. The union of contemporaneous species in a particular higher level taxon, e.g., family, implies (rather too precisely) that all originated in a single ancestral species (Figure 3-3). Species in a

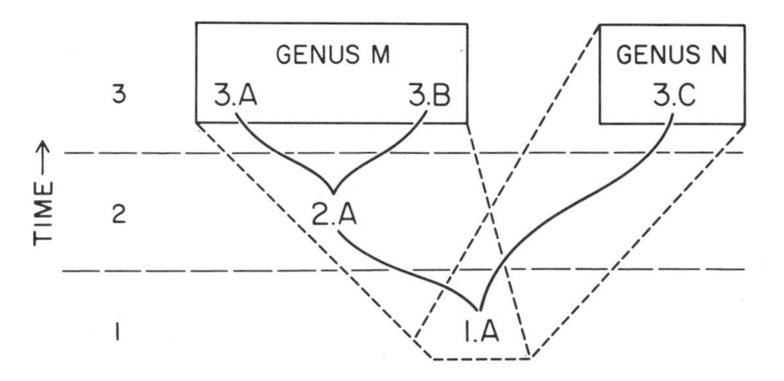

Fig. 3-3. PHYLOGENETIC RELATIONSHIPS. A lower-ranking unit, e.g., **1.A**, "belongs" to two higher-ranking units, Genus **M** and Genus **N.**

different family did not derive from this same ancestral species, although their union in the next highest taxon would imply a common ancestry at a more remote time.

Phylogeny vs. classification

The "family tree" of Figure 3-3 is superficially an inversion of the hierarchy of Figure 3-2. So long as one deals, like biologists, with only contemporaneous taxa, i.e., the intersection of phyletic lineages with a time plane, this similarity is sufficient. Species 3.A and 3.B have a proximal common ancestor, 2.A and are included in the same genus; 3.C has a somewhat more distal connection through 1.A and so is placed in a separate genus but the same family.

But what is the relation of 2.A and 3.A? Separate genera? If so, what of a point midway between the two? Or do they belong to the same genus? Then is 1.A in the same genus or, if not, why not? And 1.A as the common ancestor of 3.A and 3.C must belong to both genera (or families). But in a hierarchy, a species can belong to only one genus. In consequence, simplifying transformations are necessary to map phylogeny on a hierarchy—since these involve evolutionary processes and patterns they are considered in detail in Chapters 6 and 8.

Phylogenetic inference

The rules of the hierarchic classification and mapping of phylogeny are procedural problems, but the inference of genetic and phyletic systems from the observed organisms or fossils are substantive ones. A biologist who infers membership in a genetic system can test this inference—though typically he doesn't (doesn't need to?) do so.

A phylogenetic system is a flux observable only by samples drawn from the system. For a biologist these samples are from a single time slice, and he must infer prior states of the system entirely from the present state. Paleontological samples are scattered through the system, but the distance between most samples is great and the knowledge of the state sampled is limited. Thus connections between samples involve considerable uncertainty.

The principal data of phylogenetic inference derive from fossil form, although distribution in space, in time, and in environment is also important (and for the biologist, physiology and behavior). These data are weighted on a scale one can call "evolutionary transformation." This transformation involves development of an evolutionary model—presented in Chapter 6.

Classification vs. identification

The process of identification is essentially the reverse of classification. A classification and the taxa forming it are derived from study of three or more (typically, many more) fossils about which one can conclude that several are more like one another (belong to the same taxon) than they are like the others in the set. Identification places a fossil within a previously recognized taxon on the basis of previously defined criteria. Inability to identify a fossil may indicate that the classification is unsatisfactory and necessitate a significant redefinition and/or creation of new taxa. But, and the distinction is important though subtle, the new classification should derive from a restudy of the total set of fossils—not from the inability to identify.

Concluding thoughts

A classification should perform three important functions:

a). Provide communication about significant relationships (both differences and similarities) with an appropriate audience.

b). Incorporate maximum information about the biological systems and, for geologists, about their temporal distribution.

c). Organize apparently heterogeneous data to further research and interpretation.

A classification is a tool honed to fit a need. Several different classifications can exist simultaneously to serve alternate purposes. Computers can organize data, transform from one classification to another, and retrieve information for identification in any chosen classification; they open new potentials in classification and identification if we are willing to think radically enough about the problems.

The major animal groups

Zoologists recognize approximately twenty phyla in the animal kingdom (Table 3-1), the number varying with the allocation of certain small groups to the major phyla or to a phylum of their own. These phyla differ in complexity of organization, in symmetry, in the presence and kinds of spaces within the body, in the absence or presence of an anus, in metamerism, in the possession of limbs of one sort or another, in excretory, respiratory, and skeletal systems, and in the pattern of development of these features.

Some phyla are highly diversified in form and in mode of life; others are very limited in diversity. The phylum Chordata, for example, includes three subphyla and one of these subphyla, the Vertebrata, comprises eight classes. Each class is a variation upon the plan of the phylum, and these variations are related to an adaptation to a way of life: the birds (Class Aves) to flight; the bony fishes (Class Osteichthyes) to swimming; the frogs, salamanders, and their extinct relatives (Class Amphibia) to existence in transition between aquatic and terrestrial life. Although within a phylum some classes are adapted to a similar mode of life, the adaptations either involve different levels of organization (like the Mammalia and the Reptilia) or different approaches to the same adaptations (like the sharks, Chondrichthyes, and the bony fishes, Osteichthyes).

The subclasses and orders, of course, show still finer variation in detail within the plan of phylum and class and consequently more limited adaptation. Some of these subdivisions are worth closer attention, but this can best be done in the concluding chapters. Table 3-1 therefore includes only the larger divisions in animal classification. The groups most important in the fossil record are shown in boldface. (See pp. 54-59.)

REFERENCES

Anderson, E. 1957. "An Experimental Investigation of Judgments Concerning Genera and Species," *Evolution*, vol. 11, pp. 260-263. On the psychology of classification.

Beckner, M. 1959. *The Biological Way of Thought*. New York: Columbia University Press. Analysis of philosophy of classification.

Buchsbaum, R. 1948. *Animals without Backbones*, 2nd ed. Chicago: University of Chicago Press. Brief but excellent on the invertebrate phyla.

Easton, W. H. 1960. *Invertebrate Paleontology*. New York: Harper and Brothers. A summary of morphology and stratigraphic occurrence.

Gregg, J. R. 1954. *The Language of Taxonomy*. New York: Columbia University Press. Consideration of the logic of taxonomy.

Hyman, L. H. 1940. *The Invertebrates*. New York: McGraw-Hill. A series that will, when finished, cover in detail all invertebrate groups.

Mayr, E., E. G. Linsley, and R. L. Usinger. 1952. *Methods and Principles of Systematic Zoology*. New York: McGraw-Hill. A description of taxonomic practice.

Moore, R. C., C. G. Lalicker, and A. G. Fisher. 1952. *Invertebrate Fossils*. New York: McGraw-Hill. A text dealing with morphology and stratigraphic occurrence.

Piveteau, J. (Ed.). 1952. *Traité de Paléontologie*, 7 vols. Paris: Masson et Cie. An authoritative treatment of all fossil animal groups.

Rescigno, A. and G. A. Maccacaro. 1961. "The Information Content of Biological Classifications," *Information Theory*, Fourth London Symposium, 1960, pp. 437-446.

Romer, A. S. 1966. *Vertebrate Paleontology*. Chicago: University of Chicago Press. The standard textbook on vertebrate fossils.

TABLE 3-1

The animal phyla

Unless otherwise noted, the members of each phylum retain the characteristics of the preceding phylum as well as possessing their own unique structures and organization. A question mark before a geologic age implies that the identification of the group in rocks of this age is uncertain. The more important fossil groups appear in **bold** type.

Phylum	*Characteristics*	*Modern representatives*	*Mode of life*	*Geologic range*	*Importance as fossils*
Protozoa	Unicellular animals of various grades of complexity. Some colonial types. Some have skeletons of calcite, of silica, or chitinoid material. Skeleton may show spherical, radial, or bilateral symmetry; may be coiled or asymmetric.	*Euglena*, *Amoeba*, *Paramecium*, *foraminifers*, *radiolarians*, and so on.	Aquatic, parasitic.	(?)Precambrian, Ordovician to Recent.	Types with skeleton, particularly *Foraminifera*, of great importance in stratigraphy of well cuttings.
Porifera	Many-celled animals of low grade of complexity. Cells show incipient organization into tissues; lack organs. Most have skeleton of spicules of calcite or silica or of organic fibers; spongin. Some have radial symmetry; others asymmetric. System of canals and pores.	Sponges. Approximately 3000 recent species.	Aquatic. Attached to bottom. Primarily in waters with little mud. Feed on microorganisms filtered from water.	(?) Precambrian, Cambrian to Recent.	Spicules common; complete skeletons less so. Of relatively small importance except for a few genera.
Mesozoa	Many-celled animals of low grade of complexity. Incipient development of tissues but lack organs. No skeletons. Elongate mass of slightly differentiated cells.	Small number of obscure species.	Parasitic.	Recent.	Unknown as fossils, but of interest as possible connection of protozoans to the many-celled animals.

Coelenterata	Tissues developed and include three layers of cells, but organs only slightly developed. Middle cell layer develops largely from external layer. Large central body cavity with single opening, the mouth. Mouth encircled by tentacles. Some have skeleton of $CaCO_3$. Radial or biradial symmetry; a few with bilateral symmetry. Some colonial.	Jelly fish, hydra, various types of corals. Approximately 10,000 recent species.	Aquatic, predominately marine. Attached to bottom or free swimming and/or floating. Primarily predaceous.	(?) Precambrian, Cambrian to Recent.	Those with calcareous skeletons common and interesting fossils. Soft bodied types rare.
Ctenophora	No skeleton. Tentacles do not encircle mouth. Bear plates of cilia for swimming. Biradial symmetry. None colonial. Medial cell layer (mesoderm) developed from internal cell layer.	Comb jellies	Aquatic. Free swimming. Primarily predaceous.	Recent.	Unknown as fossils.
Platyhelminthes	Tissues developed; some organs differentiated. Body cavity opens only through mouth. Bilateral symmetry; flattened; head slightly differentiated.	Flat worms. Over 6000 recent species.	Aquatic, primarily crawling or swimming over bottom. Parasitic.	Recent.	Unknown as fossils.
Nemertinea	Anus present as well as mouth. Extrusible proboscis.	Nemertine worms. About 500 recent species.	Largely marine. Predaceous.	(?) Jurassic.	Two genera of doubtful affinities.
Entoprocta	Cavity between wall of gut and body wall; cavity develops as remnant of hollow in "hollow-ball" stage of development and is not completely lined by medial cell layer (p. 267). Gut looped so that anus is near mouth. Mouth encircled by ciliated tentacles. No proboscis. Most are colonial.	Small number of obscure species.	Fresh water, attached to bottom. Collect microscopic animals with ciliated tentacles.	Recent.	Unknown as fossils.

TABLE 3-1 (continued)

Phylum	*Characteristics*	*Modern representatives*	*Mode of life*	*Geologic range*	*Importance as fossils*
Aschelminthes	Gut not looped but straight; no ciliated tentacles. Not colonial.	Large number of parasitic and free living "worms." Rotifers.	Aquatic; parasitic.	(?) Eocene.	Five fossil genera.
Bryozoa	Body cavity between wall of gut and body wall; cavity formed within medial cell layer and is lined completely by cells from this layer. Gut looped to bring anus near mouth. Circle or horseshoe of ciliated tentacles about mouth. Skeleton of $CaCO_3$. Colonial.	Moss animals. Approximately 3000 living species.	Aquatic. Attached to bottom. Nearly all are marine. Feed on microscopic animals filtered by ciliated tentacles.	Ordovician to Recent.	Important in numbers and variety. Useful in correlation and study of ancient environments.
Phoronida	Ridge bearing tentacles is horseshoe shaped. Simple circulatory system of closed vessels. Solitary. No skeleton or chitinous tube.	Approximately 15 species.	Aquatic, marine. Burrowers. Feed on microscopic organisms collected by ciliated tentacles.	Recent.	Unknown as fossils.
Brachiopoda	Ridge bearing tentacles is drawn into elongate coiled arms. Solitary; bivalve; shell chitinophosphatic or calcareous; valves dorsal and ventral; each valve symmetrical about midline.	Lamp shells. About 120 recent species.	Aquatic, marine or brackish water. Feed on microscopic organisms collected by ciliated tentacles.	Early Cambrian to Recent.	Common; particularly abundant and varied in Paleozoic rocks. Very important in correlation and in studies of fossil environments.

Mollusca	Without ridge bearing tentacles. Most have calcareous shell of one or more pieces. If bivalve, valves are right and left. Muscular foot (in some modified to arms) for locomotion. Head differentiated; well developed circulatory, nervous, respiratory and excretory systems. Spaces of internal body cavity reduced in size. Circulatory system includes spacious cavities about organs. Differentiated dorsal skin, the *mantle*.	Chitons, tusk shell, snails, clams, squids. Approximately 70,000 living species.	Aquatic, marine and fresh water; terrestrial. Attached to bottom, burrowing, crawling, swimming. Predators, scavengers and herbivores. One group (the clams) filters microscopic organisms from water passing over gill system.	Early Cambrian to Recent.	Probably most common and varied group of fossils. Very important in correlation and in studies of fossil environments.
Siphunculoidea	Spaces of body cavity large; without shell or mantle. Proboscis. Anus dorsal.	Small number of obscure species.	Marine	Cambrian to Recent.	Only six genera.
Priapuloidea	Anus terminal. Otherwise similar to siphunculoids.	Small number of obscure species.	Marine	Recent.	Unknown as fossils.
Echiuroidea	Without proboscis. Otherwise similar to Priapuloidea.	Small number of obscure species.	Marine	Recent.	Unknown as fossils.
Annelida	Body divided into segments. In some the segments may bear pairs of appendages. Well developed nervous, excretory, circulatory and respiratory systems. Head not strongly differentiated. Internal body cavity relatively large. Anus terminal. No proboscis.	Earth worms, leeches, and so on. About 7000 recent species.	Aquatic; some terrestrial. Burrowing, crawling, swimming. Some parasitic. Predators, herbivores, scavengers.	(?) Precambrian, Cambrian to Recent.	Chitinous jaws found occasionally as micro-fossils. Otherwise very rare.

TABLE 3-1 (continued)

Phylum	*Characteristics*	*Modern representatives*	*Mode of life*	*Geologic range*	*Importance as fossils*
Arthropoda	Segments typically differentiated into head and one or two other regions. Paired, jointed appendages. External skeleton of chiton; the skeleton in some is impregnated with $CaCO_3$. Circulatory system includes large spaces around organs. Internal body cavity reduced in size.	Insects, shrimp, lobster, spiders, barnacles, and so on. Over one million recent species.	Aquatic; terrestrial. Burrowing, crawling, swimming, flying. Predators, herbivores, scavengers, parasites.	(?) Precambrian, Cambrian to Recent.	Trilobites common and significant in early and middle Paleozoic faunas. Ostracods important since Ordovician.
Chaetognatha	Non-segmented. No paired appendages. No skeleton. Nervous system poorly developed; no definite excretory and circulatory organs. Bilateral symmetry. Anus subterminal on ventral surface.	Arrow worms; about 30 species.	Aquatic, marine. Swimming. Predators.	(?) Middle Cambrian. Recent.	Questionable fossils in Burgess shale.
Echinodermata	Secondary radial, typically pentamerous, symmetry. Torsion of viscera in development changes position of mouth and anus. System of radial canals and associated structures called water vascular system. Skeleton of calcareous plates and spines developed in medial cell layer just below skin.	Sea lilies, sea-cucumbers, starfish, brittle stars, sea urchins. Approximately 5000 species	Aquatic, marine. Attached to bottom or crawling. Feed by collecting micro-organisms on ciliated tracts or are predators, herbivores, and scavengers.	Cambrian to Recent.	All groups except starfish and sea-cucumbers common and varied.

Protochordata	Bilateral symmetry. No water vascular system. Gill slits between anterior gut and external surface. Nervous, excretory, circulatory systems moderately well developed. Some colonial. Chitinoid skeleton in some. Anus terminal.	Acorn worms.	Aquatic, marine. Burrowing, attached to substrate or floating.	Cambrian to Recent.	Only a group of Paleozoic colonial forms, the graptolites, are common.
Chordata	Notochord and gill slits well developed, at least in embryo. Nervous, excretory, circulatory, and respiratory system well developed. Segmented. Most have internal skeleton of bone and/or cartilage. Anus subterminal.	Sea squirts, fish, amphibians, reptiles, birds, mammals. About 65,000 species.	Aquatic, marine and fresh water; terrestrial. Burrowing, crawling, swimming, flying. Predators, herbivores, scavengers. Some filter microorganisms out of water passing over gills.	Ordovician to Recent.	Rare as fossils but of great interest to their fellow vertebrates.

4 The species

A species is . . . ? A collection of similar fossils? An abstract construct based on such a collection? A population? A genetic system? The answer must be all of these things—and many more. Just as a taxon comprises individuals overlapping in a large number of characteristics but never identical, so the category "species" comprises a large number of taxa with overlapping but not completely congruous properties. The phenomena are neither simple nor static; definition of the concept "species" merely sets a standard and helps one decide that one taxon is a species and another is not. Therefore what I say here simplifies an extremely complex reality.

The biological concept

A few acres of woodlot shelters several dozen adult deer mice. Any male in the woodlot has an equal chance of breeding with any female; the population clearly represents a unitary genetic system. It is, therefore, a part of a species population. Another woodlot five miles away shelters another such deer mouse population. No individual in this woodlot has a finite chance of breeding with an individual in the first and distant woodlot; it has, however, a large chance of mating with an individual in an adjacent overgrown pasture; an individual in the pasture with one in a swampy valley-bottom forest; one from the forest with one in a hedgerow, and that one with an individual in the original woodlot. This network is a larger part of a species population; if one follows out all such probable breeding relationships the entire species population is defined. Other individuals who coexist in the same or in distant areas but who lack any connection with this breeding network belong to other species populations.* You will note I have not defined "probable" with precision; this reflects the existence of examples in which interbreeding is very infrequent but may occur. Such cases must be argued out individually.

As a consequence of the reproductive network, a species population shares a common pool of hereditary material. Or to put it another way, the hereditary complements of the individuals in one generation represent a reassortment or *recombination* of those in the preceding generation.

A species, though defined as a genetic system, possesses other distinctive properties. The individuals, typically, are similar in physiology and form—partly as a consequence of their similar genetic composition and partly as a result of similar environments of development. Even if local populations within the system differ, intermediate populations intergrade. Very obviously, organisms with similar physiology and form perform similar activities and survive in similar environments. As will be demonstrated in a succeeding chapter, this correspondence of genetic, physiological, and ecologic systems results from evolution. In contrast, local populations from different systems lack intergrades in genetics, form, and physiology, have distinct ecologies, and have had different evolutionary histories.

As defined, a biologic species has no extension in time, since the "probable breeding relationship" either exists or does not exist at any instant—regardless of how long it may take to test for the relationship. A biological species is a time-parallel slice across a phyletic continuum, a continuum engendered by continuous operation of a genetic system. In a very strict sense, two genetic systems represent distinct species only if contemporaneous. If separated in time, the genetic definition becomes meaningless because no probability of interbreeding exists.

In practice, a biologist works with samples distributed through time, so that "contemporaneous" approximates a few decades or centuries. From his samples he constructs a species "model" which presumably conforms to the genetic system. Typically, the data for these models are morphologic and geographic, although more and more physiologic, ecologic, and genetic

* This definition excludes obligatory asexual organisms whose only hereditary relationship must be phyletic. "Species" of such organisms are analogous to a supraspecific category in sexually reproducing organisms.

information is being used. Such a model permits the estimation of properties, e.g., rate of genetic interchange, range of ecological tolerances, etc., not observed in the original data.

The paleontologic problem

The average paleontological species model does not differ qualitatively from the average biological one. It differs quantitatively because the samples are smaller, bear a more complex relationship to the population, and are distributed through a longer (and partly unmeasurable) period of time. The basic data consist almost entirely of skeletal morphology, stratigraphic position, and geographic distribution, with some contributions from rock association and position with the rock. The model therefore uses a large number of "normic" statements, i.e., those that assert that a relationship holds "normally" or "under certain conditions (p. 11)." For example, a paleontologist might state that the ecologic properties of a species include survival in rough-water environments (model inference) because *a*) its morphologic properties include massive skeletons (observation) and *b*) massive skeletons are typical of modern species living in rough-water environments (a normic statement, because the property occurs, albeit rarely, in quiet-water species).

The fossil sample

With these restrictive properties of the paleontologic species model, the relation of the fossil sample to the living population assumes great significance. Figure 4-1 displays the history of a brachiopod population

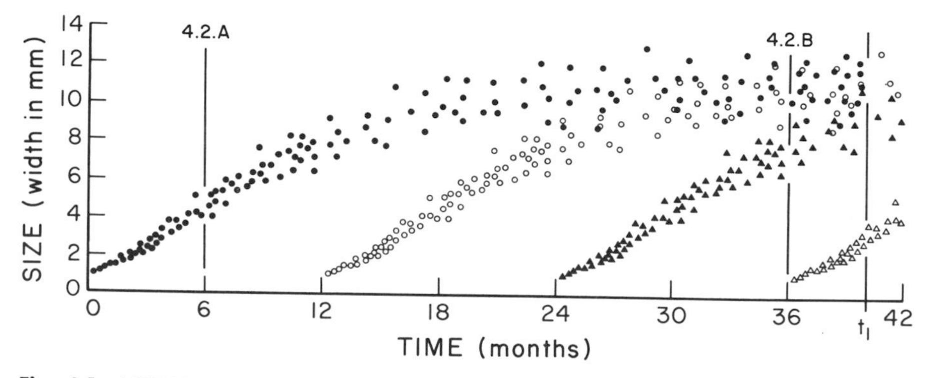

Fig. 4-1. HISTORY OF A BRACHIOPOD POPULATION. Each point plotted corresponds to the width of the shell at the death of the individual. Four distinct generations spaced twelve months apart are shown. Census results at 6 months and 36 months are illustrated at 4.2.A and 4.2.B.

through several years as growth of shell width plotted against time. Figures 4-2A and 4-2B are slices of this history and show the frequency of various widths at successive intervals for a single generation. Any such slice may be a *target population* for the paleontologist even though it is

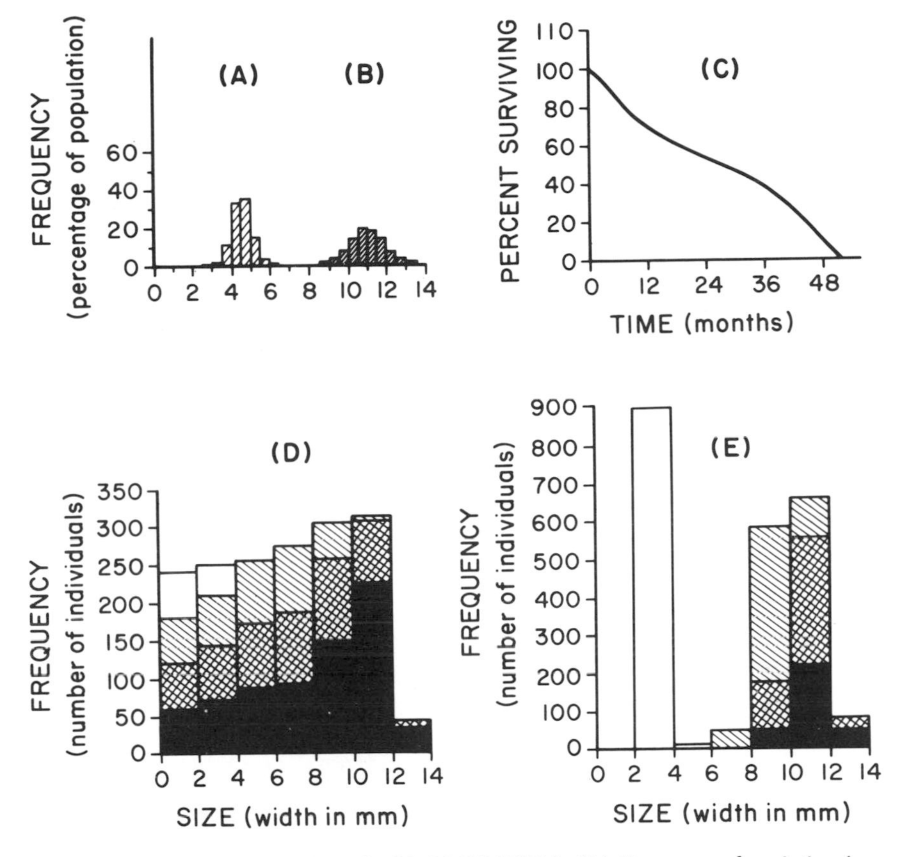

Fig. 4-2. CENSUS OF A BRACHIOPOD POPULATION. **(A)** Histogram of variation in width at 6-months census of population in Fig. 4-1. **(B)** Variation in width at 36 months. **(C)** Survivorship (%) of single generation. **(D)** Cumulative census of brachiopods dying during 40-month interval (normal mortality). Contribution of each generation shown by distinct pattern. **(E)** Instantaneous census of brachiopods by catastrophic mortality at 40 months. Contribution of each generation shown by distinct pattern.

only a sample of a more inclusive target, the temporal sequence of populations.

A fossil sample derives from a population in a series of steps commencing with the death of the organisms and terminating with the classification of the fossils. The incidence of mortality controls the fossil sample initially; death is the census-taker. Figure 4-1 displays two distinct modes of mortality. From time t_0 to time t_1, the deaths are a consequence of a large number of factors, environmental stress, predation, disease, etc. A *survivorship curve* (Figure 4-2C) expresses the percentage of an age group surviving at successive intervals; this is *normal mortality*. At time t_1, however, *catastrophic mortality** eliminates the surviving individuals at once. Normal and catastrophic mortality deliver quite different samples (Figures 4-2D and 4-2E) for fossilization, but many fossil collections probably mix normal and catastrophic censuses.

With death, the organic remains shift into a realm of factors and processes that I will call geologic. They include transportation, burial, chemical dissolution, and modification before and after burial, mechanical stresses imposed during compaction and deformation, and the effects of erosion and weathering on the outcrop. Figure 4-3 illustrates the effects of these events on the mortality samples delivered in Figure 4-2.

I'll take two extreme cases as illustrations. First assume: 1) the census is a combination of normal mortality ($t_0 - t_1$) and catastrophic mortality at t_1; 2) transportation is nil; 3) burial occurs at time t_1; 4) skeletons of all individuals are preserved; 5) mechanical deformation is insignificant; 6) destruction by erosion and weathering is random with respect to size, form, and location of the fossil. The fossil population generated and available for collection is, in consequence, a random sample of the individuals living at the site during the interval t_0 to t_1 (Figure 4-3A). Very young individuals without skeletons are, of course, lost, so that any inferences about the biologic population require normic statements (see above) of the relationship between such nonpreservable individuals and the preservable ones in modern biologic systems.

The other extreme is represented in Figures 4-3B and 4-3C. At t_1 the living population is transported from the life site after catastrophic mortality and deposited elsewhere, is sorted by size and shape and is modified by breakage of very large skeletons (Figure 4-3B). Burial is again complete, but solution destroys small skeletons (Figure 4-3B). Erosion breaks some larger skeletons, and weathering dissolves many smaller ones. Therefore, the fossil population is a highly biased sample (Figure 4-3C) of the target, and reconstruction of the original requires a large number of normic statements. Since such statements operate only in some conditions (say 90% of the time) the final depiction of the population is rather crude (five superposed statements each with 90% likelihood yield a conclusion with approximately 40% likelihood). Most fossil populations lie between these extreme examples—but this makes them no easier to interpret.

But the fossils are still in the rock—not classified in museum trays. A paleontologist who stops for a few minutes at a road cut outcrop which contains a fossil population like that of Figure 4-3A will almost certainly collect only the most obvious specimens—typically the larger ones (Figure 4-3D)—so that the sample distribution is not that of Figure 4-3A but of

* Catastrophic for the local population being sampled but normal in the total species survivorship curve.

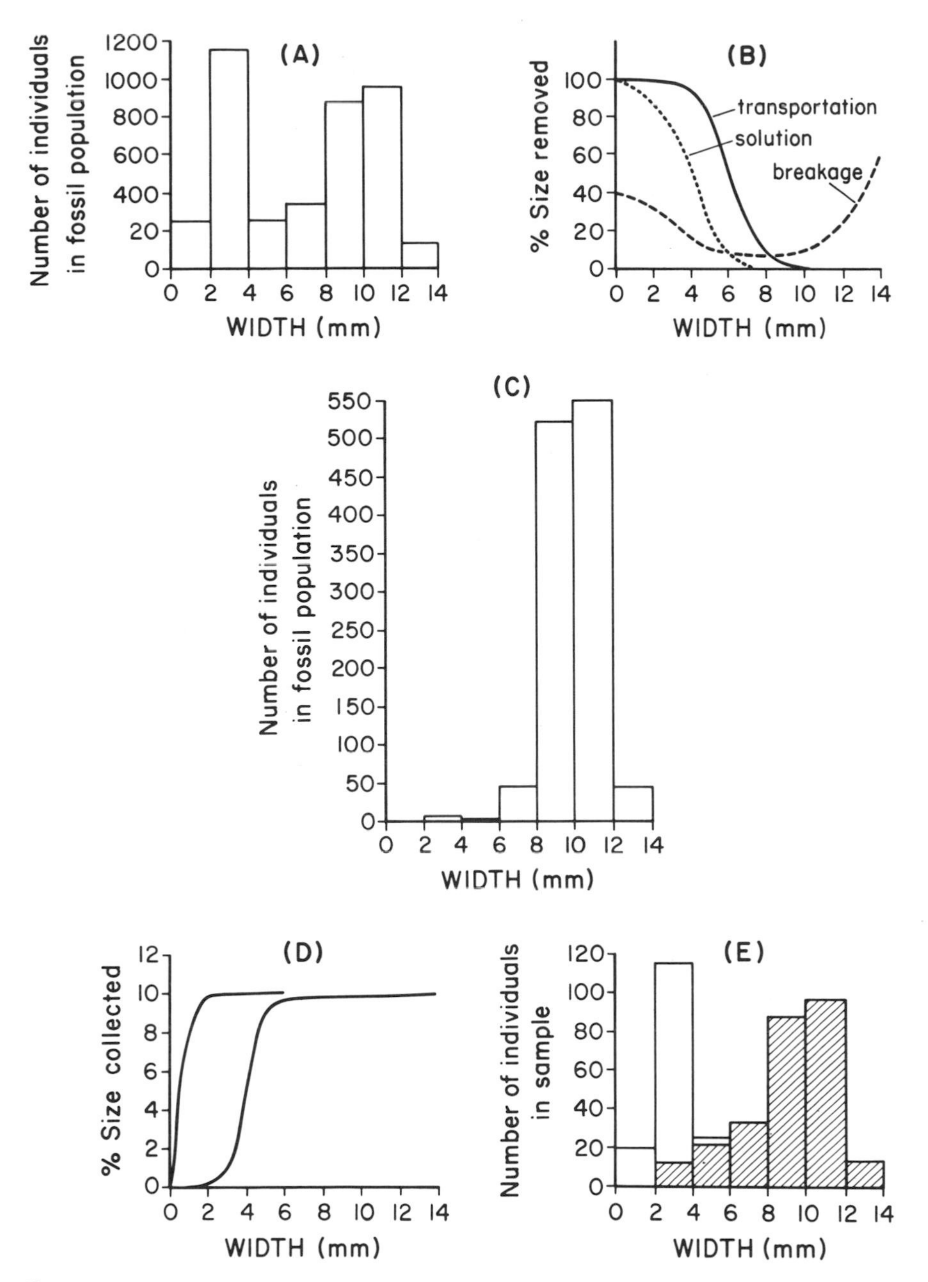

Fig. 4-3. FOSSIL SAMPLES OF A BRACHIOPOD POPULATION. **(A)** Fossil population formed from cumulative and instantaneous censuses at 40 months **(4-2(D)** and **4-2(E))** with no subsequent loss by mechanical or chemical destruction. **(B)** Selective preservation rates for various shell sizes. **(C)** Fossil population formed from instantaneous census at 40 months with subsequent selective modification as shown in **4-3(B)**. **(D)** Collection bias with size of fossil by casual collecting (curve at right) and exhaustive collecting (curve at left). **(E)** Fossil sample generated by application of collection bias curves to fossil population of **4-3(A)**. Patterned bars from casual collecting; patterned and open bars by intense collecting.

Figure 4-3E—which has quite a different significance. Another paleontologist works slowly and carefully so that his collection approaches a random sample of the population, but concludes from the bimodal distribution that two distinct species populations are present. Such distortions by *operator error* introduce additional uncertainty into a reconstruction of the living population.

Problem choice and sampling design

And where is scientific rigor in this welter of uncertainty? There is no foolproof answer. If two or more completely or partially independent statements about the population can be made, they serve to test and confirm each other. Equally important, the nature of the problem specifies the degree of rigor. Paleontologists make a choice. They may ask, looking at a particular set of rocks and fossils, what meaningful statements they can derive from it. Or they can formulate a significant question and then search for the proper sets of rocks and fossils to answer the question. The questions form a spectrum from the easily answerable but essentially trivial to the highly significant but unanswerable.

In any case, the first question concerns the geologic bias entered in the fossil sample. Typically, quantitative evaluation is difficult or impossible; at best one can say that transportation was large or small, differential chemical destruction considerable or slight. Table 4-1 summarizes some criteria for geologic modification.* The greatest amount of biologic information derives from samples in which geological modifications have been relatively slight and uniform. Valuable geologic and biological information can emerge however from biased samples if one asks the proper questions or is satisfied with low levels of likelihood. For example, the comparison of morphological variation between two samples such as those defined in Figure 4-2A yields rather precise statements about differences between the living populations. Comparison of relative growth rates between samples like Figure 4-2C would be less precise but equally valid. On the other hand, the trace element composition in a recrystallized Paleozoic coral compared with that of a living coral representing a different order produces a nonsense answer.

Control of operator bias is equally critical. Control begins with the choice of an appropriate sampling design and concludes with application of logical, uniform, thoroughly tested methods of classification and measurement. Sampling modes (following Krumbein and Graybill, 1965) include *a*) *search* in which any relevant specimen is collected without conscious pattern, *b*) *purposeful* in which a chosen class of specimens is col-

* These are themselves "normic" statements and so enter, at the beginning of a study, a certain degree of uncertainty.

TABLE 4-1

Criteria for geologic modification of fossil samples (*based in part* on Johnson, 1960)

Feature	*Unmodified*	*Modified*	*Processes*
1. Individual fossils			
a. Surface	Delicate spines, etc., preserved.	Surface smooth, worn.	Abrasion. Solution.
b. Composition	Similar to living relatives; microstructure and unstable minerals preserved. Chemical heterogeneity.	Dissimilar to living relatives; microstructure not preserved; no unstable minerals; pitting; homogeneous.	Chemical diagenesis.
c. Microstructure	Similar to living relatives; consistent with biological nucleation of growth.	Dissimilar to living relatives, consistent with random or physically oriented nucleation of growth.	Chemical diagenesis.
d. Physical condition	Unbroken.	Fragmented.	Transportation. Reworking.
e. Articulation	Separate parts articulated.	Parts separated.	Transportation. Reworking.
f. Position	Not stable hydrodynamically; life position indicated by interpretation of function and/or position in living forms.	Hydrodynamically stable.	Transportation. Reworking. Burrowing.
2. Aggregates			
a. Faunal composition	Consistent with inferred ecologic overlap of taxa.	Inconsistencies in ecologic limits.	Transportation. Reworking.
b. Morphologic variation	Wide variety of shapes, delicate specimens intact.	Limited variety of shapes, delicate specimens missing or broken.	Abrasion. Transportation. Diagenesis.
c. Size distribution	Consistent with census of living population; different taxa show different distributions.	Inconsistent with census of living population; taxa with similar size, shape, etc., show similar distributions.	Transportation. Diagenesis.
d. Orientation	Random or biologically functional.	Nonrandom and nonfunctional.	Transportation. Reworking.
e. Dispersion and association	Similar to comparable modern biologic systems; inconsistent with transportation.	Consistent with transportation; dissimilar to comparable modern biological systems.	Transportation. Reworking.
f. Frequency of disassociated parts	Frequency of disassociated parts appropriate for frequency in articulated individuals.	Frequency of disassociated parts deviates from frequency in articulated individual.	Transportation. Diagenesis.
3. Lithologic associations			
a. Structures	Indicate low flow velocity relative to most easily transported fossils.	Indicate flow velocity appropriate to transport all fossils.	Transportation.
b. Size frequency of grains	Inconsistent with size frequency distribution (hydrodynamic) of fossils.	Consistent with size frequency distribution (hydrodynamic) of fossils.	Transportation.
c. Crystalline texture	Little evidence of diagenesis.	Diagenetic.	Diagenesis.
e. Environmental inferences	Consistent with environmental limits of taxa.	Inconsistent with environmental limits of taxa.	Transportation.

lected, and *c*) *probability* in which any specimen encountered in a designed random pattern is collected. Probability sampling permits inference from fossil sample to fossil population by rigorous statistical methods; search and purposeful sampling require application of "normic" statements to relate fossil sample and fossil population; they, in effect, add another stratum of uncertainty.

Probability sampling methods include collection at points chosen by some random method, e.g., grid intersections selected from a table of random numbers. Alternatively, collection may be systematic, e.g., every five feet along a layer or up a section; such collection will be random because the sampling interval is unrelated to any features in the rocks to be sampled. If inspection indicates a marked variation within the layer or section, scattered coral banks within a shale, for example, *stratified sampling* may be desirable. In such cases, the different parts are sampled separately—either systematically or randomly. Thus, in the example selected, separate sampling grids would be used on the coral banks and on the shale; the shale grid could be used to define the frequency of coral banks in the total outcrop.

Genetic systems

A biological species is a model of a genetic system. A paleontological species is a model of a genetic system continuum. What then is a genetic system—beyond the interbreeding relationship?

To answer, one must examine the material basis of heredity. Biologists have demonstrated that most if not all inheritance corresponds to a coding within giant helical molecules (DNA) located in cell nuclei. This coding yields discrete hereditary units, *genes;* each gene is defined by its position, the *gene locus,* on a ribbon-like body, a *chromosome,* within the cell nucleus. Typically, a chromosome bears one of several variant genes at a locus; the series of variants are called alleles. The set of genes within an organism provides a particular genotype. The DNA code determines, in ways not yet fully understood, the activities of the cells and thus the activities and structure of the organism, i.e., its phenotype. Allelic genes modify cellular activities and produce alternate phenotypes. Further inquiry into gene activity and structure is unnecessary here, but I should emphasize *a*) that the genetic material consists of units inherited as discrete particles, *b*) that a gene effects a process operating in an environment, *c*) that several genes may contribute to a complex process system, *d*) that, conversely, a single gene may affect several different process systems, and *e*) that few if any structures except molecular structures stand for, represent, a single gene.

The hereditary mechanism

The behavior of the chromosome in reproduction controls gene transmission. Most cells possess duplicate sets of chromosomes, i.e., chromosome pairs which bear the same gene loci. In ordinary cell duplication

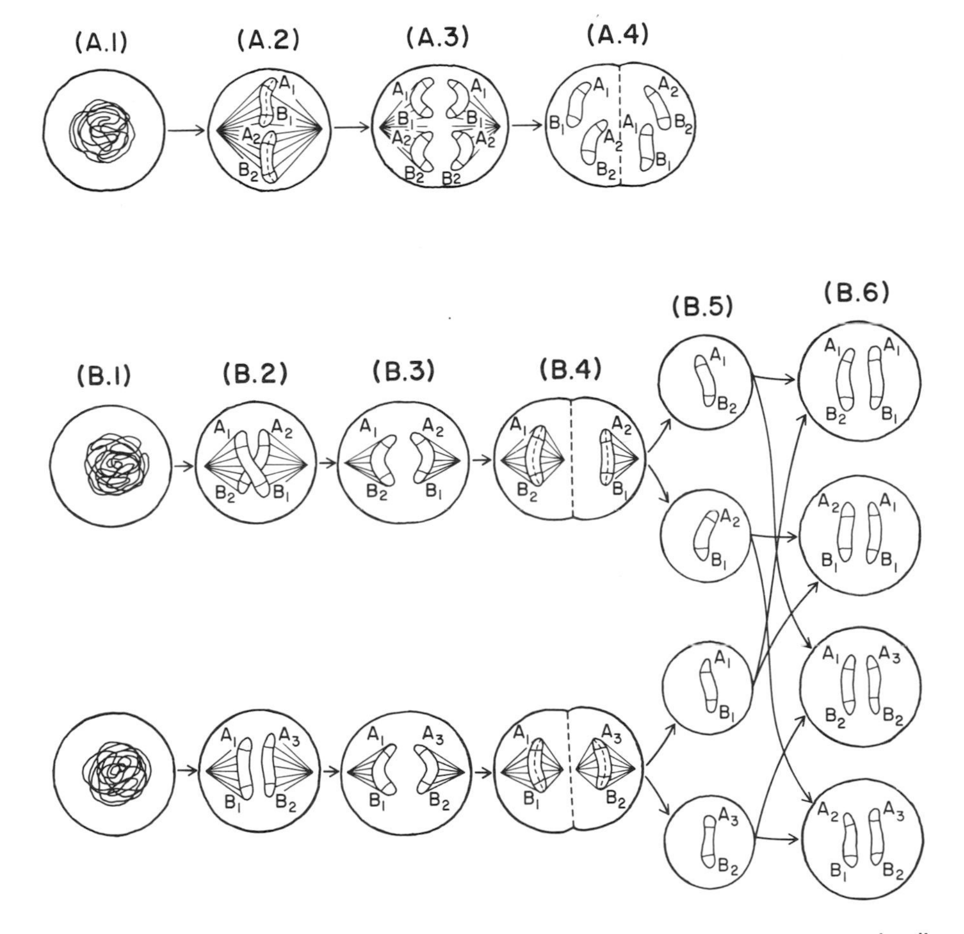

Fig. 4-4. CELL REPRODUCTION AND HEREDITY. **(A)** Successive stages in asexual cell division. The rod-shaped chromosomes split longitudinally and each daughter cell has the same hereditary complement as the original. $\mathbf{A}_1$ and $\mathbf{A}_2$, $\mathbf{B}_1$ and $\mathbf{B}_2$ are variant genes (alleles) occupying the same position (locus) on the chromosome. **(B)** Successive stages in sexual reproduction. Daughter cells after division bear only half the original hereditary complement. Fertilization combines hereditary complements from the two parents. **(B.2)** shows the interchange of material between chromosomes (crossing over).

(Figure 4-4A), each chromosome produces an identical copy of itself, and, as the cell divides, the copy is split off and passes into one of the new cells.

The "daughter" cells, therefore, contain the same chromosome pairs and same genes as the parent, and nearly every cell in an organism possesses an identical genotype. If one or more of the duplicate cells splits off to constitute a new individual, the process is called *asexual reproduction.* All the individuals descended from a single parent form a *clone* and possess, at least initially, the same genotype.

Some cell duplications, however, display a different pattern, one that ends in *sexual reproduction.* At the time of division, the chromosomes of a pair join side by side (Figure 4-4B). Typically, there is some breakage and interchange (*crossing-over*) of parts of the chromosomal material so that each member of the pair becomes, in effect, a hybrid of the original pair. As division continues, one chromosome of each pair passes to each daughter cell. These are the sex cells (or gametes) and each possesses half the normal chromosomal and gene complement (n rather than $2n$). The assortment of chromosomes in each gamete is random, i.e., a particular chromosome of a pair has one chance in two of being in a particular gamete, the chance of two particular chromosomes is one in four, etc. At fertilization, two gametes unite to form a new individual (zygote) with the normal amount of genetic material ($2n$) reconstituted. Note that crossing-over, random assortment, and fertilization *recombine* genes from the original parental chromosomes in a new arrangement.

If all the individuals in the interbreeding population possessed identical sets of genes, recombination would lack significance; it would simply put back together that which was torn asunder. But no such populations are known, for the DNA code is not completely stable and may, infrequently, change. Such changes (or *mutations*) produce *mutant genes;* these become the variant alleles found at a particular gene locus. These alleles are switched into new combinations by crossing over and fertilization to produce novel genotypes (Figure 4-4B).

As an example, a particular gene A_1 changes in structure to A_2. In a few generations some individuals will have a complement of A_1 A_2 (the *heterozygous state*); others A_2 A_2 (the *homozygous* state); and the remainder will be homozygous for the original A_1 allele. Another gene B_1 mutates to B_2. Different individuals in the population can now carry one of nine combinations: $A_1\ A_1\ B_1\ B_1$; $A_1\ A_2\ B_1\ B_1$; $A_2\ A_2\ B_1\ B_1$; $A_1\ A_2\ B_1\ B_2$; etc. Furthermore A_2 might mutate to A_3; A_3 to A_4 and so on. If each cell contained only one hundred different loci with four alleles at a locus, recombination could yield, theoretically, 10^{100} genotypes.

Observation demonstrates, however, that the phenotypes in a species population are quite similar. Either a relatively small number of gene systems predominates in the population, and/or most genotypic differences produce only small changes in phenotype. Experimental studies of gene frequencies and gene action indicate that both of these alternatives are true: Some genes are much more common than others, and only a very few of the observed genotypic differences result in major phenotypic differences.

Genes and populations

But why do some genes have high frequencies and others low? To answer that I'll have to go back to the description of genetic recombination and exhume some buried assumptions. First, mutation affects gene frequencies. For example, if one out of every million A_1 genes mutates to A_2 in every generation and if A_2 does not mutate at all, the population will eventually lose all of its A_1's, and the frequency of A_2 will be 1.0. A different mutation rate changes the time for A_2 to replace A_1 but not the result. In contrast, if A_2 mutates back to A_1 at an equal rate, the frequencies of A_2 will increase to .5 but go no further. In a population of one million individuals at this equilibrium point, there would be one million A_1 genes and one million A_2 genes. Mutation would produce, on the average, one new A_2 in every generation, but this would be counterbalanced by one mutation from A_2 to A_1. Therefore, gene frequencies are proportional to net mutation rates. Unless something interferes!

One obvious kind of interference arises from environmental selection of certain genotypes. To examine this "favoritism," assume that genes A_1 and A_2 occur in guinea pigs. The genes function in the developmental system by the kind of enzymes they produce (or cause to be produced). When cells containing $A_1\ A_1$ and $A_1\ A_2$ are present at the anterior pole of the embryo, they tend to develop by growth and cellular differentiation into a head. When the cells contain only $A_2\ A_2$, an insufficient amount of the enzyme necessary to this growth and differentiation is produced. The head fails to develop, and a guinea pig bearing only A_2 genes survives only a few hours after birth. Only individuals bearing $A_1\ A_2$ will contribute A_2 to the succeeding generation, because $A_2\ A_2$ individuals fail to reproduce. The result will be a progressive reduction of A_2 to a level where loss from selection and by mutation away from A_2 is exactly balanced by additions by mutation to A_2.

Both mutation and selection operate within the framework of a breeding population. The Hardy-Weinberg law predicts gene frequencies for large populations in which any male gamete has an equal chance of fertilizing any female gamete and in which mutation and selection are effectively zero. If $a_i(A_i)$ is the proportion (a_i) of an allele (A_i) in the population and $a_1 + a_2 + \ldots + a_n = 1$ then:

$$\left[\sum_{i=1}^{k} a_i(A_i)\right]^2 = [a_1(A_1) + a_2(A_2) + \ldots a_n(A_n)]^2$$

For a simple case of two alleles present in equal proportions:

$$[.5(A_1) + .5(A_2)]^2 = .25(A_1A_1) + .5(A_1A_2) + .25(A_2A_2)$$

More generally, if several loci are considered and if cross-over occurs, the Hardy-Weinberg law assumes this form:

$$\prod_{l=1}^{k}\left[\sum_{i=1}^{k} a_i(A_i)\right]_l^2 = \left[\sum_{i=1}^{k} a_i(A_i)\right]_1^2 \cdot \left[\sum_{i=1}^{k} a_i(A_i)\right]_2^2 \cdot \ldots \cdot \left[\sum_{i=1}^{k} a_i(A_i)\right]_n^2$$

The departure of genotypic frequencies from the predictions of the Hardy-Weinberg model measures the influence of sampling accidents in small populations, the effects of mutation, the incidence of selection, and the results of migration; thus it provides a framework for development of theoretical, quantitative evolutionary models.

The minimum conditions for application of the Hardy-Weinberg law include random fertilization and a large population size. Many species populations cannot meet these conditions. In the general case, the randomly interbreeding population, the *deme,* is smaller than the species population, i.e., a species population comprises several demes. If a deme is small, some genes are likely to be eliminated by chance, e.g., if the effective breeding population numbers ten and the proportion of an allele, A_1, is only one-twentieth, then accidental death or chance failure of fertilization will remove it from the system. This random elimination fixes a small number of gene combinations in small populations.

But small demes may be parts of large species populations. If interbreeding between demes is high, random additions by migration will offset random loss. On the other hand, if individuals migrate only slowly, the parents of an individual may be drawn from a breeding population of only a few dozen. Local fixation of gene frequencies thus can occur within species populations numbering in the billions. Consequently, the individuals in a local unit may display a unity of form that distinguishes them from individuals of other units. Such modifications of local populations do not, however, change the frequency of genes in the population as a whole.

Gene and form

The genotype "expresses itself" in the properties of an organism, the phenotype. I have suggested above that this expression is accomplished through a complex developmental sequence in a particular environmental context. The "expression" of a particular gene depends on the genetic combination to which it belongs and the environmental context within which development occurs. Since in normal cells each locus is duplicated, expression of a gene is controlled first by the allele at the other locus of the pair. An allele may be expressed by a phenotypic property even though a different allele is present at the other locus, i.e., the condition is heterozygous, and one may then state that the first allele is *dominant.* The second allele, expressed only if present at both loci, is *recessive.* Thus, in fruit fly genetics, the gene R is dominant to its allele r in the expression of eye color; if a fly carries $R\ R$ or $R\ r$, its eyes are red; if it has only the recessive, $r\ r$, then the eyes are white. Perhaps more commonly dominance is incomplete, so that the heterozygote differs phenotypically from both homozygotes.

Interactions among genes further complicate the relationship of phenotype to genotype. One gene may produce identical phenotypic effects with all the gene combinations normally available in a population and in all environments in which development occurs; another gene may produce different effects in every genetic combination and/or with every environmental variation. In most populations, the biochemical and developmental effects of various genes overlap so that quite similar phenotypes occur regardless of genotypic variation. A gene may apparently express itself only in the presence or size of a single characteristic; careful analysis, however, reveals that most genes are *pleiotropic,* i.e., have manifold effects, and that most characteristics are influenced by several genes, are *polygenic.*

If one begins at the other end of the system, phenotypic variation reflects in some complex fashion genotypic variation. It is convenient to think of variance in phenotype as comprising a genetic variance component and an environmental component. The genetic variance, in turn, includes 1) an *additive component* representing differences between the average value of the different alleles in all genetic combinations, and 2) a *nonadditive component* including interaction between alleles at a given locus (dominance) and interaction between genes at different loci (epistasis).

Fossils and genetic systems

These, then, are the genetic characteristics of the species population: an array of gene combinations controlled by gene frequencies and modified by gene mutation, by selection, by population size, and by the amount of inbreeding. But the paleontologist cannot study fossil chromosomes under a microscope nor conduct breeding experiments on dinosaurs, although study of amino acids in fossils offers some hope of such genetic fingerprinting (p. 23). If he is to determine the genetics of fossil populations, he must do so indirectly—through morphology. As a further complication, the relation between gene systems and form is neither a straightforward nor a one-to-one relationship, since different gene combinations can produce quite similar morphological characteristics. The correlation between sample measurement (fossil morphology) and the characteristic of the living system (the gene system) is too low for comfortable armchair hypotheses.

In spite of this difficulty, studies of "fossil genetics" are of considerable importance. The genetic characteristics of populations are a critical factor in their evolution (Chapter 6). Two approaches have been tried in order to estimate the gene system from fossils. The first involves study of discrete differences among individuals of a fossil species, that is, the third body segment of a trilobite may bear 1) a long spine, 2) a short spine or 3) no spine.

Kurtén (1955), for example, attempted such an explanation for tooth size relationships in Recent and Pleistocene bears. In the two species of bears studied, he found that the height of the crown of the first upper

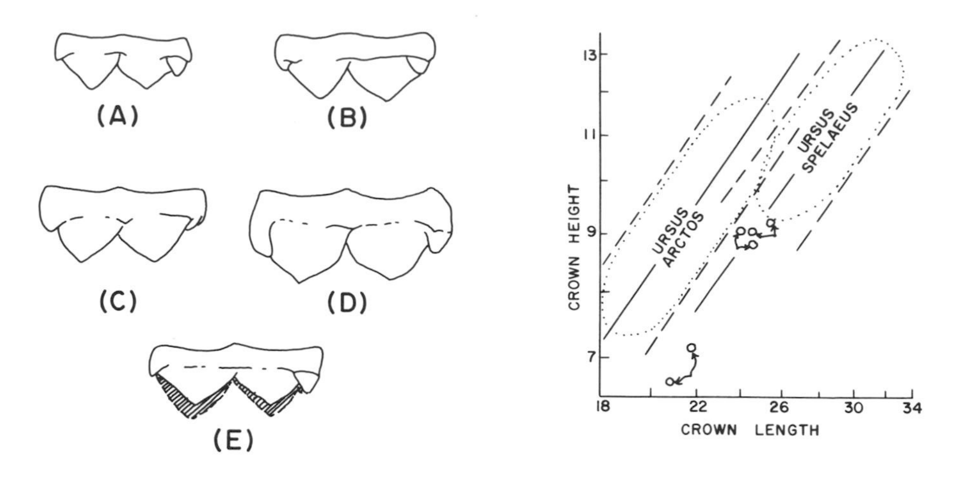

Fig. 4-5. RELATIVE CROWN HEIGHT IN BEAR TEETH. Side view of crown of first upper molar. **(A)** and **(B)** Small and large specimens of the European brown bear, *Ursus arctos.* **(C)** and **(D)** Small and large specimens of the cave bear, *Ursus spelaeus.* **(E)** Tooth of *U. arctos* with exceptionally low crown (*U. spelaeus* type). Shaded outline indicates crown height of normal *U. arctos* of the same crown length. Diagram at right summarizes measurements of crown height and length of 96 brown bears and 109 cave bears. Dotted line bounds distribution of points; solid lines are regression lines for each distribution; dashed lines indicated expected range of variation in a sample of 1,000. Circles are measurements on three brown bears that fall in the cave bear range. Specimen **(E)** at left is one of these bears. (*After* Kurtén.)

molar is relatively greater in longer teeth. Short teeth have relatively low crowns; long teeth relatively high. Figure 4-5 shows this by a plot of crown height against crown length. The points on the graph fall in two clouds. All those in one cloud belong to the species *Ursus arctos,* the European brown bear. The other cloud includes all specimens of the Pleistocene cave bear, *Ursus spelaeus,* and a very few (four individuals) of *Ursus arctos.* The problem arises with these four variant *U. arctos.*

To quote Kurtén:

> "But in the height-length relation these specimens may be said to 'imitate' the cave bear. The conclusion that these are mutants seems to have much to say for it . . ."

His reasoning is somewhat like this:

a). Part if not all the variation *within* each population resulted from variation in environment of development or from continuous minor variation in the genetic system.

b). The difference between the populations resulted from a major genotypic difference.

c). Since the difference is discrete—the bears have either an "arctos" type or a "spelaeus" type tooth—a pair of alleles—call them A_a and A_s, can explain it. Most *U. arctos* carried either A_a A_a or A_a A_s which provide an "arctos" tooth, but a few were homozygous for the recessive A_s A_s. All *U. spelaeus* bears, however, had the latter combination.

As Kurtén recognizes, the interpretation depends on a number of assumptions. The sharp distinction between the "spelaeus" and "arctos" teeth within the *U. arctos* population suggests the operation of a single discrete factor—otherwise variation would be continuous over the "spelaeus-arctos" range. This factor could be either environmental or genetic, although the latter seems more probable. If genetic, however, the phenotypic similarity of the teeth in *U. spelaeus* to the "spelaeus" variants in *U. arctos* does not demonstrate a genotypic similarity. A third allele might produce "spelaeus" in the *U. arctos* genotype; a variant allele at a different locus might simulate the effect of the *U. spelaeus,* or a combination of alleles at several loci might be the requisite code for "spelaeus." The data fits a two-allele expansion of the Hardy-Weinberg law but this is only the simplest of several reasonable explanations—the only limit is that the genotypic variation must produce a discrete, nongradational phenotypic difference.

The alternative approach emphasizes the resolution of phenotypic variance into environmental and genotypic components (see p. 73) rather than the correlation of an allele with a specific phenotype. Experimental studies indicate that environmental variance can be a rather major component of total variance, e.g., Bader found between 63% and 53% of variance in tooth dimensions of wild mice populations contributed by the environment. Thus a difference in phenotypic variance between two populations might arise from either a change in environmental variance or in genotypic inheritance or both. If, however, one can reasonably assume for the population in question either 1) that environmental variance is constant through the range of environmental variety or 2) that the populations inhabited a similar effective range of environmental variety, the *difference in variance* must result from differences in genotypic variance. The first of these conditions has yet to be tested thoroughly in recent populations; experience suggests that closely related species typically have similar environmental variance components. The second condition requires that "effective" be defined since a 2°C temperature difference may be "effective" for reptiles but "ineffective" for mammals.

In either case, application of this approach to fossils requires stringent control of sampling. What kind of population has been collected?

Are the fossils from a single contemporaneous population? Or are part of them from one population, part from another later in time, and a third part from a still later population? If contemporaneous, were they all members of a single small interbreeding unit or of a much larger breeding unit or derived from several partly isolated units? The answers come from careful field work in which the stratigraphic position and the spatial relation of the fossils are defined very precisely.

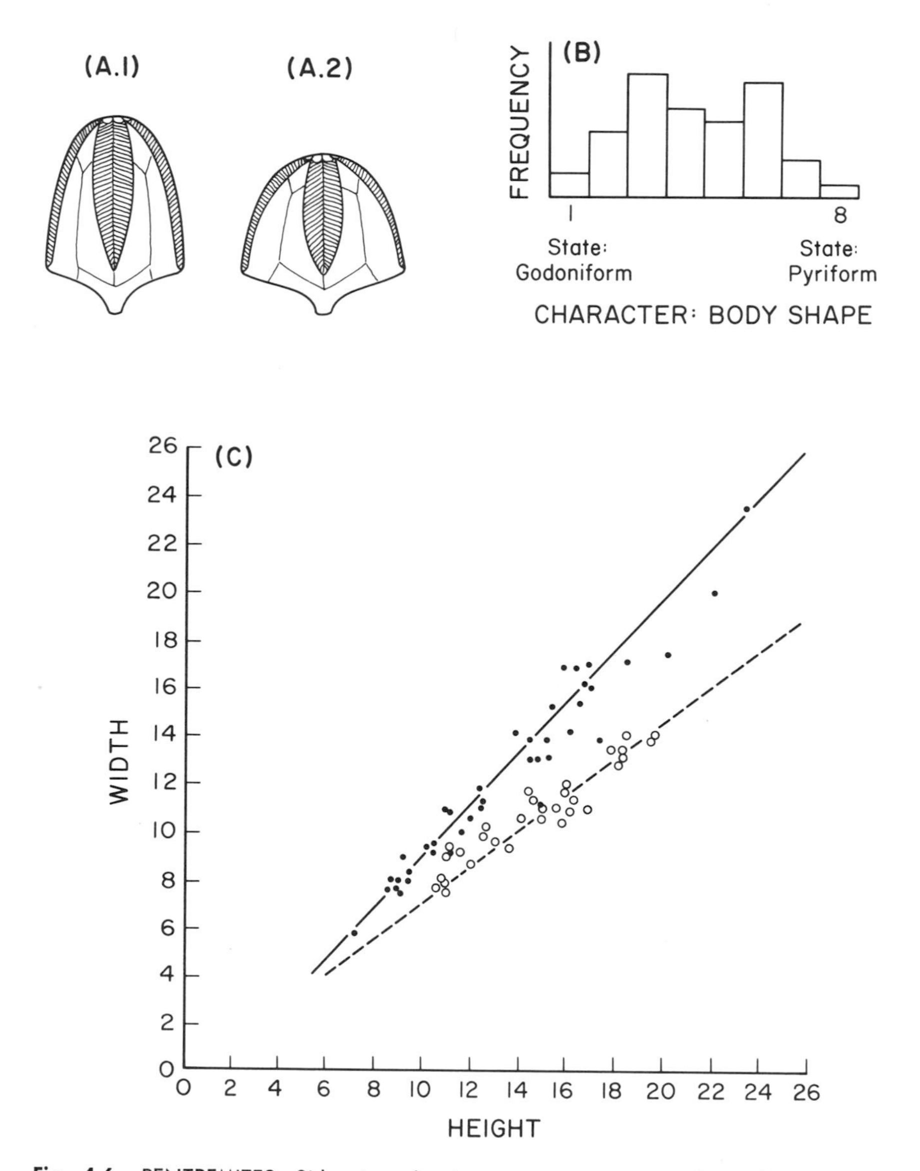

Fig. 4-6. PENTREMITES. Side view showing extreme states, pyriform, **(A.1)**, and godoniform, **(A.2)**, of the character, body shape. **(B)** Frequency distribution of character states. **(C)** Plot of width *vs.* height on 71 *Pentremites.* Dots represent specimens assigned to the godoniform group; circles represent specimens assigned to the pyriform group. Lines approximate reduced major axes of two groups. (*Data from* Olson *and* Miller.)

Morphological characteristics of species populations

For the paleontologist, interpretation of species populations returns always to morphology regardless of attempts at genetic—or any other biologic—analysis. He has the fossils, not the genes. Chapter 2 dealt with the aspects of individual morphology; fossil populations are an assemblage of individual morphologies that as an assemblage have morphologic properties not associated with any individual. These properties arise from the varied array of phenotypes in the population and include as abstracts an "average," "normal," or "typical" phenotype for the array and a "range" or "variance" of phenotypes. For a species population, the phenotypic array is distinct from any other such array.

What of the practicalities?

Let's study a hypothetical paleontologist as he examines a real sample of blastoids (a group of extinct echinoderms) from a single formation, the Paint Creek, of Mississippian age in southern Illinois (Olson and Miller, 1958, pp. 104-105, 300-301). Examination of a single property—body shape—indicates the presence of two rather distinct forms within the sample (Figure 4-6). Body shape is a *character;* the alternate expressions of the character are sorted into *states,* with the two extreme states "godoniform" and "pyriform." An individual fossil "possesses" a particular state of a character; the sample of fossils has an array of states expressed here as a frequency distribution (Figure 4-6B). Is this distribution an expression of a single genetic system? Or of two distinct and isolated systems, i.e., different species, represented by the extreme states "godoniform" and "pyriform?"

Our fictional fossil scientist asks himself: "Are there numerous intermediates in shape which might suggest genetic recombination? Do they differ in the state of other characters?"

After further study he concludes that intermediates are relatively few, that other differences are associated, and that the two morphological groupings, "godoniform" and "pyriform," deserve recognition as different species. Since similar species of blastoids have already been described under the generic name *Pentremites,* he places his species with them and provides distinguishing specific adjectives, *godoni* and *pyriformis*—thus the new species, *Pentremites godoni* and *P. pyriformis.* Finally, he publishes a description consisting of:

1. the name.
2. the designation of a single specimen as the species "type"—the name bearer. The locality and stratigraphic position are also given.
3. a list of other specimens referred to the species and used in establishing the limits of specific variation.

4. a diagnosis of the features by which *P. godoni* and *P. pyriformis* may be distinguished from each other and from the other species of *Pentremites.*

5. a general description of the morphology of the two species and of variations in morphology among the referred specimens.

The statistical study of species morphology

Does this satisfy all the requirements of the species concept? Consider that he had only seventy-one specimens for study, not the millions that composed the species populations. The samples cannot cover the entire limits of variation for the populations nor is the sample "norm" likely to be the population "norm." The paleontologist must, therefore, estimate the properties of the populations and compare populations rather than samples. What is the chance of getting, from the same population, two samples with the particular distribution of character states observed here?

The problem is partly a biological one concerning the pattern of variation expected in biological populations, but it also involves the chance differences between samples taken at random from a population and with the relation of the fossil population to the biological one. The latter part of the problem is more straightforward and less subject to constraints imposed by "normic" statements, so it can be dealt with first. Let's begin by considering the individuals placed in *P. godoni* to be one sample and those in *P. pyriformis* the second. We now ask:

> If one drew very many pairs of samples, say a thousand, from a population consisting of individuals ranging from godoniform to pyriform in shape, how frequently would one get a pair in which all the individuals in one sample were *pyriform* and all those in the other were strictly *godoniform*?

If such a pairing would happen only once in a thousand or even in a hundred times, the odds would favor a bet against a single source population, i.e., the difference is probably *significant.* What makes a difference significant? Quite plainly, if each sample comprised only one individual, a bet for the two-population hypothesis would be extremely risky unless the difference was extremely large. As the sample size increases, the significance of a particular difference also increases. Further, the amount of variation is important in this game. If the variation is quite large, then many sample pairs are likely to be quite different and one should bet against the two-species hypothesis; if variation is very small, even a small average difference becomes significant.

The study, description, and interpretation of variation is, of course,

TABLE 4-2

Height and width measurements on pentremites

No.	*Height*	*Width*	No.	*Height*	*Width*	No.	*Height*	*Width*
1	12.3	11.0	26	10.0	9.4	51	14.9	11.0
2	13.7	14.1	27	15.0	13.9	52	18.1	13.3
3	16.9	16.1	28	7.1	5.9	53	11.9	8.8
4	9.0	9.0	29	16.2	16.9	54	12.5	10.3
5	18.3	17.7	30	16.0	14.1	55	14.2	11.7
6	15.7	16.9	31	11.9	10.6	56	11.0	9.1
7	17.1	13.9	32	12.2	11.9	57	12.9	9.6
8	10.9	11.0	33	8.8	8.7	58	16.1	11.4
9	23.2	23.6	34	16.7	17.0	59	15.8	11.8
10	16.7	16.1	35	14.6	13.0	60	18.0	12.9
11	10.3	9.2	36	8.9	7.5	61	10.9	7.5
12	22.0	20.0	37	12.4	11.3	62	12.4	9.9
13	8.5	8.0	38	16.3	15.4	63	18.2	13.1
14	9.2	8.4	39	8.3	7.7	64	14.0	10.6
15	11.0	9.1	40	14.4	13.9	65	19.5	14.0
16	8.8	8.0	41	16.8	11.0	66	13.5	9.3
17	10.2	9.6	42	18.1	13.1	67	10.7	8.0
18	9.2	8.0	43	10.8	7.9	68	15.4	11.0
19	14.8	11.1	44	19.4	12.8	69	15.9	12.0
20	20.0	17.5	45	17.7	13.5	70	11.3	9.2
21	15.3	15.3	46	15.6	10.3	71	16.0	10.9
22	11.0	10.9	47	14.9	10.6			
23	15.1	13.1	48	14.5	11.3			
24	11.5	10.0	49	11.9	9.0			
25	14.4	13.0	50	10.3	7.6			

1. Numbers 1 through 40 = godoniform; 41 through 71 = pyriform
2. For godoniform specimens:
   reduced major axis = $y = -0.999436 + 1.0070x$
   correlation = 0.967
   standard error of slope = 0.0016
3. For pyriform specimens:
   reduced major axis = $y = 1.083 + .6598x$
   correlation = 0.936
   standard error of slope = 0.0017
4. Comparison:
   z (diff. of slope) = 6.05 Prob ($z = 6.05$) < 0.0001
   z (diff. of intercept) = 12.98 Prob ($z = 12.98$) < 0.0001

the province of statistics, and various texts (Simpson, *et al.*, 1960, Miller and Kahn, 1962, Krumbein and Graybill, 1965) cover their application to paleontology. I can, however, introduce a brief example using the information on *Pentremites*. Table 4-2 reports height and width measurements on the collection described above. Since they show continuous skeletal growth without a definite adult stage, size comparisons are not meaningful; more useful is the relationship between height and width plotted in Figure 4-6. The character is the height-width ratio; the alternate states are its values. The scatter of points for each sample defines a line (the *reduced major axis*) of the general form, $y = b + kx$. This line represents an average or, better, a *central tendency*. The distances of the various points from the line express the variation or dispersion. The difference between samples is represented by 1) the difference between lines and/or 2) the difference in variation about the lines. The lines may differ in slope and/or in the vertical distance between them at some meaningful point on the x-axis.* The dispersion is taken as the *correlation coefficient,* i.e., the ratio between a) the calculated dispersion in the state of one character if all the observations were exactly on the line and b) the observed dispersion of the same measure. If we drew a large number of samples at random from a large population, each sample would display a different reduced major axis. Most "sample axes" would approximate that of the population, but a few would be some distance away. The question stated above can now be reformulated:

> How often would sample axes as different as this pair be observed in samples drawn from a single population?

The variation of sample axes is related to the variation in the population; if there are many points far from the line, i.e., a low correlation coefficient, a large number of samples will be composed in large part by such distant points; widely different sample pairs will be rather likely. The probable sample variation is stated as a *standard error;* in general, about 2/3 of the sample lines will be within two standard errors of each other.† A ratio of the difference between samples to the variation within samples expresses the relative importance of that difference. If the frequency distribution(s) approaches normality, the probability of getting a particular ratio is calculable. For Figure 4-6 the ratio for slope differences is 6.049; the tables appended to statistical texts indicate a ratio this large would occur in fewer than 10 sample pairs in 100,000 trials. Therefore, one risks very little in asserting that the samples represent two distinct fossil populations.

But do these populations represent distinct species? Generally in paleontology, one concludes so, but several alternatives exist. Specifically they are:

a). Distinct fossil populations drawn by selective (nonrandom)

* A note to this chapter summarizes the mathematics used in this analysis.

† For a mathematical exposition, see the note, p. 97.

preservation or by periodic sampling of growth series (p. 65) from a single biologic population.

b). Distinct phenotypic arrays representing separate genetic systems (species).

c). Distinct phenotypic arrays from distinct genotypic arrays in a single genetic system.

1) genotypic differences within a deme (polymorphism, sexual dimorphism).
2) genotypic differences between successive demes.
3) genotypic differences between contemporaneous demes.

d). Distinct phenotypic arrays due to distinct differences in environment of development.

1) differences between geographic sites.
2) variation in time at a geographic site.

Bias introduced by periodic sampling or selective preservation is unlikely in this case. They both include growth series and have a similar range of sizes so that one cannot represent an earlier growth stage. Two types of individuals occur in many of the same thin depositional units and have been subject there to the same geologic factors. They are similar in skeletal structure and composition; as far as sedimentary sorting is concerned, they are not very different, and intermediates in shape would hardly be sorted out leaving only the end members. Samples from different beds, however, or from different localities may bear the subtle impress of selective preservation. This prospect becomes more likely if the depositional environment of the beds is markedly different and if evidence of mechanical sorting or chemical destruction is present. If a number of samples representing different preservational environments are observed, variation between samples would probably be bridged by intermediates in intermediate environments. In any case, examination of characteristics unaffected by selective preservation would be desirable.

The remaining alternatives require some sort of estimate of the biological characteristics of the populations. In the *Pentremites* case, the "godoniform" and "pyriform" types occur in the same place, so that differences in environment (*4.a.*) and/or in demes between sites (*3.c.*) can be rejected. The others are more difficult. It is usually difficult if not impossible to determine whether the individual fossils even on a single bedding plane lived at exactly the same time (*3.b.*). Nor is it possible to eliminate the possibility that distinct genotypes within the population produce distinctly different phenotypes (*3.a.*) without intergrades—as in the example of Kurtén's bears. In fact, *sexual dimorphism* appears to be quite common in some groups of organisms, e.g., Westermann's work on fossil cephalopods. If the variant forms occur together in every reasonably sized sample then polymorphism is highly probable if not certain. The converse, however, is not true, for polymorphism varies in space or time.

One returns finally to "normic" biologic statements for decision:

a). Polymorphism within a deme is *generally* limited to relatively few characters or to a set of functionally related characters;

b). Large differences between contemporaneous demes of the same species are *generally* bridged in intervening demes;

c). Relatively few groups of organisms display distinct temporal polymorphism over short time periods (months or a few years).

When these statements are applied to *Pentremites,* one can note that: a) *P. godoni* and *P. pyriformis* do not invariably occur together and, therefore, are not necessarily dimorphs; b) they differ in a considerable number of properties (see Olson and Miller, p. 217); and, c) they occur together in the same sedimentary units deposited over a short interval. Consequently one can suggest with *reasonable certainty* that *P. godoni* and *P. pyriformis* represent distinct genetic systems.*

The *Pentremites* example by no means exhausts the techniques of population study nor the problems in interpretation. If a definite adult stage can be determined, or if growth occurred in a series of discrete stages as in the molts of arthropods, absolute size comparisons are valid, and analysis of variation in single characteristics is meaningful. A ratio, t, is formed between the average (mean) and the sum of the standard errors of the averages.†

As t becomes large, the probability that the samples were drawn from the same population becomes small. Techniques are also available for simultaneous comparison of a number of different characteristics; these are discussed under the heading of multivariate analysis and factor analysis in statistics texts—see in particular, Simpson, Roe, and Lewontin, 1960, Miller and Kahn, 1962, and Sokal and Sneath, 1964. None of these techniques, however, eliminates the need for evaluation of geological factors and the biological relationships underlying the statistics; they do however provide additional tools for such evaluation.

* With the proviso of contemporaneity *P. godoni* occurs by itself in the earliest samples of the series and may have split into two phyletic systems, "conservative" godoniform and the "progressive" pyriform. Conventionally, godoniform phenotypes would be assigned to the paleospecies, *P. godoni;* all pyriform phenotypes to a separate paleospecies, *P. pyriformis.* Only contemporaneous representatives of each would be a biological species.

† See note to chapter.

Species as ecological systems

With or without statistics, observation of species morphology is an essential of paleontology, but no species (or paleontologist for that matter) survives on form alone. Somewhere back of the abstractions, back of type specimens, means, and analysis of relative growth, animals must exist. And existence depends upon eating and not being eaten, upon heat and light, upon all the relations of the animal to the world in which it participates. The interpretation of fossil species depends, in great part, upon the discovery of these relations, the *ecological system,* from fossil form, occurrence, abundance, and association. A sea urchin on the bottom of Puget Sound has a specific ecology, which is described by measurement of the environmental factors acting on it and of its reactions to those factors. If one measured the ecology of a series of urchins of a single species, e.g., *Strongylocentrotus drobachiensis,* he would find a general similarity though each one differed slightly. One individual might live in 10°C water and another in 16°C but none in 4°C or 22°C. The action-reaction system varies, but only within limits. Experimental manipulation of the environment measures the tolerance of an individual urchin, and the action-reaction systems of the individual in nature are within these limits of tolerance. One might expect, offhand, that the limits of tolerance for the species would coincide with the tolerance limits of an individual. The population, however, since it includes variants in form, physiology, and behavior, tolerates a wider environmental range than any individual. But the important measure of these tolerances is not the laboratory experiment; it is rather the occurrence of the animal and species in nature. A complete analysis of a species population is beyond present methodology, but studies of the different aspects of individual and species ecology provide a framework for analysis. This analysis in turn offers a basis for *paleoecology,* the study of ancient ecology.

Ecology and population structure

Let me set up a hypothetical experiment. Assume that a very large number of fertilized sea urchin (*S. drobachiensis*) eggs are sown from an airplane along a stretch of the Pacific coast so that they fall randomly over a five-square-mile area (Figure 4-7). The experiment uses a sufficient number of eggs to give an average of one per square yard. The number of sea urchins that develop from these eggs is counted at intervals. The number of individuals per hundred-square-yard plot provides a convenient measure of *population density.* Density variation is plotted by "iso-density" contours on a map of the area.

A census is made immediately after distribution of the eggs, at the end

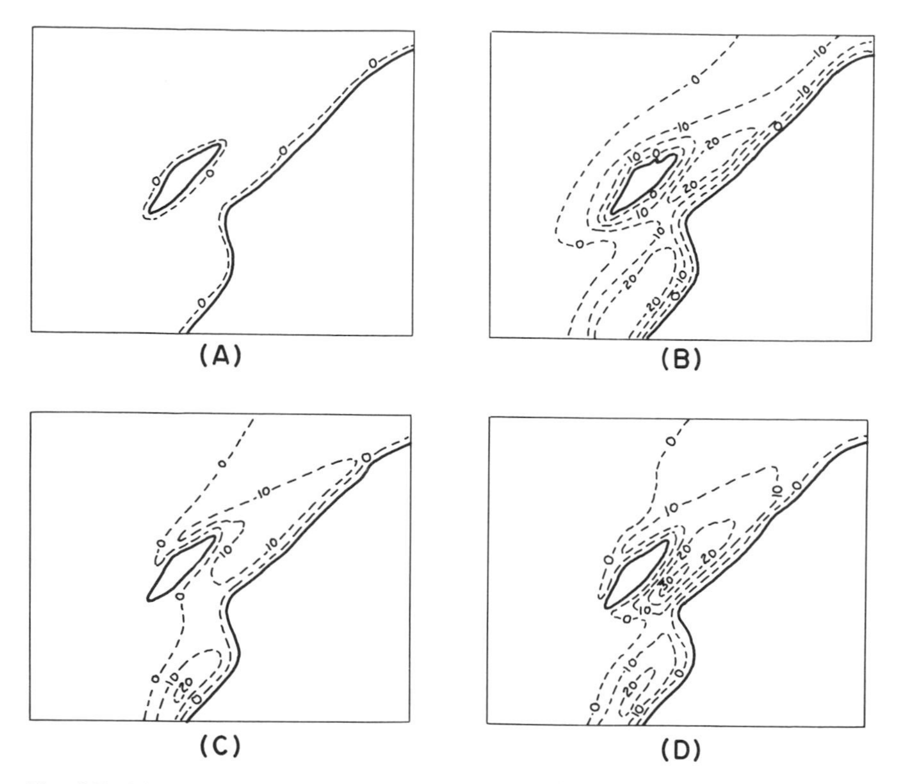

Fig. 4-7. VARIATION IN SEA URCHIN POPULATION DENSITY. Illustration of a hypothetical experiment in which urchin eggs are sown at random over a five-mile area. Land area shaded. Densities indicated by dashed contours. **(A)** Population at the end of 24 hours. Zero density line along shore lines. Essentially random densities otherwise. **(B)** Population at the end of one week. Distinct pattern of high and low density shown by contours. **(C)** Population at the end of one month. Total population has decreased, but pattern of distribution is similar. **(D)** Population at the end of one year. Reproduction has increased total population and maximum densities. Note that areas of maximum survival and reproduction do not coincide completely.

of twenty-four hours (Figure 4-7A), at the end of one week (Figure 4-7B), at the end of one month (Figure 4-7C), and at the end of one year (Figure 4-7D). At the first census, the numbers in adjacent plots vary randomly, so no significant contour pattern can be drawn. At the end of twenty-four hours, however, a pattern begins to form because all the eggs that fell on land have died. These deaths define a zero contour along the shore line, but elsewhere the distribution remains pretty much at random, though average density has fallen slightly.

Six days later the average density has fallen still more, but, more important, the distribution of densities is no longer random. Some of the squares have many more individuals than they had earlier; others have many less. Because the average density has fallen, many individuals must have died (or moved from the census area). The local populations that increased in density must have done so by immigration from the remaining plots. The number in any one of these plots comprises the original number plus the immigrants less the dead and less the emigrants. Or write it this way:

Rate of Change in Density =
Immigration Rate − Death Rate − Emigration Rate

The thirty-day census accentuates the results of the one-week census: the areas of zero density are more extensive, and the concentration of density in a few of the census plots is more marked. Average density is now only a few percent of the original. The rate of change in density for the whole population continues to be negative and quite high, though some local populations have a positive rate.

At the final census, the average density has jumped; obviously some individuals have reproduced. The situation can be summed up in an expansion of the original equation:

Rate of Change in Density =
Immigration Rate + Birth Rate − Death Rate − Emigration Rate

The birth rate for areas of zero density must, of course, be zero, so the population of these areas, if it increases, must do so by immigration. On the other hand, some populations of moderate density may have a higher birth rate than those of high density, so the pattern of the density map changes.

Numbers and distribution of the sea urchin result from individual survival, from dispersal, and from reproduction. These, in turn, must be the product of interaction between animal and environment. For a population, these processes are summarized as rates; for the individual animal they are living, mating, and dying. By measuring the environment of life, fertilization, and death, the ecologist can refer these rates back to the limits of tolerance of the individual and to the range of those limits within the population.

Factors in species ecology

Variation in population density results, then, from variation in environment. "Environment" includes the interaction of the individuals within the species population as well as factors external to that population. Animals respond to the environment-as-a-whole, but this environment-as-a-whole cannot be adequately measured. An ecologist does measure components of the environment, such as temperature and moisture, and relates these factors to the processes of population increase and decrease. Since any population capable of reproduction can potentially increase in number, these factors can be regarded as limiting, either through restriction of reproduction and of immigration or through increase of death and emigration rates. An animal species lives in a *habitat* defined by these *limiting factors.* Finally, these factors limit because they influence or are part of the basic requirements of an organism (p. 34); or because they influence the requirements of a population for reproduction.

All this is admirably theoretical, one might even say spiritual. But again the animal! What are the components of a sea urchin's environment? How do the urchin and the urchin population react to that environment?

First, the energy source. Urchins of this species browse on seaweed, diatoms, tube worms, and hydroids, hold their food with spines and tube feet and chew it by a complex jaw system called "Aristotle's lantern." They gather food in part by random search and in part by distant perception—probably largely by "smell." Movements of the tube feet carry them toward food (or away from enemies). The urchin survives by obtaining a certain minimum of food; the species survives only if a minimum number of individuals obtains sufficient additional food to reproduce. The amount of available food in Puget Sound, for example, limits the density of the urchin population—a limit set by the total mass of suitable food less the amount used by other species less the amount not found by urchins in their search. If the number of urchins increases, the amount of food not found is reduced. Ultimately, increased predation on the food species (plants, worms, hydroids) will exceed their reproductive increase and reduce the density and the total amount of food. More urchins starve, and fewer individuals reproduce. The death rate goes up; the rate of reproduction down, and the population density decreases. In this circumstance, the population of *Strongylocentrotus drobachiensis,* and of its food species constantly adjust to one another.

Physiochemical factors that affect death and/or reproduction rates also limit population density. For example, water temperature determines the activity rate of the urchin, modifies internal processes, and may induce new processes. The cessation of necessary activities or mechanical disruption by freezing limits survival at the lower end of the temperature spectrum; at the upper end, the induction of harmful reactions culminates in coagulation of the proteins. Heat, therefore, is a limiting factor, the effects of which vary with different individuals, so that the population has a range of tolerance greater than that of the individual. Rather, the existence of protected and exposed areas modifies the effects of temperature on survival. The urchins avoid light and either burrow into the bottom or retreat beneath rocks and shells. In a shallow pool, this behavior protects them—in part—against temperature changes. The number of such protected locations is limited, and the number of suitable rocks and places to burrow sets a population density limit. One might consider competition for hiding places the factor responsible for the limitation, but this does not detract from the limit imposed by the number of hiding places nor from the fact that the temperature changes produce death.

The burrowing habits of the urchin also affect survival in storms, since unprotected individuals are broken in the surf. Again the availability of protected burrows sets a maximum density. In this case, other behavior might modify survival. If the food species were rare, some urchins would be forced farther from protected areas in search of food and exposed to greater risk of wave and turbulence. Several factors interacting in a complex fashion would control the density of the population.

The complexity of the relationship between population and environment increases if the action of one individual on another and the effects of density on reproduction are considered. Ecologists know practically nothing of these factors in sea urchins, but they have been studied in other animal species, so I can shift to another example, muskrat populations in the central United States. In a large marsh inhabited by muskrats, low densities reduce the chance of mating and in consequence the birthrate. Each individual, though, has a greater chance of survival because of relative abundance of food and good hiding places. If the death rate for this reason is lower than the birthrate (assuming immigration = emigration), the density will increase; and as it increases, the birthrate will likewise increase, i.e., not only will more young be born but the number of births per adult will increase. But, as the number of individuals increases, a greater percentage of them must occupy less protected places exposed to attack from owls, hawks, and foxes. Further, fighting for the available space increases; individuals entering the marsh from outside are driven away or drive away the original inhabitants, and many immature or senescent individuals are driven out by vigorous adults. The birthrate decreases as a result of changes in the physiology and behavior of the individuals affected by crowding and a lesser supply of food. The death rate increases as does the emigration rate until, together, they equal or exceed the rates of birth and immigration. The population density now assumes a steady state or decreases. Since it consists largely of aging matures with only a few young it is very vulnerable to further deterioration. Changes in the marsh complicate the situation. High-water levels result in more food and more dwelling places; densities increase as does the rate of increase. Several years of favorable conditions would result in great numbers and high densities, and the death rate in the succeeding low-water years would increase correspondingly. Because of these fluctuations in the environmental factors, most, if not all, natural population densities fluctuate more or less widely about an equilibrium.

Probably no one factor limits a population throughout its existence; almost certainly they fluctuate as the population changes in density and in character of component individuals. The limiting factors change also as the environment changes. Although the ecologist isolates certain factors or combinations, no animal exists in isolation with temperature, with a parasite, or with its siblings. The animal is rather an intersection in the network of environment. The strands of that network may be classified in many ways but most conveniently in terms of isolated measurable factors, light, heat, predation, food supply, and so on (Table 4-3). The effect

TABLE 4-3

Factors of the environment

1. Primarily Physical or Chemical.
   - A. *Solar radiation.* High intensity damages living organisms. Low intensity limits production by photosynthetic plants. Visible light affects perception and thus periods and places of activity.
   - B. *Heat.* High and low temperatures can damage living organisms. Temperature controls the rate of animal activity by modifying rate of chemical processes.
   - C. *Gravity.* This limits the bulk and orientation of animals that live on a land surface or on the sea bottom. It is related to density, bulk, and body form of swimming, floating, and flying animals.
   - D. *Pressure.* High pressures in oceanic depths influence marine organisms.
   - E. *Currents of air.* This is important in determining climatic patterns. Directly influences animals through winds.
   - F. *Currents of water.* Currents in marine and fresh water affect distribution of aquatic animals, modify their activities, and may damage them if intense.
   - G. *Substrate.* This is the surface on which animals live or are supported. It includes surface of water, land surface, and surfaces beneath bodies of water. It influences organisms by stability, firmness, roughness, and so on. It is important to paleoecologists because it is related to sedimentary environment.
   - H. *Physiochemical and chemical phases.* These are viscosity, diffusion, and osmosis. Viscosity affects rates of sinking of swimming and floating organisms. Osmosis involves the exchange of ions through cell membranes and so influences survival in water of low or high ion concentrations.
   - I. *Water.* Water is an essential constituent of protoplasm. The dry environments of the atmosphere and land surface restrict those environments to animals that can store and conserve water.
   - J. *Atmospheric gases.* Free oxygen is essential to energy utilization in most organisms. Dissolved gases, particularly oxygen, limit the distribution of aquatic organisms.
   - K. *Dissolved salts.* Some elements are essential to life processes. The concentration of $CaCO_3$ and other shell-building materials assumes particular importance to the paleontologist.
   - L. *Fluctuations in environmental factors.* Regular or irregular variation in any of the above factors is an important modifying influence in itself.
   - M. *Combinations of environmental factors.* The availability of essential materials and the effects of individual physical and chemical factors are modified by other factors in the environment. Thus, temperature is related to water vapor, and calcium carbonate to pH.
2. Primarily Biological.
   - A. *Predation.* This is the effect of a predator on a prey species.
   - B. *Food.* This includes animal food for predators and parasites and plants for herbivores.
   - C. *Parasitism.* This is the effect of a parasite on the parasitized species.
   - D. *Modification of physical and chemical factors affecting several species.*
   - E. *Symbiosis.* This is the interaction of two or more species without injury to either and with benefit to at least one partner.

of variation in any one factor is modified by the total combination, by the flexibility of individuals in adjusting to the variation, and by the range of species tolerance. The flexibility of the individuals and the variation in their tolerance define the range of species tolerance.

The fossil and its environment

Since an animal is an intersection in the biologic net, it has its fullest meaning only if considered in the environmental context. A paleontologist must seek beyond form and association to this wider meaning and trace out the environmental strands or at least some part of them. He may conclude that a limestone in central Nebraska was deposited in a shallow, warm sea of essentially normal salinity because of an abundance and diversity of fossil corals. He may argue for a warming trend near the end of the Mesozoic because of the oxygen isotope ratio in fossil cephalopod shells. Or he may recognize that Carboniferous sharks inhabited fresh water rather than marine because they occur in river channel deposits. Even these few examples demonstrate the possibilities and the complexities, the diverse aims and the differing techniques of paleoecology.

An analysis of the environmental relations of the fossil oyster *Cubitostrea lisbonensis,* made by H. B. Stenzel (1945), illustrates the development of a paleoecologic study. This species (Figure 4-8) is known from the

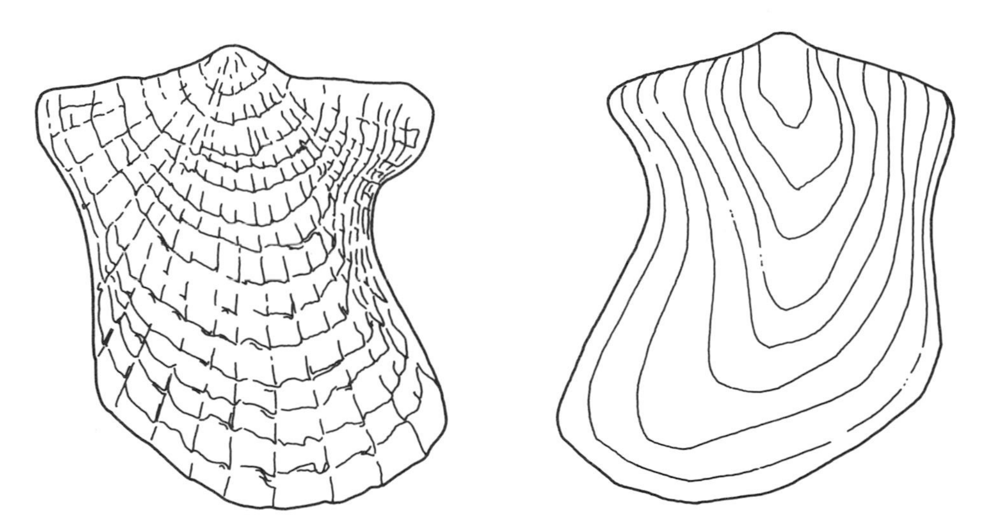

Fig. 4-8. *CUBITOSTREA LISBONENSIS.* Left and right valves of this middle Eocene oyster. (*After* Stenzel.)

Weches formation of middle Eocene age and its equivalents on the Gulf Coastal Plain in Alabama, Mississippi, Louisiana, and Texas (Figure 4-9). The sample comprises a series of collections from different stratigraphic levels and different localities in Texas and includes data on associated species and rock characteristics (Table 4-4). The fossil population includes all possible collections of this oyster with the association data throughout the area of outcrop. All individuals of this genetic continuum

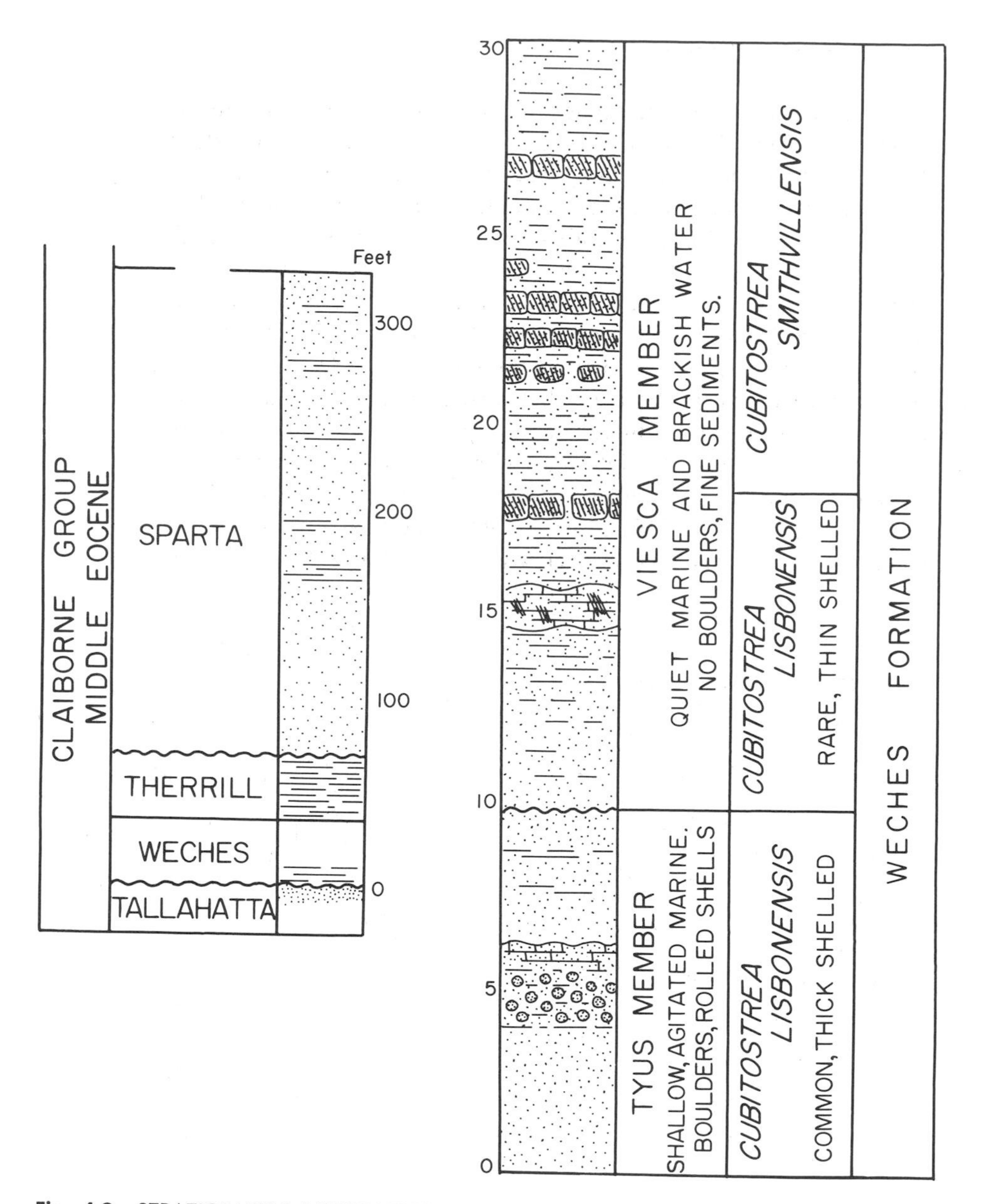

Fig. 4-9. STRATIGRAPHIC DISTRIBUTION OF *CUBITOSTREA LISBONENSIS*. The section at the left shows the general position of the Weches within the Claiborne group. The section at the right shows the character of the Weches at Hurricane Shoals, western Houston County, Texas. Farther west, the Tyus member becomes finer grained, loses the boulders and rolled shells, and contains rare, thin-shelled *Ostrea lisbonensis*. (*After* Stenzel.)

that lived in the present outcrop area during early and middle Weches time form the corresponding biological population. The sampling method was a combination of search and purposeful design.

TABLE 4-4

Data on form, occurrence, and associations of *Cubitostrea lisbonensis* (*information from Stenzel 1945, and 1949*)

Parameters	C. *lisbonensis Type 1*	C. *lisbonensis Type 2*
Stratigraphic occurrence	Early Weches only.	Throughout Weches time.
Geographic occurrence	Central Leon County, Texas, and eastward.	Central Leon County, Texas, and westward. Transgressive eastward in later Weches time.
Form of C. *lisbonensis*	Large heavy shells heavily ribbed and fairly regular form predominate. Thin and irregular shells rare or absent. Intergrades with Type 2 westward.	Smaller, thin shells, with low ribs and irregular form predominate. Thick and regular shells rare or absent. Intergrades with Type 1 eastward.
Fragmentation, wear, etc.	Some broken shells; many show evidence of abrasion and rolling.	Very few broken shells; no evidence of wear or rolling.
Population characters	Mature individuals predominate; immature specimens rare.	Many very young and immature specimens.
Association with other fossils	?	Thin-shelled clams, delicate bryozoa, and some rare corals, sea urchins, and crabs.
Character of rock	Limestone boulders, layers of rolled and worn shells, coarse glauconite grains.	Shaly, thin bedded, fine glauconite grains.

Comparison of samples of *C. lisbonensis* shows a gradual change from thick-shelled (Type 1) to thin-shelled (Type 2) forms along an east-west trend in the lower Weches strata (Table 4-4). Some variation in shell thickness among local samples probably corresponds to differences among local biologic populations, since the heavy-shelled individuals, the most likely to be preserved, do not occur with the thin-shelled. The absence of the thin-shelled types in a sample does not demonstrate, however, that such individuals were absent from the living population. They may have been removed or destroyed by wave action or never fossilized. Differential preservation could also explain the decreasing percentage of immature oysters from east to west. One might test by comparison of thickness and size of the broken and/or leached shells with intact shells from the same sample. These two portions of the sample should be similar in these characteristics if no bias exists.

Shell thickness in many species of living pelecypods correlates with the vigor of wave action or strength of currents, but the correlation is imperfect. Stenzel infers turbulence to be greater in the eastern portion of the "Weches Sea" than in the western, but the inference is not very firmly established; other explanations are equally probable.

His field observations, however, include information on sediment size, a parameter rather highly correlated with turbulence. An estimate of turbulence on the basis of sediment size agrees with the estimate derived from a study of shell thickness.

In the westernmost outcrops, *Cubitostrea* is found in fewer samples and is rarer where found than to the east. What factors restricted *Cubitostrea* densities here? Studies of sediment size and shell thickness indicate lower turbulence in this area. This reduction could result from protection by bars and from shallower water. Other sedimentary features demonstrate that the western outcrops were near the shore line of the Weches Sea. Stenzel concluded that the local environments in the west consisted of broad shallow lagoons. Their environment probably differed significantly from that of the open sea in temperature, in light, in substrate, in gas concentrations, and in salinity.

Study of the fossils occurring with *Cubitostrea* helps to sort out the influence of these several variables. These associates include other pelecypods (pectinids), bryozoans, corals, and sea urchins. Their closest modern relatives are marine and cannot tolerate *brackish* (low salinity) water for prolonged periods, nor are closely related species known from ancient freshwater or brackish environments. Further, the oyster beds formed by *Cubitostrea lisbonensis* always include several other animal species in fairly large numbers. This variety contrasts with modern brackish water oyster beds, which contain few organisms and few species of organisms other than oysters. Because *Cubitostrea* distribution coincides with that of salinity sensitive species, Stenzel concluded that lowered salinity limited the westward distribution of this oyster species.

Notice, however, that I have not said that lower salinity causes the excess of deaths over births and immigration. The effect might be more subtle and, for example, reflect control of food sources or of parasites by salinity. With this restriction, one can conclude that salinity changes limited, either directly or indirectly, the distribution of *Cubitostrea lisbonensis*.

What can one learn of paleoecologic method from this example? Stenzel began with variation in particular characteristics, e.g., shell thickness. He concluded that some of this variation reflected variation in the biologic population, although a large component might also be related to postmortem geologic factors. He then noted a general (though not invariable) relationship between shell thickness and turbulence in modern marine animals and concluded that the variation in fossil shells corresponded to variation in turbulence. He tested this conclusion by reference to characteristics of the sediments which are also correlated with turbulence. The interpretation of salinity was more complicated since the variation in abundance of *Cubitostrea* simply demonstrated variation in some unknown environmental property. He then identified this property with correlative variation in other taxa. The most striking correlation was with organisms which do not at present tolerate low salinities so he argued for salinity control of *Cubitostrea.* Regardless of the complexity of the argument, the logical chain is similar:

> Variation in samples is a function of variation in the fossil population; variation in the fossil population is a function of variation in the biologic population; variation in the biologic population is a function of variation in environmental parameters.

Fossils for paleoecologists

Paleoecologic analyses derive from two sources, fossils and associated rocks; environmental interpretation of sediments is beyond the scope of this book, so further discussion will be limited to the fossils. As pointed out above, much of the "information" in a fossil collection is entered by processes geological and operator, which have little or nothing to do with the environment of the living population. In summary, these biases are the result of:

a). Selective chemical modification or destruction

b). Selective mechanical destruction

c). Selective sorting with transportation

d). Selective exposure of variant rock types in outcrop

e). Inappropriate sampling methods

f). Nonrandom errors in measurement and classification.

In the *Cubitostrea* example, solution and/or mechanical abrasion might well have destroyed more thin-shelled than thick-shelled oysters at some localities; sorting in transport may have tended to form lag "pavements" of thick shells at some places and bar deposits of thin shells at others; since the lithology enclosing the oysters varies from place to place, the resistance of lithologies might have controlled relative outcrop frequency; the collector may have tended to pick up only the largest and best-preserved fossils; the thin-shelled and thick-shelled forms might represent different species rather than environmental variants. This particular study rejects all these alternatives with some degree of certainty, but they must be considered and tested to attain reliability and precision in paleoecology. Table 4-1 provides some criteria for evaluation of geologic bias in fossil samples.

But you have established the correspondence of fossil sample and biologic population. Now what?

A single fossil obviously has some paleoecologic significance. I described above (pp. 36-37) the interpretation of fossil form in terms of specific functions, e.g., swimming, that are possible only in specific environments. Therefore, a functional analysis yields environmental interpretation; briefly, it depends on comparison with living relatives, on comparison with form-function correlations in modern organisms, and on interpretation of mechanical relationships (including orientation in rocks).

A second source of paleoecologic information is the variation of form among samples—the thick shell of a *Cubitostrea* has little significance until the thin-shelled form is observed in other localities. Study of variation among localities and among adjacent beds at a locality measures the association with depositional environments—*Cubitostrea* with turbulent water sediments—and of association with other fossil species. In the latter case, a few associated occurrences of several species mean some overlap of environmental requirement; if these requirements can be defined, even grossly, the overlap may be defined with greater precision. On the other hand, frequent association indicates response to the same limiting factors, e.g., the association of *Cubitostrea* with corals, bryozoans, and sea urchins.

Skeletal composition and environment

Modification of form, of shell mineralogy, etc., by the environment provides a third source of paleoecologic data. Responsive variation in form may be difficult to separate from genetically controlled adaptive variation. For example, are the elongate forms developed by oysters in crowded beds genetic, or are they a response to crowding during growth? On the other hand, the varied forms of stromatolites (calcite-depositing algae) appear to have little adaptive significance but are simply permissive form variants molded by the environment (Kaufmann, 1964). Variations in mineralogy and in elemental and isotopic composition appear to be almost entirely passive, although some may have significance in chemical buffering of organic systems.

Studies of skeletal composition have dealt almost entirely with temperature and/or salinity variations in the environment. Composition varies as a function 1) of biological partition of elements (or isotopes) within the organism prior to their deposition in the skeleton, 2) of the chemical concentrations in the environment, and 3) of the chemical fractionation during skeletal deposition. The latter in turn is a function of temperature and pressure, but the pressure effect is so negligible that only temperature need be considered. With three unknowns, physiology, concentration, and temperature, and a single known, composition, valuation of one unknown requires assumption of some constant value for the others.

One can take as an example the employment of oxygen isotope ratios, O^{18}/O^{16}, for paleotemperature determination. No recent molluscs are known to show any physiological partition of oxygen isotopes, although algae and echinoids do fractionate them. One begins then with the assumption that the isotope ratios in fossil mollusc skeletons reflect isotopic equilibrium with the surrounding water. Second, studies of recent environments demonstrate that normal-salinity ocean water has a rather constant O^{18}/O^{16} ratio, although fresh, brackish, and hypersaline waters are quite variable. One therefore limits isotopic studies to fossils which are undoubtedly (?) normal marine—particularly a fossil cephalopod group called belemnites. Two of the variables have now been eliminated, so one can solve for temperature. With a fixed mineral, calcite, and a fixed initial composition, the ratio of isotopic composition between the calcite and the water defines an equilibrium constant. This constant is a function of temperature (in earth surface pressure ranges) and the equilibrium constant can be determined experimentally for various temperatures. The O^{18}/O^{16} ratio in a particular fossil belemnite therefore determines an equilibrium constant, and that, in turn, a temperature.

But what of geologic factors? Could the belemnite skeleton reequilibrate much later during or after rock formation? No single answer is conclusive, but a) only belemnites with apparently fresh, unaltered skeletons are used; b) the O^{18}/O^{16} ratio within the skeletons varies in a regular pattern consistent with seasonal variation in temperature; c) the temperatures are reasonable in terms of modern marine environments; d) the regional pattern of temperature variation is reasonable.

Other somewhat similar techniques include:

- **a).** C^{13}/C^{12} ratios to distinguish marine from freshwater environments
- **b).** Aragonite/calcite ratios as temperature indicators
- **c).** Strontium/calcium and magnesium/calcium ratios as temperature indicators

All of these offer similar possibilities, require similar kinds of assumptions, and pose similar problems in interpretation. The latter two tend to show very strong vital effects. For example, Pilkey and Hower (1960) found no discernible relationship between $MgCO_3$ percentage and temperature among individuals of the same species of the recent sea urchin *Dendraster,* but discovered a higher average percentage in warm-water species (Figure 2-8, p. 24). In contrast, individuals within a species showed a high correlation between strontium content and water temperature, but no two species showed the same relationship.

NOTE

Statistical techniques of morphologic comparison

A. Single variable.

1. *Variable.* Some measurable property observed in sample.

$$x = \text{variable}; \ x_i = \text{the } i^{\text{th}} \text{ value of } x$$

2. *Sample.* The set of objects observed or measured.

$$N_j = \text{number of objects measured in the } j^{\text{th}} \text{ sample}$$

3. *Mean of Sample.* The average of a set of measurements.

$$\bar{x}_j = \frac{x_1 + x_2 + \cdots + x_i + \cdots x_n}{N_j} = \frac{\sum_{i=1,2}^{K} x_i}{N_j}$$

4. *Standard Deviation of Sample.* A measure of the dispersal (*variance*) of measured values about the mean.

$$s_x^2 = \frac{(x_1 - \bar{x}_j)^2 + (x_2 - \bar{x}_j)^2 + \cdots + (x_n - \bar{x}_j)^2}{N_j - 1} = \frac{\sum_{i=1,2}^{K} (x_i - \bar{x}_j)^2}{N_j - 1}$$

5. *Standard Error of the Mean.* An estimate of the dispersal of sample means for the chosen variable over the population.

$$s_{\bar{x}} = \frac{s_x}{\sqrt{N_j}}$$

6. *t-test for Comparison of Sample Means.* Ratio of differences between sample means to variance.

$$t = \frac{\bar{x}_1 - \bar{x}_2}{\sqrt{s_{\bar{x}_1}^2 + s_{\bar{x}_2}^2}}$$

Probability that t will equal or exceed a particular value in two samples from the same population with $N_1 + N_2 - 1$ observations is given in tables in statistics books.

7. *F-test for Comparison of Sample Variances.* Ratio between sample variances.

$$F = \frac{s_1^2}{s_2^2}$$

Probability that F will equal or exceed a particular value in two samples from the same population with $N_1 - 1$ and $N_2 - 1$ observations is given in tables in statistics books.

B. Relationship of two variables

1. *Reduced Major Axis.* The linear relation between variation in two variables (x and y) measured on the same sample.

   a. Reduced major axis: $y = \mathbf{b} + x\mathbf{k}$

   b. The slope $= \mathbf{k} = \dfrac{s_y}{s_x}$

   c. The intercept $= \mathbf{b} = \bar{y} - \bar{x}\mathbf{k}$

2. *Regression Line.* The linear variation of one variable with variation in the other measured on the same sample.

   a. Regression line: $y = \mathbf{b}_y + x\mathbf{k}_{yx}$

   b. The slope $= \mathbf{k}_{yx} = \dfrac{\sum_{i=1,2}^{K} (x_i - \bar{x})(y_i - \bar{y})}{\sum_{i=1,2}^{K} (x_i - \bar{x})^2} = \dfrac{\sum_{i=1,2}^{K} x_i y_i}{\sum x_i^2}$

   c. The intercept $= \mathbf{b}_y = \bar{y} - \bar{x}\mathbf{k}_{yx}$

3. *Correlation.* The intensity of the relationship between two variables expressed as the ratio between calculated and observed dispersion from the regression line.

$$r_{xy} = \frac{s_{y_c}}{s_{y_0}}$$

where s_{y_c} is the variance of the calculated values of the variable y corresponding with the values of the variable x along the line $y = b + xk$. Or for calculation:

$$r_{xy} = \frac{\sum_{i=1,2.}^{K} (x_i - \bar{x})(y_i - \bar{y})}{\sqrt{\sum_{i=1,2}^{K} (x_i - \bar{x})^2 \sum_{i=1,2}^{K} (y_i - \bar{y})^2}} = \frac{\sum_{i=1,2.}^{K} x_i y_i}{\sqrt{\sum_{i=1,2.}^{K} x_i^2 \sum_{i=1,2.}^{K} y_i^2}}$$

4. *Comparison of Reduced Major Axis Between Samples.*

   a. *Slopes.* The ratio of difference in slopes to the standard errors of the slopes

$$z = \frac{k_1 - k_2}{\sqrt{s_{k_1}^2 + s_{k_2}^2}}$$

where k_1, k_2 are the slopes for the two samples and the standard error of the slope is equal to:

$$s_{k_j}^2 = \frac{s_y^2}{s_x^2}\left(\frac{1 - r_{xy}^2}{N_j}\right)$$

The probability that z, the normal standard variable, will be exceeded in two samples from the same population is given in tables in statistics texts.

b. *Intercept.* The ratio of the vertical distance between the lines at some selected value of x (x_0) to the standard error of the vertical distance.

$$z = \frac{x_0(k_1 - k_2) + (b_1 - b_2)}{\sqrt{s_{k_1}^2(x_0 - \bar{x}_1)^2 + s_{k_2}(x_0 - \bar{x}_2)^2}}$$

where x_0 is chosen at the greatest *observed* vertical difference between the two lines.

See note under 4.a. concerning z.

REFERENCES

Ager, D. V. 1963. *Principles of Paleoecology.* New York: McGraw-Hill.

Alee, W. C. *et al.* 1949. *Principles of Animal Ecology.* Philadelphia: Saunders.

Bader, R. S. 1965. "A Partition of Variance in Dental Traits of the House Mouse," *Jour. of Mammalogy,* vol. 46, pp. 384-388.

Cock, A. G. 1966. "Genetical Aspects of Metrical Growth and Form in Animals," *Quart. Rev. Biology,* vol. 41, pp. 131-190.

Craig, G. Y. 1967. "Size-Frequency Distributions of Living and Dead Populations of Pelecypods from Bimini, Bahamas, B. W. I.," *Jour. of Geology,* vol. 75, pp. 34-45.

——— and G. Oertel. 1966. "Deterministic Models of Living and Fossil Populations of Animals," *Quart. Jour. Geol. Soc. of London,* vol. 122, pp. 315-355.

Ehrlich, P. R. and R. W. Holm. 1963. *The Process of Evolution.* New York: McGraw-Hill. First nine chapters provide extended discussion of genetic systems and some of their morphological and ecological attributes.

Hecker, R. F. 1965. *Introduction to Paleoecology.* New York: Elsevier.

Hedgpeth, J. W. and H. S. Ladd (eds.). 1957. "Treatise on Marine Ecology and Paleoecology," 2 vols. *Geol. Soc. of America, Memoir 67.*

Imbrie, J. and Norman Newell. 1964. *Approaches to Paleoecology.* New York: John Wiley.

Johnson, R. G. 1960. "Models and Methods for Analysis of the Mode of Formation of Fossil Assemblages," *Bull. Geol. Soc. of America,* vol. 71, pp. 1075-1086.

Kaufmann, W. L. 1964. "Diverse Stromatolite Forms from the Upper Devonian of Saskatchewan," *Bull. Canadian Petrol. Geol.,* vol. 12, pp. 311-316. Environmental modification of form.

Keith, M. D. *et al.* 1964. "Carbon and Oxygen Isotopic Composition of Mollusk Shells from Marine and Fresh-Water Environments," *Geochimica et Cosmochimica Acta,* vol. 28, pp. 1757-1786.

Krumbein, W. C. and F. A. Graybill. 1965. *An Introduction to Statistical Models in Geology.* New York: McGraw-Hill.

Kurtén, B. 1955. "Contribution to the History of a Mutation During 1,000,000 Years," *Evolution,* vol. 9, pp. 107-118.

Lloyd, R. M. 1964. "Variations in the Oxygen and Carbon Isotope Ratios of Florida Bay Mollusks and their Environmental Significance," *Jour. of Geology,* vol. 72, pp. 84-111.

Lowenstam, H. A. 1963. "Biologic Problems Relating to the Composition and Diagenesis of Sediments," in Donnelly, T. W. (Ed.), *The Earth Sciences: Problems and Progress in Research*. Chicago: University of Chicago Press. Pp. 137-195. Among other things a review of the environmental significance of skeletal mineralogy and composition.

McAlester, A. L. 1961. "Some Comments on the Species Problem," *Jour. of Paleontology*, vol. 36, pp. 1377-1381. Biological and/or paleontological species concepts.

Miller, R. L. and J. S. Kahn. 1962. *Statistical Analysis in the Geological Sciences*. New York: John Wiley.

Olson, E. C. 1957. "Size-Frequency Distributions in Samples of Extinct Organisms," *Jour. of Geology*, vol. 65, pp. 309-333.

——— and R. L. Miller. 1958. *Morphological Integration*. Chicago: University of Chicago Press. Model and methods for quantitative study of morphological variation.

Pilkey, O. H. and J. Hower. 1960. See references Chapter 2.

Simpson, G. G. *et al.* 1960. *Quantitative Zoology*. New York: Harcourt, Brace.

Slobodkin, L. B. 1961. *Growth and Regulation of Animal Populations*. New York: Holt, Rinehart, and Winston.

Sokal, R. R. 1965. "Statistical Methods in Systematics," *Biol. Reviews*, vol. 40, pp. 337-391.

Stenzel, H. B. 1945. "Paleoecology of Some Oysters," *Rept. Comm. on Marine Ecol. as Related to Paleontology*, Nat. Research Council, No. 5, pp. 37-46.

———. 1949. "Successional Speciation in Paleontology: the Case of the Oysters of the *Sellaeformis* Stock," *Evolution*, vol. 3, pp. 34-50.

Weber, J. N. and D. M. Raup. 1966. "Fractionation of the Stable Isotopes of Carbon and Oxygen in Marine Calcareous Organisms—the Echinoidea," *Geochim. et Cosmochim. Acta*, vol. 30, pp. 681-736.

Weller, J. M. 1955. "Fatuous Species and Hybrid Populations," *Jour. of Paleontology*, vol. 29, pp. 1066-1069.

5

The organic web

A dozen different fossil species may occur in a few cubic centimeters of rock collected from a single thin stratum; dozens of different plant and animal species live in a single bunch of grass. The stratum, followed along a hillside, may continue to yield, sample after sample, the same species; every bunch of grass on a sand dune will have a majority of species in common. But the same rock layer twenty kilometers away may contain quite different fossils, just as the grass bunches in a swampy meadow yield quite different organisms from those on the adjacent dunes. One notes a double pattern, variation or diversity in a single place and variation between places. The latter phenomenon has an obvious explanation: different species have different limiting environmental factors; *ergo,* different environments contain different species. But why should a single environment contain a multiplicity of types?

Study of the species from any modern environment demonstrates that they are bound together in a web of biological connections—prey and predator, parasite and host, shelter, competition, and so on. In effect, the environment is parceled out among the various types of organisms, and each species participates in a dynamic system within the environment. Experiment and observation indicate that systems with more species maintain a higher organic mass relative to energy input and in this sense are more efficient (Margalef, 1963). This efficiency presumably "explains" the diversity of species in an environment, although it does not explain how the diversity originates. But regardless of explanations, complex ecological systems exist; they have influenced the character of the fossil record (and, as will be seen, the evolution of life); and they are one of the richest, most interesting, and most difficult subjects for paleontologic research.

Ecologic systems and the fossil record

Paleontologists must treat models of ecological systems, because they have only models of the species which compose such systems. Obviously, the next question must be the relation of system to model. From a consideration of species models (pp. 62-66), one can generalize:

Information in Fossil = Information in Biologic System
+ Geologic Changes
+ Modifications in Collection and Observation

Typically, the geologic changes and observation errors reduce the information; they also introduce new information difficult to separate from the biologic. For example, an oyster bank association might easily comprise forty species, including microorganisms. Of these forty, only ten, perhaps, have preservable skeletons. Of the ten preserved, two are very small and washed out by wave action; one has an aragonite skeleton destroyed in diagenesis. Small clams living on the adjacent level bottom are added by wave transport, and other clams, after burial of the bank by mud, burrow down into it. The fossil assemblage would include two "aliens" and seven "natives." The collector, in turn, mistakes variant oysters for distinct taxa and recognizes three species rather than one.

A large portion of paleontologic information is, in consequence, "noise" so far as reconstruction of ecological systems is concerned—though it may be of very real geologic (or psychological) significance. Preceding sections dealt with the recognition and elimination of "noise" in individual form (pp. 19-21) and in variation among samples (pp. 62-66, 93-94). I need only add here that some ecologically significant information may be detected in spite of a high noise level. For example, diversity (number of species) is affected by a very large number of geologic and operator factors; the variation induced by selective preservation, by differences in sampling, and by bias in classification may exceed ecological variation. If, however, the diversity entered by these factors is random or can be partitioned into nonrandom components, residual patterns in sample variation reflect ecologic controls. Thus Stehli (1965) has defined environmental trends from diversity patterns, using statistical analysis to remove random variation and to separate nonrandom components.

One can acquire, by one method or another, a quantity of ecologically significant information from fossil samples. But this in itself is not a model of an ecologic system. It requires modification into meaningful statements about activities within the system.

Examples of such statements include:

a). Brachiopod species *Y* and *Z* lived in clumps on a muddy sea bottom, and the intensity of clumping decreased as silt content increased.

b). Trilobite species *R* fed on organic detritus accumulating around brachiopod clumps.

c). The species *Y* and *Z* fed on algae and small animals floating in the water.

These three statements display different levels of inference. The first is a simple translation of an observation. Three conditions are necessary for its validity: a) the clumping is not of geologic origin; b) the clumping is not a result of biased collection and/or identification; c) the clumping is greater than one would expect in a random distribution. If these conditions are met, the statement is a powerful description of the structure of the system, although a weak one about its dynamics.*

The second statement is more involved. First, the trilobite occurs more often within the brachiopod clumps than chance provides. This inference has the same form as **a).** and the same conditions. Second, the trilobite lacks crushing or piercing mouth parts—a functional analysis from form. Third, modern arthropods with similar mouths typically feed on detritus. Fourth, organic detritus is now associated with many clumps of bottom organisms. To use Tasch's analogy (1965), a paleontological message decodes into a biological one; the decoding depends in part on a "memory" which associates from observation of modern organisms preservable data, e.g., jaw parts, shell clumps, with nonpreservable activities or characteristics, e.g., detritus feeding, accumulation of detritus. The "memory" consists of "normic" statements (discussed on pp. 77-82). "Memory" is subject to "misrecollection," so this type of inference is less likely to be correct than the first, even though in many ways a more interesting description of the dynamics of the system.

* To be completely rigorous, *Y* and *Z* must be accepted as traces of biological activity through analogy with the structures produced by recent organisms; the clumping, however, need not be present in recent ecologic systems.

The third statement depends entirely on "memory" and either its likelihood or its precision may be so low that it has very little value. Because of the traditional emphasis on reconstruction of dynamic connections, statements of the second and third type have been considered the most important. One wonders if the possibilities of the first type of statement have been fully realized.

Neighbors, foundlings, and stray waifs

The critical step in interpretation of ecologic systems from fossil evidence is the inference that fossils lived together. Modern systems show three levels of intensity in association:

a). **Obligatory.** The presence of a species depends completely on the presence (or absence) of other species, e.g., a herbivore species that lives on one and only one kind of plant.

b). **Direct.** The species interact and are at least partially dependent, but alternative species can substitute in this relationship, e.g., a predator species which requires a prey species but which feeds on one of several. (See p. 116.)

c). **Facultative.** The species "tolerate" each other without interacting and overlap in environmental tolerance.

The first two types express the dynamics of the system. Even a facultative relationship, however, is a significant structural property and indicates environmental characteristics if nothing else. The distribution of organisms reflects all three aspects.

The association of fossil species, however, reflects to a very large extent the factors of preservation and even of collection. R. G. Johnson, who has been deeply concerned with this problem, has attempted a deductive analysis (1960) of three alternate modes of fossil accumulation: I. Catastrophic census (see above, p. 64) primarily *in situ;* II. normal census primarily *in situ* with some disturbance by removal of some skeletons; III. normal census with major transportation to and from site, and has suggested criterial (Table 5-1) for recognition of these models. In a study of Recent and Pleistocene shell accumulations (1962, 1965), he found that a considerable percentage of those from low turbulence (fine-grained sediment) sites represented life associations (Modes I and II). In another study of Pleistocene assemblages, Valentine and Mallory (1965) found that fossil species associations based on many samples corresponded to recent living associations in spite of evidence that some transportation had occurred and that individual samples were, characteristically, derived from several living associations. Physical association of fossils seems a reasonable basis on

TABLE 5-1

Criteria for faunal accumulations (*after* Johnson, 1960)

Feature	*Expression* I *Little disturbed*	II *Some disturbance*	III *Highly disturbed*
1. Faunal Composition	Ecologically coherent.	As in *I*.	Not necessarily coherent.
2. Morphologic Variation	Wide variation in shape and delicacy.	As in *I*.	Little variation—consists of most durable elements.
3. Density of Fossils	Wide range possible.	As in *I*.	High.
4. Size-Frequency Distribution	Consistent with census of biological population.	Inconsistent with biologic census in some species.	Consistent with sediment sorting.
5. Disassociation	High proportion articulated; disarticulated parts in appropriate biologic ratios.	Intermediate between *I* and *III*.	Most disarticulated; disarticulated parts not in appropriate ratios.
6. Fragmentation	Low proportion.	Moderate proportion.	High proportion.
7. Surface of Fossils	Little abrasion.	Some abrasion.	Much abrasion.
8. Position	Some in life position; not hydrodynamically stable.	Majority in hydrodynamically stable positions.	As in *II*.
9. Dispersion	Consistent with pattern in living pop.	Random or consistent with mechanical reworking.	As in *II*.
10. Sedimentary Structure and Grain Size	Imply subcritical velocity for smallest fossils.	Imply critical velocity for smallest fossils.	As in *II*.
11. Sedimentary Environment	Consistent with inferred tolerances of fauna.	As in *I*.	May be inconsistent with faunal inferences.

which to begin interpretation of interspecies relationships—just as association with rock type is critical in interpretation of species ecology.

Characteristics of ecologic systems

A system, biological, social, or otherwise, is characterized by the transfer of energy, information and/or material and by a structure. An ecologic system requires for survival an input of energy (since it shows negative entropy) and certain amounts and kinds of material. These constitute the extrinsic dynamics of the system. Within the system, energy and material are transferred from unit to unit. These transfers constitute the system's intrinsic dynamics. In turn, system structure, extrinsic and intrinsic, must parallel dynamics. If some part of system structure remains in the fossil record, reconstruction of dynamics becomes feasible.

Extrinsic properties

The organisms living together form a *biocoenosis;* a biocoenosis inhabits a specific place with a specific set of environmental properties, a *habitat* (see also p. 86). The properties of a habitat define the input of energy and material to a biocoenosis; these properties form the extrinsic framework of an ecological system. The environmental properties and the organisms of two adjacent rocky shores are strikingly similar; even an Arctic headland and a tropic clifty coast resemble each other closely in habitat properties and in the features of their biocoenoses. On the other hand, a clifty coast and an adjacent inlet diverge sharply. Obviously, one can classify habitats in a rather small number of distinctive types; one can then map the spatial distribution of habitats as a part of ecological structure. Conversely, the association of particular biocoenotic types with particular habitat types permits recognition of habitat without measurement of the environment. This implies for paleontology the definition of ancient habitats, ancient environments, and ancient geography by study of fossil assemblages and their distribution.

Habitat classifications are based on differences in environmental factors that appear most important in determining animal distributions ("most important" factors meaning those which affect the most species). Cross-classifications involving several different factors are also made.

In this way, the ecologist distinguishes the terrestrial and aquatic environments by the character of the medium; classifies habitats within the aquatic environment as fresh, brackish, and marine, according to salinity; and further subdivides the marine habitats by depth of water (which is correlated with turbulence or light intensity). A cross-classification on temperature yields even finer subdivision; for example, temperate, aquatic, marine, shallow, rocky, and a further cross-classification on geographic location as North Atlantic temperate, aquatic, marine, shallow, rocky.

Neither the classification of habitats used here (Table 5-2; Figure 5-1) nor any classification I know of is completely consistent within itself. Moreover, the division between habitats is, in part, arbitrary. This classification, however, yields a useful visualization of habitat differences and establishes a standard terminology.

Habitats and environments of deposition

In the fossil record, loss of diagnostic criteria such as depth and temperature, difficulties in distinguishing life environments from those of

TABLE 5-2

Classification and characteristics of the more important habitats

Classification of marine habitats follows Hedgpeth, 1957. Classification of sedimentary environments is based on Krumbein and Sloss, 1963. Description is not intended to be exhaustive.

Habitat, biome or biotope	*Primary ecologic factors*	*Sedimentary environments*	*Sedimentary characteristics*	*Organic characteristics*
1. Terrestrial A. Forest	Light intense to moderate; moisture, high to moderate; substrate, soft to firm; succulent vegetation; abundant O_2.	Fluvial—alluvial plain, channel.	Rapid fluctuations—coarse to fine; oxidation; laterization; channeling; structureless shales; interlensing beds of different rock types; interbedded swamp and lacustrine lithotopes.	Terrestrial adapted; arboreal animals; browsing herbivores; vertebrates—tetrapods; fluvial, swamp, and lacustrine associates.
B. Grasslands	Light intense; moisture moderate to low except on river borders; substrate firm; grass; abundant O_2.	Fluvial. Eolian. Lacustrine—ephemeral.	As above except more oxidation; caliches; some laterization; pond ls. and playa facies. Eolian deposits with well sorted and rounded, frosted grains.	Terrestrial adapted; cursorial vertebrates; grazing herbivores; tetrapod vertebrates; fluvial and lacustrine associates.
C. Deserts	Light intense; moisture low; substrate firm or rocky; scrub-grass vegetation; abundant O_2.	Eolian. Fluvial. Playa.	Fanglomerates; eolian sands and silts; evaporite facies.	Terrestrial adapted; cursorial vertebrates; few fish or snails.
D. Swamps, marshes, tundra	Light intense to moderate; moisture abundant; substrate soft; vegetation; bottoms acidic, anerobic.	Swamp.	Fine sediments; high amount of carbonaceous material; reducing; usually lensing into lacustrine and fluvial beds.	Aquatic and semiaquatic animals; fish, amphibians, snails, small pelecypods, arthropods of various types. Admixture of terrestrial adapted types.
E. Caves	Little or no light; generally abundant moisture; no vegetation; O_2 variable.	Cave.	Clay or silt fillings in fissures or pockets.	Mixture of animal types washed in or seeking shelter in caves. A few specially adapted types: blind fish, amphibians, insects.
F. Polar	Low temperatures; H_2O difficult to obtain; tundra vegetation; abundant O_2.	Glacial.	Till and outwash; solufluction.	Terrestrial adapted animals; some insects, homiothermal vertebrates.

G. Alpine	Low to moderate temperatures; moisture high to low; substrate firm, irregular; intense light; O_2 partial pressure reduced.	Glacial. Fluvial.	Till and outwash; fanglomerates.	Terrestrial adapted animals; some insects, homiothermal vertebrates.
2. Aquatic A. Fresh water (1) Running waters	Salinity low; abundant oxygen; high turbulence; substrate variable—usually firm to moderate; light intense to low (controlled by turbidity); turbidity variable; vegetation slight.	Fluvial—channel. Deltaic—distributaries.	Channel fills in fluvial facies; relatively coarse; fossils commonly broken or worn.	Fish, amphibians; crustacea and other aquatic arthropods; snails, pelecypods; intrusive terrestrial animals.
(2) Standing waters a) Lakes	Turbidity moderate to low; salinity low; light intense to low; substrate firm to soft; low turbulence; O_2 abundant to anerobic; abundant plant life.	Lacustrine.	Fine-grained silts and shales; laminated; interfingering with delta swamp, and alluvial deposits.	Fish; amphibians on margins; aquatic arthropods; snails and pelecypods; intrusive terrestrial animals on margins.
b) Ephemeral ponds	Turbidity moderate to high; light intense to moderate; substrate firm to soft; low turbulence; O_2 abundant; salinity low except in playas.	Lacustrine. Swamp.	Restricted extent; mudcracks; lime mud breccias; occasionally playa facies. Lenses in alluvial and/or swamp.	Some fish, amphibia, crustacea snails, pelecypods; intrusive terrestrial animals.
c) Swamps and marshes	See above.	Swamp.	See above.	See above.
B. Brackish	Low salinity, otherwise as in running waters (estuaries), lakes (lagoons), and swamps and marsh (lagoons).	Deltaic. Lagoonal. Littoral. Possibly sublittoral in ancient epicontinental seas.	As in Fluvial (channel), Lacustrine and Swamp.	Few species—fish, gastropods, pelecypods, few brachiopods.
C. Marine (1) Benthic a) Supralittoral	Light intense; moisture slight to moderate; vegetation varied —grasses, forest, swamp; substrate firm to soft; above high tide but may be affected by storm waves; much O_2.	Supralittoral intergrades with Fluvial and Deltaic. In many cases Eolian.	Fluvial and deltaic sediments and associations. Dune structures with well sorted, rounded grains of sand.	Variety of terrestrial invertebrates and vertebrates.

TABLE 5-2 (continued)

Habitat, biome or biotope	*Primary ecologic factors*	*Sedimentary environments*	*Sedimentary characteristics*	*Organic characteristics*
b) Littoral	Between high and low tides; O_2 typically high; moisture and light fluctuate. Algae. Water saline; substrate moderate to soft. Wave action strong except in estuaries and lagoons.	Lagoonal. Littoral. Deltaic.	Beach structures—ripple marks, cross-bedding, rain drop prints, beach cusps. Mud flats—ripple marks, mudcracks; swamp sediments. Sediments typically well sorted.	Burrowing and crawling invertebrates; some echinoderms. A few types of sessile animals with external shells.
c) Sublittoral 1. Inner	Depths to 100m; light moderate to low intensity; wave action moderate to slight; O_2 typically moderate. Saline—occasionally hypersaline; substrate firm to soft; algae; turbidity moderate to high.	Infralittoral. Circalittoral.	Moderate to fine-grained sediments; sorting moderate to poor. Typically reduced. Ripple marks; even bedded, some cross-bedded; "sheet" deposits. Carbonates. Evaporites.	Many types of attached invertebrates; brachiopods, corals, bryozoans, pelecypods. Burrowing and crawling echinoderms, molluscs, and arthropods.
2. Outer	Depths to 200m; light low to absent; wave action slight to none; O_2 moderate to low; saline—occasionally hypersaline; substrate soft; a few algae; turbidity moderate to high.	Circalittoral.	Fine-grained sediments. Sorting moderate to poor. Typically reduced. Graded beds from turbidity currents. Even bedded —"sheet" deposits. Many carbonates. Evaporites.	As in inner sublittoral.
d) Bathyal	Depths between 200 and 4000m; light and wave action absent; O_2 moderate to low; substrate soft; temperatures low; low velocity currents.	Bathyal.	Very fine sediments—carbonates subordinate to clastic, glauconitic, and siliceous. Evidences of turbidity currents. Even bedded —"sheet" deposits; reduced.	Organisms rare with increasing depth—mostly scavengers and predators. Sessile types limited.
e) Abyssal	Depths between 4000 and 5000m. No light or wave action; O_2 moderate to low; substrate soft, temperatures quite low; low velocity currents.	Bathyal. Abyssal.	Very fine sediments—calcareous and siliceous oozes and red and blue muds. Possibly some coarser sediments brought by turbidity currents.	Organisms relatively rare—mostly scavengers and predators. Sessile types limited.

f) Hadal	Depths below 5000m. Otherwise similar to abyssal.	Abyssal Hadal	Very fine sediments. No calcareous deposits.	Organisms as in abyssal.
(2) Pelagic a) Neritic	Waters above sublittoral benthic; characteristics the same except lack of substrate; turbidity moderate to low.	Epineritic. Infraneritic.	As given above for these sedimentary environments.	Floating and swimming organisms. Abundant microscopic plants and animals.
b) Oceanic 1. Epipelagic	Depths to 100m±. Characteristics as in inner sublittoral except lack of substrate; low turbidity.	Bathyal Abyssal	Same as above.	Same as above.
2. Mesopelagic	Depths between 100 and 1000m. Characteristics as in outer sublittoral and bathyal except lack of substrate; low turbidity.	Bathyal	Same as preceding.	Floating and swimming organisms. Few plants; animals primarily larger scavengers and predators.
3. Bathypelagic	Depths 1000-4000m±. Characteristics as in bathyal except lack of substrate.	Bathyal	Same as preceding.	Floating and swimming organisms. No photosynthetic plants; animals primarily larger scavengers and predators.
4. Abyssopelagic	Depths below 4000m. Characteristics as in abyssal except lack of substrate.	Abyssal	Same as preceding.	Same as preceding.

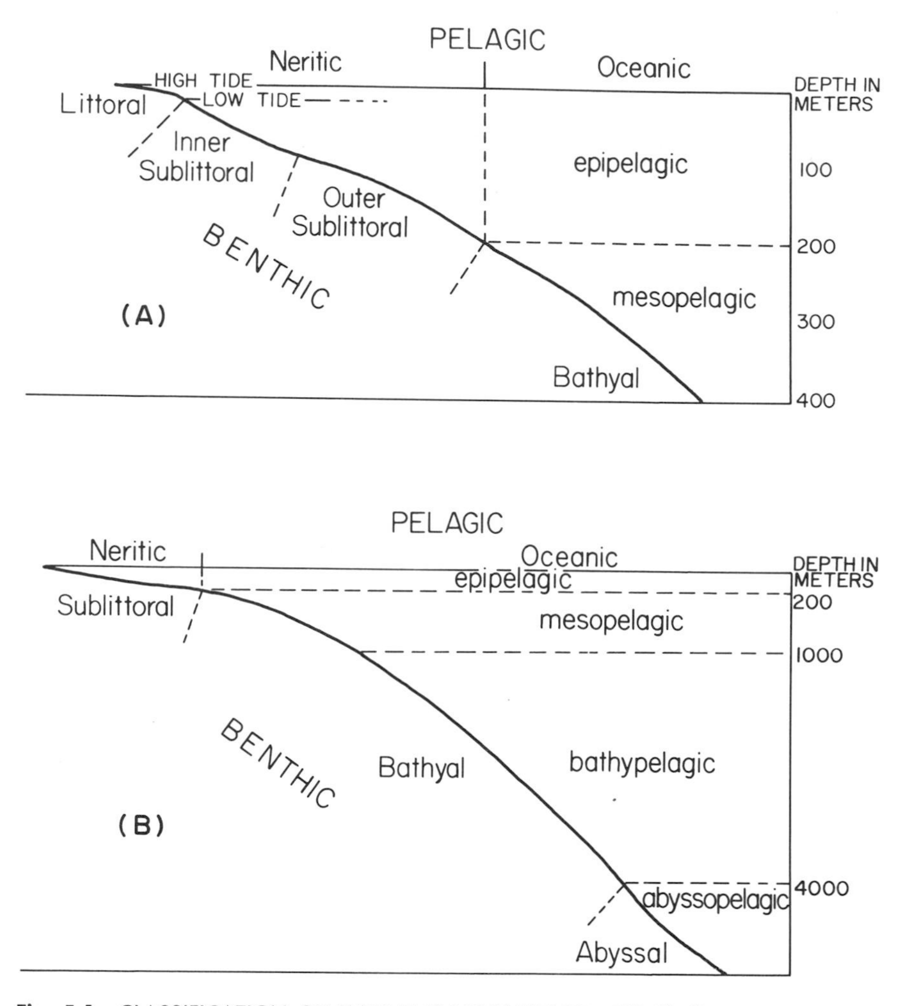

Fig. 5-1. CLASSIFICATION OF MARINE ENVIRONMENTS. **(A)** Shallow water zones. **(B)** Deep water zones. The pelagic environment and its subdivisions are water zones. The benthic environment and its subdivisions are bottom zones. The upper portion of the epipelagic and the inner sublittoral and littoral environments are lighted; the others receive no effective amount of sunlight. (*Based on* Hedgpeth, 1957.)

burial, and admixture of foreign species handicaps recognition of habitat types. Otherwise, the interpretation proceeds much like that of single species populations, with the use of 1) the adaptations of the organisms, 2) the taxonomic affiliations of the fossils to modern animals, and 3) the sedimentary characteristics to establish critical parameters. In the sedimentary rocks deposited in a single sedimentary basin, a paleontologist can typically recognize several distinct animal associations. Each association presumably occupied a distinct habitat, the same one throughout the basin.

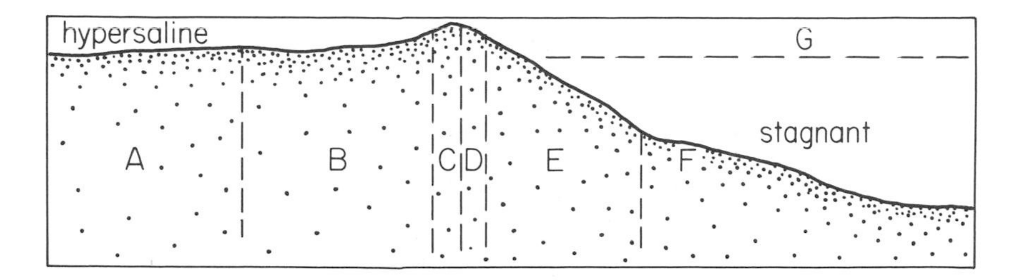

Fig. 5-2. MARINE ENVIRONMENTS IN THE WEST TEXAS PERMIAN BASIN. A section from the open basin across the barrier reef and back reef lagoon. **(A)** Lagoons behind barrier reef, sublittoral (inner) and neritic environments. Salt concentrated by evaporation. Fossils very rare; a few bachiopods, pelecypods, and gastropods occur. **(B)** Bank behind reef; inner sublittoral and neritic; Foraminifera, crinoids, some brachiopods, bryozoans, and algae are characteristically found. **(C)** Reef top, sublittoral and possibly littoral. The fossils include massive calcareous sponges, bryozoans, and algae, as well as cemented brachiopods. **(D)** Reef front, inner sublittoral. Brachiopods and bryozoans of more delicate structure than those in **(C)**. Calcareous algae. **(E)** Upper talus slope, sublittoral, and upper bathyal environments. Water intermittently stagnant with high concentration of hydrogen sulphide. Sponges with siliceous skeletons are most common fossils. **(F)** Basin floor, upper bathyal. Water stagnant. Very few fossils other than pelagic types are found in rocks representing this environment. **(G)** Epipelagic and neritic environments. Radiolaria, calcareous planktonic algae, and ammonoids are the most common fossils. (*Information from* Newell, 1957.)

The Permian rocks in the basins of west Texas contain several associations and habitats, as shown in Table 5-3 and Figure 5-2 (Newell, 1957). The basin floors were low in oxygen. The high organic content of the sediment and the small numbers and slight variety of bottom-dwelling organisms indicate this. The fine grain size of the sediments and their even bedding demonstrate deposition in quiet water, probably at considerable depth. In addition to scattered fossils of bottom-living animals, the basin floor deposits contain numbers of radiolarians, calcareous algae (simple plants), and cephalopods. These resemble recent floating and swimming organisms and presumably lived in the aerated water near the surface. Here a single sedimentary environment contains fossils from two distinct habitats.

On the margins of the basins are reefs (*bioherms*), built largely by calcareous algae but also containing masses of calcareous sponges, bryozoans, and hydrocoralines. The accumulation of such an abundance and variety of organisms could occur only in relatively warm, well-aerated waters. The breakage of the fossils indicates that the upper part of the reef was subject to intense wave action, but delicate bryozoans occur on the lower parts of the reef. The basinward side and shoreward side of the reef also contain different species. Thus, the reef consists of several different but related habitats, each with characteristic animal species, most of them bottom-living.

TABLE 5-3

Interpretation of marine habitats in Permian rocks of west Texas (*data and interpretations based on* Newell, 1957)

General environment of deposition	*Rocks*	*Organisms*	*Probable habitats*
1. Basin Phase—infraneritic and bathyal.	1. Fine-grained quartz sandstones; black, bituminous limestones; black, bituminous, platy siltstones. Thin uniform beds.	1. a. Brachiopods, pelecypods, small gastropods, (?) sponges.	1. *a*) Outer sublittoral, bathyal. Environment probably anerobic at times.
		b. Pelagic protozoans, algae, ammonoid cephalopods.	*b*) Neritic, epipelagic, mesopelagic.
2. Reef and Bank Phases—epineritic and littoral.	2. Limestone composed chiefly of fossils and fossil fragments. Typically dolomitized, some detrital sand and mud.	2. Calcareous algae, calcareous sponges, benthic foraminifers, corals, bryozoans, brachiopods, a few ammonoid cephalopods.	2. Inner sublittoral, (?) littoral, neritic.
3. Shelf Phase—epineritic and littoral.	3. Thin dolomitic limestones, many pisolithic; thin wedges of quartz sandstone; limestones arenaceous. Some anhydrite and red sandstones.	3. Calcareous algae, benthic foraminifera. A few species of brachiopods, pelecypods, and gastropods.	3. Inner sublittoral, (?) littoral, neritic. Environment probably hypersaline.

The sediments and fossils of the lagoons behind the reefs are those characteristic of shallow, well-aerated waters. The fauna was limited to those animals that could tolerate high salt concentration. Apparently the reefs prevented free circulation of water into the lagoons from the basins, and evaporation produced highly saline conditions. The bottom-living animals were types adapted to support themselves on oozy mud—represented now by micrite.

Some associations occur in several different lithologies; obviously they were not limited by factors of sediment deposit. Other sedimentary environments encompassed several different habitats. These occurrences confuse the paleoecologist, because identification of habitat depends, in part, on identification of sedimentary environment. They also confuse the sedimentary petrologist, since his interpretations of sedimentary environment rest, in a measure, on interpretation of organic habitat. A stratigrapher-paleoecologist therefore makes constant cross reference from rock to fossil and back. He joins the fossil assemblage to a critical suite of sedimentary lithologies; the strata to a critical suite of fossil assemblages.

This interplay sometimes yields more detailed interpretations of ecology (Figure 5-3, adapted from Phleger). The foraminifer assemblages cross the boundaries between different environments of deposition. They do not, however, live in all the sedimentary environments of the area, nor do they inhabit all of any one sedimentary environment. Obviously, the sedimentary characteristics (turbidity and substrate) do not limit their distribution, because they occur on both mud and sand bottoms. On the other hand, depth is not the critical factor; they are always found in relatively shallow water but not on all shallow bottoms. They live near shore but not in all near-shore environments.

Inspection of the map suggests a relation to the main distributary channels of the Mississippi. The water in these channels freshens the ocean water near the point of discharge and bears a load of suspended and dissolved organic nutrients. Almost surely the distribution of this foraminifer assemblage is related to one or the other of these factors. The ecologist can now explain the anomalies of distribution. Both sand and mud bottoms occur off the distributary mouth; therefore this assemblage lives in both, but only at the distributary. The depths at the distributary mouth are shallow, but shallow waters elsewhere are too saline or lack the necessary organic nutrients.

But how would a paleoecologist recognize the control of the river distributary on animal distribution? The sedimentary environments are not unique in this particular point; sands are found elsewhere and so are muds.

But the configuration of environments, the interrelation of sand in channel fills and bar fingers and of silts and clays in the natural levees and delta front deposits, establishes the presence of the distributary mouth. The relation of the assemblage to this brackish habitat becomes an environmental index for other deposits. A lagoon might be brackish, normal, or

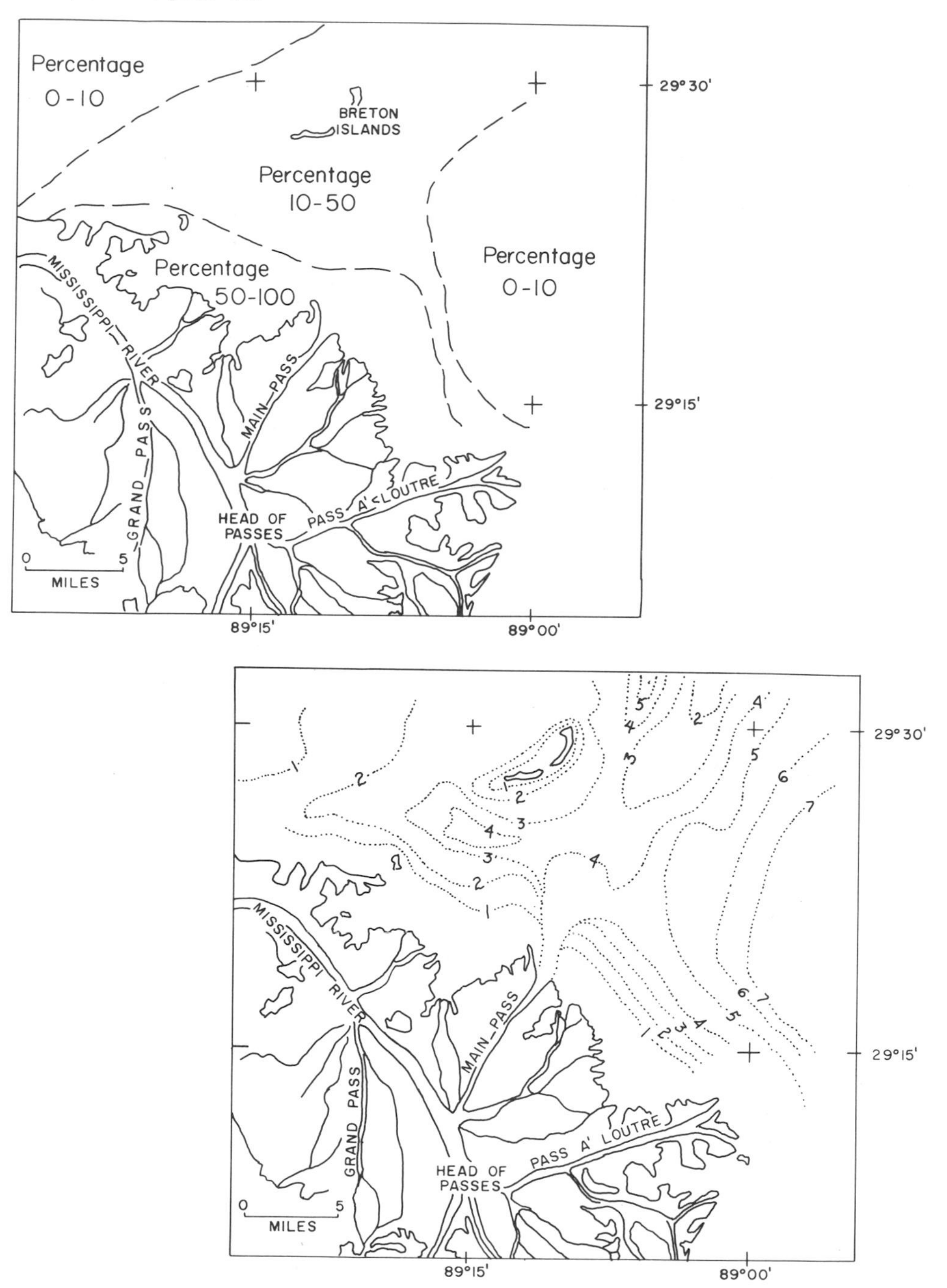

Fig. 5-3. DISTRIBUTION OF A FORAMINIFER SPECIES AND RELATION TO ENVIRONMENT. The upper map shows the relative abundance of the species *Ammobaculites salsus* in percent. The lower map shows the depth of water in fathoms. The sediments, not shown here, display a rather complex pattern across abundance trends and, in part, across water depth. Note relation of distribution to distributaries (passes) of Mississippi. (*Data from* Phleger, 1955.)

hypersaline without any indication in the sediments or their interrelation. If a paleoecologist found this assemblage of foraminifers, he would logically conclude that the lagoon was brackish.

A first summation

To recapitulate, the paleontologist recognizes differences and similarities between fossil samples of the same age collected in different localities. He separates features that correspond to those of the living populations from those that result from transportation and selective preservation. Next, he isolates the environmental factors that controlled the distribution of the animals. Typically, he observes distinct fossil associations whose occurrences correlate in part with sediment differences. By interpretation of the organic adaptations and of the rocks, he recognizes ancient habitats and their characteristic species. This analysis differs from that of a species population (see pp. 89-95) only in the variety of animal types.

But interpretation of habitat fails to explain all differences between fossil samples. For example, the early Pleistocene mammals of Australia differed radically from those of southeastern Asia, even though they were adapted to similar environments. Further, they are found with other fossils and rocks that indicate similar habitats. Nevertheless, they differ in species, genera, families, and orders. Only a historical explanation can solve this puzzle—so it seems proper to put it aside until a basis in evolutionary theory can be laid. I'll simply note that biological variation between fossil samples may represent differences in space and time (Chapter 8) as well as in habitat.

The architecture of biocoenoses

Although the relationships within ecological systems are many, complex, and varied, one can classify them in terms of 1) consumption of nutrients, 2) modification or conditioning of the environment, 3) competition for limited resources, 4) spatial (or temporal) subdivision of the habitat, and 5) movement within the habitat. These characteristics define both the overall structure and the role played by each species, i.e., its *niche*. This section will treat each of them (and their paleontologic aspects) in turn.

Consumption

The production and utilization of food provides the basic structure of a biocoenosis. Organisms include 1) producers (*autotrophs*), which manufacture needed complex organic molecules from simpler inorganic ones, 2) consumers (*allotrophs*), which require an external supply of organic molecules, and 3) the rarer *heterotrophs,* which function as both producers and consumers. Autotrophs comprise the photosynthetic plants which rely on sunlight as an energy source and those bacteria which utilize exothermal chemical reactions. Allotrophs include a) *parasites* (many bacteria and fungi; some animals), b) *saprophytes* and *scavengers* (many bacteria and fungi; some animals) which consume dead organic material, c) *herbivores* (many animals) which eat plants, and d) *predators* (many animals) which prey upon other animals. Heterotrophs include a considerable group of microscopic organisms ordinarily included in the animal phylum Protozoa; they are capable of photosynthesis in the presence of light but are also carnivorous.

A plant-herbivore link is the basic element of a multispecific ecologic system, but most systems add successive predators and parasites, linked in a *food chain* (Figure 5-4A). In most systems, several chains are interconnected to build a *food web* (Figure 5-4B).

To survive, a population in a food web must produce sufficient material to maintain itself, to provide for reproduction, and to offset losses to other populations. Only part of the material supplied to that population is incorporated in building animal tissue; some is metabolized, and direct loss back to the extrinsic environment may also occur. In consequence, the mass of organic material produced at the lowest level in a food web averages about ten times that in the second level—varying from five to twenty times. Similar relationships appear between successive levels to define a food pyramid (Figure 5-4C).

The food pyramid as defined here applies only to the rate of production or to the total amount of organic material produced. Under rather special conditions one observes corresponding pyramids of numbers or of *biomass* (weight of organic tissue) at one time. If the sizes of organisms in successive levels are the same or larger and if the rate of production is similar or higher on successively higher levels, then the number of individuals and the biomass diminishes upward (Figure 5-4D). On the other hand, if size is less and/or rate of production lower the number and biomass pyramids are inverted (Figure 5-4E). The inversion of the pyramid emphasizes the distinction between the production of organic material during a unit time (the *productivity*) and the amount of material existing at any one time (the *standing crop*). This distinction is particularly striking in some marine environments. There the standing crop of plants is relatively low, but they grow and reproduce rapidly so that productivity is high; the bottom organisms that feed on these plants grow and reproduce very slowly so that productivity is low. In consequence, a score of generations of plants supports one generation of animals. If ten grams of plant material supply one gram of animal material, the standing crop of animals would be twice that of plants.

Nutrient cycles are closely related to productivity and the consequent turnover of organic material. In all systems, the supply of nutrients, whether organic compounds or inorganic salts, is limited; in any one system, some critical nutrient imposes a limit on the system (or some level within it). If productivity is high and turnover rapid, the limiting nutrient

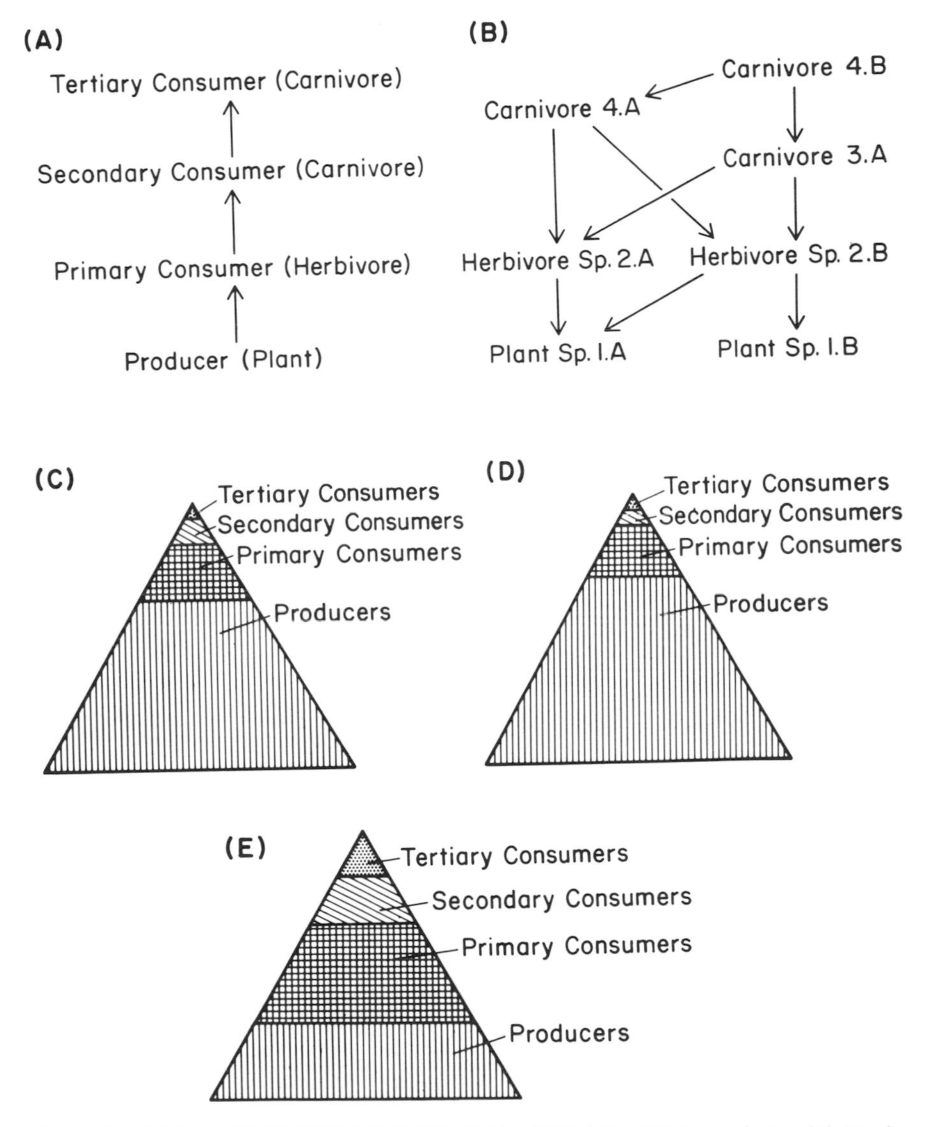

Fig. 5-4. TROPHIC STRUCTURE OF ECOLOGICAL SYSTEMS. **(A)** Food chain. **(B)** Food web; trophic levels indicated by numbers. **(C)** Food pyramid. Area indicates production at each trophic level. **(D)** Food pyramid. Area indicates number (or mass) of individuals with size constant or increasing and productivity constant or increasing upward. **(E)** Food pyramid—biomass or numbers. Size and/or productivity decreasing upward.

circulates rapidly through the system; if productivity is low and turnover slow, a large part is banked in "standing crop."

The requirements and adaptations of particular organisms determine the mode of nutrition (p. 33), but ecologic systems display persistent nutrition patterns. In land environments, the plant stratum consists of relatively large, immobile individuals. The herbivores, therefore, actively seek out and select a plant for consumption. In turn, the predators attack individual herbivores after selective search. Only a few terrestrial animals feed on undifferentiated masses of organic particles and inorganic material (earthworms) or collect masses of small organisms (anteaters perhaps). In contrast, in water environments—particularly the sea—the plants are principally very small individuals floating in the water. In consequence, the herbivores develop diverse feeding modes. Very small animals live among the plants and feed on them actively and selectively. These in turn are prey for slightly larger animals. Superposition of successive predation levels results in successively larger predators—up to the sharks, giant squids, and carnivorous whales. Another mode involves mass collection of the small plants and the accompanying animals. The process may involve active search as in whalebone whales and in whale sharks, but most collectors are fixed or sluggish bottom animals. Some extract small organisms and organic debris from the water by a filter system; these are the *suspension feeders*. Others remove organic particles selectively from the bottom sediments; these are *selective deposit feeders*. Only a short step further are the *nonselective deposit feeders* which ingest undifferentiated masses of sediment and organic material. The herbivores that feed on large masses of attached plant material resemble land herbivores in their active search and selective consumption. All the bottom-living herbivores, whether motile or fixed, are prey for active carnivores of various types and with varied feeding mechanisms. One group, the coelenterates, contributes a large number of carnivores which remain fixed and prey on swimming animals. Plants, herbivores, and predators are also parasitized—typically by individuals of very small size.

The reconstruction of nutrient systems is obviously very important in the study of ancient biocoenoses; it is also difficult and rarely very satisfying except at a general level. Assignment as auto- or allotrophs is done primarily on affinity to modern photosynthetic plants on one hand or to an animal phylum on the other. The deduction is a normic law: plants are not required to be autotrophic—witness the Venus flytrap or the pitcher plant—or animals to be allotrophic—*viz.* the heterotrophic protozoans. The normic statement, however, is so general and powerful that the testimony of adaptive structures, e.g., leaves or teeth, is essentially trivial.

The distinction of predators from herbivores typically derives from functional interpretation of the feeding mechanism, where that involves skeletal structures (p. 36). Assignments based on feeding habits of recent relatives are, at best, good guesses projected into Mesozoic and Paleozoic biocoenoses. As a geochemical alternative, Toots (1965) has demonstrated a lower strontium content in fossil carnivore bone—related presumably to chemical fractionations along the food chain.

Reconstructed food chains and webs are also relatively generalized and derive 1) from inferences about feeding mechanisms and modes, 2) from relative size, e.g., a weasel can hardly be a regular predator on elephants, and, 3) from association in occurrence. The food web reconstructed by Olson (1952, 1961) for an early Permian land biocoenosis (Figure 5-5),

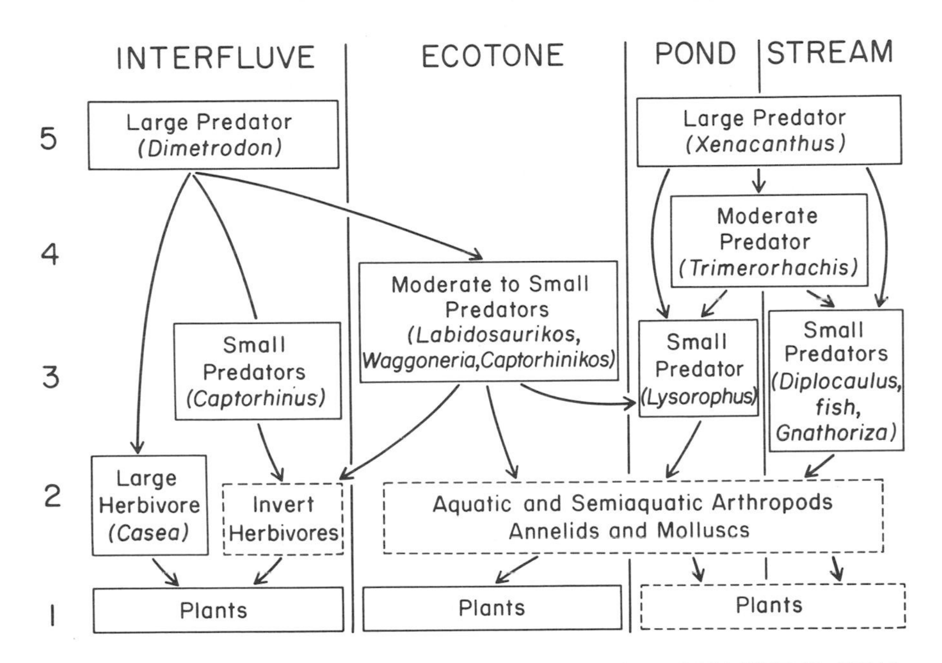

Fig. 5-5. ANALYSIS OF THE EARLY PERMIAN FAUNA OF NORTH-CENTRAL TEXAS. Arrows indicate probable predation. Species in any one block are in part competitive. The plants, though poorly known as fossils, must have furnished the ultimate food source for whole society. The herbivores included one large reptile genus, *Casea*, and, less well known from fossils, a variety of small invertebrates. The latter were consumed by the small- to moderate-sized predators, including a variety of amphibians and reptiles. The master predator was a reptile, *Dimetrodon*. Three distinct habitats—interfluve, pond, and stream, are included in this area, with a transition, an *ecotone*, between the interfluves and the aquatic environments. The probable life occurrence of the various genera in these habitats is indicated. (*Based on* Olson).

is almost certainly qualitatively correct, although some interpretations, e.g., *Edaphosaurus* and *Diadectes* as invertebrate feeders, may be wrong. Note that the producers and the primary consumers are poorly known. Construction of food web models seems more useful for terrestrial vertebrate assemblages than for marine biocoenoses. In the latter, the producers and several levels of consumers are rarely if ever preserved, and inferences of diet tend to be less precise.

Standing crop and productivity are even less amenable to reconstruction. The relative abundance and/or mass of various fossil taxa from a single depositional unit at a single sample site approaches an estimate of *relative* standing crop. Since standing crop and productivity show no necessary correlation, the latter seems beyond the paleoecologist's direct grasp. One angle of attack as yet unexplored is the quantity and variety of organic compounds in fine-grained sedimentary rocks. If geologic bias could be removed, these might prove an important index to nutritive structure—Margalef points out (1963) that the variety of assimilatory pigments used by plants is proportional to the complexity of the system.

Environmental modification

Organisms modify or condition the external environment in their vicinity. The modification may be very large and quite extensive geographically, as in coral reefs or forests, where a large number of organisms unite in collective alteration of the environment. It may be very restricted and quite local where on a muddy substrate a clam shell serves for attachment of a sponge. It may be physical modification of the environment or subtle chemical alterations.

Although the varieties of environmental modification are as numerous as the varieties of organisms, they form three broad categories. The first would include gross environmental modifications. Within a forest, light, temperature, and humidity differ from that in an adjacent meadow; the turbulence, salinity, and dissolved gases of the water behind a coral reef will be very different than they would be in the absence of a reef.

The general modifications of the environment grade into more specific ones as a second category: A coral reef creates a new general environment, but it also creates a large number of "places to live," i.e., to hide, to attach, to lay eggs, to feed, etc. The shells, living or dead, on the sea floor provide a similar set of *microhabitats,* although the gross environment remains unchanged. Both general modifications and microhabitats occur regardless of the species that participate, e.g., a bryozoan encrusting a sea urchin test would be equally content on a different species of urchin or on a barnacle or clam. Participation in an association is neither obligatory nor exclusive for the species.

The third category (broadly, *symbiosis*) comprises obligate relationships and nonobligate ones limited to particular species. An extreme appears in the special barnacles that live only on other barnacles that live only on certain whales. These relationships may be of mutual benefit (*mutualism*) or of value to only one (*commensalism*). The latter grades into parasitism, which is deleterious to one of the partners.

The reconstruction of environmental modification depends on its nature, e.g., links established by biochemistry or behavior are inaccessible. Gross physical modifications such as that of the back reef environment of fossil bioherms (see above, p. 111) are likely to be obvious. Symbiosis if it requires physical attachment is equally obvious, as when specimens of the fossil snail *Platyceras* commonly are found over the anal opening of a crinoid, or skeletons of the worm *Spirorbis* occur attached on a clam shell near the exhalant current position. Aside from direct physical association (see Ager, 1963, pp. 259-266, for other examples), environmental modification is inferred from analogies with modern systems and from association in occurrence. A coral framework will modify wave action, will alter the

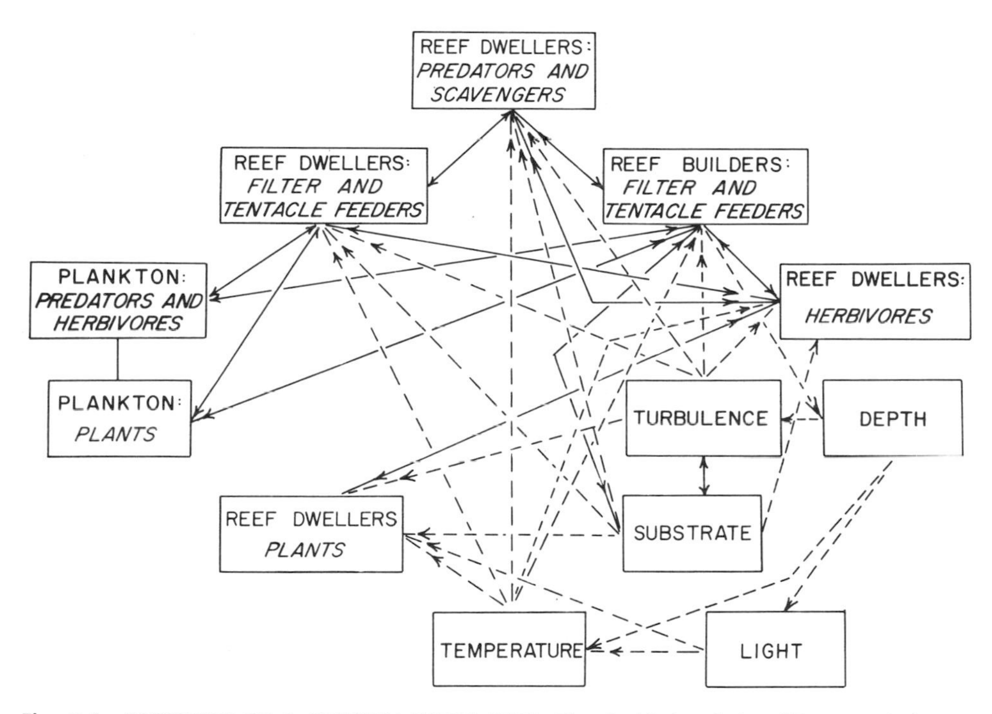

Fig. 5-6. ECONOMY OF A SILURIAN CORAL REEF. The double-headed, solid arrows indicate mutual interaction; the single-headed, dashed arrows indicate unilateral action. The physical factors of the environment are grouped at lower right. The critical relations in reef formation are 1.) the supply of plankton for the reef builders; 2.) the interaction of the reef builders and substrate, and 3.) the effect of reef-building on depth and turbulence. As these change, the physical character of the environment changes and, consequently, the other species are limited by these factors. This reconstruction is based on Lowenstam, 1957. The reef builders include stromatoporoid, tabulate corals, and *Stromatactis,* a form of unknown affinities. Among the herbivorous reef dwellers are the gastropods, possibly some of the trilobites, and a variety of "worms" unrepresented as fossils. The filter and tentacle feeders include crinoids, cystoids, brachiopods, bryozoans, corals, and pelecypods. The scavengers and predators comprise trilobites, cephalopods, and possibly annelid worms. The other components of the society were not preserved.

environment to the leeward in regular fashion, and will create a large number of microhabitats. Thus the interpretation of Silurian coral reefs by Lowenstam (1957) (Figure 5-6) depends on these analogies and on the association of species in reefs of this age. Measurement of environmental modification depends in turn on variation in adaptations, e.g., an increase of skeletal size and thickness parallels increased turbulence from back to forereef.

Consistent associations without obvious environmental modification pose more difficult interpretive problems. If two fossil brachiopod species, *A* and *B*, occur together more often than one would expect by chance, does this imply that one modifies the environment for the others? Or that both respond similarly to the modifications created by a third species? Or that they simply have rather similar limiting environmental factors? One approach is by definition of microhabitats: Where *A* and *B* occur on the same bedding surface or within the same thin sedimentation unit, are they more likely to be close than distant neighbors? In other terms, is the nearest neighbor of any *A* more likely to be a *B* than any other species? Since the general environment is probably constant over any small area, clusters within the area suggest microhabitats created by organic activities.

Competition

Competition is easily handled in theory:

a). If a resource is limited, then organisms requiring this resource compete for it.

b). If two organisms are similar, they require similar resources and potentially are competitors.

c). The density of the limiting resource fixes an upper limit to competitor density.

d). Since individuals from different species differ in their ability to obtain a resource, one species will be superior in a specific competition.

e). In a stable homogeneous environment, if species compete for the same limiting resource, only one survives, because density is limited and one species excels in the resulting competition.

As a consequence, an ecological system should consist of species that compete little if at all.

For the paleontologist, the general applicability of the theory would simplify interpretation by removing competition as a factor. Unfortunately, natural environments are neither stable nor homogeneous. *A* utilizes the resource more efficiently at one place (within a microhabitat) or at one time, but *B* has an advantage in an adjacent microhabitat or at a slightly later time. This variation balances competition among species, and competition may be large. Similarly, variation within a species may be large in spite of the competitive superiority of one variant at one time and place.

Measurement of competition even in modern ecologic systems is extremely difficult, although qualitative recognition of its existence is rather easy. The paleontologist recognizes competition by similarity in requirements for food or places to live—reconstructed by means described earlier. A rather special type of "competition" exists among organisms preyed upon by the same species—a competition to avoid or reduce predation. Within microhabitats, interspecies competition should be at a minimum, but the fossil assemblage may mix several "generations" of biocoenoses and juxtapose strong competitors. In general, microhabitats and homogeneous macrohabitats do not contain closely similar species (members of the same genus) because of competitive exclusion; this holds for many fossil assemblages. But exclusion of a species can result from factors other than competition. In all, a paleontologist reconstructs the competitive relationship in his model only in very broad and general terms.

Stratification and temporal fluctuations

Ecological systems even in a single habitat show a regular ordering in space or stratification. Stratification may correspond to extrinsically determined gradients such as turbidity variation above the bottom, or it may reflect gradients created by organic activities. Thus the effect of trees and shrubs on light, temperature, and humidity produce vertical stratification in a forest. Each stratum becomes a microhabitat populated by a somewhat different biocoenosis than adjacent strata.

In marine environments, the major stratification is between the *benthic* organisms living within or on the bottom and the *pelagic* ones living in water above the bottom. In benthonic biocoenoses, the *infauna* living within and the *epifauna* living on the substrate form obvious strata, but even within each of these there may be further stratification—for example, between assemblages deep within the bottom sediment and those just below the surface, between organisms that feed at the water-substrate interface and those that extend some food-collecting device well above this interface.

Recognition of stratification in the fossil record depends on observation of position, e.g., burrowing clams within a layer, or upon adaptations for life in different strata, e.g., the stratification in Mississippian crinoid assemblages reconstructed by Lane (1964) and shown in Figure 5-7. If stratification models are based on strong evidence—particularly unambiguous life positions in the rocks or direct mechanical analysis of adapta-

Fig. 5-7. STRATIFICATION IN ANCIENT ECOLOGICAL SYSTEMS. The height of the various types of animals above the bottom defines three rather distinct strata. All are suspension feeders; presumably each stratum feeds on food particles in distinct water masses. (*After* Lane.)

tion they provide a rather powerful measure of organization within the ancient ecologic system.

Environmental gradients extend in time as well as space. One can eliminate both irregular fluctuations and long-term trends in the extrinsic environment because they do not permit development of any regular periodicity or sequence; they will be dealt with, however, in the three succeeding chapters. Regular, short-term, extrinsic fluctuations (*periodicity*) and trends determined by intrinsic sequences of environmental modification (*seres*) remain for discussion. Very short cycle periodicity, e.g., tidal rhythms, daily light variations, etc., divides a habitat among species but is difficult to recognize in the fossil record. Wells (1960) has distinguished daily growth increments in Devonian corals; the incremental laminae in modern calcareous algae are controlled by tidal variation and/or by day-night cycles. The difficulty becomes obvious, however, if one asks: Did the brachiopod species in a fossil assemblage feed continuously, feed only part of the day, or feed at various times during a day? Longer term periodicity controlled by seasonal fluctuations also subdivides the habitat. Seasonal periodicity is known in the fossil record from growth variation (see p. 29) and has been recorded also in oxygen-isotope paleotemperature determinations (p. 95). Polymodal size distributions (Olson, 1951) and periodic variation in shape (Tasch, 1958) also suggest seasonal cycles. Variations in trace element composition and in the organic compounds within skeletal material might trace seasonal periodicity.

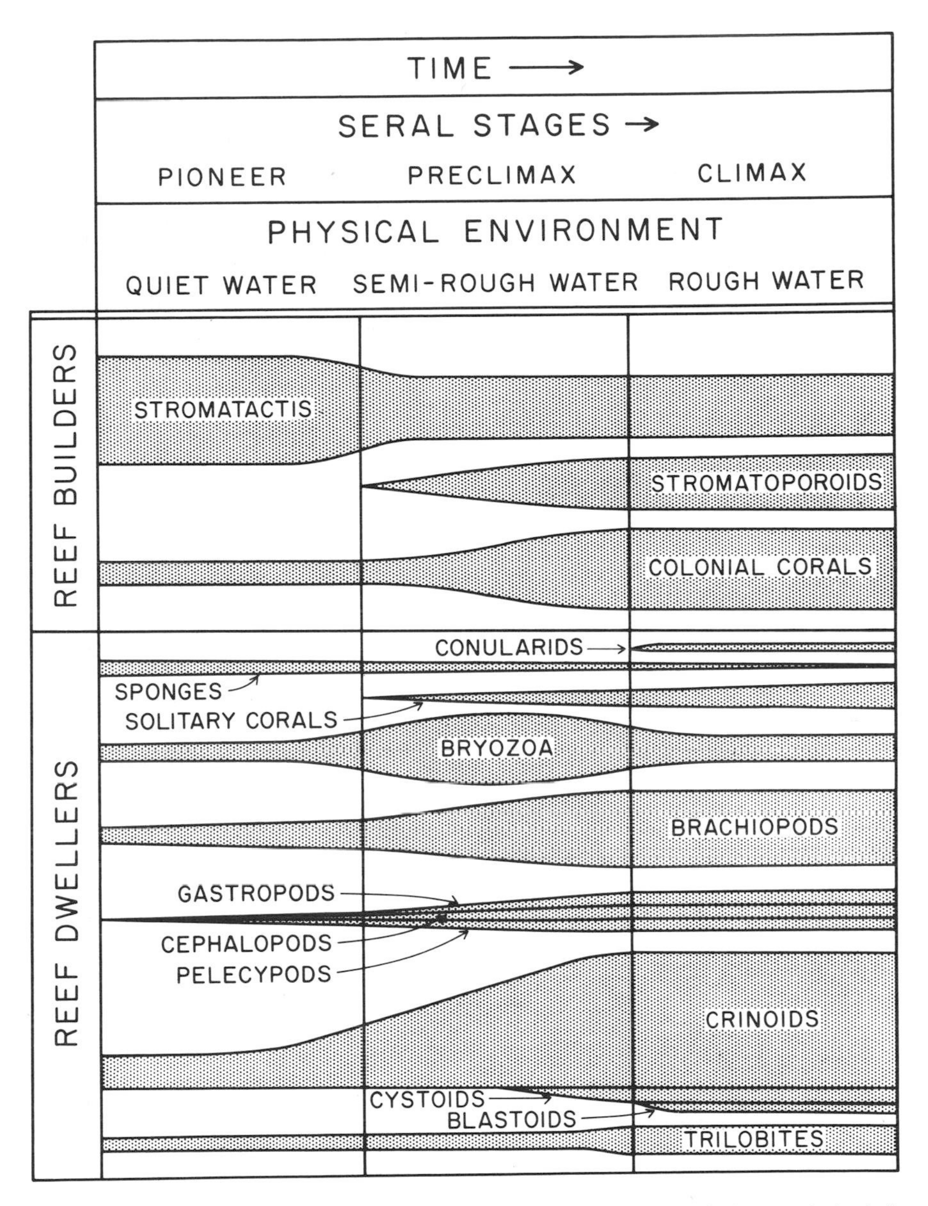

Fig. 5-8. ECOLOGICAL SEQUENCE IN SILURIAN CORAL REEFS. Thickness of shaded band indicates abundance of group. (*After* Nicol, *interpreting* Lowenstam.)

Progressive environmental modification induces a sere, e.g., the sequence of aquatic, swamp, and forest plants and animals in filling a lake or the sequence of shrubs, grass, and various trees in replacing a burnt-over forest. In such sequences, one species modifies the habitat so that others may enter—thus an accumulation of oyster shells on a muddy substrate creates a hard surface on which various sponges, corals, and bryozoans attach. Because many pioneers in the succession do not survive in the modified environment a rather complete turnover of species occurs.

In fossil successions, the same sequence of assemblages may appear in many places suggesting seral replacement. Changes in the extrinsic environment, e.g., a marine transgression, produce, however, similar sequences without intrinsic habitat modification. Lowenstam (1957) has interpreted the succession from the base to the top of Silurian coral reefs as a sere in which rough-water species successively replaced quiet-water ones as the reef built upward (Figure 5-8 and also Nicol, 1962). On the other hand, Lecompte has argued that similar sequences in the Devonian of Europe result from increasingly rapid subsidence, i.e., an extrinsic environmental change. Other examples (Ager, 1963, pp. 276-288) have similar alternate interpretations. The problem is to establish that the extrinsic environment did not change—an almost insoluble problem unless one can couple control samples to the sequence. For example, if one assumes that a coral reef succession is seral (intrinsic), then the contemporaneous succession adjacent to the reef and in the same general environment should remain constant or show a different trend; if the change were extrinsic, then they should change in a similar direction. The loaded terms, of course, are "contemporaneous" and "same general environment"; demonstration of these conditions is difficult. Missing such controls, fossil successions may be accepted as seres if the sequence makes sense in terms of habitat modification, e.g., growth of a coral bank carries it upward into the zone of wave action.

Movement

The role of an organism within an ecologic system is defined to considerable extent by the amount and kind of movements it undertakes—a whalebone whale browsing through a cloud of floating microorganisms has a different ecologic significance from a brachiopod filtering food particles from the portions of the cloud that drift into its vicinity. Benthic and pelagic marine organisms are categorized separately. *Sessile* benthic organisms live continuously in one place on or in the substrate; *vagrant* organisms swim or crawl over the substrate or burrow actively within it. *Planktonic* pelagic organisms drift with the movements of a water mass; *nektonic* organisms swim through or among such masses. Recognition of the modes of locomotion in fossils commonly derives from a functional analysis of form. The categories are sufficiently broad and distinct that they are marked by quite different adaptations. But difficulties arise in distinguishing vagrant benthic organisms from nektonic pelagic forms and some large planktonic species from nektonic ones.

Biocoenotic structure and the niche concept

Within an ecological system, the individuals of a species play a characteristic role. Conversely, the system includes a number of roles, each with distinctive relationships to the other roles and played uniquely by individuals from a single species. A particular role is a *niche* occupied by one and only one species. From this concept a number of philosophical problems arise. Can there be unoccupied niches? Is the niche structure independent of the species that occupy it? Does a species define a niche by its presence? Do niches overlap? These and similar questions are important in the analysis of the structure and genesis of ecologic systems but involve too many difficulties to argue out here. Therefore, I will simply observe that niches and niche structure are models and that one may choose different models to express different things—so long as the model is not used to demonstrate the a priori statements on which it is based.

Since niches are defined by their ecological relationships—nutrition, competition, stratification, periodicity, etc.—reconstruction of these relationships describes the niche structure of a fossil ecologic system. Thus in Figure 5-6 the gross structure of a Silurian coral reef is modeled in terms of 1) nutrition, 2) environmental modification, 3) stratification, and 4) movement. Competition is omitted, since competitors are placed in the same niche category, e.g., all brachiopods in niches grouped as sessile benthonic reef-dwelling suspension feeders. This example also illustrates the limitations of paleontologic interpretation. From the existence of several species, one recognizes the presence of several niches, but one cannot define precisely the adaptive differences between similar species. In consequence, a paleoecologist cannot specify the individual niches, but he can, as in Figure 5-6, recognize categories of niches.

A model of niche structure defines the structure of an ecological system, i.e., it expresses those relationships through which energy, material, and information are exchanged. As an aggregate it posseses properties that are independent of any particular relationship or for that matter of the particular species. A model of niche structure includes 1) total number of species, 2) number of species relative to total number of individuals, 3) relative abundance of species, 4) absolute or relative abundance of species in various niche categories, e.g., infaunal vs. epifaunal species, or 5) the relative or absolute abundance of individuals in these categories.

Although considerable difficulties pertain to these measurements from fossil evidence, some inferences are possible. For example, Van Valen (1964) analyzed two mammalian assemblages and found that species abundances fit a niche model (proposed by McArthur, 1957) consisting of contiguous nonoverlapping niches. Unfortunately, the abundance distributions generated by random factors in preservation and collection are unknown, but Johnson (1965) found low correlation between life and death abundance of species in samples of recent marine biocoenoses, although the species composition was quite similar. Several paleontologists have reported an increase in diversity, i.e., in number of species relative to number of fossils, in marine transgressive sequences (e.g., Ferguson, 1962). Trechmann (1925) demonstrated a reverse trend in diversity related to progressive salinity increases in a restricted marine basin. Selective preservation is almost certainly a factor in some of these observations, since 1) a changing environment may systematically reduce or increase the probabilities of preservation for all elements of the biocoenosis and 2) different biocoenoses have different proportions of preservable species, e.g., mud bottoms have a high proportion of soft-bodied infauna, but silty and sandy substrates have a high proportion of epifauna with hard skeletons.

Ecologic analysis of fossil assemblages

In recent years, many paleontologists have published complex reconstructions of ancient biocoenoses—for example, Shotwell (1963) on mid-Cenozoic mammalian faunas of the Great Basin, Gekker and others (1948) of a late Jurassic lake, and Zangerl and Richardson (1963) of a Pennsylvanian black shale assemblage. One of the most instructive of these and a fitting summary for this chapter is Ferguson's study (1962) of a thin Mississippian shale in Scotland.

Ferguson collected a continuous series of slabs, 45 in all, from the 2.8 meter thick shale in a single locality. Then he split the slabs to yield a total of 125 vertical samples. The average thickness of the slabs was approximately 6 centimeters; of the individual samples, 2.2 centimeters. He removed stubs from each sample for thin sections and for preparations of microfossils. He disaggregated the shale by the kerosene-water method and examined, identified, and recorded all the specimens in each sample. Since the samples varied somewhat in size, he recalculated the number of individuals in each sample as number of individuals per kilogram of shale. He recorded the microfossils from 10 gram sub-samples of each sample. The total fossils included approximately 3000 macrofossils, representing fifty species, and 8000 microfossils, representing seventy species.

Results

From the species distribution patterns, Ferguson defined four successive assemblages or "topozones." The major characteristics of each topozone and of the included fossils are displayed in Table 5-4. He concluded that the assemblage represented a biocoenosis *in situ* because of 1) the wide size range including microstages of large animals, 2) the high frequency of life orientations, and 3) the high percentage of articulated bivalves. He noted a progressive increase in number of species from the basal to the upper topozone, particularly in the number of species in taxa whose modern representatives require normal marine salinity (32-36 parts/thousand dissolved salts). Of the two genera still surviving today, the brachiopod *Lingula,* which is abundant in Topozones I and II, typically occurs in shallow water and intertidal habitats 10°C and warmer; the ostracod *Bairdia* from Topozone III now requires normal marine salinities

TABLE 5-4

Summary of data for "topozones" in Mississippian shale sequence (*data from Ferguson, 1962*)

Topozone	*Lithologic parameters*	*Paleontologic parameters*
IV	Shale; laminae indistinct or absent; Fe approx. 1-25%; Mn .13-.11%.	Fauna: *Eomarginifera, Echinoconchus, Pustula, Dielasma, Chonetes, Spirifer* (brachs), *Amplexizaphrentis, Clisiophyllum, Aulophyllum, Lithostrotion* (corals), crinoids, variety of bryozoans, 11 spp. of forams, 7 spp. of ostracods. Size Distribution: Brachs show wide size range; *Eomarginifera* young abundant. Position: *Amplexizaphrentis* in life position. Disassociation: Many bivalves articulated.
III	Shale; laminae indistinct or absent; Fe approx. 3%; Mn approx. .17%.	Fauna: *Crurithyris, Eomarginifera, Schizophoria* (brachs), *Bucaniopsis, Glabrocingulum* (snails), *Paraparchites, Bairdia* (ostracods), *Stacheioides, Plectogyra* (forams) predominate. Size Distribution: *Schizophoria,* 0.29-36 mm., young predominate; *Eomarginifera,* 4-60 mm.; many gastropod young; microshells of *Crurithyris* and *Eomarginifera* rare. Position: 12% of *Crurithyris,* most *Eomarginifera* in life position. Disassociation: Most bivalves articulated.
II	Interlaminated shale and silt; laminae less distinct than in I; granules and aggregates of pyrite; approx. 4% Fe; Mn 0.15 to 0.19%.	Fauna: *Lingula, Crurithyris, Eomarginifera* (brachs) and *Paraparchites* (ostracod), *Glamospira* (foram) predominate. Size Distribution: *Crurithyris,* 0.2-10 mm., microshells 55x abundant as adults. Position: Many *Lingula,* 17% of *Crurithyris* in life position. Disassociation: 96% of *Crurithyris* articulated.
I	Interlaminated shale and silt; laminae streaky and lenticular; granules and aggregates of pyrite; 4.5 to 7% Fe content; Mn 1.4 to 1.7%.	Fauna: *Lingula* (brach), *Streblopteria* (clams), *Paraparchites* (ostracod), predominate. Size Distributions: *Lingula,* 1.25-25.0 mm., mode between 9 and 15 mm.; *Streblopteria,* microscopic to 30 mm.; young predominate; *Paraparchites,* 0.136-1.870 mm. Position: Many *Lingula,* clams, and snails in life positions. Disassociation: Most bivalves articulated.

(30-40%) and is most abundant where water temperatures are between 25° and 31°C. The size-frequency distributions for individual species change strikingly among and within topozones: in Topozone II, very small individuals dominate populations of the brachiopod *Crurithyris,* but in Topozone III, adults are predominant. In Topozone III, the number of small *Schizophoria,* another brachiopod, decreases progressively from bottom to top.

Ferguson interprets these changes as a consequence of a shift from tidal and possibly brackish habitats in Topozone I to normal marine habitats in Topozone IV. Because the tidal habitat is highly variable and because brackish water limits the distribution of many organisms, the ecologic system was relatively simple, comprising only a few niches and thus few species, although the number of individuals may be high. The abundance of infaunal species, *Lingula* and *Nuculopsis,* implies the preponderance of the more protected infaunal niches in the system. In contrast, the deeper water, the more stable environment, and the normal salinity of Topozone IV permitted population of the substrate by a larger variety of species in a more complex niche structure and survival of a large epifaunal component.

Animals, natural societies, and stratigraphers

The ecological interpretation of fossil assemblages provides a synthesis of "two-dimensional" paleontology. The first step in this sort of paleontology is the observation of individual form. Next, the paleontologist interprets growth and development of form and the relation of form to function. He proceeds to a classification of animals based primarily on their form; from this classification he evolves the concept of the species. He recognizes that a species has genetic and ecologic characteristics as well as morphologic ones and that species distributions map environmental factors. He utilizes adaptations to define niches and the association of species to characterize niche structure. He reconstructs an ecological system by integration of all these observations and inferences.

Since paleontologists and stratigraphers, whether they interpret evolutionary patterns, draw paleogeographic maps, or correlate rocks, deal with organisms once parts of natural societies, they must also deal with such societies. If each fossil bore a convenient label, "I was a scavenger on the bottom of a 30-meter-deep sea from 231,612,907 BC to 231,612,906 BC," there would be no need for paleoecologic studies of fossil assemblages. So long as they are not so labeled, or until a satisfactory time machine is available, paleontologists are limited in their study of fossils by their restricted knowledge of the paleoecologic framework.

REFERENCES

Ager, D. V. 1963. See ref. for Chapter 4.

Broadhurst, F. M. 1964. "Some Aspects of the Palaeoecology of Non-marine Faunas and Rates of Sedimentation in the Lancashire Coal Measures," *American Jour. of Sci.*, vol. 262, pp. 858-869. Deals with effects of selective preservation.

Fagerstrom, J. A. 1964. "Fossil Communities in Paleoecology: Their Recognition and Significance," *Geol. Soc. of America Bull.*, vol. 75, pp. 1197-1216.

Ferguson, L. 1962. "The Paleoecology of a Lower Carboniferous Marine Transgression," *Jour. of Paleontology*, vol. 36, pp. 1090-1107.

Gekker, R. F. *et al.* 1948. "A Fossil Jurassic Lake in the Kara-Tau Chain," *Trud Paleont. Inst. Akad. Nauk. S.S.S.R.*, vol. 15. (In Russian.)

Gutshick, R. C. and T. G. Perry. 1959. "Sappington (Kinderhookian) Sponges and Their Environment," *Jour. of Paleontology*, vol. 33, pp. 977-985.

Hazen, W. E. (Ed.) 1964. *Readings in Population and Community Ecology*. Philadelphia: W. B. Saunders.

Hedgpeth, J. W. 1957. "Classification of Marine Environments," vol. 2, *Geol. Soc. of America, Memoir 67*, pp. 93-100.

Johnson, R. G. 1960. See ref. for Chapter 4.

———. 1964. "The Community Approach to Paleoecology," in Imbrie, J. and N. D. Newell (Eds.), *Approaches to Paleoecology*. New York: John Wiley. Pp. 107-134.

———. 1965. "Pelecypod Death Assemblages in Tomales Bay, California," *Jour. of Paleontology*, vol. 39, pp. 80-85.

Lane, N. G. 1963. "The Berkeley Crinoid Collection from Crawfordsville, Indiana," *Jour. of Paleontology*, vol. 37, pp. 1001-1008. Stratification in Paleozoic marine biocoenosis.

Lecompte, M. 1959. "Certain Data on the Genesis and Ecologic Character of Frasnian Reefs of the Ardennes," *International Geol. Review*, vol. 1, pp. 1-23.

Lowenstam, H. 1957. "Niagaran Reefs in the Great Lakes Area," vol. 2, *Geol. Soc. of America, Memoir 67*, pp. 215-248.

Margalef, R. 1963. "On Certain Unifying Principles in Ecology," *American Nat.*, vol. 97, pp. 357-374.

McArthur, R. H. 1965. "Patterns of Species Diversity," *Biol. Review*, vol. 40, pp. 510-533.

Newell, N. D. 1957. "Paleoecology of Permian Reefs in the Guadalupe Mountains Area," vol. 2, *Geol. Soc. of America, Memoir 67*, pp. 407-436.

Nicol, D. 1962. "The Biotic Development of Some Niagaran Reefs—an Example of an Ecological Succession or Sere," *Jour. of Paleontology*, vol. 36, pp. 172-176.

Olson, E. C. 1952. "The Evolution of a Permian Vertebrate Chronofauna," *Evolution*, vol. 6, pp. 181-196.

———. 1961. "The Food Chain and the Origin of Mammals," *Int. Colloq. on Evol. of Mammals. Kon. Vlaamse Acad. Wetensch. Lett. Sch. Kunsten Belgie*, part I, pp. 97-116.

Phleger, F. B. 1955. "Ecology of Foraminifera in Southeastern Mississippi Delta Area," *Amer. Assoc. of Petr. Geologists, Bull.*, vol. 39, pp. 712-752.

Reyment, R. A. 1963. Multivariate Analytical Treatment of Quantitative Species Associations: An Example from Palaeoecology," *Jour. of Animal Ecol.*, vol. 32, pp. 535-547.

Rhoads, D. C. 1967. "Biogenic Reworking of Intertidal and Subtidal Sediments in Barnstable Harbor and Buzzards Bay, Massachusetts," *Jour. of Geology*, vol. 75, pp. 461-476.

Shotwell, J. A. 1964. "Community Succession in Mammals of the Late Tertiary," in Imbrie, J. and N. D. Newell (Eds.), *Approaches to Paleoecology*. New York: Wiley. Pp. 135-150.

Speden, I. G. 1966. "Paleoecology and the Study of Fossil Benthic Assemblages and Communities," *New Zealand Jour. of Geol. and Geophy.*, vol. 9, pp. 408-423.

Stehli, F. G. 1965. "Paleontologic Technique for Defining Ancient Ocean Currents," *Science*, vol. 148, pp. 943-946.

Tasch, P. 1958. "Permian Conchostracan-bearing Beds of Kansas. Part 1. Jester Creek Section-Fauna and Paleoecology," *Jour. of Paleontology*, vol. 32, pp. 525-540.

———. 1965. "Communications Theory and the Fossil Record of Invertebrates," *Trans., Kansas Acad. of Sci.*, vol. 68, pp. 322-329.

Toots, H. and M. R. Voorhies. 1965. "Strontium in Fossil Bones and the Reconstruction of Food Chains," *Science*, vol. 149, pp. 854-855.

Trechmann, C. T. 1925. "The Permian Formation in Durham," *Proc. Geol. Assoc., London*, vol. 36, pp. 135-145.

Valentine, J. W. and B. Mallory. 1965. "Recurrent Groups of Bonded Species in Mixed Death Assemblages," *Jour. of Geol.*, vol. 73, pp. 683-701.

Van Valen, L. 1964. "Relative Abundance of Species in Some Fossil Mammal Faunas," *American Nat.*, vol. 98, pp. 109-116.

Wells, J. W. 1963. "Coral Growth and Geochronometry," *Nature*, vol. 197, pp. 948-950.

Zangerl, R. and E. S. Richardson, Jr. 1963. "The Paleoecological History of Pennsylvanian Black Shales," *Fieldiana, Geol. Memoirs*, vol. 4.

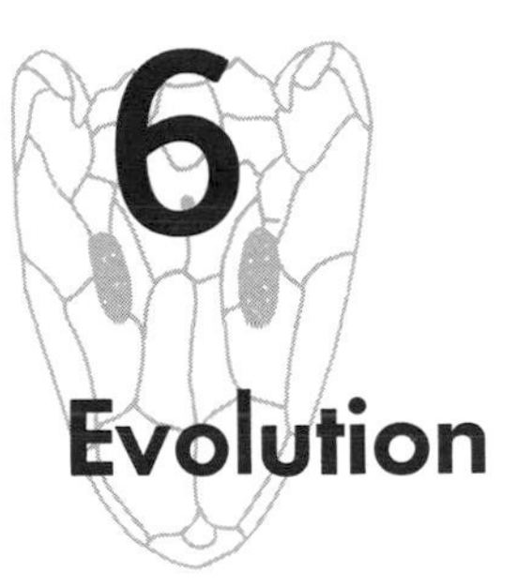

6 Evolution

Although plant and animal species (genetic systems) live only in a "now" of breeding, of death and survival, of predation and competition, they exist as a consequence of an evolutionary "past" and most of them predicate a "future." A genetic system operating through time generates a phylogeny; the evolutionary processes impinging on the genetic system control this phylogeny. Evolution appears in the fossil record in several guises. It connects fossil species in a sequence of form and explains the succession of fossil assemblages in the stratigraphic column. Less obviously, it explains the similarities and differences between contemporaneous species, the geographic diversity of species, and the diversity within natural societies. This chapter and Chapter 7 will discuss the mechanics of evolution, and Chapter 8 will explain the significance of evolution to the stratigrapher and paleogeographer. Evolution runs as underlying thread beneath all paleontologic phenomena; without an evolutionary perspective, the paleontology is a poor, half-crippled thing.

Organic change in the geologic record

In any considerable thickness of sedimentary rocks, successive layers contain different species. If the oldest beds were deposited in a marine environment and later ones on a floodplain, a fauna of terrestrial vertebrates, insects, and snails succeeded a fauna of clams, fish, and crabs. The change in environment controlled the change in species. When the sea again encroached on the floodplain, the original species reappeared. The succession of faunas resulted from the same limiting factors that determined the distribution of a species at any particular time.

Most fossil species, however, disappear for good as the environment changes—disappear from the area and, indeed, from the fossil record. New species, unknown in older rocks, take their places, so that the total number of species does not change greatly. This succession of

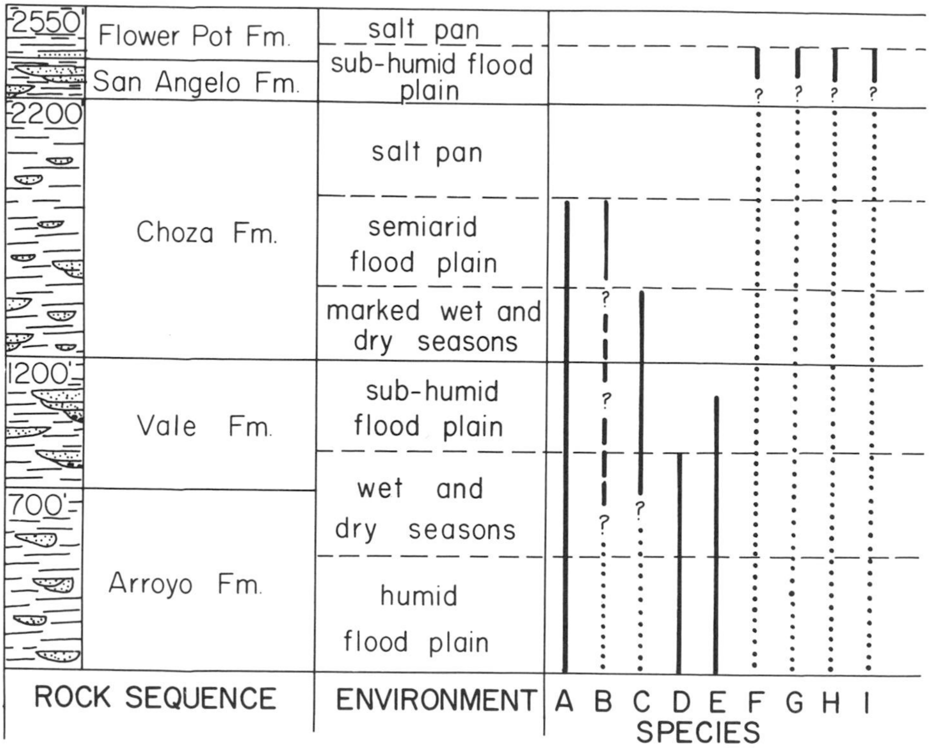

Fig. 6-1. SUCCESSION OF FOSSILS IN A LOCAL STRATIGRAPHIC SEQUENCE. The faunal and lithologic sequence in the early and middle Permian, north-central Texas. As the environment changed back to a more favorable type in San Angelo time, none of the old species reappeared. In place of them appeared several new but ecologically similar species. **A** and **F** are large carnivorous reptiles; **B, C, G, H,** and **I** are moderate to large herbivorous reptiles. The two amphibians, **D** and **E,** apparently had no ecological successors. (*After* Olson, 1952, *and* Olson and Beerbower, 1953.)

faunas is, in some examples, correlated with environmental change, but the early species do not recur if the environment switches back (Figure 6-1).

Occasionally a paleontologist finds a third type of faunal succession. He discovers distinct species from the top and bottom of the sequence but finds that intermediate beds contain populations intermediate in character, some of which might be logically placed in either species (Figure 6-2). Again, if fortunate, he can relate the change in morphology to environment changes. The only reasonable conclusion seems that of a gradual change in genetic composition of an interbreeding population (genetic system) through time—formally, a phyletic system produced by evolution.

Obviously, he forces the term *species* on this succession, for the successive populations were never reproductively isolated in the same sense that contemporaneous species populations are. Nor are the successive populations distinct morphologically. If he divides the phyletic lineage into species, he provides arbitrary assignments (this practical problem is discussed at greater length on page 158) so that the differences in "average morphology" between temporal species is about the same as that between similar contemporaneous species. A few rock sequences, representing nearly continuous deposition through a considerable period of time, include long successions of "temporal species," successions sufficiently long so that the extremes can be classified in different genera.

These three modes of faunal succession are characteristic not only of species but also of higher categories. Genera, families, and so on, may succeed each other temporarily or permanently and, if permanently, either abruptly or by gradual transformation. Since the species are the basic units in the construction of these categories, this should be expected; arguments for the special origins of higher categories will be considered later.

Abrupt faunal changes

The first mode of faunal succession—a direct and reversible consequence of environmental change—has been disposed of; the second and third, which are, in any definition, evolution, require extended discussion.

The sudden appearance of new species, families, orders, and so on, might ensue either a) from the sudden origin of these groups, b) from their migration into the area after slow evolution elsewhere, or c) from an extended period of local nonpreservation during which gradual change took place. The second and third possibilities are just special cases of the gradual succession described earlier and can be shown to be valid explanations for some sudden faunal changes. For example, cats, deer, elephants (mastodons), dogs, horses, peccaries, camels, and several other mammal groups are unknown from the Cenozoic rocks of South America until the very top of the sequence. Then these groups appear without any inter-

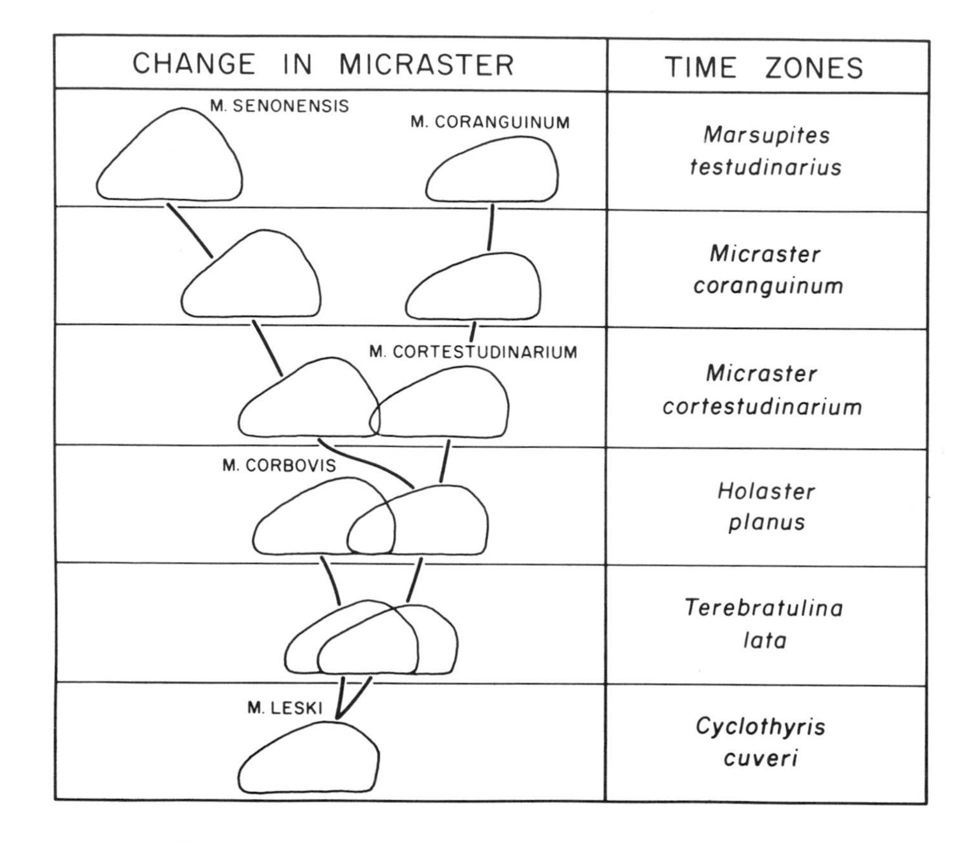

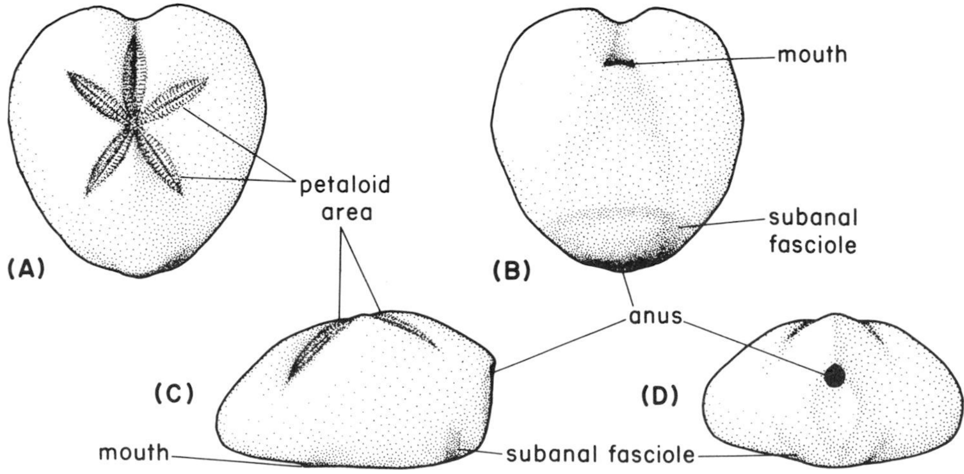

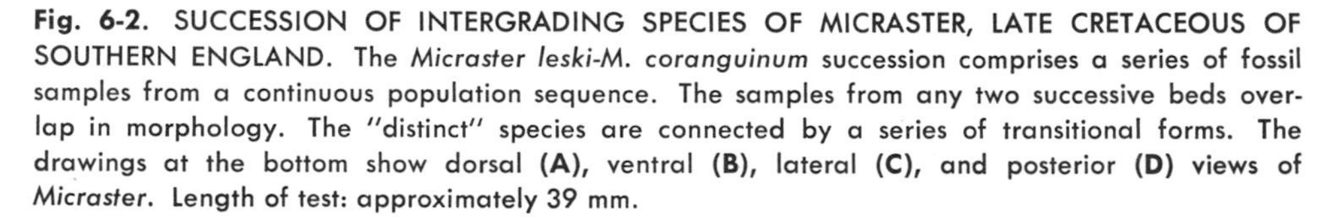

Fig. 6-2. SUCCESSION OF INTERGRADING SPECIES OF MICRASTER, LATE CRETACEOUS OF SOUTHERN ENGLAND. The *Micraster leski-M. coranguinum* succession comprises a series of fossil samples from a continuous population sequence. The samples from any two successive beds overlap in morphology. The "distinct" species are connected by a series of transitional forms. The drawings at the bottom show dorsal **(A)**, ventral **(B)**, lateral **(C)**, and posterior **(D)** views of *Micraster*. Length of test: approximately 39 mm.

mediate links to the earlier mammalian fauna. The antecedents of these mammals, however, exist in older rocks of North America and Eurasia. Similarly, the Cenozoic mammalian faunas in the deposits of the intermontane valleys of Montana change abruptly with no transitions between genera and families. But the transitional species and genera are known from rocks in South Dakota, Colorado, Wyoming, and Nebraska. Although these transitional forms certainly lived in Montana, the deposition of sediments was spasmodic (as can be shown by physical geologic evidence), and the fossil sequence lacks continuity.

Because some abrupt faunal changes resulted from migration or gaps in preservation, because the fossil record is discontinuous, and because the sedimentary rocks of any one geologic age cover only a small part of the earth's surface, most paleontologists argue that this explanation holds for all sudden appearances of taxa in the fossil record. Some disagree, however, and the magnitude and nature of some of these "jumps" are admittedly puzzling. Since a general theory of evolution cannot depend on a lack of data, I'll postpone discussion of abrupt evolutionary changes to a later page.

Gradual changes

Let me return to an example of gradual transformation and examine its details. A sea urchin, *Micraster* (Figure 6-2), occurs through several hundred feet of the Cretaceous section in southern England. Specimens from the top of the section differ notably from those at the base; transitional forms appear in intermediate layers (Rowe, 1899, Kermack, 1953, Nichols, 1959). Urchins from any one level vary considerably; this variation presumably represents differences within the population at one time. Successive samples overlap so that some individuals in the later samples display characteristics like some of those in the earlier. I've expressed this in Figure 6-3 by frequency curves showing variation in several characteristics for successive populations. These and other characteristics indicate three phyletic lineages, *M. leski* → *M. coranguinum, M. leski* → *M. senonensis,* and *M. leski* → *M. corbovis.*

The three characteristics included in the illustration appear to have functional significance. Low body form is correlated with burrowing habits in modern sea urchins; a flattened form makes mechanical sense also for a burrower. The number of pores in the petaloid area (see Figure 6-2) is related to respiration, since these serve in modern sea urchins for passage of long slender respiratory organs, i.e., specially modified "tube feet." In modern sea urchins, these are most numerous in burrowing forms. Finally, the depressed tract around the anus corresponds to the subanal fasciole in modern urchins. Currents generated by cilia on the fasciole sweep debris from the surface of the body, renew the water for respiration, and

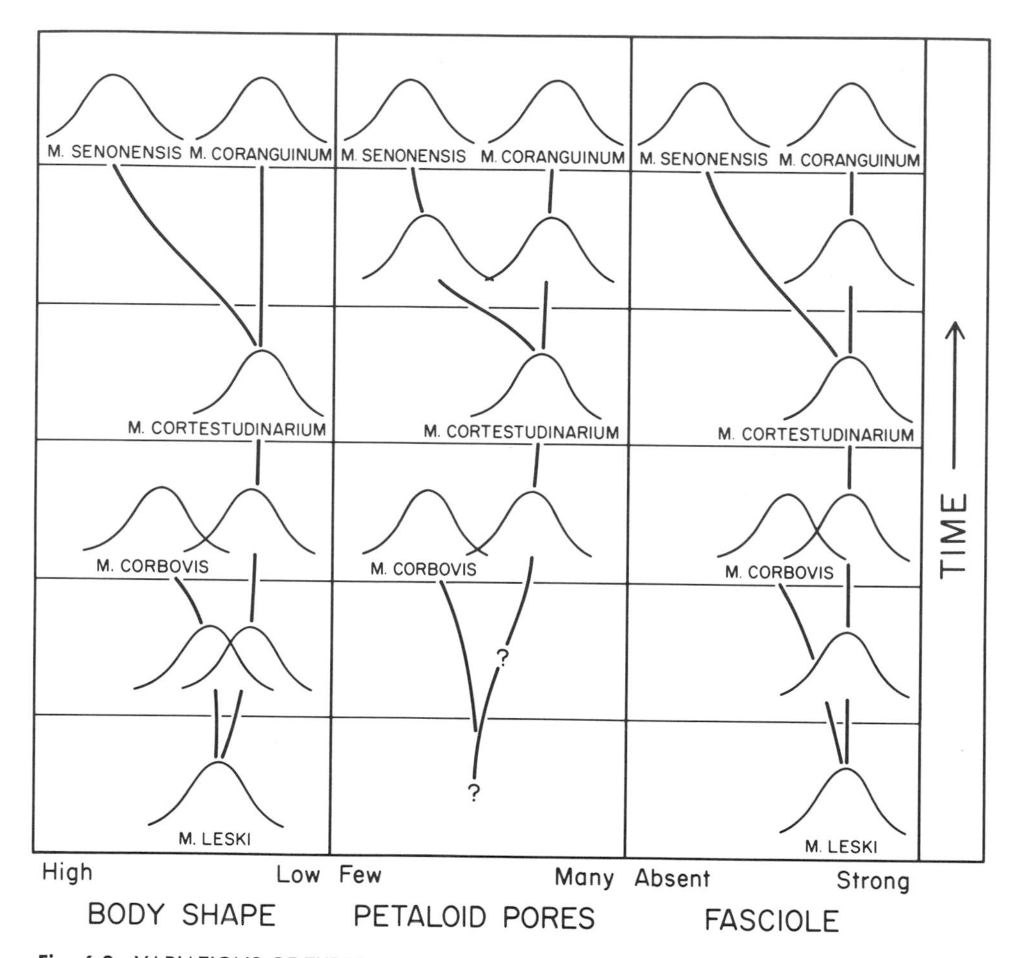

Fig. 6-3. VARIATIONS OF THREE CHARACTERS IN MICRASTER EVOLUTION. The frequency curves show, in generalized fashion, variation of body shape, number of petaloid pores, and development of the subanal fasciole (see *also* 6-2. on body shape).

remove wastes. In surface-living types, however, the normal movements of water about the urchin handle these functions, and the subanal fasciole is absent or weakly developed.

When one applies these functional interpretations to the phyletic lineages, one finds that a shallow burrower, *M. leski,* gave rise to a line of shallow burrowers, *M. corbovis,* a line of deep burrowers, *M. coranguinum,* and a line of surface dwellers, *M. senonensis.** The changes in

* *M. coranguinum* and *M. senonensis* intergrade in contemporaneous samples and may represent partly isolated populations (? subspecies) within a single biologic species. This alternative does not affect the evolutionary argument in essentials, so I continue to use the distinct species names in the discussion.

structure correlate with occupation of new niches with different environmental properties.

But does the environment "cause" the evolutionary change? Or does the morphologic change require a shift into new niches? This question is essentially unanswerable from the geologic record, and one must turn to the dynamics of modern populations to provide an explanatory model.

The mechanism of evolution

The form (phenotype) of an organism results from the operation of genetic information, the genotype, in an environment (p. 25). Obviously, the array of phenotypes within a population derives from an array of genotypes developing in an array of environments. Further, the genotypic array (pp. 69-73) is the consequence of recombination, of mutations, of selection of genotypes through differentials in survival and/or fecundity, and of random loss and fixation. Finally, the phenotypic array defines the habitats and niches within which individuals may survive. Therefore, correlation between environmental and population changes might arise from:

a). Change in the environment of development

b). Direction of mutation by the environment

c). Alteration of selection values

d). Displacement in habitat and/or niche preference

Biological processes

I think it fairly obvious that item **d),** niche preference, requires some sort of prior phenotypic change in population. If induced by items **a), b),** or **c),** these become explanatory in themselves and **d)** is not needed. Change in niche preference, therefore, is an autonomous explanation if and only if random mutations or random loss and fixation of genes supplies the phenotypic change (p. 72). The mathematics of populations indicate that the known range of mutation rates cannot induce genotypic shifts when opposed by selection. If favored by selection, mutational shifts again become trivial as an evolutionary explanation. This leaves a single possible case for directive mutation: shifts when selection is neutral. The consensus of biologists rejects this possibility as an important factor because a) most mutations appear to be unfavorable, b) a shift affecting niche preference is unlikely to be selectively neutral, c) successive mutations at a locus do not, typically, induce unidirectional phenotypic variation.

Although mutational drift is intellectually respectable because it must operate, it has probably contributed little to evolution, since continued drift against selection leads to extinction (see page 72).

Changes in developmental environment (item **a**) must underlie some observed changes in the fossil record but cannot be accepted as an evolutionary mechanism for several reasons. First, they are not unique to a particular time, i.e., they would recur at any time or place that a particular genotype was subject to a particular environment; second, they are not transferred in time, i.e., the genotypic array remains constant, so phenotypic changes can be either renewed or reversed in any successive generation. Some evidence exists that genotypic evolution may copy environmentally induced modifications through selection of genotypes that develop more strongly in this particular way. For example, if a preponderant genotype A_1 produced very high profiles in *Micraster* living on the surface and moderately low forms in burrowing individuals, and if increased predation encouraged burrowing, then the population would shift phenotypically from very high to moderately low. A rare genotype A_2, for consistently low profile, however, would be favored by selection and in a number of generations would become the most common in the population.

Nearly all biologists reject environmental direction of mutation (item **b**) as a significant cause of evolution. With a very few and somewhat doubtful exceptions, phenotypic response to mutation is unrelated to the cause of the mutation, and, as already noted, sequential mutations at a locus do not induce linear phenotypic changes. Thus, cosmic rays produce mutations, but only by chance would the mutation affect phenotypic response to ionizing radiation. The exceptional cases mentioned above involve biochemical responses in acellular organisms (bacteria, protozoans) to chemical stimuli. Some of these, apparently, induce a specific alteration in the genetic code, but experimental difficulties make alternative explanations plausible.

By exclusion, environmental selection (item **c**) directs evolution within the limits of genetic variation arising from mutation and recombination (pp. 71-72). The direction of mutation and the products of recombination are at random with respect to selection; the rate of mutation and the amount of recombination are, however, influenced by selection. As defined, evolution by natural selection is *not theory* but a logical certainty, given a population limited in size by the environment and containing genotypic variation that has phenotypic expression.

Model for *Micraster*

With this general model in view, one may develop a specific model for *Micraster* evolution. Consider the early population of *Micraster* to include

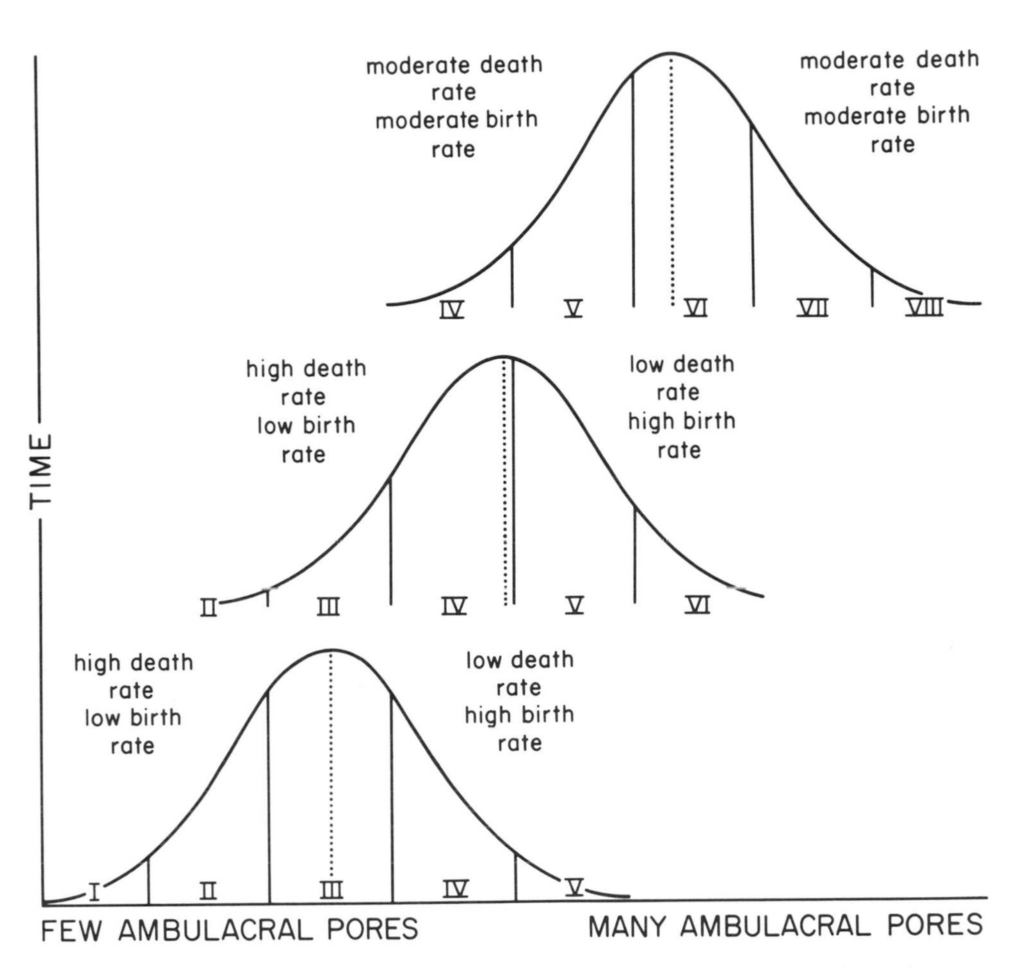

Fig. 6-4. TRANSFORMATION OF A MICRASTER POPULATION SUCCESSION. The early population has a large number of individuals with few ambulacral pores (Groups I, II, III). The other groups, however, have higher survival and reproduction rates so that the population frequency distribution shifts to the right in time.

five genetically controlled, phenotypically distinct subpopulations (Figure 6-4), the most common type III, the less common types II and IV, and the rarest types I and V. Each subpopulation has its own specific fecundity and survivorship curves and, as a product of these curves, net birth and death rates. If the population is stable, i.e., changing neither in number nor in morphologic or physiological characteristics, the birthrate in each subpopulation equals the death rate. If birthrates exceed death rates, the total population will increase in number. It will change in average morphology if the births are disproportionate among the subpopulations.

If *Micrasters* with relatively few ambulacral pores (subpopulations IV and V) have the lowest net rate, the morphological characteristics of the species will change, the average shifting from "few ambulacral pores" toward "many ambulacral pores." These rates also determine the range of variation, for morphologic types with negative net rates will diminish and disappear. For reasons not yet understood, the amount of variation may remain large or even increase during intense selection for one phenotype. Ultimately, however, variation must stabilize at some value determined by the differences between net rates of increase—little variation if net rates are quite different, much if they are very similar.

Selection and variation

So far the explanation is simple enough, but I have hidden complexity beneath this simplicity. Genes produced enzymes that manifested themselves in the development of the characteristic "ambulacral pores." These enzymes also affected other features of the animal. A number of different gene sets probably induced "few ambulacral pores" but produced different results in other parts of the body. Gene set *A* might have reduced the death rate by improving the "ambulacral pores," but at the same time decreased the birthrate by adverse effects on the reproductive organs. The recombination of genes complicates the matter further, for crossing a male with gene set *A* and a female also with *A* might have yielded only a few offspring with set *A* and many with other gene sets. On the other hand, a male with set *B* and a female with set *C* might have produced 90 percent set *A* young.

Analysis in terms of specific matings would bog down in millions of computations, but one can lump all these specific matings and individual deaths in terms of averages. The geneticist prefers to do this with respect to single genes.

The sum of rates of fecundity and of survivorship can be stated as an adaptive or selective value. In theory, one could determine the average selective value of each gene and each gene combination in a living natural population at any time. Since no combination of researchers could measure fast enough to actually do this, biologists accept the theoretic results supported by limited experiments and by mathematical deduction.

An evolving population shifts in the direction of the gene combination or combinations possessing the highest selective value at a rate determined in part by this value. In the *Micraster* example, each of the variant subpopulations (the products of different gene combinations) presumably had different selective values. The direction of evolution indicates that the "many ambulacral pores" subpopulation had the highest selective value. As the population changed, the probability of getting some gene combinations decreased (sets for "few ambulacral pores"), and others increased (sets for "many ambulacral pores"). Individuals with "very many ambulacral pores" appeared in the population; individuals with "very few ambulacral pores" disappeared. Since the new subpopulation "very many ambulacral pores" had a higher survival value than the "few ambulacral pores," the population average continued to shift. Note that this did not reduce the variety of recombinants but rather the chances of getting a particular recombinant. In the earliest population, individuals carrying a gene combination for "very many ambulacral pores" had a chance of occurrence of, perhaps, 10^{-12}, but with the shift of the population mean the probability jumps to about 10^{-1}. In practice, "very many ambulacral pores" individuals would not appear in the original population but would turn up fairly often in the later one. In this way evolution generates new variation without new gene mutation, but it also deprives the population of other variants.

Mutation and recombination

The Hardy-Weinberg Law (page 71) predicts the chances of getting a particular gene set in any generation. Because morphology is correlated with the gene system, the law also predicts the chance of getting an animal with a particular morphology if the animal develops in a particular environment. The frequency of a gene or set of genes is constant for a population in equilibrium. If the factors acting on the population are not in equilibrium, the frequency of genes and gene combinations changes. In addition to selection, these factors include 1) mutation rate of genes involved in the various combinations, 2) size of the breeding population, 3) immigration, 4) emigration.

If gene *A* mutates to *a* more often than *a* mutates back to *A*, the frequency of gene *A* in the population decreases until the gene frequencies are in equilibrium. This change results in increasing numbers of individuals bearing the favored gene. If this were the only process acting, the preponderant genes in the population would be those with the greatest chemical stability. Once this point were reached, the population would cease to change. As pointed out above, the direction of mutation cannot, however, direct evolution, but the rate, as well as direction, does supply variation on which selection can act.

The preceding discussion of selection and mutation was based on simplified assumptions about the randomly breeding population, i.e., that it is large and is closed for either immigration or emigration, is *isolated*. In fact, all species populations are conveniently thought of as being divided into demes, i.e., local, freely interbreeding units (p. 72). The genetic variation (and possibly the morphologic) within these local populations will be somewhat less than that of the species, because no single local population includes all the less frequent gene combinations. As long as each local population breeds freely with adjacent populations, the genes circulate freely among these local units, and the probability of a particular gene combination in a local population is the same as that for the entire population. If, however, no breeding occurs between local populations, the gene frequencies within a local population determine the probability of that combination. Since the frequencies differ from those in the species as a whole, each local population differs from all the others in morphology, behavior, and so on. Most species populations probably fall between these extremes of fully interbreeding or of completely isolated local populations and consist of partly isolated units.

As pointed out above (pp. 72, 139), if the effective breeding population is quite small and if the system is very nearly isolated, accidents of sampling eliminate genes and gene combinations. The frequencies drift even against rather strong selection; the local population loses its genetic variance, and many populations will become extinct because of fixation of combinations with low fecundity and/or high death rates. Species populations with such a structure lose variability with the fixation of a relatively few gene combinations. Such trends are likely to be unfavorable for the species—extinction is hardly favorable—but might, on rare occasions, be of advantage if they, by accident, resulted ultimately in a favorable combination. Thus a population could drift for a time against selection and shift from one niche to another through a period of nonadaptive change.*

If the effective breeding population is large or completely open, rare favorable gene combinations tend to be swamped and lost by continuous recombination. Such combinations are much more likely to survive intact in small, partly isolated populations where the number of recombinations possible is much smaller.

If immigration and emigration are random with respect to the array of gene frequencies, they reduce differences between demes, "homogenize" the species population and, in effect, increase the effective breeding size of the demes. On the other hand, if immigration or emigration of particular genotypes is nonrandom, differences between demes would increase, and average effective breeding size decrease. For example, *Micraster* with few ambulacral pores might aggregate where food was abundant in upper sediment layers and those with many pores where food was found at depth.

In consequence, population structure controls the amount of variability and its availability for evolution. Population structure does not, however, exist in isolation but expresses the environmental limits and the powers of dispersal of the local populations. Ecological barriers reduce

* Genetic drift would account for the sudden appearance of new animal groups, since the ancestral populations might be so small that they do not appear in the fossil record. For this reason, paleontologists have found the concept tempting—but dangerous, since such conclusions are based on the lack of evidence, the lack of fossils.

migration between local populations. Conversely some species and phenotypes within species possess greater ability to cross these barriers. Therefore the selective advantage of a genotype includes the environmental tolerance and dispersal powers of those that bear it.

Effect of local environments

Finally, different local populations seldom, if ever, have identical environments. For this reason, the selective value for a gene combination varies from population to population. In a freely interbreeding population, recombination overcomes this tendency to local differentiation, and the frequency of a combination remains the same everywhere. On the other hand, if interbreeding is restricted, the local populations evolve in different directions. In species consisting of demes that interbreed freely only with adjacent units, the adjacent populations, inhabiting somewhat similar environments, resemble each other more than they do nonadjacent populations. Each deme has an average phenotype determined by the equilibrium between local selection and immigration from adjacent demes. The average changes over the geographic range of species as local selection differentials change. A gradual change produces a *cline;* sharp changes divide the species into relatively distinct *subspecies* or *races*.

In summary of neo-Darwinism

The ideas set forth in the preceding paragraphs form the basic concepts of neo-Darwinism. Some paleontologists and biologists argue for other evolutionary "models" or for major revisions in the neo-Darwinian "model." I've described some of these suggested mutant concepts above and will treat some in the following chapter, but most workers in the field of evolution believe that the primary mode of organic change is by selection of variant genotypes within a species population. Mutation and recombination supply the variants; the structure of the population modifies the effects of recombination and of selection.

To what extent are these factors observable and measurable? In experiments with rapidly breeding organisms, e.g., fruit flies, simulation of selection and population structures has confirmed the deductive arguments. The best documented case from natural populations is probably that of melanism in moth populations (Kettlewell, 1965). In environments darkened by deposits of industrial grime, the frequency of dark moths has increased strikingly over the past hundred years; experiments with marked individuals of one species indicate a 30% advantage of a dark phenotype over a light one; experiments with bird predators demonstrate differential predation on light and dark forms related to background color; the experiments also indicate that replacement of the light type by the dark type can occur in as few as fifty generations.

The characteristics of the fossil record preclude similar rigorous tests of the evolutionary mechanism. If, however, much or all evolution derived from the processes of the neo-Darwinian model, some fossil sequences should be consistent with this explanation, and some specification of selective values should be possible. Several examples are considered in the next chapter, on pages 163 to 167.

Isolation and the diversity of species

The transformation of one species into another accounts in part for the succession of fossil species in the geologic column. This transformation, however, does not in itself explain the diversity of contemporaneous species nor the apparent family resemblances among species.

Do the differences between local populations of a species foreshadow differences between species? The gene frequencies in local populations reflect an equilibrium between local selection, genetic drift if any, and limited sampling of total species variability on one hand and immigration on the other. Complete isolation ends gene flow and permits further divergence of the local populations. If isolation persists, the diverging populations may lose their interbreeding potentialities and form new species. When and if isolation ends, the two intermingle without interchange of genes.* Because the environments were initially different the "daughter species" evolve differently, each adapted to its local environment. In plain words, the two species differ in form (and mode of life), but resemble each other in characters retained from their common ancestor. Thus the divergence of isolated populations explains the origin of diversity of species between areas.

Can this mechanism extend to explain diversity in a single area? Consider the moment when the daughter species come in contact. In general, they utilize many of the same environmental resources. Therefore numerous individuals in the two species compete for the same food, seek similar hiding places, suffer the same predation, etc., but by the nature of their origin, individuals from different species differ in their ability to exploit food, utilize hiding places, and tolerate predation. In these circumstances, one or the other of the species will be forced toward extinction in areas of overlap, because of inherent inferiority in a particular local ecologic

* In some groups, plants particularly, interspecific crosses are fairly common, and in some cases the recombinants are adaptively superior to the original populations. Since this process, evolution by introgression, is only a particular case of the general evolutionary model, there is no point in discussing it at length in a short treatment of evolution.

system (p. 127). Those individuals most different from the other species in requirements will be those most likely to survive. In fact, if both species survive in the same area, selection will force continued divergence. The ultimate result will be diversity of species in the same area with diverse adaptation.

This forced divergence of species must have occurred frequently in the geologic past. Species split and resplit, each division carrying the species form and adaptation further from the ancestral type. Given time, the results could be as different as tapeworms, ants, and elephants. Each daughter species adjusts to a distinctive mode of life, and those species that occupy the same habitat do so because they occupy different niches (see further, p. 178).

The search for ancestors

Genetic systems in time become phyletic systems (p. 47); phyletic systems involve both the ancestor-descendent relation of noncontemporary populations and the common ancestry of several contemporaries. The fossil record consists of biased, irregular, infrequent, and geographically scattered samples of the morphological attributes of phyletic continua. A paleontologist in reconstructing evolutionary history must first attempt to develop a phyletic model that has some valid relationship to these continua. Only if he restricts his interests to single strata at single localities can he avoid this challenge; phyletic models form a skeleton for integration of paleontologic (and for that matter, biological) explanations.

Some theoretical scaffolding

One has a set of points, samples of local biologic populations scattered geographically and stratigraphically. How does one decide that a particular pair of points should be connected as ancestor-descendent? Or that *A* and *B* have a proximal common ancestry but that *C* and *B* have only a remote phyletic connection?

Begin with a single picture of a phyletic system comprising individual genotypes and genetic interchange. The genotype is one "input" into an organism; the organism through its total existence with all the various inputs is a phenotype. A phenotype is sampled by a set of objectively defined characters, each one of which exists in one of several possible states, e.g., the number (state) of ambulacral pores (character). The same characters observed in several individuals provide a sample of the array of character states in the population. This array "maps" the inputs into the system, including that critical one, the genetic.

When two arrays are compared, the similarities and differences in form, i.e., the *phenotypic distance,* are a consequence of differences in genotype, i.e., the *genotypic distance,* and of the environmental effects on phenotype. Environmentally controlled variance, unless related to genetic difference, is effectively random. Even if it produces large phenotypic effects, they should be distinguishable by their random pattern in time.* Genetic distance, unfortunately, displays no simple linear relationship to phenotypic distance. Genetic similarity, of course, demands developmental, morphologic, and physiological similarity, but quite different genotypes may underlie similar character states and slightly different genotypes may produce a large phenotypic distance. Different rates of genetic displacement in different phyletic lines complicate the problem further. In consequence, a direct approach to phylogeny through genetic models seems unsatisfactory.†

Consider this alternative: Phenotypic variation in modern organisms includes three components: 1) random developmental and genetic variation, 2) obligatory correlation with a particular mode of life, 3) nonobligatory association of phenotype and mode of life. Random variation is significant only in comparison of very similar forms, i.e., in defining species boundaries; it can be removed in part by statistical analysis. The obligatory component requires a one-to-one relationship between a character state and a specific function, i.e., an organism performs x in habitat y if and only if it has z, and z is meaningful only in performing x in y. Thus, an organism flies in the air only if it has wings; wings are significant only in flying in the air. If two phenotypes are similar in possessing wings, the similarity is *fully* explained if they both fly. In contrast, nonobligatory association requires no one-to-one correspondence, i.e., an organism performs m in habitat o if it has either r, s or u and conversely, either r, s or u are compatible with alternate functions m or n in different habitats o and p. Thus animals fly whether skin membranes or feathers form the airfoil, and the arrangement of bones in the forelimbs of birds, antelopes, and whales is similar in spite of different functions and habitats.

Can this be related now to a phyletic model? A population at a particular time represents a unique accumulation of prior selective events and of "accidents" in mutation and recombination (extending back to the origin of life). A later population on the phyletic continuum differs only as a consequence of intervening selection and "accidents." Selective change corresponds to functional change; accidental differences reflect population size and structure and need only be compatible with functional divergence. The later population "inherits" from the earlier ones characteristics compatible with the difference in mode of life but not necessary to

* If the samples are very small and few in number, or if the genetic distance is small, environmental variance obviously assumes more importance.

† But see p. 159 in relation to "numerical taxonomy."

it. Therefore "inheritance" explains the nonobligatory component of phenotypic distance; selection the obligatory one. Similarly, two contemporaneous populations possess obligatory similarities and differences as a consequence of similarities and differences in selective history; they also possess nonobligatory similarities acquired from the common ancestor. Obligatory differences express phyletic divergence; the ratio of nonobligatory similarities to nonobligatory differences expresses the proximity of the common ancestor. The distances within this phyletic model are not necessarily proportional to genetic distances and reflect rather environmental (adaptive) divergence and time. But the phyletic pattern conforms qualitatively to the underlying genetic one.

I can now proceed to a consideration of the working rules by which one derives a phyletic model from the fossil record.

Interpretation of overlapping fossil successions

When fossil samples from successive layers of rock overlap phenotypically, the paleontologist usually infers an overlap of genotypes. Although the same genotype could arise in different species, this is most improbable. The samples are then regarded as segments of an evolving population continuum, and their differences, if significant, are regarded as the consequence of that evolution. Morphologic similarity here means near identity; some specimens from sample *1* can hardly be distinguished from specimens of sample *2*. The *Micraster* population sequence described on p. 137 exemplifies this sort of relationship.

Interpretation of discontinuous fossil successions

If evolution ran on tracks, directly from fish to men or from worm to trilobite, interpretation of phylogeny would be simple. The paleontologist could finish his business and leave the geologist to fit new forms into a neatly drawn chart. Fortunately for the careers of paleontologists, though not so fortunately for their ease of mind, nature lacks rails, crossties, and block signals. A species population may evolve in one direction for a while under the impetus of a single selective factor, but few such simple trends can long continue. If a paleontologist has but a few fossils from isolated spots along this cow path of evolution, how is he to tell that they are segments of a single phyletic lineage?

I will try to illustrate the answer by discussing the relationships of Devonian lungfish, lobe-finned fish, and amphibians. For the sake of brevity and simplicity, I will limit the example to two middle Devonian fish genera, *Dipterus* and *Osteolepis* (Figure 6-5), and to three late

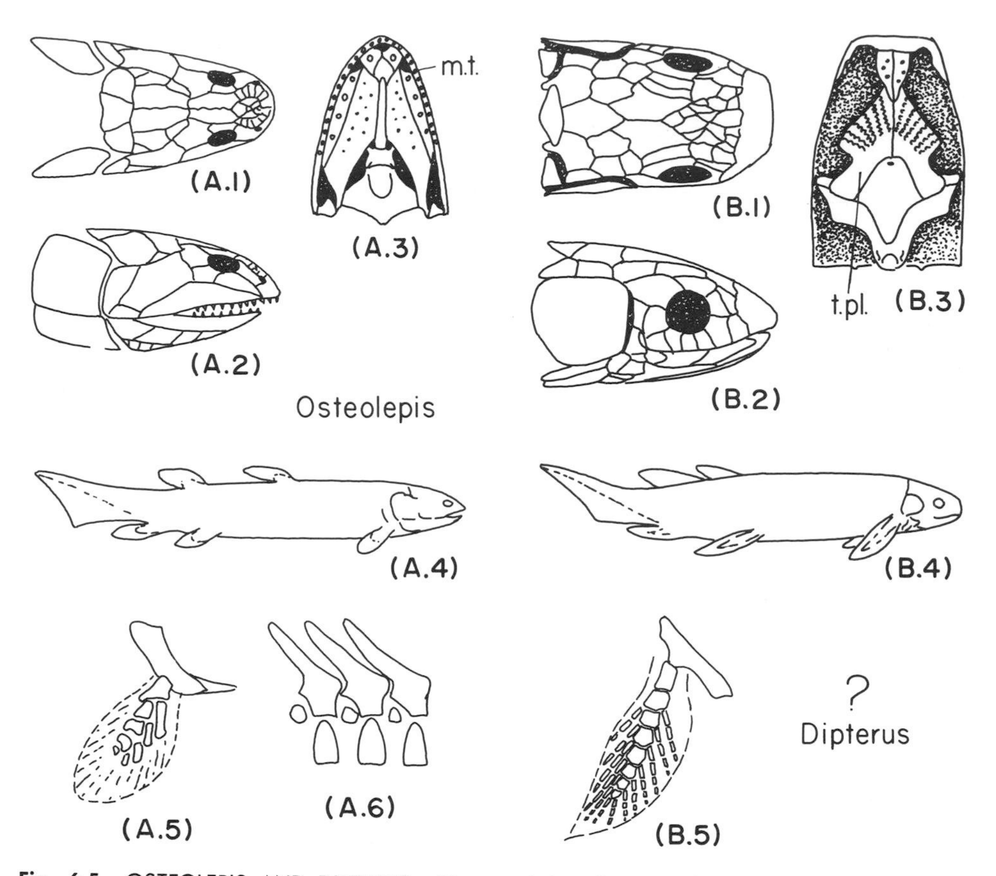

Fig. 6-5. OSTEOLEPIS AND DIPTERUS. Characteristics of two middle Devonian fish genera. **(A)** *Osteolepis;* **(A.1)** skull roof; **(A.2)** side view of skull; **(A.3)** palate; **(A.4)** general body form; **(A.5)** one of anterior paired fins articulated with bony plate in body wall; **(A.6)** three vertebrae. Structure of fin and palate based in part on later genera. **(B)** *Dipterus;* various views as in *Osteolepis;* vertebrae probably similar to Osteolepis but not known certainly. *Abbreviations:* mt, marginal tooth; tpl, tooth plate. [**(A.1-3)** *after various sources;* **(A.4)** *after* Traguair; **(B.1-3)** *after* Westoll; **(B.4)** *after* Traguair.]

Devonian genera, the fishes *Scaumenacia* and *Eusthenopteron* (Figure 6-6), and the amphibian *Ichthyostega* (Figure 6-7).

Dipterus and *Scaumenacia* are similar in character. Their bodies are relatively slender, compressed from side to side, elongate, and covered by thick bony scales. Two dorsal median fins are present, and the tail is heterocercal, i.e., the vertebrae bend into the upper half of the tail fin. As in all advanced fish, there are two sets of paired fins, one pair just behind the head, the other just in front of the anus. Each fin is leaf-shaped, with a bony, segmented axis and numerous short side branches. The inner end of the axis articulates with a small bony plate in the body wall. The backbone is ossified, although the individual vertebrae are not

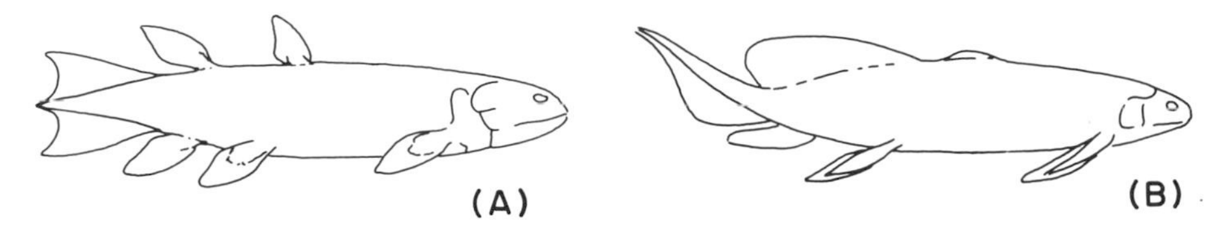

Fig. 6-6. EUSTHENOPTERON AND SCAUMENACIA. Late Devonian fish genera. **(A)** *Eusthenopteron,* general view. **(B)** *Scaumenacia,* general view. Details of skulls and fins quite similar to *Osteolepis* and *Dipterus,* respectively. [**(A)** *After* Gregory *and* Raymond; **(B)** *after* Hussakof.]

tightly articulated. The ribs are cartilaginous. The skull is moderately short with an unusually short cheek region. The skull comprises a large number of plates in the skull roof, a well-ossified brain case, and several large elements in the roof of the mouth. The lower jaw consists of several unfused bones. The bones in the roof of the mouth bear radial rows of large teeth, but marginal teeth are either absent (*Dipterus*), or rudimentary (*Scaumenacia*). The bones of the upper and lower jaws that bear marginal teeth in other vertebrates are also lacking or very small.

The two genera differ in only a few characters. The first dorsal fin of *Scaumenacia* is smaller and the second dorsal larger than that of *Dipterus.* The paired fins of *Scaumenacia* are somewhat more slender. Finally, the arrangement of the skull bones and character of the teeth on the palate are slightly different.

Osteolepis and *Eusthenopteron* resemble the preceding in general form. The axial portions of their paired fins, however, are quite short, the bones heavier, and the number of side branches smaller. The skulls

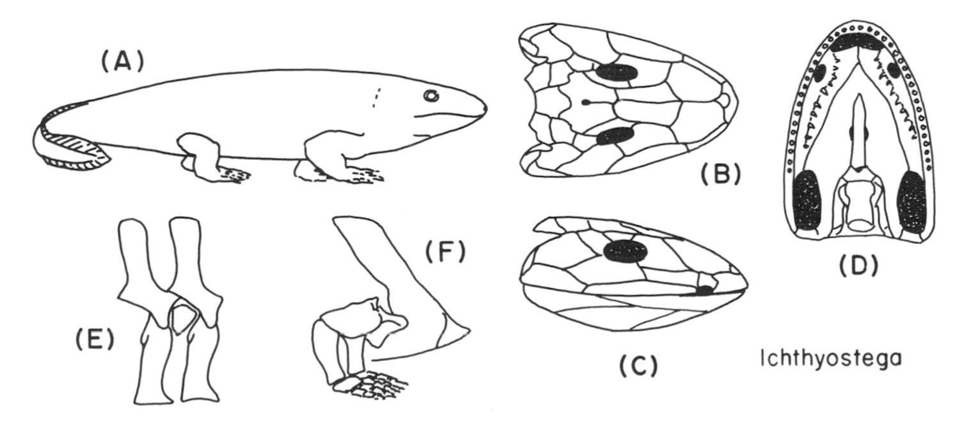

Fig. 6-7. ICHTHYOSTEGA. Late Devonian amphibian, *Ichthyostega.* **(A)** General form. **(B)** Skull roof. **(C)** Side view of skull. **(D)** Palatal view. **(E)** Vertebrae. **(F)** Forelimb and bony plates (pectoral limb girdle) with which it articulates. Compare with Fig. 6-5. (After Jarvik.)

are relatively longer. The marginal bones of the jaws bear large pointed teeth, but the teeth on the palate are small and lack a radial pattern. The bones of the skull roof are larger and are fewer in number. Their brain cases are well-ossified but, unlike those of the first two genera, consist of separate front and rear elements joined by a movable articulation. The two genera differ from each other chiefly in the structure of the tail—the vertebrae in *Osteolepis* turn into the upper lobe; in *Eusthenopteron* they extend straight backward to form a central lobe of the tail—and in minor details of proportion and bone arrangement.

The amphibian, *Ichthyostega,* is quite different from any of these fish. The body is flattened from top to bottom rather than from side to side. The long slender tail bears low fins rather than the expanded ones of the fish. The vertebrae are large, heavy, and closely articulated, and, except for those in the tail, each bears a pair of ribs. In place of the paired fins are legs articulated with broad plates, the pelvic and pectoral girdles, on the sides of the body. The skull resembles those of *Osteolepis* and *Eusthenopteron* to some extent and comprises a number of large elements in the skull roof, a well-ossified brain case, and a series of marginal plates on both upper and lower jaws bearing large teeth. Some of the palatal bones bear large teeth, but these are never in radial rows as they are in the lungfish. As in the fish, the nostrils open internally into the mouth.

What sort of pattern appears in these observations? What are the evolutionary relationships of the five genera? I could now attempt an analysis of similarities and differences, but discussion may be clearer with a hypothetical reconstruction of the evolutionary changes.

Imagine that two local populations of a fish species live in the rivers of adjacent drainage basins. The two populations differ slightly in average form and physiology. The differences are results of genetic adaptation to differences in environment and local genetic drift (p. 145). A barrier arises between basins; the local populations diverge in isolation; two species appear where there was one before. The relationship of these species is obvious, for the differences are slight (recognized by placing them in the same genus), and their relationship to the ancestral species is equally obvious. Each retains most of the characteristics of its ancestors; their differences are consequent upon minor variations in the same basic mode of life.

The species, however, continue to diverge.* The original one ate other fish and occasionally varied its diet with worms and crayfish. One of the new species sticks to this diet, but the other shifts entirely to crayfish and worms. In the first species, individuals with large marginal teeth have an advantage in selection and the average phenotype shifts in this

* If the two subsequently occupied the same rivers, their competition may have forced greater divergence.

direction. In the second species, individuals with large, closely-spaced palatal teeth have higher survival and fecundity rates.

The resultant species differ chiefly in skull and jaw structures adaptive to different diets (obligatory differences). They retain generally similar characters of fins, body, and skull so far as these are not affected by dietary adaptation (compatible and nonobligatory similarities). You can recognize *Dipterus* and *Osteolepis,* the middle Devonian genera. They differ to the extent that they had different modes of life, but they retain many characteristics of their common ancestral species. One cannot be certain that the fish-eating habit was the ancestral type, but our knowledge of early Devonian fish suggests it was.

The analysis, however, has hardly begun, for the paleontologist must connect the three late Devonian genera to the middle Devonian forms. *Scaumenacia* resembles *Dipterus* most closely. Did *Dipterus* give rise to *Scaumenacia?* The modification of the palatal teeth occurred once in response to selection. Could it occur again in an isolated population of *Osteolepis?* If it did, several improbable events must have happened. First, the new adaptive line arising from *Osteolepis* recreated the precise form of tooth and jaw structure. Selection might choose a similar structure to fit similar needs (an obligatory similarity), but production of an identical structure seems less likely. Second, the new line developed paired fins and other characteristics like those of *Dipterus,* although these are not adaptations to the same selection pressures as the jaws and teeth. It seems extremely improbable that the same selective factors that acted on the ancestor of *Dipterus* would again be combined in the same way and that they would find the same genotypes among which to select. Finally, if *Scaumenacia* evolved from *Osteolepis,* the truncated axis of the paired fins and the regular arrangement of skull roof bones disappeared. These features, however, are compatible with the different mode of life, and their loss is inexplicable by adaptation. Because of these improbabilities, the paleontologist makes *Scaumenacia* a descendent of *Dipterus.*

Now for *Eusthenopteron.* Purely on degree of resemblance, it descended from *Osteolepis.* But could it have arisen from *Dipterus?* Again it depends on probabilities. Origin from *Dipterus* requires that truncation of the fin axes, rearrangement of the skull bones, and jointing of the brain case evolved twice—a double coincidence of selection and available variation. Also, the peculiarly adapted dentition of *Dipterus* would be lost and marginal teeth regained—a clear and minute reversal of an evolutionary trend. The probabilities, therefore, indicate a *Eusthenopteron-Osteolepis* connection as the correct one.

These cases are not too difficult, since the ancestor and descendent are quite similar, but bridging the stream between fish and amphibian is far more difficult. Since, however, the amphibian structures evolve from those available in the fish, some piscine features should remain. Since the chief adaptation is for life on land, the limbs, limb girdles, tail, and backbone should show the largest change. The modifications of the skull should be less marked.

Could *Ichthyostega* have evolved from *Dipterus*? The bony axis of the paired fins might change to the bones of a foot and leg, although the change might be troublesome. The plates supporting the paired fins become the limb girdles. The vertebrae need only firmer articulation—not a major change. The ribs change to bone from cartilage. The tail, no longer used for a scull, becomes relatively slender and narrow. The small plates of the skull roof either fuse or are lost to leave a series of large regular plates. The bones covering the gills are reduced or lost. The brain case and the internal openings of the nostrils require no modification. The teeth and jaws—there's the difficulty. The amphibian has large marginal teeth and jaw plates; *Dipterus* has few or none. Either the trend exhibited by *Dipterus* must have reversed or the connection of *Dipterus* and *Ichthyostega* is impossible.

Let me try *Osteolepis* as an ancestor. The tail and vertebrae would need no more modification than for *Dipterus*. The change in paired fins is a lot simpler—and thus more probable. The fin of the fish and the leg of the amphibian share a number of compatible and nonobligatory properties; they differ largely in features related to function, i.e., obligatory differences. The skull roof plates of both are large and display a similar pattern: in number, in relation to eyes, nostrils, and jaw articulation, and in openings for nerves and blood vessels. Palate, jaw, and teeth are similar, although these may be obligatory similarities with similar dietary function. The brain cases are different in that the distinct joint between anterior and posterior portions in the fish is closed in the amphibian (a nonobligatory difference?). This could disbar *Osteolepis* from its job as ancestor, but *Ichthyostega* is intermediate between *Osteolepis* and later amphibians in this character. It might in fact be an adaptive difference. Of the two choices, *Osteolepis* is the more probable ancestor.

The paleontologist in his imagination has stripped away the tetrapod terrestrial adaptations and restored the pristine piscine form. Since the restored form agrees with that of the middle Devonian fish, he has found the ancestor—maybe. Actually, he may have as fossils a species off the direct evolutionary line. Reconstruction seldom can be so precise as to pick an ancestral population from among similar species, or, if the gap is large, from among similar genera.

A final consideration is the taxonomic expression of the amphibian-fish relationship. The morphologic difference between the primitive amphibian and its ancestor are not much greater than those between orders of fish in the same subclass. Should the paleoichthyologist establish a new order in the same subclass to receive the amphibian? Certainly a logical step. But he displays illogic, for the amphibian becomes representative of a new class. His reasoning was like this: "this animal (he should have thought 'population') opens new prospects in evolution; if given a chance, he will evolve into something as remarkable as a frog or even a dinosaur or a monkey. The differences between frog and fish are great; therefore, I will begin a new class with this animal. I'll classify him not for what he is but for what he'll become."

I'll end my piscatorial research here. But you may say: "I'm not interested in fish except as food. Trilobites (or brachiopods) are what excite me. How do I work out their phylogeny?" Even if you don't ask this, I'll give some general rules anyway.

1) The paleontologist must have some assurance he's comparing the right things—tail with tail, paired fin with leg. Since evolution acts by building on or by reshaping the ancestral structure, ancestor and descendent have a common structural plan defining a common set of characters, and descendent lineages should likewise share a basic organization of parts. Studies of development of the embryo assist here, since the organization may be laid out in simpler form in the embryo.* Thus the embryos of fish and of humans have similar gill arches, though the adult fish gill bars and the human hyoid bones seem to have little in common. Parts occupying the same position in the structural plan are termed *homologues,* e.g., the fish hyomandibular gill bar is homologous with the stapes of the amphibian ear.

2) If all else is equal, phyletic distance is equal to phenotypic distance. Some taxonomists hold that "all else is equal" if a large number of logically independent characters (~60) are taken into consideration. If they are only partly right (as I believe), or if only a few characters are available, then further criteria must be sought.

3) If possible, phenotypic distances must be scaled by reference to adaptive explanations. Similarly adapted (obligatory) properties are least important, because they may evolve independently in separate lines, whereas similarities in structures that serve different functions (nonobligatory properties) indicate relationship. The similarity between the leg of the amphibian and the paired fin of the fish cannot be laid to similar selective pressures; it can be found in a common ancestry. Several characteristics are better than one in establishing relationship—unless they are adaptations to similar modes of life or features related to a single adaptation. The paleontologist gives most weight to the "conservative" features held in common by diversely adapted species.

4) He must infer a phylogenetic model which reconstructs the evolutionary sequence in adaptive terms. The model should predict hypothetical ancestors or descendents to be matched with the fossil record.

* Sometimes ontogeny recapitulates phylogeny when structures are added or reformed near the end of development. But the adult may retain embryonic structures or the embryo (or better, larva) may evolve its own peculiar adaptations. In general, phylogeny adopts the possibility inherent in development. Therefore, developmental stages of ancestor and descendent or of descendents of a common ancestor resemble each other, though the resemblance may be between different stages.

Ideally, this reconstruction fulfills four conditions: *a*) reconstructed structures are compatible in form with their probable function; *b*) the proposed modification of a structure provides continued adaptation during the period of evolution; *c*) the modifications are compatible with what is

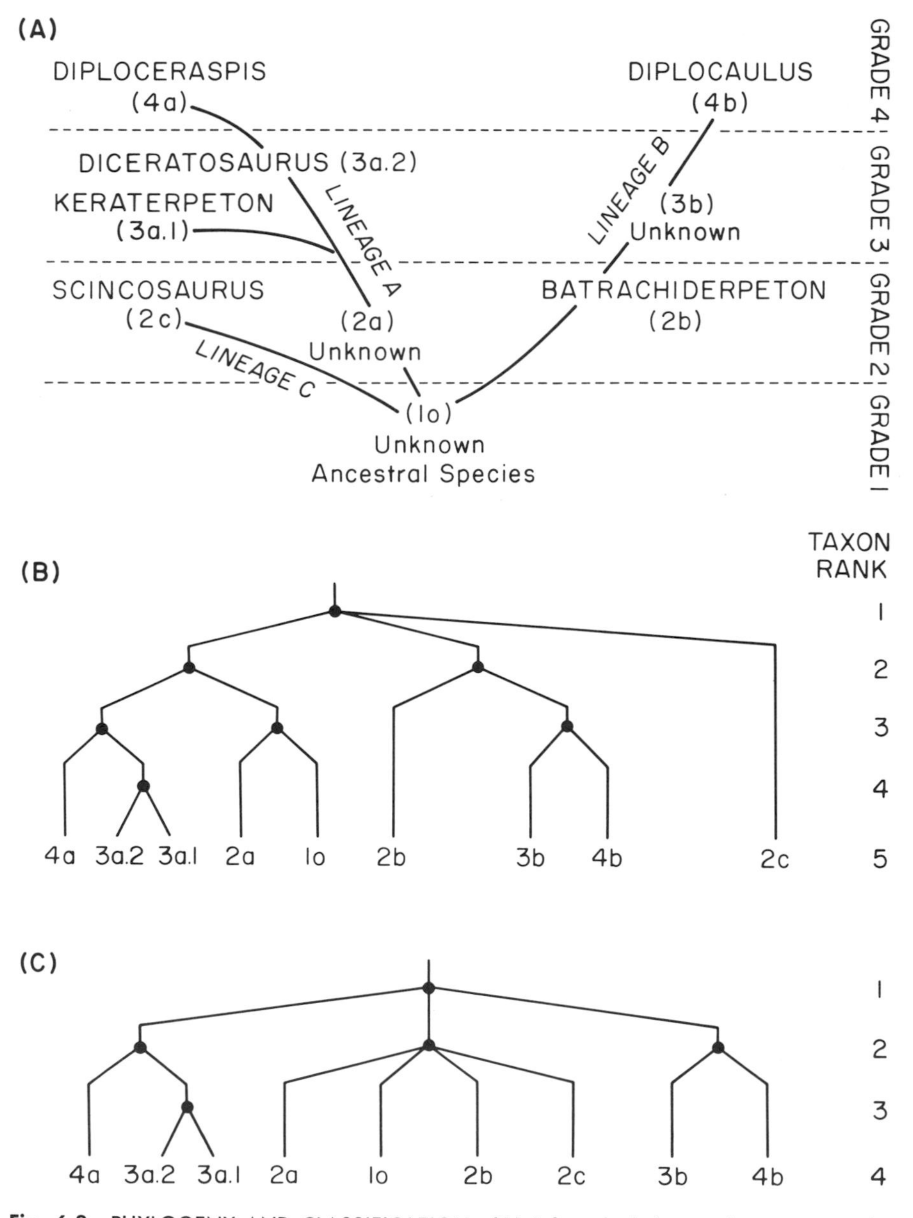

Fig. 6-8. PHYLOGENY AND CLASSIFICATION. **(A)** Inferred phylogeny for a group of late Paleozoic amphibian genera. Numbers indicate structural levels (grades); letters indicate evolutionary lineages (clades). **(B)** and **(C)** are alternate classifications of the same genera. Numbers correspond to those in **(A)**.

known of the ontogeny of the structure, and *d*) the modifications involve the fewest possible changes between ancestor and descendent.

5) If the presumed ancestor reversed its evolutionary trend to produce the descendent, it's unlikely to have been the ancestor. This bars two-toed horse species from evolving into three-toed horses, or a species of large elephant from producing a species of small elephant. In detail, structures lost will not be regained; complex structures will not become simple; large forms will not give rise to small. *Unfortunately, this "law" [Dollo's] only works a six-day week—on Sunday anything can happen.* Certainly selection can reverse an evolutionary trend. This reversal probably does not retrieve the lost genotypes and the precise form of the ancestor, but it can produce a close facsimile of that form. It is, then, only as a working principle to be disregarded on the basis of other evidence.

6) The evolutionary relationships should be compatible with geographic distribution. The connection between South American and Old World porcupines was suspect for this reason. One couldn't have given rise to the other without a ferry between South America and Africa. Restudy of the morphology suggests that the similarities are due to similar adaptations developed in two lineages that were closely related to begin with.

7) The shorter the distance across which the paleontologist must interpolate, the smaller is the chance of error. Missing links, when found, serve not only to confirm a hypothesis but also to render it more precise.

8) Conservative groups—those that change slowly—provide knowledge of ancestral types not available directly as fossils. The paleontologist depends on Permian amphibians for evidence of reptile origins, because little of the true, late Mississippian and early Pennsylvanian, ancestry is preserved.

Some phylogenies cannot be fully justified by these criteria, e.g., the adaptive significance of cephalopod septal variation is uncertain. Those that meet all these criteria are objective, i.e., could be repeated by anyone employing the same premises. Unfortunately, the premises, e.g., that the palatal teeth of *Dipterus* indicate an invertebrate diet, are normic statements of limited (and sometimes disputed) likelihood. Alternate phylogenetic models cannot be ruled out on objective grounds, and the choice of a particular model involves an arbitrary decision.

Phylogeny and taxonomy

Biologists and paleontologists generally attempt to express their phyletic models in the arrangement of taxa in a classification (p. 51). Phylogenies are nonhierarchic continua; the Linnean classification is a discontinuous hierarchy; these discrepancies require discussion of the rules by which a classification is brought to map a phylogeny.

The continuum problem

A phylogenetic model (Beerbower, 1963) for a group of Paleozoic amphibians is shown in Figure 6-8A. The terminal species on one of the branches, *4a,* is quite different from the initial species, *1o* and, therefore, requires placement in a different taxon. Where shall the boundary be placed? The only possible nonarbitrary break lies at the split of phyletic lines (clades) *A* and *B*—but a population sample just above this break is still quite different from *4a,* so one cannot avoid the problem. Quite simply, one arbitrarily breaks the continuum at some convenient point, e.g., at a break in the fossil record, at a place where the rate or direction of evolution changes rapidly, etc. In this case the difference between *4a* and *3a* is rather large, so I'll draw a line there.

The hierarchy problem

A phylogeny and a hierarchy (Figures 3-2 and 3-3) both "branch" and to this extent parallel each other. Species populations occupy only the tips of hierarchic branches, however, so that the branches and their dichotomies are pure abstracts. In contrast, real organisms occupy the phylogenetic branches. One must therefore decide, arbitrarily again, whether to place the population occupying a dichotomy, *1o* for example, with one or the other taxon defined by the corresponding hierarchy branches (Figure 6-8B). Once that decision is made, a descendent population on the other branch, e.g., *2c,* is placed in a distinct taxon, even though it may be closer in all properties to the population at the dichotomy than the latter is to later populations (*4a*) in the same taxon. Phyletically, *1o* "belongs" with *4a*; *1o* "belongs" with *2c,* but *2c* does not "belong" with *4a.* Therefore, *1o* is a member of two categories of equal rank—a logical impossibility in a hierarchy. The alternative is fairly simple: *1o, 2a, 2b,* and *2c* are placed in the same taxon, a logical hierarchic alternative (Figure 6-8C). By this arbitrary exclusion, *3a* and *4a* must form a second equivalent taxon; *3b* and *4b* a third. The first taxon includes species with close common ancestry, i.e., slightly divergent phyletic lineages. In contrast, the other taxa comprise populations along single lineages. Other logical alternatives are possible, *so long as the only objective dividing point, the dichotomy, is not used as a division between taxa.* A specialist on late Paleozoic stratigraphy might emphasize the "horizontal" classification that divides early and late species; a specialist on amphibian evolution might emphasize "vertical" taxa that group *1o* with *4b* and *2a* with *4a*.

Problems of scale

The rank assigned a taxon and the amount of subdivision of a taxon varies with the purpose (and the psychology) of the classifier. The recent cats, for example, are often placed in three genera, one for the hunting leopard, one for the large cats, lion, tiger, and so on, and one for the smaller cats, wildcats, ocelot, and others (Simpson, 1945). Such a grouping emphasizes the similarities of most cat species. Some mammalogists, however, emphasize the differences and distinguish twenty genera. In the latter classification, the relationships among the genera are shown by grouping them into subfamilies. I can only give a personal reaction to lumping vs. splitting of categories: that it is better to lump than to split. Splitting helps those few paleontologists who specialize in a particular group but injures the majority who are not specialists. An effective compromise might be to use the chief steps in the classification (genus, family, and so on), to group animals, and the intermediate steps to divide them. Thus, one could recognize three cat genera (which all could know) divided into twenty subgenera (which only the specialists need know). Of course, the arrangement of supraspecific taxa, whether lumped or split, carries the burden of phyletic relationship, and so the classifier is not entirely free to lump or split as he chooses.

Numerical taxonomy

The general philosophy of phyletic modeling as discussed here has been opposed recently by the growth of a new approach called *numerical taxonomy*. This opposes not only a methodology—elaborate statistical analysis of groups of characters *vs.* semiquantitative or qualitative analysis of single characters—but also, and more significantly, a new rationale to the traditional approach. Because the level of statistical understanding prerequisite to a study of the methodology goes beyond the scope of this book, I can only refer you to sources, particularly Sokal and Sneath, 1963.

The rationale, however, is critical. The argument is thus: If a very large number of independent characters are considered in their totality, the statistical phenotypic distance provides a better representation of phyletic distance than the weighting of character displacement by adaptive value. The various statistical techniques cluster samples; members of a cluster belong to adjacent phyletic lines or are near neighbors on a single line. Such clusters possess a hierarchic, metrically ordered pattern and, in effect, decide the problems of continuum, hierarchy, and scale. Practical difficulties aside—the few characters available on many fossils, the small sample sizes, etc.—the approach offers some intriguing possibilities—particularly where the adaptive values are extremely dubious. In fact, many fossil classifications are little better than morphologic keys set in hierarchic form; numerical taxonomy could do no worse. As to the more ambitious and general claims, the comparative values of the two philosophies have yet to have an adequate test.*

The flux of phylogeny

Fossils are found, phylogenies established, classifications published. Truth is graven eternally into the literature of fossil fishes, trilobites, or brachiopods. And then someone turns a slab of rock and finds a new species that fits into a phyletic sequence—but not the accepted one—or new evidence accrues from studies of adaptive significance or embryos or comparison of proteins in the blood serum of lions and antelope or a classification of the parasites of kangaroos and South American monkeys. Or stratigraphic research discloses that the "ancestor" was really the descendent. The remarkable thing is not that phylogenies are often found in error, but that they frequently have been confirmed by new evidence. This is of importance not only to the reputation of the phylogeneticist but also to the soft-rock geologist who would rely on interpretation of evolutionary trends to establish relative age and on degree of divergence to determine paleogeography.

This discussion has touched but superficially on the methods of phylogenetic interpretation. I have omitted much and simplified the remainder. It is, however, sufficiently detailed to permit in the next chapter a description of the patterns of evolution as they show in the phyletic reconstructions.

REFERENCES

Beerbower, J. R. 1963. "Morphology, Paleoecology, and Phylogeny of the Permo-Pennsylvanian Amphibian *Diploceraspis*," *Bull. Mus. of Comp. Zoology*, vol. 130, pp. 31-108.

Camin, J. H. and R. R. Sokal. 1965. "A Method for Deducing Branching Sequences in Phylogeny," *Evolution*, vol. 19, pp. 311-326.

* "Non-numerical" taxonomists can and do employ statistical methods. In such cases, the paleontologist evaluates the statistical distances in terms of an adaptive model—see, for example, the *pF* approach developed by Olson and Miller (1959).

Dobzhansky, I. 1951. *Genetics and the Origin of Species*, 3rd ed. New York: Columbia University Press.

Ehrlich, P. R. and R. W. Holm. 1963. *The Process of Evolution*. New York: McGraw-Hill.

Kermack, K. A. 1954. "A Biometric Study of *Micraster coranguinum* and M. (*Isomicraster*) *senonensis*," *Roy. Soc. of London, Philos. Trans., Ser. B*, vol. 237, pp. 375-428.

Kettlewell, H. B. D. 1965. "Insect Survival and Selection for Pattern," *Science*, vol. 148, pp. 1290-1296. Evolution in action.

Kurtén, B. 1963. "Return of a Lost Structure in the Evolution of the Felid Dentition," *Soc. Sc. Fennica, Comment. Biol.*, vol. 26, pp. 1-11.

Mayr, E. 1963. *Animal Species and Evolution*. Cambridge, Mass.: Harvard University Press.

Nichols, D. 1959. "Changes in the Chalk Heart-Urchin *Micraster* Interpreted in Relation to Living Forms," *Roy. Soc. of London, Philos. Trans., Ser. B*, vol. 242, pp. 347-437.

Olson, E. C. 1952. "The Evolution of a Permian Vertebrate Chronofauna," *Evolution*, vol. 6, pp. 181-196.

Orton, G. L. 1955. "The Role of Ontogeny in Systematics and Evolution," *Evolution*, vol. 9, pp. 75-83.

Rowe, A. W. 1899. "An Analysis of the Genus *Micraster* as Determined by Rigid Zonal Collecting from the Zone of *Rhynchonella cuvieri* to that of *Micraster coranguinum*," *Quart. Jour. Geol. Society of London*, vol. 55, pp. 494-547.

Simpson, G. G. 1952. *The Major Features of Evolution*. New York: Columbia University Press.

———. 1961. *Principles of Animal Taxonomy*. New York: Columbia University Press. Phylogenetics as well as systematics.

Sokal, R. R. and P. H. A. Sneath. 1963. *Numerical Taxonomy*. San Francisco: Freeman and Co.

Tax, S. (Ed.) 1960. *Evolution after Darwin, Vol. I, The Evolution of Life*. Chicago: University of Chicago Press. A collection of very important critical and review papers.

Zangerl, R. 1948. "The Methods of Comparative Anatomy and Its Contribution to the Study of Evolution," *Evolution*, vol. 2, pp. 351-374. Significant review of the rationale of morphologic comparison.

7 Patterns of evolution

In an ideal world, the fossil record would provide a continuous motion picture of evolution, each frame in the sequence a successive layer of fossil-bearing rock. Since this world is not ideal, the evolutionary record consists of scraps on the cutting-room floor. Some are single frames, still-pictures in fact, the actors frozen in silence. A few are continuous strips in which the principals scurry into motion—and then the lights go up; the reel is changed, and we have a new strip with a different plot, old actors disguised, and new ones on the scene.

The paleontologist has two tasks in restoring this ancient movie: to discover the actors in their disguises and to work out the vagaries of plot that require the disguise. He has a third task as a critic of the reconstructed film; to discern the common features of the various subplots. The preceding chapter dealt with the reconstruction. This one must synthesize the unities of the evolutionary drama. These unities include process, rate, and mode of evolution; they involve tests of neo-Darwinian theory as well as its application to phylogenetic interpretation.

Fossils and evolutionary mechanisms

Can paleontologists measure the various evolutionary processes and factors from the fossil record?

The parameters of an evolving population are 1) genetic variation, 2) morphologic variation, 3) reproductive structure, and 4) the selection values for various genotypes or the corresponding phenotypes. To the extent paleontologists measure these, they can explain the course of evolution illustrated in the various phyletic lines.

A warning is necessary, however; since the paleontologist uses evolutionary models to work out phylogeny, he may devour his own intellectual flesh. Thus, a theory of linear evolution yields linear phylogenies which support a linear theory of evolution. Some major works on evolutionary theory should bear the sign *cave canem*.

Genetic and morphologic variation

I have argued elsewhere (pp. 73, 75) the difficulty of measuring either individual genes or genetic variation from fossil evidence. Such efforts as that of Kurtén (1955) on Pleistocene bears demonstrate that interpretations of genetic change from fossil sequence are consistent with our knowledge of genetics—but it is doubtful that any fossil sequence could be shown rigorously to be inconsistent with modern genetic theory.

Paleontologists, therefore, treat the second parameter, morphologic variation, without resolving its genetic basis. They make qualitative observations of variation, or, since the fossils form a sample, they estimate population characteristics in the normal fashion (p. 78), and describe changes through time. Sounds simple enough, but, as usual, there is a booby trap buried with the bones.

Paleontologist Doakes collects mammal teeth from a single stratigraphic horizon at a single locality and concludes the average and standard deviation of molar length for the population to be thus and so. Paleontologist Smith collects from twelve horizons, over a ten-thousand square mile area. A third man, Jones, compares their date and says: "Aha! Smith's species was more variable than its descendent species (Doakes'). This proves something." What it probably proves is that Doakes sampled a local population at a single time and Smith sampled several different local populations at different times. It's not much use comparing unless one compares commensurates. Morphologic variation can be used (must be, of course) in evolutionary studies but it should be used with due caution and precaution.

Population structure

The paleontologist finds determination of population structure and of variation in morphology related to that structure very difficult. In some cases, he may be sufficiently certain of the ecological limits of the species to draw conclusions about distribution patterns, ecologic barriers, and the resulting effects on that population. For example, an aquatic amphibian species will consist of small to medium-size populations distributed linearly along water courses and nearly isolated within single drainage basins, whereas a single active carnivorous mammal (a mountain lion, perhaps) will range over an area wider than that same watershed. If he can establish synchroneity of samples from different areas, he can measure directly the variation between local populations in a species. These occasions are extremely rare, and even then he can't separate variation resulting from isolation from that which is due to different local selection values.

Direct measurement of deme size is theoretically feasible; such measurements would provide an estimate of the size of the breeding population. The limiting conditions for such measurement however are stringent; they include: 1) complete preservation of all mature individuals, 2) absence of transportation or reworking, 3) a census including only one generation, 4) easily recognized individuals of breeding age. "Nests" of the brachiopod *Mucrospirifer* found in the Devonian of western New York apparently meet these requirements: They consist of undisturbed clumps of apparently mature individuals representing a single spat-fall (attachment of planktonic larvae to bottom). Therefore, the brachiopods in a clump are a sample of a deme, and the number of breeding individuals in the deme must have approximated or exceeded the number in a clump.

Selection

Selection consists of two components, the direction (defined by the favored phenotypes) and the pressure (defined by the difference between selective values for the various phenotypes). Direction is established by comparison of characteristics of succeeding populations. Usually, it is desirable to infer causes for the superiority or inferiority of a given phenotype. This is done either on the basis of its presumed advantages in terms of the environment (as in the interpretation of *Micraster*, p. 141), or correlation of morphological and environmental changes. The common factor yielding the correlation may be complex, e.g., change in rainfall affects the vegetation, which affects the herbivores, which affects the carnivores, but one can relate some adaptive change directly or indirectly to the environmental change.

The amount of difference between successive populations relative to the time elapsed gives a measure of selection for the phenotype but even this relationship is not a simple one. Mutation rates and population structure also control evolutionary rates, and measurement of elapsed time is difficult. Since the selective significance of a phenotype is determined only within an environment, the physical difference between two phenotypes bears no direct, constant relation to their selective values. In this respect, the various studies by Kurtén (1958, 1964) are among the most interesting and enlightening. Table 7-1 shows the variation in two tooth characters in Pleistocene cave bears. The relative height of the large cusp on the second molar was measured by a "paracone index" as "cusp height" over "total tooth length." The depression or valley into which this cusp fits on the lower tooth was measured by "valley index" as "space across valley" over "total tooth length." The age of the individual represented by a particular tooth was estimated by the amount of wear. The sample represents deaths, probably during hibernation in the cave, from a local population continuum for several thousand years (several hundred generations).

TABLE 7-1

Tooth variation and mortality in cave bears (*after Kurtén 1957*)

	Survival in percent			
	Paracone Index		*M_2 Valley Index*	
Age in years	*<Mean*	*>Mean*	*<Mean*	*>Mean*
0.4	100	100	100	100
1.4	48	29	17	38
2.4	39	6	4	20
4.4	26	4	—	—
6.4	16	2	—	—

$$\text{Paracone Index} = 100 \times \frac{\text{length of paracone}}{\text{length of crown}}$$

$$\text{Valley Index} = 100 \times \frac{\text{distance between summits of protoconid and hypoconid}}{\text{length of crown}}$$

The table demonstrates a striking differential in survival (Figure 7-1A): ninety-six percent of the individuals with unusually high cusps died before sexual maturity; only seventy-four percent of those with low cusps died before maturity. The survivorship curves for "valley index" show a much higher death rate for those with a narrow valley. Kurtén argues that individuals with high cusps and narrow valleys suffered from

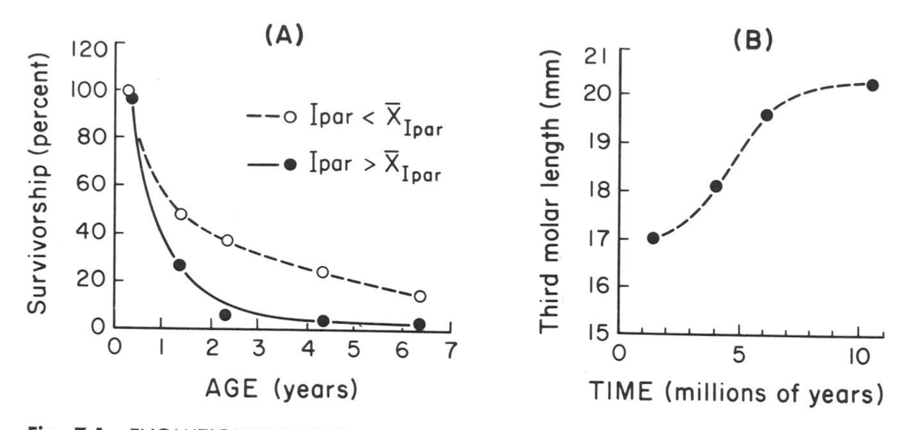

Fig. 7-1. EVOLUTIONARY MECHANISMS AND RATES. **(A)** Survivorship rates for cave bears. The paracone index (I_{par}) is the ratio of the height of the paracone to the length of the 2nd upper molar. The percentage of individuals with high paracones (dots) and small paracones (circles) surviving to particular ages is shown. (After Kurtén.) **(B)** Rate of change in length of third upper molar in lineage of Miocene mammalian herbivores, *Merychys crabilli-M. relictus.* (Data from Bader.)

severe malocclusion, were unable to feed adequately, and, therefore, were less likely to survive hibernation.

According to the selection model, this situation appears unstable, and the frequency of genotype(s) for high cusp and for narrow valley should have been reduced rapidly. In fact, these deleterious genotypes persisted over several hundred generations. At first inspection the selection model fails to operate.

Unfortunately for a test of the neo-Darwinian model the results are not as simple as they look. Kurtén argues a modified neo-Darwinian interpretation:

a). Malocclusion is most frequent in the homozygous genotypes, i.e., either A_1A_1 or A_2A_2 produces malocclusion.

b). The correct occlusal relationship occurs most frequently in the heterozygote, A_1A_2.

c). In consequence, the genetic system represents a balance (*balanced polymorphism*), because the less favored homozygotes are a necessary by-product of the favored heterozygote, i.e., A_1A_2 individuals in crossbreeding produce 50% homozygotes.

Other alternatives also save the phenomena. For example, is the malocclusion genetically controlled? In experiments on mice, the greater part of tooth variation arises from variation in the environment of development. If identical genotypes produced bears with normal occlusion or malocclusion according to variation in environment, direct selection of genotypes would be excluded.

Van Valen (1963) has argued a somewhat similar example for fossil horses and has determined selective values from the differences in milk teeth between individuals that died in the first year and those that died as yearlings. The yearlings have, on the average, wider lower milk teeth and narrower upper ones; the amount of variation is also significantly less. Here again, the problem is to relate phenetic and genetic variation. It is further complicated because one can determine tooth form only in those horses that died—not in those that survived. Van Valen suggests that individuals with narrow lower and wide upper milk teeth had a higher death rate over the first year and are "over-represented" in that mortality group. Only individuals with wide lower and narrow upper teeth remain alive to be included in the second mortality census. Alternatively, however, the first census might represent a random sample of tooth form, i.e., tooth form did not affect mortality. In the second year, higher mortality rates for individuals with wide lower and narrow upper teeth would yield a mortality sample in which these characteristics predominate. Those surviving the year would be principally optimum phenotypes with narrow lower and wide upper teeth but, because of the loss of the milk teeth, they would not be recorded as such in later mortality.

Such difficulties plague all rigorous tests of the sufficiency* of the neo-Darwinian model from the fossil record. They may be absolute and insoluble or they may yield to an entirely new approach—see Rudwick, 1964, for an analysis of the problem.

The past recaptured

But, in spite of this pessimistic view, paleontology can make—and has made—worthwhile contributions to the study of the dynamics of evolution. Remember, it deals with a very large number of examples and very long periods of time. The individual interpretations may be incorrect or, at best, dubious, but taken in bulk they suggest questions and some conclusions. The consistencies, though only qualitative, provide patterns that must test the theories of evolutionary process.

The patterns themselves are varied. The brachiopod *Lingula* persists unchanged for several hundred million years; the evolutionary line leading to mammals has passed through six classes since the Ordovician. What combination of factors produced this extraordinary difference in rate of evolution? The Ordovician fish resemble some arthropods remarkably,

* Note that "sufficiency" and not "validity" for the selection concept is, as indicated above, p. 140, a tautology, given the nature of genetic systems and of the environment.

though the respective ancestral species must have been quite different. What chance of mutation, recombination, or selection brought about this resemblance?

Or does it matter? One would like to know out of sheer alarm-clock-tinkering curiosity. One needs to know to fulfill those geologic functions of paleontology, to date rocks and to interpret their environment of deposition. One should know in order to set the modern biologic world in historical perspective.

The tempo

Measurement of evolutionary rates. Obviously, animals have evolved at different rates. There seem nearly as many ways of measuring these rates as there are different kinds of animals. The first problem is that of a time scale. If a close estimate can be made from radioactive dates, the researcher can use an absolute scale of years (Figure 7-1B). Otherwise, he must employ relative time units. Many studies include something of both, i.e., generalized and, therefore, crude, absolute time scale *and* the comparison of contemporaneous evolutionary lines in that time span (Figure 7-2). If he uses the latter method, he may plot several lines

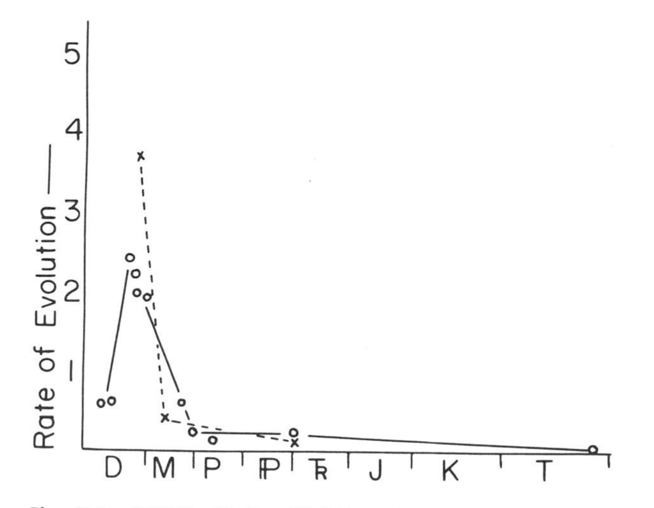

Fig. 7-2. RATES OF EVOLUTION. The rates of evolution of the lungfish (circles and solid line) and of the coelacanth fish (crosses and dashed line) plotted against time. The ordinate scale is in number of character changes per million years; the abscissa is divided into segments proportional to approximate duration of geological periods. Abbreviations: **D**, Devonian; **M**, Mississippian; **P**, Pennsylvanian; **P**, Permian; **Tr**, Triassic; **J**, Jurassic; **K**, Cretaceous; **T**, Tertiary.

against each other on an arbitrary scale (Figure 7-3A), or plot the change against the thickness of the fossil-bearing rocks.

The second, and more difficult, problem is what to plot against the time scale. Variation of a single morphologic character, e.g., the width of the third molar tooth (Figure 7-1B)? Or some group of characters? The first yields a value that may be set, in theory, in terms of specific functional

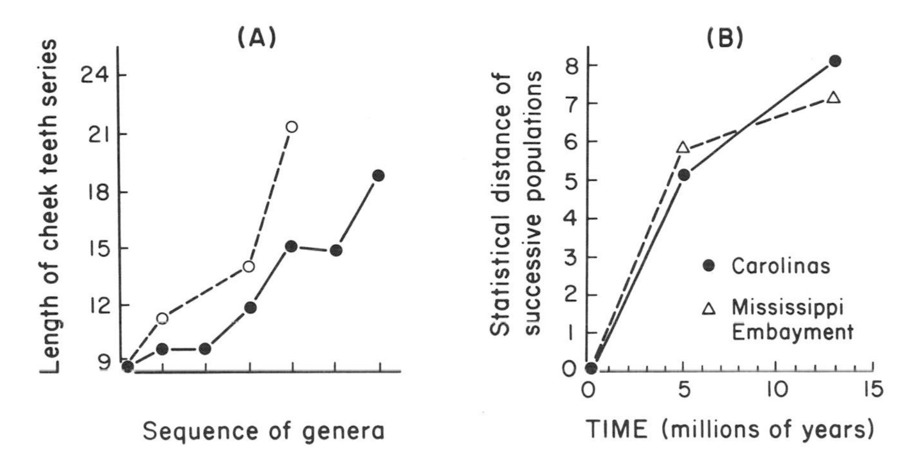

Fig. 7-3. RATES OF EVOLUTION. **(A)** Relative evolution of cheek tooth length in two horse lineages *Miohippus-Megahippus* (circles) and *Miohippus-Equus* (dots). Cheek tooth lengths are generic means; the abscissa is an arbitrary scale matching contemporaneous genera. Data from Romer. **(B)** Rate of evolution in the Cretaceous oyster *Exogyra*. The morphologic change between successive samples is represented by a summary index, *the statistical distance*. See also Fig. 8-7., p. 200. (*Data from* Lerman.)

value, e.g., better nutrition with a longer tooth, faster running with a longer leg, etc. The second can yield a similar sort of answer if restricted to the elements of a single functional system but, so restricted. does not provide an overall picture of changes in phenotype or in the niches, occupied by successive populations. One approach to overall measurement is in terms of *taxonomic rates*. Inclusion in the same taxon implies overall similarity of populations; inclusion in different taxa measures difference. Thus the number of taxa of equal rank, e.g., genera, along a phyletic line (or in a group of related lines) measures the amount of change (or diversification). This approach has limitations, since no objective criteria can assure that two taxa of equal rank are of equal morphologic scope and since the classification itself involves feedback from phylogenetic interpretation, e.g., the taxonomic rate for the lineage, *A-B-C-D,* is reduced 25% if *C* is placed in another lineage. The taxonomic data used in Figure 7-4 were weighted—subjectively—by Simpson to remove these inequalities.

More objective though still limited are measures of statistical distance between populations. If the populations are defined completely by statis-

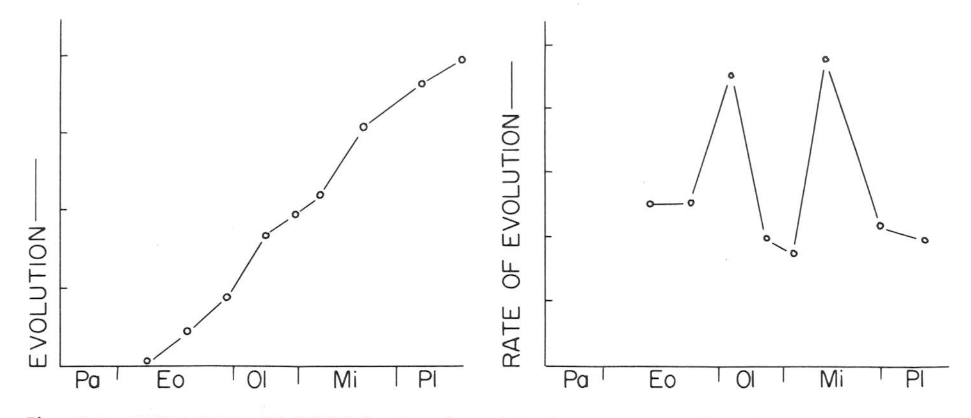

Fig. 7-4. EVOLUTION IN HORSES. Graph at left shows amount of evolution. The ordinate scale is purely relative and partly subjective—it indicates Simpson's (1952) estimate of the relative amount of difference between successive genera. The abscissa shows relative duration of the Cenozoic epochs. The graph at right indicates the rate of change per million years and is therefore a plot of the slope for each segment of the other graph. The peaks in early Oligocene and mid-Miocene are striking. Abbreviations: **Pa,** Paleocene; **Eo,** Eocene; **Ol,** Oligocene; **Mi,** Miocene; **Pl,** Pliocene.

tical distances as in numerical taxonomy, and if the lineages are defined by minimal distances, then the problems of equality of ranks and of interpretive feedback are resolved. On the other hand, the characters measured are not necessarily either a "reasonable" or a random sample of the total

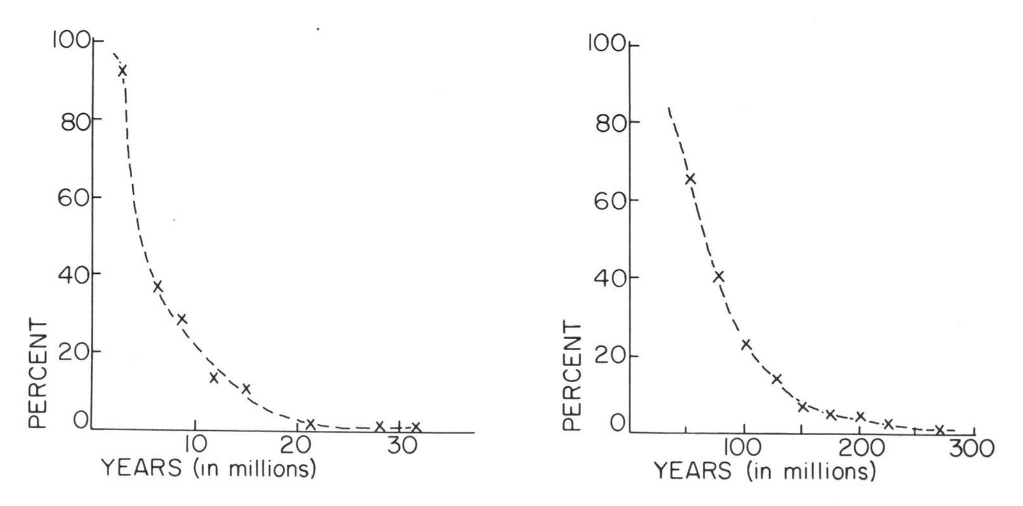

Fig. 7-5. SURVIVAL OF PELECYPOD AND MAMMALIAN CARNIVORE GENERA. Graph at left shows length of survival of carnivore genera. Less than 20 percent survived more than 10 million years. Over 20 percent of the pelecypod genera (graph on right) persisted 100 million years or more. If the genera in the two groups are of similar morphologic scope, the rate of transformation of the carnivore genera was many times as rapid. (*After* Simpson.)

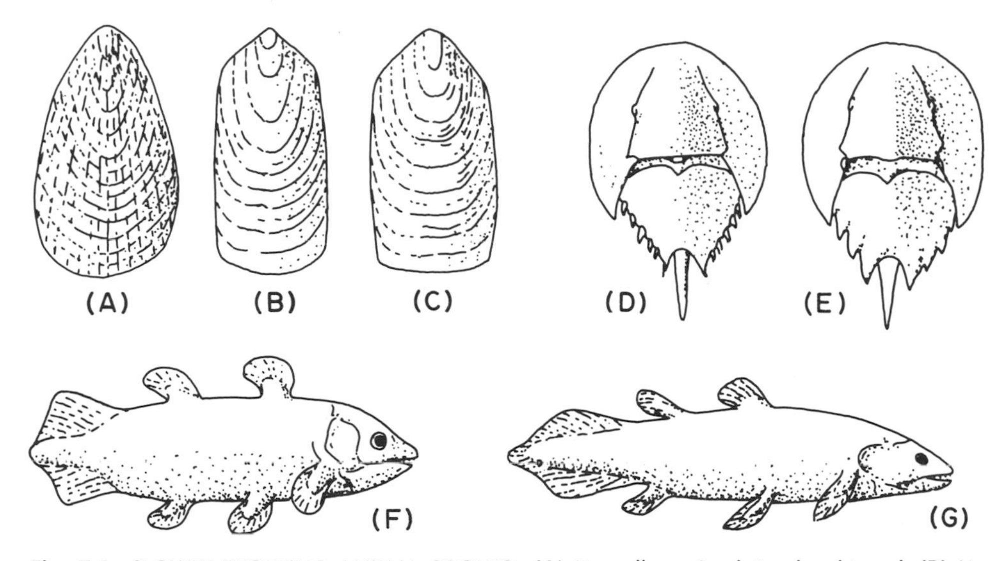

Fig. 7-6. SLOWLY EVOLVING ANIMAL GROUPS. **(A)** *Linguella,* a Cambrian brachiopod, **(B)** *Lingula münsteri,* Silurian; **(C)** *Lingula anatina,* Recent. **(D)** and **(E)** Jurassic and Recent representatives of the horseshoe crab, *Limulus.* **(F)** *Macropoma,* a Cretaceous coelacanth fish; **(G)** *Latimeria,* a Recent coelacanth fish. [**(A)** *After* Walcott; **(B)** *After* Piveteau, ed., *Traité de Paléontologie,* Copyright, Masson et Cie, Paris, used with permission; **(D)** After Störmer; **(F)** and **(G)** *after* Schaeffer.]

morphology.* In any case, the shortest statistical distance by numerical methods or the simplest path by classical ones is only a minimum phyletic path—not necessarily the true one, which may be, and probably is, longer. Figure 7-3B employs statistical distances to display differences between contemporaneous and successive populations (Lerman, 1965).

Paleontologists have worked out the precise lineages in only a few fossil groups, although the broader relationships of genera or families may be evident. Simpson has suggested an approach to determining evolutionary rates based on average "longevity" of genera (Figure 7-5). A rapidly evolving group will have genera with brief "life spans," as they appear by evolutionary transformation and disappear by further transformation or by extinction; a slowly evolving group includes genera with long life spans.

Kinds of rates. In spite of the frustrations in measuring rates of evolution precisely, paleontologists can at least distinguish between the very slow, the moderate, and the very fast. Before I discuss each kind, let me first post warnings.

Slow rates or moderate rates are averages over a long span of time. These rates may be constant through the full period, or they may fluctuate

* The same statement, of course, holds for nonnumerical methods, so this is not a comparative deficiency.

from fast to very slow. With only two or three samples in a line, interpolation between samples is questionable.

Slow and moderate rates are slow or moderate rates of morphologic change and changes in only a small part of the animals' morphology. Physiological changes or changes in soft anatomy appear only to the extent that they modify the skeleton.

Most examples of rapid change are associated with known or probable stratigraphic gaps. To the extent that the time involved in these gaps is unknown, the magnitude of the rate remains unknown. Further, the sedimentary record is so spotty that a particular line may be unknown for much of its existence and therefore may show a high apparent rate.

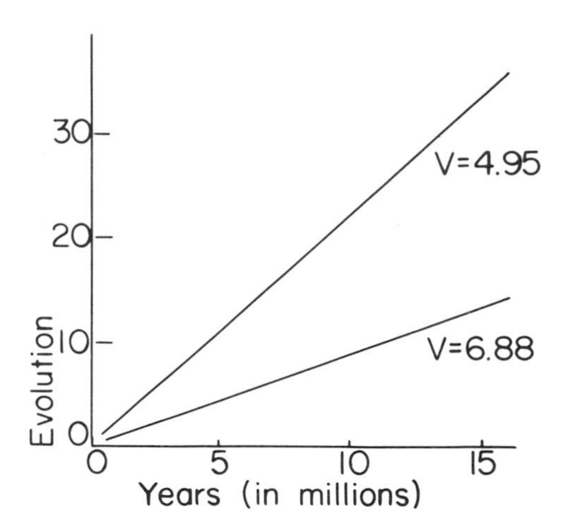

Fig. 7-7. EVOLUTION IN TWO MIOCENE OREODONT LINEAGES. The upper line represents the subfamily, *Merycochoerinae*, the lower the subfamily *Merychyinae*. The ordinate scale indicates the difference between the natural logarithms of the mean width of the third molar in the initial and in the final populations of a lineage. The slope of the lines (2.11 and 0.90, respectively) indicates rate of evolution. The V-value is the average amount of variation within species populations of a lineage. Note that the more rapidly evolving line varies less at any one time. (*Data from* Bader, 1955.)

The laggards. A brachiopod, *Lingula,* is a classic example of extremely slow evolution. Nearly every phylum and class, however, contains forms that have persisted over long periods with little change (Figure 7-6). Some paleontologists theorize that extremely slow rates of evolution result from a limitation of available variance. The population does not change because it has nothing with which to change. Very little reliable data have been collected, but Bader has shown (1955) that in a group of extinct herbivorous mammals the populations in slowly evolving lines varied more at a particular time than did the rapidly evolving lines (Figure 7-7). Guthrie (1965) found, however, greater variance in rapidly evolving lineages of mice. Sampling difficulties and the lack of linear correlation of form with genetics makes inquiry difficult.

An alternate thought would be that slowly evolving populations were large, freely interbreeding units, because such populations have great evolutionary inertia. Unfortunately, recent species of conservative lines have about the same sorts of population structure as representatives of progressive lines. Was population structure effective in some cases? If it was, the information is not now available, but then very little precise information is.

If variability was normal or even relatively great, and population structure no factor, slow rates of evolution imply very low selection pressure. You should immediately ask why a constantly changing environment fails to induce strong selection.

Some paleontologists have suggested that conservative lines had some sort of "generalized adaptation" that permitted them to survive in different and changing environments. The concept is interesting, but I find it difficult to imagine any sort of rigorous test. On a priori grounds, one might suspect that the line achieved a maximum adaptation to a niche which persisted over a long period of time. This hypothesis sounds good but is also difficult to test.

The problem may be verbal; perhaps each example has a different explanation. Certainly, the problem of conservative groups requires some radical thinking.

The moderates. If the tortoise and the hare pose refractory questions, the horse does not. The horse lineage provides a classical example of moderate evolutionary change (Figure 7-4). So far as detailed quantitative studies have been made, such lines had moderate to low variance at any one time. Selection chopped off the extremes, and new, favorable, genetic combinations appeared as the population mean shifted. Presumably, selection pressures were moderate. Though population structure may have influenced—either accelerated or decelerated—the rate of evolution, there would seem to be no reason to think it a controlling factor.

The impetuous. I approach the subject of rapid evolution rather gingerly. Many of the major animal phyla, classes, and orders pop abruptly into the fossil record. Their ancestry, if known at all, is vague, and they are much more advanced than the presumed ancestral types. So far as the fossil record goes, their evolution was extremely fast.

One faction, now the minority, claims some special sort of mutation, or extremely rapid mutation rates account for these saltations. Some of the members of this school hold what seems an extreme view about the nature and origin of taxonomic categories. I have no wish to bog down here in philosophical or semantic problems, but briefly they hold that the major categories (phyla, classes, and so on) are separated by unbridgeable gaps. Therefore, they arose in a sudden "macromutation" in a single individual, a "hopeful monster."

Such mutations have not yet been conclusively identified in recent populations. If they do occur, most would not be viable without commensurate buffering changes in the gene systems. If they do occur in viable form, they must be selected in the normal fashion from the gene combinations in an interbreeding species population. Even if selection pressures are very high, the macromutation will change in frequency at a finite rate. A "mutant" coral with sixfold radial symmetry would still belong to a species with fourfold symmetry. Is a man with six fingers on each hand a "macromutant?"

The other party to this controversy holds that major evolutionary steps occurred through high selection pressures on a population with favorable reproductive structure. Since in some cases the initial steps (usually hypothetical through lack of fossils) seem to be inadaptive, Simpson suggested they were the result of drift in very small populations, a drift that moved against selection until a new adaptive significance was established. This latter idea is interesting, but no one has yet demonstrated positively that it happened or even that it is likely to happen. An increasing amount of evidence indicates that the Class Mammalia, at least, originated by small, adaptive steps (see p. 476), and rare fossils (*Archaeopteryx,* for example) nearly bridge the gaps for several other classes.

As this is written, the controversy has quieted—principally from lack of data to substantiate various theories. A decisive series of observations must be sought, but there seems no reason to accept a theory of special origin or to dispute over definitions of macromutation or taxonomic category. Studies of evolutionary rates mean little unless they can be shown to be free of sampling difficulties, can be referred to the biology of the animals, and can be placed in an environmental context. Paleontologists have made few such complete studies.

The trends

Adaptation. I have discussed the tempo of the evolutionary drama; now what of the story line itself, the trends in evolution? Many of the well known examples of evolution, the reduction of toes in the horse lineage for one, have obvious adaptive significance. They improve the fit of the population to a particular environment. Other sequences lack obvious adaptive significance and have been interpreted as a) nonadaptive trends that neither help nor injure their bearers and b) inadaptive trends that are injurious.

Some have offered the initial evolution of rhinoceros horns as an example of nonadaptive change. They argue that the horns had no selective value until they were of considerable length. Therefore, they could not evolve by selection of short-horned variants in an otherwise hornless population. Since short horns were of no value, their appearance was nonadaptive. Only after they became long did adaptive evolution begin.

Simpson (1952) demolished a sufficient number of such interpretations to cast doubt on all. Some are simply the result of arbitrary selection of morphologic characters. The configuration of bones in the skull roof of early lungfish may have been a developmental adaptation, but whether a particular bone is large or small had little to do with protection of the skull. The character has in itself no adaptive significance (Westoll, 1949).

Other "nonadaptive trends" reflect *preadaptations.* Some scallops swim by clapping the two valves of the shell. The form of shell and musculature are adaptive to this life, but how could a primitive burrowing clam with different form and musculature develop this adaptation? The answer would seem to be that the scallop ancestors were attached to the bottom by a threadlike structure. The musculature and shell form adapted to this mode of life have proved preadaptive to swimming.

Finally, many examples of nonadaptive evolution are based on a lack of imagination. The person making the study cannot conceive of any adaptive significance. The absence of an explanation for short horns in rhinoceros evolution does not prove the absence of an adaptation. Such conclusions will not support any new and drastic theories of evolution.

The fossil record demonstrates some apparent inadaptive (deleterious) trends (Figure 7-8). For example, a pelecypod lineage developed a

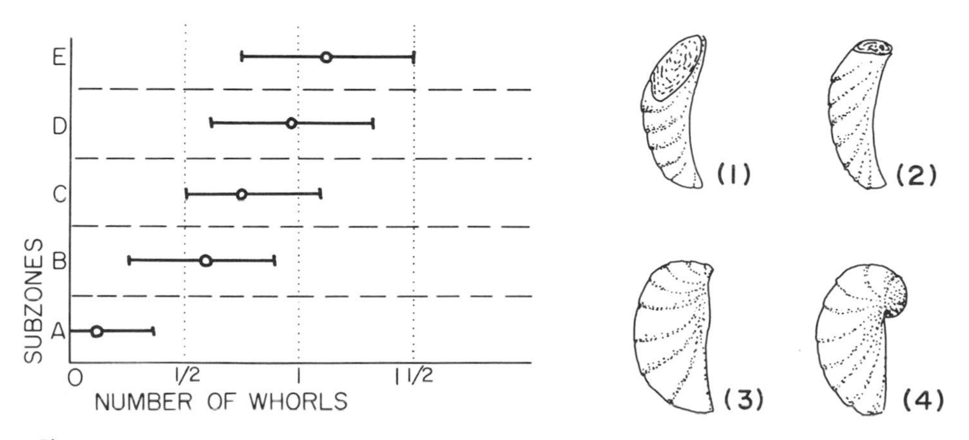

Fig. 7-8. INADAPTIVE EVOLUTION. Evolution of coiling in the Cretaceous pelecypod *Gryphaea*. The shift of the population in time is shown at left; length of the heavy lines indicates range of variation in sample; circle indicates average. Drawings at right show change from non-coiled **(1)** to strongly coiled **(4)**. A few of the strongly coiled individuals in the later populations may not have been able to open their shells at the end of their life. (After Trueman.)

coiled shell. Late populations in this line had the shell so coiled that old individuals *may not* have been able to open their shell and perforce died of starvation. The trend is deleterious, but, as Simpson pointed out, only to old individuals. These individuals lived to excessive age because they were successful. Other interpretations of inadaptive changes like the great size of horns and antlers in some mammal lines are shaky because no one can demonstrate that they were actually disadvantageous.

Until otherwise shown, all evolutionary changes should be considered adaptive or correlated with adaptive changes. If you doubt this abrupt conclusion, then prove it wrong.

Orthogenesis. One mode of evolution that has been a source of much controversy is the apparent linearity of some phyletic lines, or, more particularly, of the evolution of some characteristics. The horse line, *Hyracotherium-Equus,* has been taken as an example of this sort of evolution, and many cases of "nonadaptive" evolution are cited also. I find linear evolution difficult to define, but, in use, it refers to phyletic lines that change in a regular manner from the ancestor, e.g., a small, three-toed horse with simple teeth, to the descendent, a large, single-toed horse with complex teeth, and change without temporary reversal or divergence. Such linearity, particularly in nonadaptive and inadaptive evolution, has been taken to demonstrate a special evolutionary mechanism. Some hypotheses involve a metaphysic of purpose, either in the organism or in the organic world as a whole. Such explanations may rest on valid metaphysical grounds, but they have no scientific value. Once the operation of the evolutionary mechanism is defined, purpose becomes a superfluous hypothesis.

Other theories suppose some inherent tendency of the genetic material to mutate in one direction. Disregarding the difficulty of associating a "linear" evolution of morphology with linearity of mutation, such theories could have a valid basis in the absence of selection. "In the absence of selection" is the key; nonadaptive modifications could arise in this way, and adaptive modifications be assisted in evolution, but mutation rates sufficient to push the population against selection are not known to occur. The inadaptive and nonadaptive trends placed in evidence by

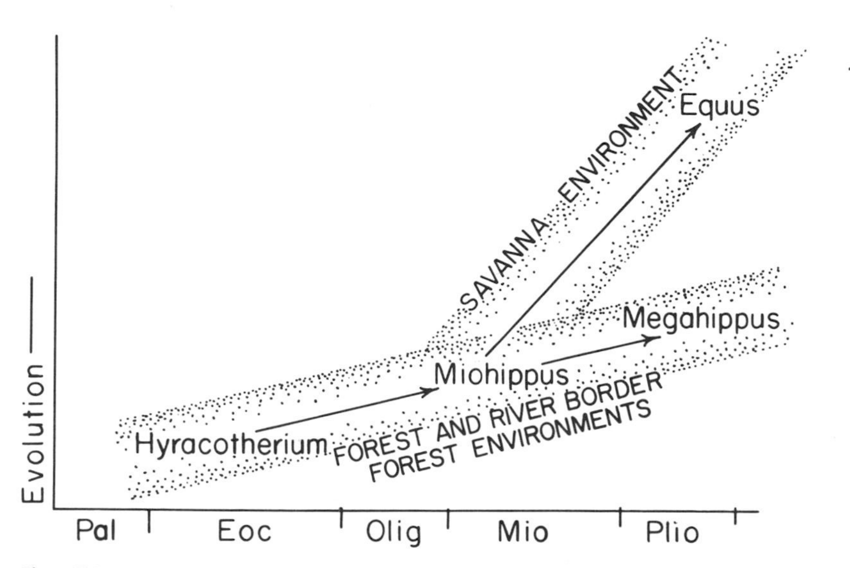

Fig. 7-9. LINEAR AND DIVERGENT EVOLUTION. The *Hyracotherium-Equus* lineage has been considered by some to be an example of straight line evolution. More careful study reveals that *Equus* is the terminus of a divergent stock that evolved in a grassland environment. Abbreviations: **Pal,** Paleocene; **Eoc,** Eocene; **Olig,** Oligocene; **Mio,** Miocene; **Plio,** Pliocene.

some orthogeneticists were kicked around thoroughly in the preceding section. In their weakened condition, they are not of much use in validating any theory.

Any serious discussion of linear evolution must include consideration of what is meant by linearity, for a basic principle used in determining phylogeny is an assumption of linear evolution. Further, I find it easier to discover linearity of trend in the absence of knowledge of the causal factors for a particular sequence; the *Hyracotherium-Equus* line looks straight (Figure 7-9) until one recognizes that the *Hyracotherium-Miohippus* portion of the line evolved in response to the limitations of a forest environment and the *Miohippus-Equus* portion in response to the limitations of a grasslands environment. However artificial some examples of linear evolution may be (as a consequence of straight thinking, perhaps), others seem valid and readily explicable by continued selection in one direction. The extrabiotic environment supplied this direction in some; in others, it was done by a feedback relationship such as prey-predator. Further, the variation available to the population and inherent structural limitations restrict the possible directions of change. A leg can become shorter or longer, but it can't sprout wheels.

Specialization. Linear evolution modifies the phenotype to perform a special set of functions. This process, *specialization,* limits the ecology of the species. The cause is simple; animals which have only to run can be much better adapted to running than those that must both run and climb. The ultimate animal would do only a very few things but do those exceedingly well.

Specialization limits diversity within a population, and thus limits possible directions of evolution. It is further restricted by the mechanical limits of a particular structural plan; bone or shell can only be so strong or a muscle so effective. Since the structural plan results from prior and current adaptations, it, too, is limited by specialization. In this sense, evolution is self-limiting.

Preadaptation. From this postulate some people have argued that evolution is a downhill process and will cease as the potential variants in the different species populations are either used or eliminated by continued adaptive modification. A different view of evolution is possible, however, for each modification of form creates potential new adaptations. In this sense, a population or variants in it are at all times *preadapted* to environments differing from the one to which the population is currently adapted. Birds, for example, probably evolved as climbing reptiles. In response to the requirement for activity and coordination of activity, they developed a highly organized central nervous system, excellent eyes, and constant-temperature mechanisms (four-chambered heart and feathers). These specializations were for climbing and thus limited the potentialities of the old reptilian morphologic plan, but they are also preadaptations for flight and thus transcend the possibilities of that same plan. Paleontologists study the preadaptive possibilities of a specializing line, as well as the course of the adaptation itself. Evolution both limits possibilities and creates new potentialities.

Adaptive radiation. Those happy few that strike upon new possibilities renew the vigor of evolution. Rather than dwindle to the spindly shoot of specialization, they root and branch (Figure 7-10).

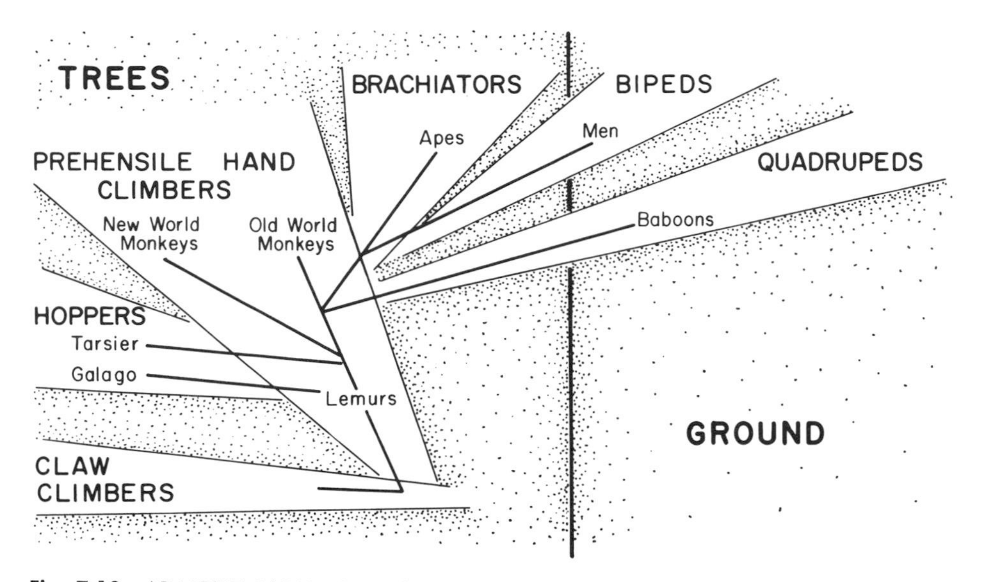

Fig. 7-10. ADAPTIVE RADIATION. The radiation of the major groups of primates as shown in their mode of locomotion.

Conceive a continent populated by species specialized variously for running, climbing, or swimming; specialized also for eating leaves, fish, or each other. Conceive also that a few species are present that are generalized with respect to these adaptations, though specialized in other ways; one of these specializations is a constant body temperature. Then, for some reason, this continent is denuded of most of the specialized species. What happens? Theorizing is unnecessary, for this happened in the early Cenozoic. The remaining generalized species can do any of these things and can evolve specializations in any one of these directions if given a chance. "Given a chance" is rather personalizing the matter, though. Actually the process must involve: first, isolation and divergence into separate though similar species in response to local environment differences, e.g., for tree-climbing on a forested island and running on a desert plain; second, breaking down of the isolating barrier, with ensuing contact and ecologic interaction between the new species—the climbers may be better adapted for eating the runners than they are for eating cactus, and the runners better adapted for eating leaves than for capturing climbers, and third, continuous selection of the phenotypes best fitted for running and eating leaves or for climbing and eating runners. The ultimate results, of course, are horses and mountain lions.

This process of diversification is called *radial evolution* or *adaptive radiation*. Since the generalized species have specializations (like hair, a four-chambered heart, and mammary glands) not present in contemporaneous species, the subsequent evolution of the generalized to the specialized produces new forms unlike their adaptive antecedents. In this way, the hippopotamus is quite different from a brontosaur, although both have (or had) semiaquatic herbivorous adaptations.

Paleontologists sometimes desire to measure the rate of adaptive radiation. Commonly, they do this by determining the rate of occurrence of new genera (or other taxonomic categories) per unit of relative or absolute time. Unfortunately, this measurement is not completely satisfactory, because: 1) the scope of a genus varies in different taxonomic groups; 2) it measures geographic diversity as well as ecologic; and 3) it

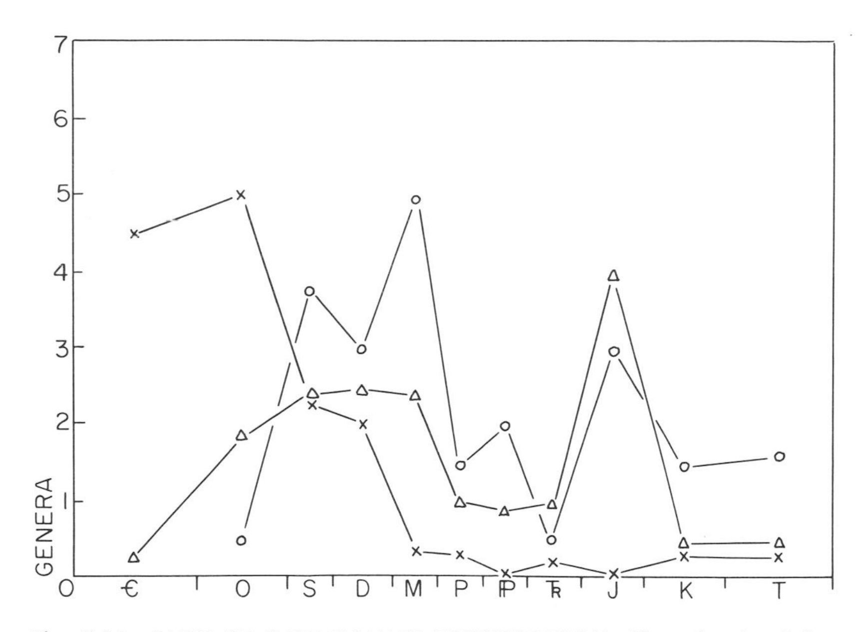

Fig. 7-11. RATES OF EVOLUTIONARY DIVERSIFICATION. The value (rate) for each period is calculated by dividing the number of genera appearing for the first time in the period by the length of the period in millions of years. Triangles represent articulate brachiopods, crosses show inarticulate brachiopods, and circles indicate corals. Note alternate periods of rapid and slow diversification: the initial bursts, the late Paleozoic decline, and the high rates again in the Jurassic. Some of the differences may result from the relative amount of work done on different groups at different ages; the Triassic decline may represent our slight knowledge of Triassic marine environments. Abbreviations: **€**, Cambrian; **O**, Ordovician; **S**, Silurian; for other abbreviations see Fig. 7-2. (*Based on* Newell, 1952.)

is dependent on the completeness of the geologic record at any particular time.

When these rates of taxonomic diversification are plotted for various groups through time (Figure 7-11), two phenomena strike the eye: rates vary within groups at different times, and some groups have higher rates than others. If these differences are real, they deserve careful interpretation—but are they real? High taxonomic rates may represent the special interest of some paleontologists in the group and, therefore, reflect the number of papers published, or they may represent a group locally diversified in a thoroughly studied rock sequence. Because of the theoretical importance of determining these rates in their relation to each other and to major geologic events, an unambiguous method of measuring diversification deserves more general interest.

Parallelism and convergence. After a group begins its radiation, several of the new species may adopt or maintain a similar mode of life, i.e., fall into the same adaptive channels. Some of these species are eliminated by their contemporaries who have evolved a better way of doing things; isolation protects others, and still others survive together by adopting variant niches of the same general adaptation. Similar environmental factors select similar individuals in the different populations. The lines

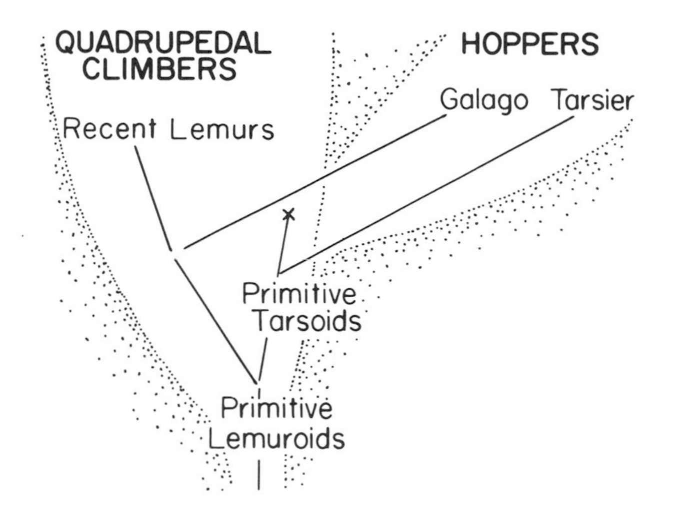

Fig. 7-12. PARALLEL EVOLUTION. The recent galagos and tarsiers evolved their arboreal, hopping adaptations independently but in parallel from a primitive lemuroid ancestry.

evolve in *parallel* (Figure 7-12); thus, the mammoth and mastodon lineages in the elephant stock paralleled each other in development of great size, tusks, and other features. In an extreme form, the concept of parallelism implies selection of the same genes or gene systems in two species populations which had diverged from their common ancestry in only a few genes. Since it is impossible to measure genetic similarity in fossil populations, the term parallelism is best used for the similarities in adaptive trends in related lines.

If species in divergent lines later adopt a similar mode of life, they, too, develop similar features, not in parallel but by *convergence* (Figure 7-13). In extreme form, this implies morphologic similarity due to selec-

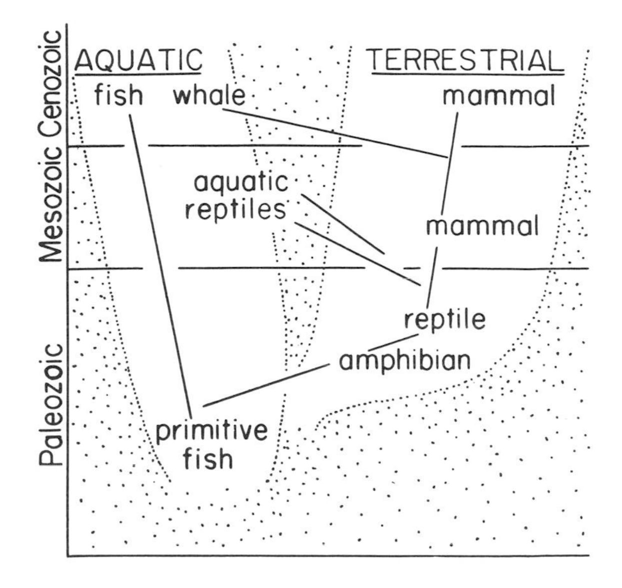

Fig. 7-13. CONVERGENT EVOLUTION. The amphibian-reptile-mammal line adapted to the terrestrial environment and diverged from the fish groups that remained in the aquatic environment. Some of the terrestrial vertebrates later readopted an aquatic habit, readapted to that environment, and converged on the fish in evolution.

tion of different genes or gene systems in species populations which have no genes in common. Thus, the similarity in general form of animals of the sessile benthos, corals, bryozoa, rudistid pelecypods, and so on, is a consequence of selection of dissimilar genetic backgrounds. The distinction between these two categories, parallelism and convergence, is, however, not so distinct as implied by the extremes, for many cases of morphologic similarity probably involve both convergence and parallelism. Among elephants the two lines initially diverged in type of tusk development as the mastodons evolved tusks in both lower and upper jaws. Later they reduced the lower tusks, and the line converged on the mammoths.

Although the numerous examples of parallelism and/or convergence are of evolutionary interest, they are commonly a nuisance in geologic paleontology. For example, the dating of rocks may depend on the interpretation of phylogeny; phylogenies are erected on evaluation of similarity and difference in form; similarity of form can result from parallelism or convergence as well as from common ancestry; therefore . . . paleontologists get confused. The weighing of characteristics in phylogenetic models (p. 148) attempts to overcome this problem by considering the function of various structures. In this scheme, similarity in structures of similar function is given relatively little weight in determining relationship. Note, however, that detailed similarity even in similarly adapted structures implies relationship. In this way the "coral-like" rudistid pelecypods resemble each other more in structures adapted for sessile bottom life than they resemble similarly adapted corals.

Evolution of ecological systems

The interrelated radial evolution of one or more groups of organisms yields a number of contemporaneous ecological systems. Because the species populations composing any one of these societies are mutually interrelated, the society itself can be said to evolve as the populations and their ecologic relationships change in character.* Just as treatment of a natural society as an ecologic unit is a higher synthesis than study of a single species in that society, so the treatment of evolving ecological systems is a higher synthesis than the study of evolution of "isolated" populations. The initial act of synthesis is more complex than the interpretation of single-population evolution, but, because it involves simultaneous approaches to the same fundamental measures of a biological system, it permits a closer estimate of characteristics of that system.

Analysis of ecosystem evolution requires the determination of association between species and environmental factors in contemporaneous systems, and of association of evolutionary trends between species and with environmental changes. Association may be measured statistically through correlation coefficients or similar techniques, and the nature and extent of ecological interaction estimated from contemporaneous association, from association of evolutionary trends, from association of species with environment, and from interpretation of ecology based on adaptations of form. Figure 7-14 is a partial analysis of the evolution of trophic structure within early terrestrial vertebrate faunas (see also p. 119, Figure 5-5). In studying analyses of terrestrial vertebrate faunas one has the impression of close linkage of evolving species by ecologic interaction

* Whether they also evolve in some holistic sense, i.e., in ways not determined solely by evolution of component populations, is a difficult and challenging problem.

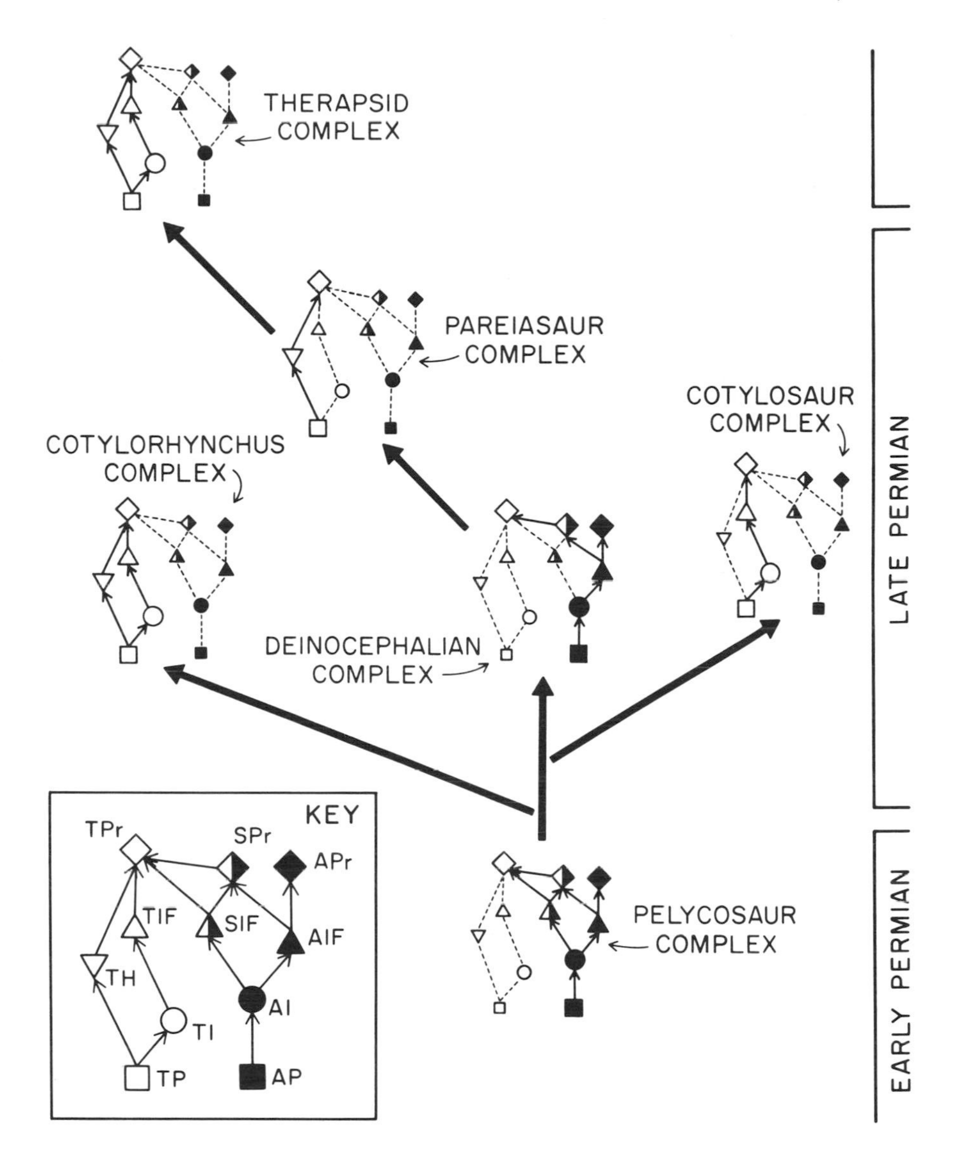

Fig. 7-14. EVOLUTION OF FOOD NET, LATE PERMIAN-EARLY TRIASSIC. TERRESTRIAL VERTEBRATE FAUNAS. Small circles and triangles and dotted lines indicate elements absent or of little importance. Large symbols, and solid lines indicate most important food links. The early Permian *Pelycosaur Complex* is based principally on aquatic plants **(AP)**; from this system arose: 1) a *Cotylorhynchus Complex* based on terrestrial plants **(TP)** through terrestrial vertebrate **(TH)** and invertebrate **(TI)** herbivores; 2) a *Deinocephalian Complex* that retained the primitive "pelycosaur" structure, and 3) a *Cotylosaur Complex* based on terrestrial plants through terrestrial invertebrates. The Deinocephalian complex evolved into the *Pareiasaur Complex* based on terrestrial plants through vertebrate herbivores and from this arose the *Therapsid Complex* with both vertebrate and invertebrate herbivore components. Abbreviations: **T,** terrestrial; **S,** semiaquatic; **A,** aquatic; **P,** plants; **I,** invertebrates; **H,** vertebrate herbivore; **IF,** vertebrate predator on invertebrates; **Pr,** vertebrate predator on vertebrates. (*Based on* Olson.)

and of a persistent tendency of rather complex natural societies to behave as units. The preliminary conclusions about the evolution of natural societies are well summarized in Olson's pioneer effort (1952), but it seems likely that there are other types of evolving systems. In particular, marine invertebrate communities must be approached by the same methods to determine their degree of integration in evolution.

Doomsday

A final aspect of evolution is the extinction of species populations. Treated simply as individual cases, extinctions are not particularly exciting (except to the group becoming extinct) and can be explained simply as continued negative rates of population change. The pattern of extinction as seen in the geologic record is considerably more mysterious and has generated sufficient controversy to excite the most blasé paleontologist.

Inadaptive evolution—the excessive antler size in the extinct Irish elk; the coiling of the shell in the fossil pelecypod *Gryphaea,* which prevented opening of the valves (Figure 7-8)—is sometimes cited as a cause of extinction. The significance of the former example depends upon the word *excessive,* which has no objective value and is thus unworthy of scientific consideration. The latter example is objective, since the coiling must lead to the animal's death, but, as Simpson points out, only very old individuals were sufficiently coiled to prevent opening of the valves. Thus, the death of such individuals was a consequence of individual senility and simply one factor in the death rate of the population. This and similar occurrences probably indicate relative success of the species, since many individuals survive through youth and maturity to die of "old age" (see Kurtén, 1954).

Theories of "racial senescence" are often linked with inadaptive linear evolution. These theories, in one form or another, carry the implication that species populations "grow old"; that the old age is likely to be associated with "excessive" ornamentation and other nonadaptive features, and that it ultimately ends in racial death (extinction). As metaphoric or metaphysic concepts, these theories are not without interest. But species populations are not metaphors, and using metaphoric terms does not make them so. The real question is something like this: Adaptive radiations, taken as a whole, tend to run out in very specialized lines. Are these lines so closely adapted to particular environmental limits that they cannot survive environmental change? There are still some undefined terms in the question (very specialized, closely adapted, and so on), but this is a little closer to an operative question.

If, as seems possible, adaptive radiations result from complex evolution in the ecologic system, the species produced by the radiation are in rather delicate ecologic adjustment to one another. If someone upsets the applecart by becoming extinct—due, say, to climatic change—the whole system is likely to be unfavorably affected. What paleontologists need is precise analysis of faunal assemblages before and during a wave of extinction—not metaphors.

Cataclysms

Some "waves of extinction" and "explosive evolution" of new groups seem to be real. Some abundant and diversified groups disappeared abruptly, and, with equal abruptness, their heirs appeared to dominate the new communities. The plots of rates of generic origin on p. 179 reflect this, even if one discounts possible sampling errors. Some workers theorize that the restriction or expansion of the epicontinental seas accelerated or damped rates of evolution and extinction. But do extensive seas result in a greater variety of habitats and thus of possibilities for diversification, or do restricted seas increase opportunities for isolation and consequent divergence? Even for those animals capable of migration, one may doubt whether the advance or retreat of the sea was sufficiently rapid to create crowding and increase competition. One can hardly envisage a clam and a brachiopod elbowing each other aside as they hurry along the retreating strand.

These bursts of extinctions and radial evolution are among the major markers in the geologic time scale. Therefore, it is of extreme importance that they be understood. Theories are plentiful; the hard facts of large collections, careful field studies of the geology, and informed synthesis of these collections and studies are excessively rare. The term *abrupt* is relative; many dinosaur lineages became extinct at various times before the close of the Cretaceous, and those species whose disappearance marks the end of that period are few in number and almost surely did not become extinct at one time or in all places at the same time.

REFERENCES

Bader, R. S. 1955. "Variability and Evolutionary Rate in the Oreodonts," *Evolution*, vol. 9, pp. 119-140.

Bock, W. J. 1959. "Preadaptation and Multiple Evolutionary Pathways," *Evolution*, vol. 13, pp. 194-211.

Guthrie, R. D. 1965. "Variability in Characters Undergoing Rapid Evolution, an Analysis of *Microtus* Molars," *Evolution*, vol. 19, pp. 214-233.

Johnson, R. G. 1964. See refs. p. 131.

Kurtén, B. 1954. "Population Dynamics and Evolution," *Evolution,* vol. 8, pp. 75-81.

———. 1964. "Population Structure in Paleoecology," in Imbrie, J. and N. D. Newell (Eds.) *Approaches to Paleoecology.* New York: John Wiley. Pp. 91-106.

Lerman, A. 1965. "On Rates of Evolution of Unit Characters and Character Complexes," *Evolution,* vol. 19, pp. 16-25.

Martin, P. S. 1966. "Africa and Pleistocene Overkill," *Nature,* vol. 212, pp. 339-342. Pleistocene extinctions by human predation argued in an ecologic context.

Newell, N. D. 1962. "Crises in the History of Life," *Sci. American,* vol. 208 (Feb., 1963), pp. 77-92. Evolutionary "explosions" and "catastrophes."

Olson, E. C. 1952. See refs. p. 161.

Pitcher, M. 1964. "Evolution of Chazyan (Ordovicial) Reefs of Eastern United States and Canada," *Bull. Canadian Petr. Geol.,* vol. 12, pp. 632-691.

Rudwick, M. J. S. 1964. See refs. p. 12.

Shotwell, J. A. 1964. "Community Succession in Mammals of the Late Tertiary," in Imbrie, J. and N. D. Newell, *Approaches to Paleoecology.* New York: John Wiley. Pp. 135-150.

Simpson, G. G. 1952. See refs. p. 161.

Van Valen, L. 1965. "Selection in Natural Populations. III. Measurement and Estimation," *Evolution,* vol. 19, pp. 514-528.

Westoll, T. S. 1949. "On the Evolution of the Dipnoi," in Jepsen, G. L., *et al.* (Eds.) *Genetics, Paleontology, and Evolution.* Princeton, N.J.: Princeton University Press. Pp. 121-184.

8

Fossils and stratigraphy: Adventures in space and time

The Cretaceous formations of the Atlantic and Gulf Coastal plains of eastern North America yield an abundance and variety of fossils from many localities. Let's conceive an imaginary field trip through this collector's paradise, from New York City southwestward to the Mississippi Valley (Figure 8-1). We stop first at Poricy Brook in northern New Jersey. An hour's collecting in the marly claystones yields brachiopods, belemnite cephalopods, small oysters very like those that now live off this same coast, and, most striking and abundant, a large oyster-like form with a strongly coiled shell. From here we drive steadily 600 miles southward to reach our next stop—along the Cape Fear River in North Carolina. There, in the chalky bluffs of the river, we encounter similar coiled oysters. Another 600-mile drive, now to the west, brings us to road-cut outcrops in northern Mississippi and an additional collection of these oysters. Subsequent comparison demonstrates significant but minor differences in the oysters from each locality, but the overlap among the collections is even more striking.

We recognize this overlap by assigning all three samples to a single

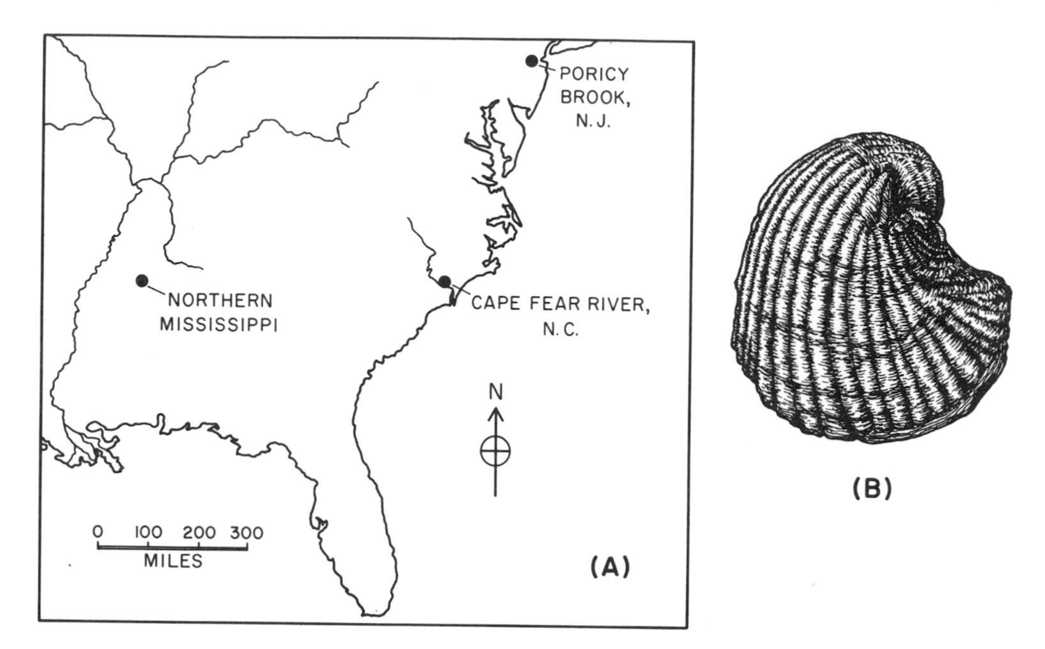

Fig. 8-1. OCCURRENCES OF *EXOGYRA COSTATA*.

taxon, the paleo-species *Exogyra costata,* described many years ago by Say. Additional collecting reveals a continuous distribution of *E. costata* from New Jersey to Mexico and northwestward as far as Colorado. The differences among collections are comparable to differences among samples from a modern interbreeding population, i.e., a biospecies. Therefore, we infer that they sample a continuum of such populations through a relatively short time interval. Collecting on the Pacific coast and the Arctic Coastal Plain fails, however, to produce samples assignable to *E. costata.* Finally, when we return to New Jersey (or any other *E. costata* localities) and make additional collections from over- and underlying beds, we find that *E. costata* occurs in a relatively small interval in the total sequence of sedimentary rocks.

The geographic distribution of *E. costata* and its occurrence through a rock sequence are strictly *stratigraphic* problems—though they obviously have ecologic and evolutionary implications. Broadly, stratigraphers are concerned with the distribution of rocks and fossils in space and time, sort of a temporal geometry. In general, they determine only one variable, either time or space, by study of one taxon, because to infer time they must assume a spatial distribution, and to infer geography they must assume contemporaneity. If the occurrence of *E. costata* is assumed to indicate *approximate* contemporaneity of all samples, the species' geographic distribution must have extended from New Jersey to Mexico at a single time. Conversely, if the geographic distribution is assumed to be widespread, then all the occurrences must be *essentially* contemporaneous. This partial circularity is the stratigrapher's dilemma; the only possible escape must be through a very thorough consideration of biologic dispersion in space and in time.

Dispersal in space

The critical datum of stratigraphic paleontology is the distribution, geographic and stratigraphic, of a fossil taxon. Similarity may be extensive—membership in the same species—or slight. The first problem is to explain similarities of form in different places. Such similarities may arise after dispersal of the ancestral population, or prior to it; these two alternatives imply quite different temporal relationships.

Dispersal and evolution

Parallel and convergent evolution produce extensive similarities between phyletic lines separated in time and/or space (p. 180). Parallelism also occurs among subpopulations of a single line. In the latter case, selection of the same genotypes in distant demes produces phenotypic similarity without the necessity of dispersal. Thus, in species of deer

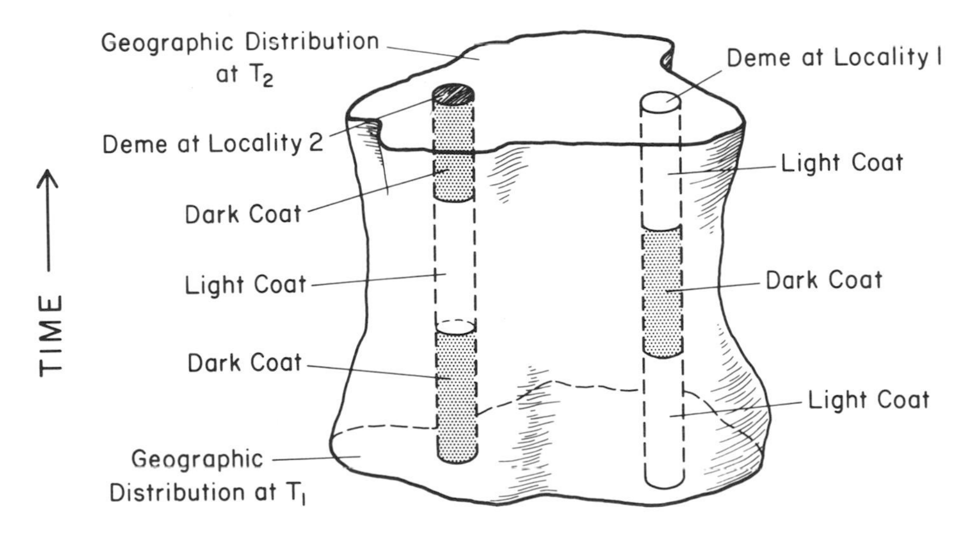

Fig. 8-2. GEOGRAPHIC AND TEMPORAL PARALLELISM WITHIN A GENETIC SYSTEM. Initially (T_1) within the distribution of the population, the deme at Locality 1 has predominantly dark coats and the one at 2, light coats. With subsequent environment shifts, the predominant coat colors change and then change back (T_2).

mice in southwestern North America, each isolated mountain forest has a population of dark-coated mice—selected from the rare dark alleles present in the typical light-coated desert populations. If the forests disappear with changing climate, the dark-coated populations disappear also; wherever and whenever a forest reappears, so does the dark phenotype—so long as the requisite alleles are still available (Figure 8-2). This reversible variation has little temporal or geographic significance, and the paleontologist has no sure way of separating it from similar small non-reversing shifts in a lineage.

Interspecific parallelism occurs when similar selection operates on two or more lineages with some genetic overlap. The common features might appear very early in one line and much later in another but can be no earlier than their separation into distinct stocks. Thus parallelism defines a lower limit in time (Figure 8-3). It also implies dispersal from some common area for geographically separated lineages.

Convergence (Figure 8-3), i.e., similarity in the absence of genetic

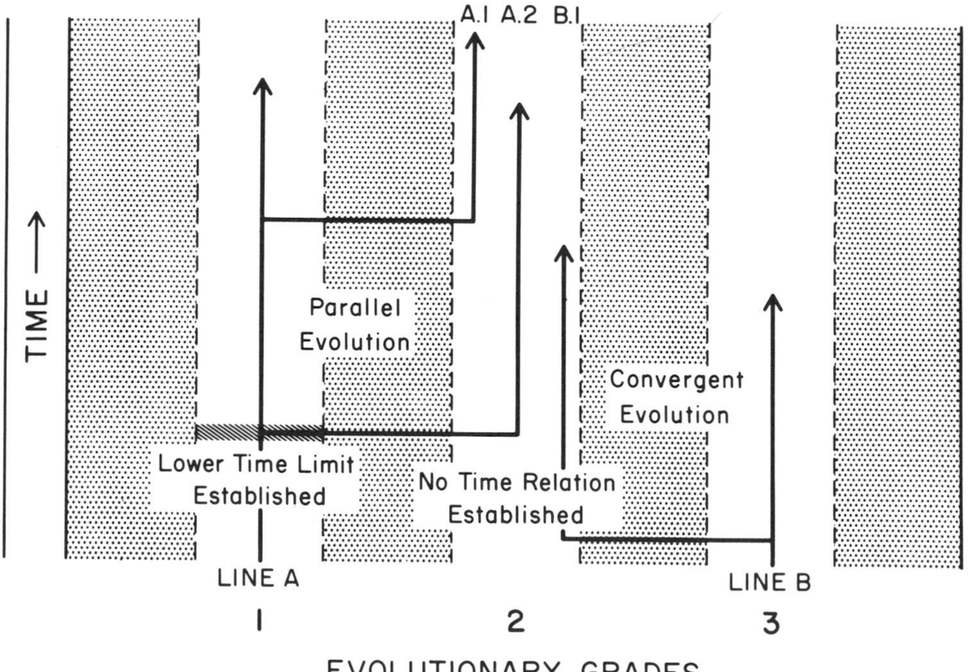

Fig. 8-3. TEMPORAL SIGNIFICANCE OF INTERSPECIFIC PARALLELISM AND CONVERGENCE. The parallelism of **A.1** and **A.2** (in structural grade **2**) can be no earlier than the separation of the two lines—but, as shown, can be considerably later. The convergence of **B.1** into Grade **2** can occur anytime—earlier, as shown, or later.

overlap, indicates neither temporal nor geographic relationships. It can arise anywhere at any time that selection assumed a similar direction—it thus possesses only environmental significance.

Evolution and dispersal

If similarity is the result of neither parallelism nor convergence, it implies dispersion of individuals through space and time; if intraspecific, it restricts possible relations in both space and time; if interspecific, it restricts relations in space and sets one limit in time, the time of origin of shared features. Since dispersal requires time as well as certain geographic and environmental connections, its interpretation contributes to temporal and paleogeographic inferences.

Consider the problem of dispersion of a particular phenotype or taxon—of a particular type of snail, for example. During a single generation, individuals of this phenotype move at varying velocities over erratic paths in various directions. What is the probability of one of the snails reaching some point A_1 several miles from A_0? High, if the phenotype is abundant, moves at high velocities (relative to distance), follows a direct path, "prefers" to move in this direction, and has a reasonable chance of surviving. Low, if any of these conditions does not hold. If, however, the phenotype endures in the population, and the net movement toward A_1 remains positive, it will ultimately appear there. An abundant type of gull might span the Atlantic several times in a single generation. Conversely, if the vector is seldom if ever selected, e.g., an oceanic path for a weak-flying land bird, or the phenotype is very rare, or the velocity is zero, e.g., plankton drift against current direction, or survival is very low, e.g., lemmings swimming the Arctic Ocean, the second site may never be occupied. Obviously, dispersion depends upon these intrinsic properties of the phenotype, as well as upon the extrinsic barriers and upon the availability of a suitable niche at the terminal point. If dispersion occurred, all of these conditions must have been satisfied. To the extent that one can define intrinsic limits of dispersal, one can infer the geography connecting the sites where the phenotype or taxon occurs. The succeeding section deals with appropriate models for several paleobiogeographic analyses.

If dispersal tells something of past geography, the failure of dispersal also yields information. Here one must distinguish between a) biological or physical barriers along the path between sites and b) environmental exclusion from the new site itself. The second case tells something of biologic or physical environmental conditions at the site, e.g., the absence of coral reefs from Iceland is presumably a consequence of low water temperature. If there is good reason for rejection of such interpretations, then (and only then) can one make inferences about environments and geography in the intervening area. Again, analysis requires construction of dispersal models—models of failure rather than of success. Dispersion, however, does not provide automatically for fossil occurrence or collection. An occasional shore bird blown across the Atlantic would be a most unlikely fossil, because of its initial rarity and because of its low chance of fossilization. In general, the probability of discovery and recognition increases with the number of individuals supplied, with their "preservability," and with their obviousness in collection and identification. Conversely, occurrence even in small numbers indicates establishment *in situ,* or quite nearby,* of a self-sustaining population. In turn, this requires, for obligatory bisexual organisms, the immigration in a single generation of a pregnant female or of at least two individuals in a single generation. In consequence, mode of reproduction and minimum population size also enter dispersal models.

Paleobiogeographic analyses

Any useful paleobiogeographic interpretations depend on imposition of temporal constraints on the model, e.g., a model for trilobite distribution in the late Cambrian of North America is hardly viable if one includes early Ordovician data. Inference of a time relationship fixes, however, the biogeographic configuration for the fossil organisms used to make the correlation. Therefore, the time constraints should be independent of the biogeographic pattern to be analyzed. For example, if graptolite taxa establish the correlation of Ordovicial strata, any paleogeographic analysis from these same taxa includes a large feedback from the correlation.

Equally, analyses of dispersal failure must include paleoecologic and collecting constraints, e.g., the nearly complete absence of advanced mammal-like reptiles from the North American Triassic becomes meaningful if and only if the sampled habitats were suitable for such reptiles. Or—the earliest Permian deposits of eastern North America have lower reptile and amphibian diversity than those of central Texas. Is this a consequence of a temperature gradient and therefore of ancient pole positions? Or is it a consequence of sampling restricted by limited outcrop to a single habitat unfavorable to reptiles and many amphibians?

Finally, paleobiogeography requires phylogenetic constraints, particularly the identification of parallelism and convergence. This particular limitation becomes quite complex where geographic distribution is employed, as phylogenetic data and/or phylogenetic inferences are used to establish temporal constraints.

* Airborne dispersal of spores and pollen and current dispersal of some planktonic organisms provide exceptions to this rule.

Paleobiogeographic models

A striking feature of the late Cenozoic mammalian faunas of both South America and Australia is the abundance and diversity of marsupials, particularly large carnivores which are unknown from the northern continents. A number of biogeographers have postulated land bridges or continental drift—across the southern Pacific or through Antarctica—to account for this distribution. Their dispersal model consists of 1) the evolution of a carnivorous marsupial lineage in either South America or Australia and 2) the mid-Cenozoic dispersion of populations from that lineage by some path other than the northern continental connections. Since the animals were *presumably* nonaquatic and *almost certainly* could not have survived a journey of several months on a natural raft—a tree or something of the sort—the path involved either a direct land connection or a series of short "hops" along an island chain through an environment tolerable to these marsupials. The geographic interpretation derives from a chain of inferences and depends upon the validity of every one. For example, could a few such animals or a pregnant female have survived trans-Pacific rafting under rather extraordinary conditions? Could sampling accidents explain the absence of such animals from the northern continents? Did the similarity evolve by parallelism *after dispersal* of primitive marsupials (which are known from the northern continents)?

The present consensus favors parallelism and quite a different biogeographic model. It postulates 1) dispersal of primitive marsupial populations throughout the northern continents in late Cretaceous—early Cenozoic times, 2) dispersal * of small parts of these populations from North to South America and from southeast Asia to Australia, and 3) parallel evolution of carnivorous marsupials in the absence of placental carnivores in both South America and Australia. This model is as logical as the first; it requires only geographic connections known from independent evidence, and it requires no a priori conclusions about ability to disperse. But it, too, depends on a phylogenetic inference that cannot be other than a probability. Note also that the only time relationship yielded by this model is "not earlier than late Cretaceous." The evolutionary similarities (parallelism or convergence) arose after dispersal.

An example of dispersal after evolution is provided by the Cenozoic of the New World. Some Cenozoic mammals of South America belong to the same species as those of North America, e.g., modern mountain lions, fossil horses and mastodons. Others belong to orders endemic (limited) to the southern continent. Since very detailed phyletic sequences are known from the North American Cenozoic and moderately good ones from the South American, since the time relationships are

* Along a "filter" bridge, i.e., an island chain or isthmus with an unfavorable environment that prevented dispersal of coexisting groups.

TABLE 8-1

Effects of migration on Cenozoic mammalian faunas of North and South America (*information taken from Simpson, 1950*)

Time of migration	*Mode of migration*	*Animals invading South America from North America*	*Animals invading North America from South America*
Late Miocene to Recent.	Along Isthmus of Panama and/or along island chains.	Deer, Camels, Peccaries, Tapirs, †Horses, †Mastodons, Cats, Weasels, Raccoons, Bears, Dogs, Mice, Squirrels, Rabbits, Shrew.	Porcupine, Armadillos, †Ground sloths, †Glyptodonts, (?) Oppossum.
Late Eocene to Oligocene.	Probably along island chains—"island hoppers."	Primitive rodents ancestral to porcupines, cavies, chinchilla, etc. Advanced lemuroid primates ancestral to New World monkeys.	(?) None
Early Paleocene.	Along land bridge connection.	Primitive herbivores ancestral to varied South American ungulates (4 orders all now extinct). Primitive edentates ancestral to sloths, armadillos, anteaters, and glyptodonts. Opossums—ancestral to marsupial carnivores (now extinct) and caenolestoids.	(?) None

† Group now extinct in area.

well established within each continent, and the paleoecologic framework is reasonably clear, the interpretation of mammal dispersion between the two continents presents a near-ideal analysis (Simpson, 1950).

Table 8-1 summarizes Cenozoic mammalian migration between the two continents. The model for late Eocene-early Oligocene includes:

a). The evolution of rodents and lemuroid primates in North America—indicated by partial phyletic sequences on that continent and their absence from South America.

b). The near-identity of many habitats in the two continents—indicated by similar adaptations of indigenous organisms.

c). The dispersal of rodents and primates to South America in late Eocene-early Oligocene—since suitable phenotypes are apparently limited to this interval.

d). The exclusion of other mammals from migration by a selective dispersal barrier—since suitable niches were available for them to inhabit. The barrier probably consisted of oceanic straits—a) since many Oligocene mammals had dispersal abilities across land equal to or greater than those of rodents and primates, b) since the latter as small arboreal forms had greater probabilities of rafting on floating trees, and c) since the phyletic unity of Oligocene South American rodents (and primates) suggests origin from a single species population—possibly a few individuals.

Items **a** and **c** possess as high a certainty as one can normally expect to attain in phylogenetic interpretation; Item **b**, however, is weaker, and therefore **d** involves considerable uncertainty.

Limitations to dispersal

The failure of North American mammals other than rodents and lemurs to reach South America in the Eocene is significant in defining paleogeography. As employed in that example, however, failure of dispersal is a relatively weak basis for analysis. It depends first on negative evidence—a lack of fossils—and it further requires the weak inference that a suitable niche existed *in the absence of the organism to fill the niche.*

Other types of nondispersing models are, however, stronger and may surpass dispersal ones, because they require no phyletic inferences.

For example, study of Jurassic ammonite cephalopod distributions (Arkell, 1956) demonstrates three distinct contemporaneous groups (faunal realms) of species (Figure 8-4). These differences derive from environmental differences and from the absence of niches, but since the realms are so widespread they indicate broad geographic variations transcending local habitats and niche structure. Similarly, Stehli (1965) has pointed out regional variations in diversity of recent planktonic foraminiferans

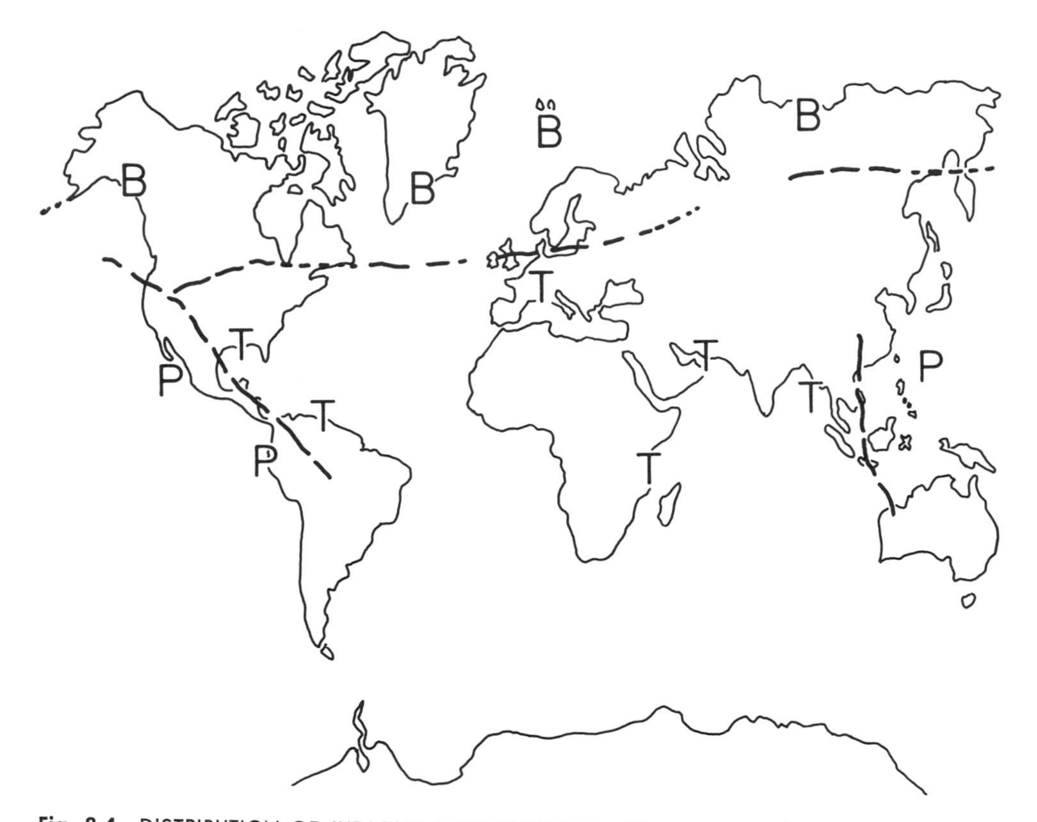

Fig. 8-4. DISTRIBUTION OF JURASSIC MARINE FAUNAL REALMS. The three realms, Boreal (**B**), Tethyan (**T**), and Pacific (**P**), are defined by distinct groups of animals. Boundaries between realms are approximate and shifted from time to time. (*Based on* Arkell, 1956.)

related to position of oceanic currents. The low diversity in a locality results from a failure of many species to immigrate and thus is a dispersion failure. The failure results from environmental differences between localities, and not barriers between them, but differences of regional scale define regional geography rather than local habitat variations. Interpretations of this sort are affected only by sampling and classification factors that can be eliminated, controlled or offset.

Time and time again

With space conquered, what of time? In the discussion that opened this chapter, we indulged in an imaginery field trip through eastern and southeastern North America. We observed at every collecting site a finite stratigraphic range for the oyster, *Exogyra costata,* i.e., only a single segment of the series of beds contained this taxon. If we had considered other fossils from these same localities, we would have observed a similar limited stratigraphic range for other taxa, e.g., for other coiled oysters, *Exogyra cancellata* and *E. ponderosa.* Further, we could have noted that *E. ponderosa* occurred only in strata below those enclosing *E. cancellata* and *costata,* that *E. costata* and *E. cancellata* overlapped but that some *E. costata* appeared higher in the stratigraphic series than any *E. cancellata* (Figure 8-5). The elemental data derive from *local sections,* i.e.,

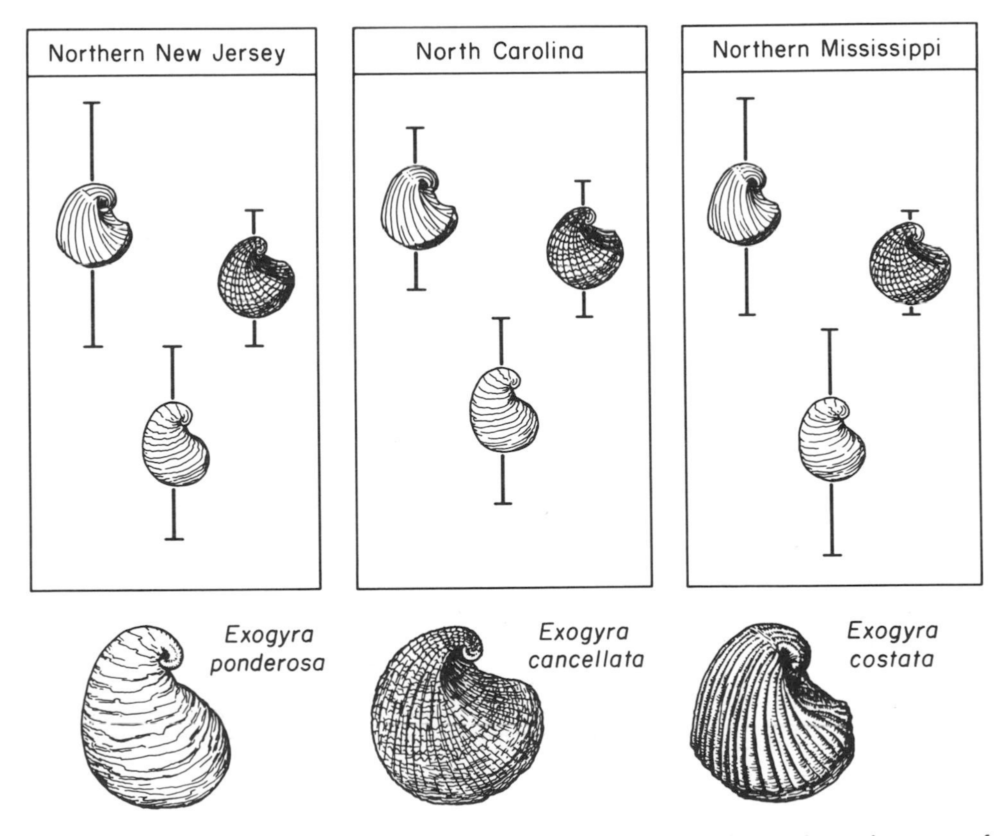

Fig. 8-5. STRATIGRAPHIC DISTRIBUTION OF *EXOGYRA.* The vertical lines indicate the range of each species in local rock sequences. *E. ponderosa* invariably occurs below *E. cancellata* and *E. costata.*

where the layers are in visible superposition. The stratigrapher is fundamentally concerned with inference of 1) physical interrelations of strata from one local section to another, i.e., rock-unit stratigraphy and 2) temporal relationships of strata from one local section to another, i.e., rock-time stratigraphy. From these, he can pass to interpretation of rock-unit genesis, paleogeography, and tectonic sequences.

Stratigraphy without time

Much rock-unit analysis is beyond the scope of this book, but it may involve models into which fossils enter as one type of environmental or geographic evidence (see pp. 83-95, 105-113, and the first part of this chapter). It may also involve use of fossils as purely physical criteria. Thus the Tichenor limestone of western New York is defined and identified in local sections by the abundance of crinoid columnals. This use extends to the definition of stratal units by fossil sequence without any necessary inference that deposition of the units was contemporaneous. For example, a geologist "sitting" a well in southern Alabama envisages completion in a permeable zone observed in outcrop in northern Alabama. At a depth of 6000 feet he recommends abandonment because he finds *Exogyra ponderosa.* Psychic stratigraphy? No, for in northern Alabama the permeable zone invariably occurs above *E. ponderosa.* His decision may be incorrect, but it is based objectively on invariable *stratigraphic succession.* Stratigraphy originated in this sort of observation, in the late eighteenth and early nineteenth centuries, and much current stratigraphy is simply a more sophisticated version. The relationships in Figure 8-6 depend solely on the succession of physical objects and one

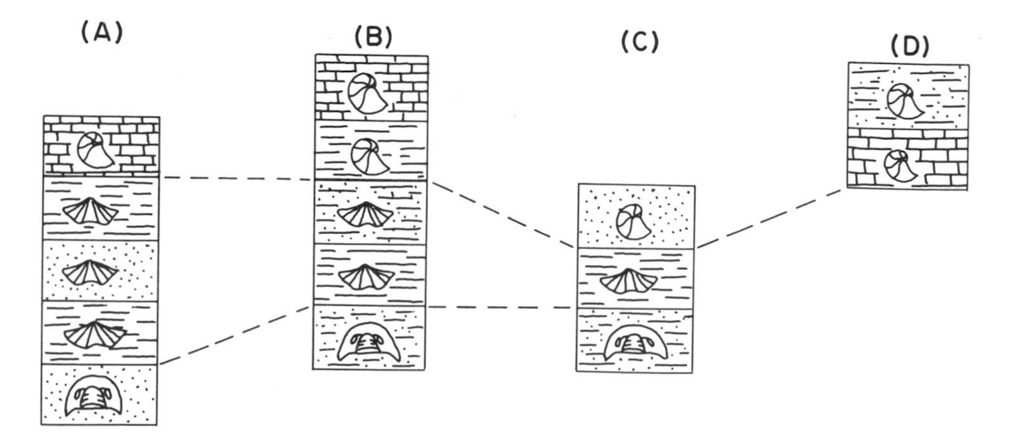

Fig. 8-6. LOCAL FOSSIL SEQUENCES. A similar sequence of fossils occurs in each of four different local sections. The occurrence of a particular type or association of fossils thus provides the basis of matching local sections.

need know nothing about these biologically (even their origin) to draw the lines.

Homotaxis vs. correlation

The geometric relation established in Figure 8-6 is *homotaxis* (same series). Homotaxy may or may not demand synchronous deposition, i.e., *correlation.** Correlation requires some sort of biological model that "explains" homotaxis and includes temporal elements.

Biologic signals and correlation

Kitts (1966) has argued that correlation is ". . . an attempt to extend the temporal ordering of events beyond the local section by means of causal chains or *signals*." Ideally such signals would *a*) be unique in time, i.e., occur only once over a very short interval, *b*) be propagated at a very high velocity, i.e., propagation time should be small relative to the intervals being ordered, and *c*) be nearly ubiquitous in space, i.e., recorded over a large part of the area in which correlation is undertaken.

What, if any, biological signals approach this ideal? Ecologic successions controlled by very widespread and rather rapid environmental changes, e.g., eustatic sea level changes during the Pleistocene, may satisfy the second and third conditions. Conclusions as to extent and rapidity derive from a model which "explains" the change, e.g., melting of continental glaciers, and defines its likely duration. Such signals, however, are not unique in time. For example, microfossils from Pleistocene deep-sea cores indicate several cold episodes, any one of which may correspond to the occurrence of a muskox in an outwash terrace in New Jersey.

Evolutionary events offer another set of choices. The transformations that occur along phyletic lines and at the splitting of lines provide the raw material of correlation. The signal consists of the transformation dividing taxa or of a taxon which had a limited existence—from initial transformation to a terminal transformation or extinction.

To what extent are such signals unique, instantaneous, and ubiquitous?

Evolutionary transformations and time

Evolutionary transformations of populations are rare in the fossil record. Since, however, they establish one or both temporal limits for a taxon, they must be considered here.

* This accords with the definition by Dunbar and Rodgers (1957) but not with that of Krumbein and Sloss (1964), who apply it to any type of equivalency—temporal or not.

The *Exogyra* sequence described above provides an example. Analysis begins with a time-parallel slice of the paleo-species *Exogyra ponderosa;* this slice is a biological species, isolated from the phyletic continuum. For simplicity this particular slice is E_0—the zero referring to the initial time slice. The population E_0 comprised demes, $E_{0.1}$, $E_{0.2}$, ... $E_{0.n}$, along the eastern and southern coasts of North America. Each deme comprised a somewhat different array of phenotypes (pp. 144-145); probably the arrays in adjacent demes were more similar than those in distant ones. Thus demes from distant localities would differ significantly. Lerman (1965), for example, reported a "statistical distance" of 0.7 between *approximately contemporaneous* samples of *E. ponderosa* from North Carolina ($E_{2.12}$) and Mississippi ($E_{2.13}$). This difference is probably not statistically significant. At a later time, another slice yields demes, $E_{5.1}$, $E_{5.2}$, $E_{5.3}$, ... $E_{5.n}$ with similar geographic variation. Purely on a chance basis, some later demes will resemble some earlier demes more than others. Thus Lerman records that $E_{5.1}$ from western Tennessee shows a statistical distance of 5.9 from $E_{2.13}$ and 6.2 from $E_{2.12}$. It follows that the similarities and differences between localities possess temporal significance if and only if geographic variation is small relative to evolutionary change. For example, $E_{5.5}$ has the same statistical distance, 1.3, from the contemporaneous $E_{5.10}$ and the later $E_{14.3}$.

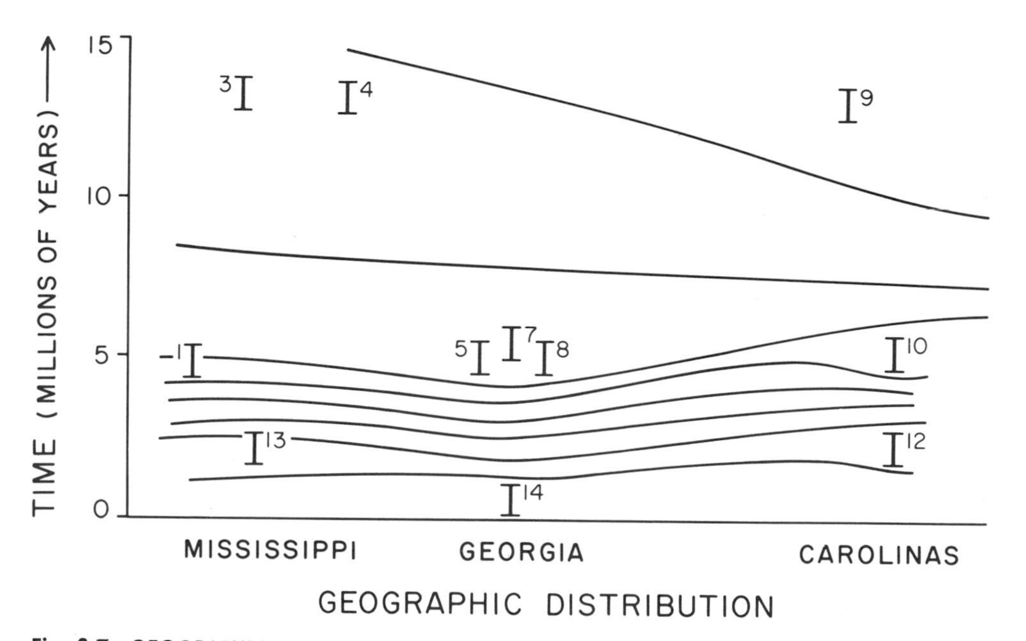

Fig. 8-7. GEOGRAPHIC VERSUS TEMPORAL VARIATION IN *EXOGYRA*. Contour lines of morphologic uniformity as measured by "statistical distance." Vertical lines accompanied by a sample number represent inferred uncertainty in correlation for sample. The number of contour lines between samples measures morphological difference between them. (*Data from* Lerman.)

Further, the interdeme variation is likely to anticipate evolutionary trends for the lineage as a whole. For example, the late Cenozoic camel lineages show increasing adaptation to arid habitats. Almost certainly, demes in subarid habitats displayed more of this adaptation than contemporaneous demes in subhumid ones; with increasing aridity, all subsequent demes will show the character states "anticipated" by the early subarid demes. I will, in fact, argue that many of the observed features of faunal succession can be best explained by migration of preexisting variant phenotypic arrays and by the evolution of these arrays during dispersal.

What are the consequences for correlation? Clearly, evolutionary transformations transect time planes, i.e., they are *diachronous*. Planes of "minimum morphologic variation," the only ones definable from fossils, approximate a plane of "maximum environmental similarity" and are limited to the time interval during which the population continuum can yield the requisite genotypes. When evolution was rapid, i.e., the limiting interval was short, and/or when contemporaneous environmental variation was small, planes of minimal morphologic variation approximated time planes (Figure 8-7). Conversely, if evolution was slow and/or environmental variation great, then they intersected time at a steep angle.

Taxonomic similarity and correlation

Since a taxon corresponds to a unique portion of a phyletic system, its representatives occupy a discrete interval of time. Strata in which they occur were deposited during this interval. Nearly all biostratigraphic correlation derives from this principle; such correlation has limits imposed by the properties of phyletic systems and by the sampling of these systems. Taxa are defined by planes of minimum morphologic variation; they originate with an evolutionary transformation; they terminate by another transformation and/or by extinction. If the taxonomic interval is long relative to the duration of the transformation, diachrony of the taxon boundary is not of practical significance. The diachrony of extinction, however, approaches the length of the taxonomic time interval, i.e., the taxon becomes extinct in some places very early in its history. The occurrence of a taxon also requires dispersal and suitable enivornments. Either a dispersal barrier or absence of a proper habitat and niche may delay its appearance (p. 191). In consequence, planes of first and last occurrence must, in general, transect time more steeply than the morphologic limits of the taxon (Figure 8-8).

The recorded fossil occurrence of a taxon requires burial and subsequent preservation and collection. Gaps in deposition (particularly obscure unconformities), changes in preservational environment, and inadequate collecting truncate the true range of a taxon in a particular

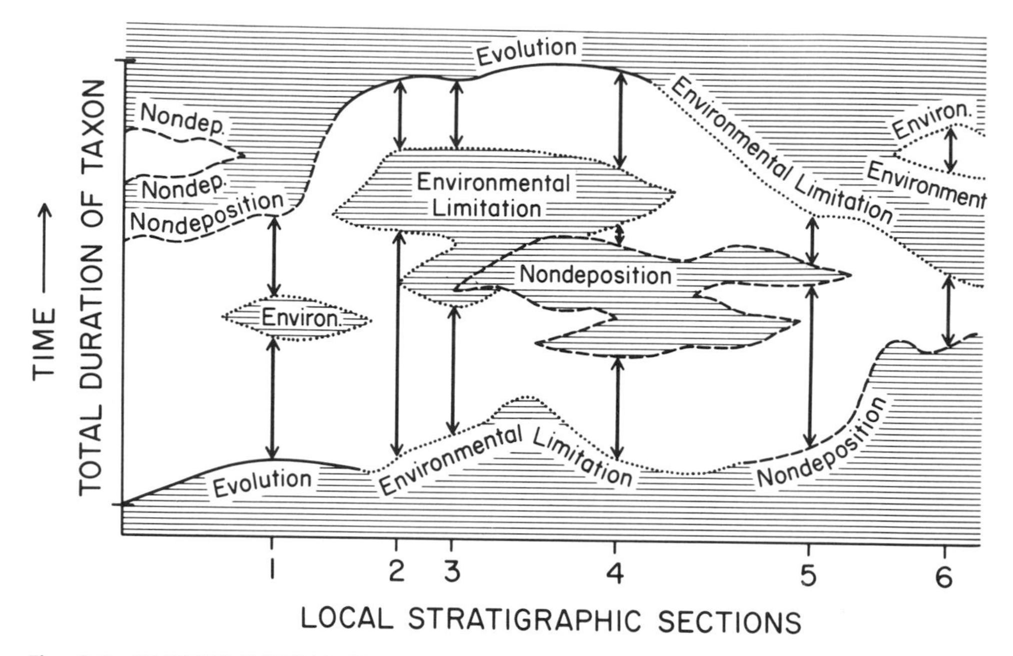

Fig. 8-8. FACTORS LIMITING STRATIGRAPHIC RANGE OF TAXON. The total potential stratigraphic range is set by evolution (lower limit) and by evolution or extinction (upper limit). In any particular local section nondeposition (or erosion) and environmental exclusion further limit the range. The ranges in any two local sections therefore may or may not indicate an overlap in time.

section—or eliminate it altogether. Grant (1962) in a study of late Cambrian trilobites demonstrated the effect of all three factors in determining range. Such truncations are, in general, diachronous even over a small area, as Grant shows.

The precision of correlation from a single taxon cannot exceed the total duration of the taxon—itself largely indeterminable—unless one can construct an acceptable model of its evolution, its dispersal, and its preservation.

A recent paper (Clark, *et al.*, 1967) justifies correlations between Cenozoic strata in South Dakota and Montana (Figure 8-9) with such a model. The lower Oligocene beds at Pipestone Springs, Montana, contain a forest fauna. Comparable animals appear in the lower Oligocene of South Dakota only in local outcrops that represent a river border forest. Most of the South Dakota deposits formed in a semi-tropical grassland and scrub forest environment and contain different animals. The only comparison is between the forest animals. Among these is a horse, *Mesohippus*. The Montana horse is closest to a species that occurs toward the top of the South Dakota section. In South Dakota, this species was preceded and succeeded by other *Mesohippus* species in a phyletic

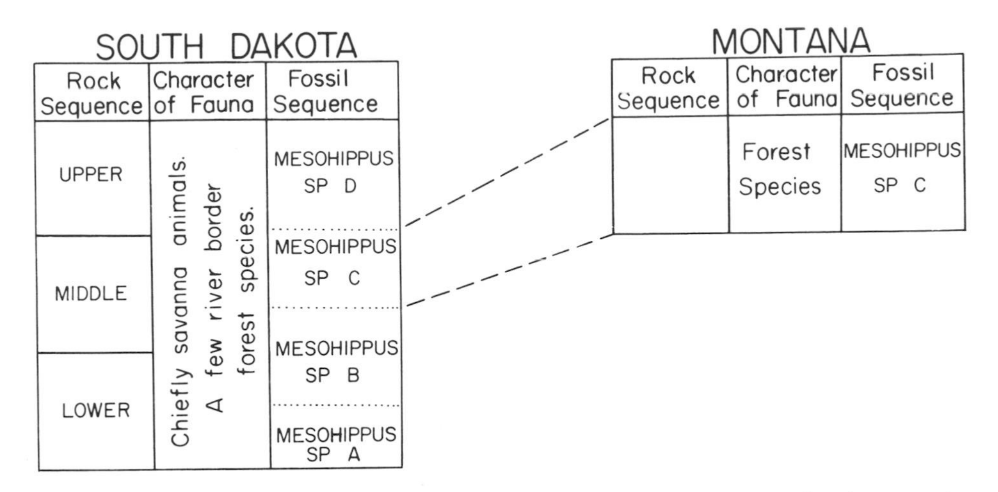

Fig. 8-9. PROBLEMS IN CORRELATION. *Mesohippus,* an inhabitant of the early Oligocene river border forest in South Dakota, provides a basis of correlation with the forest fauna of the Pipestone Springs formation in Montana. The other species, savanna and forest, are not diagnostic.

sequence. Since these are absent in Montana, the stratigrapher correlates the Montana local section with a small part of the South Dakota local section. Since he compares similar environments and he believes with some evidence that no geographic barriers separated them, he expects the same phenotypes to appear in the two areas nearly simultaneously. If evolution in the savanna-river border forest anticipated evolution in the upland forest, a slight temporal offset is probable.

In the absence of explanatory models, stratigraphers have sought other approaches to precise correlation. For example, overlapping stratigraphic ranges may subdivide the succession and provide closer correlation. In Figure 8-10, the interval marked by the A_1, B_2, C_1 assemblage is shorter (therefore provides more precise ordering) than the interval of any one taxon. In practice, these correlations also involve diachrony in evolution, extinction, and dispersal. They enter, in addition, the mechanisms of classification into the range data. If two taxa are first described from a single stratum, as is commonly the case, this action restrains the placement of taxonomic boundaries in both lineages. If their rates of evolution are similar, the upper and lower boundaries of both taxa approach contemporaneity. On the other hand, another taxon defined originally from a different stratum will overlap.

The rationale behind such correlations argues that each taxon is an independent "signal." So argued, only temporal overlap explains the coincidence of such signals in several localities, and two taxa in common are twice as important as one. Thus correlation is made between beds with the largest number of taxa in common. But are the signals inde-

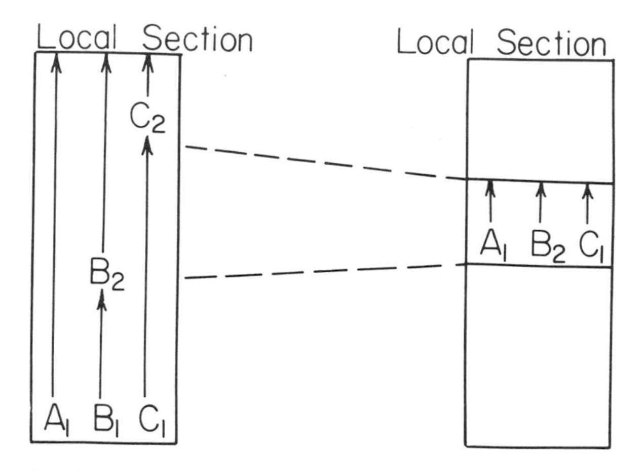

Fig. 8-10. USE OF OVERLAPPING FOSSIL SEQUENCES IN CORRELATION. In one local section, species A_1 continues through the section; B_1 is replaced by B_2; and, somewhat higher in the section, C_1 by C_2. In another local section, the occurrence of A_1, B_2, and C_1 in a unit implies a correlation with the middle of the first local section.

pendent? Some portion of intertaxon correlation results from common ecological requirements, similar dispersal patterns, and interaction in evolution. Only if the common taxa belonged to independent ecologic systems and if their definition as taxa was independent do they carry much more weight than individual similarities.

Correlation by succession

Taxonomic successions (Figure 8-11) may be compared empirically (p. 198) without regard to their temporal significance. Most stratigraphers, however, employ them as temporal indicators and as the principal basis of biostratigraphic correlation. The rationale is simple and powerful because it does not require development of a biological model for each correlation. For simplicity, assume that two taxa, *A* and *B,* succeed each other in a local section. Next ask the chances of getting the same sequence by chance in all local sections. This is like asking how often an equivalent run of heads will occur in tossing a coin. As the probability becomes small, one assumes that the succession corresponds to a real relationship, either an invariable ecologic succession or a restriction of the taxa to different time intervals. Over any large area and/or through any considerable period of time an environment suitable for *A* is likely to occur later than one for *B,* as well as earlier. Therefore the ordered sequence of taxa $A \longrightarrow B$ in every local section indicates temporal exclu-

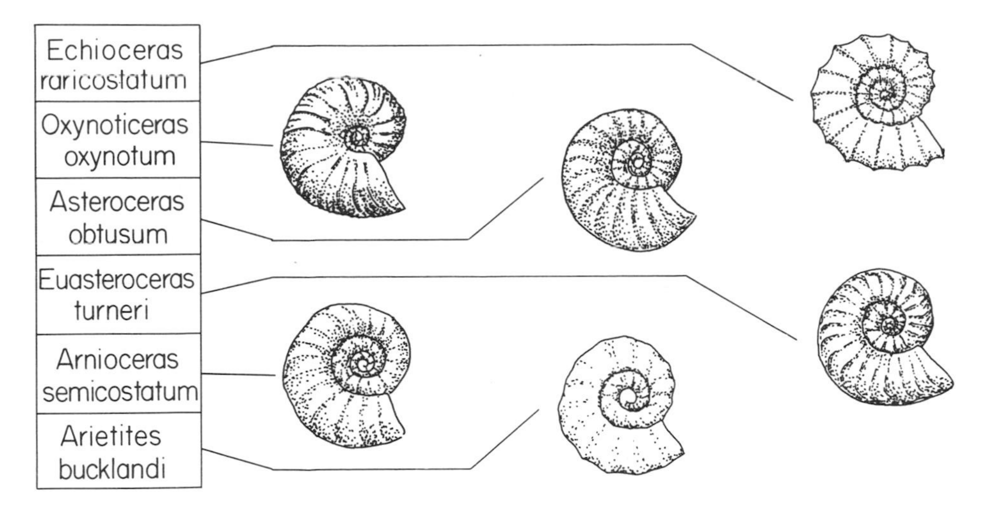

Fig. 8-11. FOSSIL ZONES AND ZONAL FOSSILS. The sequence of these six species of Jurassic ammonites provides a guide for correlation wherever they occur. The succession appears invariable, and the species have never been found to overlap in stratigraphic distribution. (*Based on* Arkell.)

sion and assures the correlation of all A-bearing strata with each other, all B-bearing strata, etc. Acceptance of the initial premises excludes biological problems—even evolution and/or extinction are not necessary processes. The method requires only that some taxa do not overlap in time; a requirement that is certainly met even if not all phyletic systems are irreversible.

The limitations of the method arise from the same source as its strengths. No biological necessity exists for any particular sequence. In consequence, the succession always remains empirical. Strict application of inductive probability is difficult, because all irregular sequences are discarded. Perhaps the most severe objection is to the premise that environmental variations are random over any geologically significant period of time. In fact, abundant evidence indicates that variations occur in an ordered sequence and that particular habitats shift laterally with time in an ordered pattern (Walther's Law). On the other hand, an invariable succession of four or five environments is unlikely because of local hydrographic and topographic variation. Grant (1962) demonstrated overlap of successive trilobite taxa, but the overlap was less than their total range.

In effect, taxonomic succession selects "signals" that are unique in time, that are propagated at a relatively high velocity, and that are relatively ubiquitous. The resolution of the method depends on the range of the taxa involved, and it is unlikely that correlation surfaces derived from such successions transect time by more than the duration of one taxon, e.g., for the rapidly evolving Jurassic ammonites the error of correlation would be between five hundred thousand and a million years.

Further refinement generally requires very weak or unjustified normic statements. For example, maximum abundance is used for some correlations. Since fossil abundance reflects both local biologic and preservational environments, it is certainly strongly diachronous. Some correlations derive from the concept of an *acme,* i.e., a maximum, in geographic distribution. Acmes existed by definition, but *cannot* be determined from the fossil record and are about as significant in correlation as Plato's heavenly chair.

What good is biostratigraphy? It is very useful for reasonably reliable ordering and precise correlation, particularly if backed by an interpretation of phylogeny, paleogeography, and paleoecology. From this view, stratigraphy is "good" if it predicts efficiently the sequence of fossils in a hitherto unknown section—even if the prediction involves an unresolved mixture of ecologic trends, geographic patterns, and evolutionary sequence. In general, resolution in fossil correlation is better than that provided by radiometric methods or physical stratigraphy, and it is probably susceptible to at least as much improvement as these alternate techniques.

The compleat stratigrapher

The stratigrapher-paleontologist observes age relationships only in a local section where visible superposition establishes a succession of species. These successions form the central pier of correlation; all else is deduction and hypothesis. Each species has a definable range within the local section. That range samples the total duration of the morphologic group classified as a species. In addition the stratigrapher has information on the association with rock characters, on the occurrence of species with each other, and on the numbers, condition, orientation, and distribution of the fossils.

He next compares adjacent local sections (Figure 8-3). Similarities of rock type and rock sequence establish rock unit equivalence. Similarities of species occurrence and sequence establish biological equivalence. Each local section, however, contains somewhat different species assemblages, and the ranges of species relative to each other and to the rock units vary. Local environmental differences produce this variation imposed as noise upon the temporal signal.

The alternatives were considered in the preceding section: evolutionary transformations, similarity of one or a few taxa, similarity of total assemblages, similarity in sequence of taxa. Evolutionary transformations are so rarely observed as to be of little stratigraphic significance. Similarity of individual taxa is largely meaningless unless specific biologic

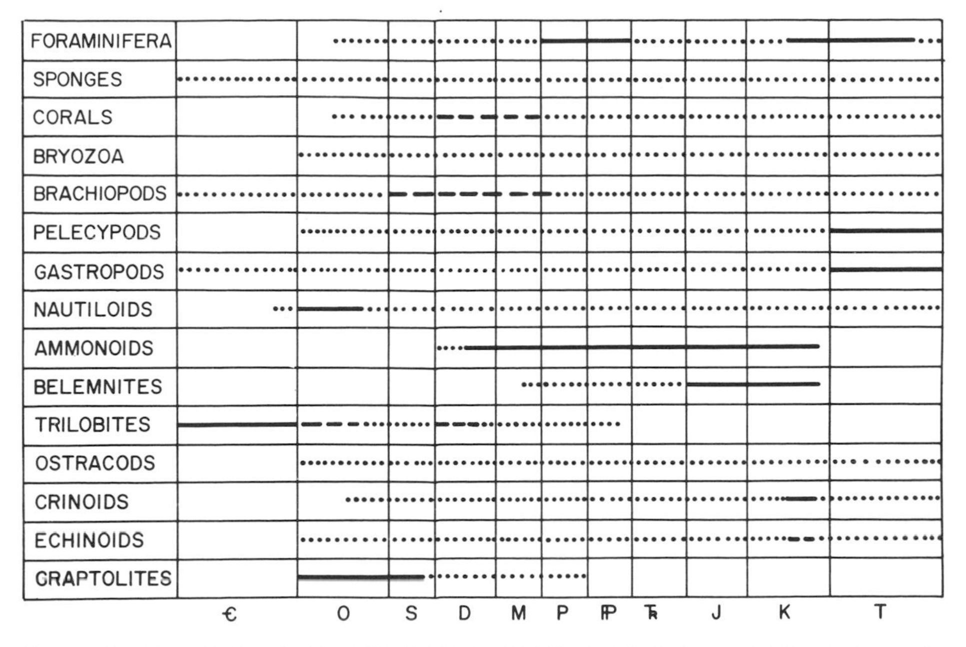

Fig. 8-12. STRATIGRAPHIC SIGNIFICANCE OF FOSSIL GROUPS. The solid line indicates the group is important for world-wide zoning and correlation; the dashed line means that it was important in regional correlation; the dotted line signifies only occasional or rare use as zonal fossils. (*After* Teichert.)

explanations remove the variation introduced by environment and dispersal. Correlation by similarity of whole faunas is generally more probable than correlation by randomly selected taxa, but the improvement is not linear, because of ecologic and phylogenetic associations. Invariable sequences of taxa imply that the component taxa have short and nonoverlapping temporal ranges. This method selects the taxa most useful for correlation without an elaborate (and therefore highly uncertain) explanation of ecologic and dispersal controls. These taxa are *guide* or *index* fossils. Please remember, however, that guide fossils are subject to revision, that not many qualify, and that their appearance in local sections will differ even though their total period of existence may be so short as to make these differences unimportant. Relatively few groups of organisms provide most of the index fossils and do so for only part of their history (Figure 8-12).

Regional and interregional correlation

Within a basin of deposition, the species distributions depend upon local environmental controls. Extension of correlation from adjacent sections to more distant ones, therefore, follows the same pattern outlined above. Invariable successions due to transgressive or regressive environments are less likely because of greater topographic, hydrographic, and tectonic variation. Shifts in climate and in water mass distribution, however, may induce uniform regional successions, e.g., a sequence of warm and cold water faunas generated by progressive cooling. Conversely, regional temperature gradients may produce strong faunal differences among contemporary deposits. For these reasons, the stratigrapher analyzes dispersal more carefully and places more reliance on a few widely distributed species. Thus he builds correlations throughout a basin and into interconnected basins.

Correlation between distinct biologic provinces is, inevitably, less precise. It depends on a few widely ranging taxa, e.g., pelagic animals such as the cephalopods and graptolites. It also depends upon migration between provinces and realms during intervals when dispersal barriers are reduced. For example, the South American terrestrial succession is distinct from that of North America throughout most of the Cenozoic. Only near the beginning and near the end of the era do the same or closely related genera and species occur in both areas.

Units of correlation

The founders of biostratigraphy, impressed by the limited time range of most fossil types, divided the geologic column into *zones* characterized by one or more genera or species. These were variously *marker, guide,* or *index* fossils. Subsequent workers evolved somewhat divergent conceptions of the zone and also found that some of the diagnostic fossils were not restricted to "their" zone, that most had limited geographic ranges, and that all were confined to relatively few environments of deposition.

In spite of this, the zone with its guide fossils provides the most practical and accurate unit of correlation. A zone represents the practical limit of correlation within a faunal province. In general, each faunal province or realm requires a separate zonal sequence. Further, many local sections lack the critical guide fossils, and correlations are built up as fossils of local significance are tied into the sections that contain zonal guide fossils.

American stratigraphers have developed an elaborate system of stratigraphic classification involving 1) rock-stratigraphic units inferred by physical similarities of strata and by similar sequences of strata among local sections, 2) biostratigraphic units inferred by similarities in fossils and by similar sequences of fossils, 3) time-stratigraphic units (the body of strata deposited in a time interval) inferred by correlation of rock units or, more typically, of biostratigraphic units; 4) time units corresponding on a one-to-one basis with selected time-stratigraphic units. In this code, a zone is a biostratigraphic unit without any necessary time value.*

The definition of a time-stratigraphic unit, the Permian system or the Paleocene series, is built up by correlation between local sections. One or more local sections that cover the entire interval are designated *standard sections,* to provide a reference for further correlation. Typically, stages have significance only within a faunal province, because they involve narrowly defined taxa, i.e., species and genera, with limited geographic distribution. Larger taxa, subfamilies, etc., which have poorer temporal resolution but broader distribution define interprovincial series and systems.

European stratigraphers, probably influenced by the relative lack of subsurface and outcrop data on rock units, have held closely to the zone as the unit of stratigraphy, disregarded rock units, and neglected distinctions between rock-stratigraphic, time-stratigraphic, biostratigraphic, and time units. Work in North America has centered more on rock and rock-time problems—perhaps because of a relative lack of such superb zonal sequences as occur in the European Mesozoic. In any case, stratigraphers now employ the term "zone" in several different ways. This confusing situation still awaits the decision of usage.

All stratigraphers would wish for even finer resolution than is possible with the faunal succession method. Evolutionary paleontologists and paleoecologists would like a very precise and accurate dating system—with a resolution of a few years—independent of the fossil data they have to analyze. As geologic methodology goes, however, the faunal succession technique is unusually effective. Radiometric correlation typically includes a two to five percent experimental error—yielding for the Jurassic, for example, a resolution limit of two to five million years. Zonal correlation for the Jurassic has a resolution limit below five hundred thousand years—within a province at least. In some cases, e.g., the correlation between South Dakota and Montana described on p. 202, the resolution is presumably below one hundred thousand years, and the theoretical limit probably approaches ten thousand years.

REFERENCES

Arkell, W. J. 1956. *Jurassic Geology of the World.* New York: Hafner.

Clark, J. *et al.* 1967. "Oligocene Sedimentation, Stratigraphy, Paleoecology, and Paleoclimatology," *Fieldiana: Geol. Memoirs,* vol. 5. See especially pp. 56-59.

Cockbain, A. E. 1966. "An Attempt to Measure the Relative Biostratigraphic Usefulness of Fossils," *Jour. of Paleontology,* vol. 40, pp. 206-207.

* I have yet to discover a contemporary American stratigrapher who does not act as if zones had a time value—in spite of his metaphysics.

Dunbar, C. O. and J. Rodgers, 1958. *Principles of Stratigraphy*. New York: John Wiley.

Grant, R. E. 1962. "Trilobite Distribution, Upper Franconian Formation (Upper Cambrian), Southeastern Minnesota," *Jour. of Paleontology*, vol. 36, pp. 965-998.

Horowitz, A. S. 1966. "A Model for Simulating Paleontologic Correlations" (Abst.), *Geol. Soc. of America*, Program of Annual Meeting, p. 98.

Jeletsky, J. A. 1965. "Is It Possible to Quantify Biochronological Correlation?" *Jour. of Paleontology*, vol. 39, pp. 135-140. An argument against efforts of Shaw (see ref. below) and others to develop quantitative correlations.

Kitts, D. B. 1966. "Geologic Time" *Jour. of Geology*, vol. 74, pp. 127-146.

Krumbein, W. C. and L. L. Sloss. 1963. *Stratigraphy and Sedimentation*. San Francisco: W. H. Freeman.

Lerman, A. 1965. "Evolution of *Exogyra* in the Late Cretaceous of the Southeastern United States," *Jour. of Paleontology*, vol. 39, pp. 414-435.

Miller, T. G. 1965. "Time in Stratigraphy," *Paleontology*, vol. 8, pp. 113-131. An exposition of current concepts in biostratigraphic nomenclature and philosophy.

Newell, N. D. 1962. "Paleontologic Gaps and Geochronology," *Jour. of Paleontology*, vol. 36, pp. 592-610.

Schindewolf, O. H. 1950. *Grundlagen und Methoden der Paläontologischen Chronologie*. Berlin: Borntraeger.

Scott, G. H. 1965. "Homotaxial Stratigraphy," *New Zealand Jour. of Geol. and Geoph.*, vol. 8, pp. 859-862.

Shaw, A. B. 1964. *Time in Stratigraphy*. New York: McGraw-Hill. Quantitative biostratigraphy—the biologic model appears inadequate.

Shotwell, J. A. 1961. "Late Tertiary Biogeography of Horses in the Northern Great Basin," *Jour. of Paleontology*, vol. 35, pp. 203-217. A study of geographic and stratigraphic variation in horse distribution in an ecologic framework.

Simpson, G. G. 1950. "History of the Fauna of Latin America," *American Sci.*, vol. 38, pp. 361-389.

Stehli, F. G. 1965. See ref. on p. 132.

——— and C. E. Helsley. 1963. "Paleontologic Technique for Defining Ancient Pole Positions," *Science*, vol. 142, pp. 1057-1059. Stimulating approach to paleogeographic analysis employing statistical techniques.

Weller, J. M. 1960. *Stratigraphic Principles and Practice*. New York: Harper. Good review of correlation and stratigraphic practice.

Young, K. 1960. "Biostratigraphy and the New Paleontology," *Jour. of Paleontology*, vol. 34, pp. 347-358. One of the few adequate analyses of the biostratigraphic rationale.

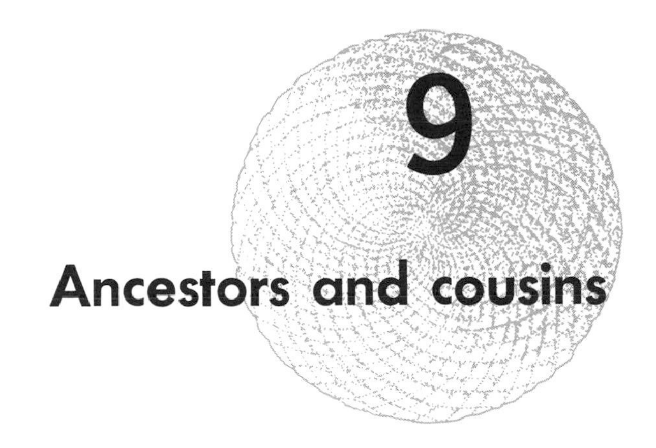

9 Ancestors and cousins

The preceding pages described the methods, principles, and conclusions of paleontology. The succeeding ones apply these generalities to specific groups of fossil animals. Unfortunately, many important events in the history of life are unrecorded and must be inferred from related fossil and recent animals. Little is known of the adaptations, ecology, and evolution of most fossil species. These areas of ignorance must be brought into perspective with that which is known, but this is too much for a few hundred pages of text and illustrations. The best to be done in the confines of this book is to mention some of the most interesting (interesting to me at least) things, and to illustrate further the methods, principles, and conclusions.

The Protista

Single-celled organisms include today a complex variety of types ranging from simple parasitic bacteria to elaborately organized "animals," and

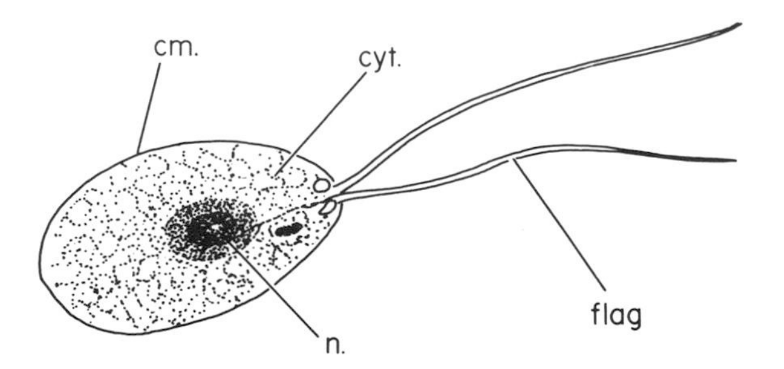

Fig. 9-1. GENERAL FORM OF THE PROTOZOA. A relatively simple flagellate protozoan, much enlarged. Abbreviations: cm., cell membrane; cyt., cytoplasm; flag., flagellum; n., nucleus.

"plants." The quotation marks are placed because the boundaries between plants and animals disappear at this level, e.g., some organisms are simultaneously photosynthetic autotrophs and predatory allotrophs. Their history is very poorly known, and only a few kinds contributed significant numbers of fossils. Phylogenetic relationships are inferred primarily from biochemical and microstructural similarities of recent species. Apparently, several lineages evolved from proto-organic levels through the organizational stages represented today by the bacteria and the simplest algae. One or more of the radiating lines adopted herbivorous and/or predatory modes of life; from these derive single-celled "animals," multi-

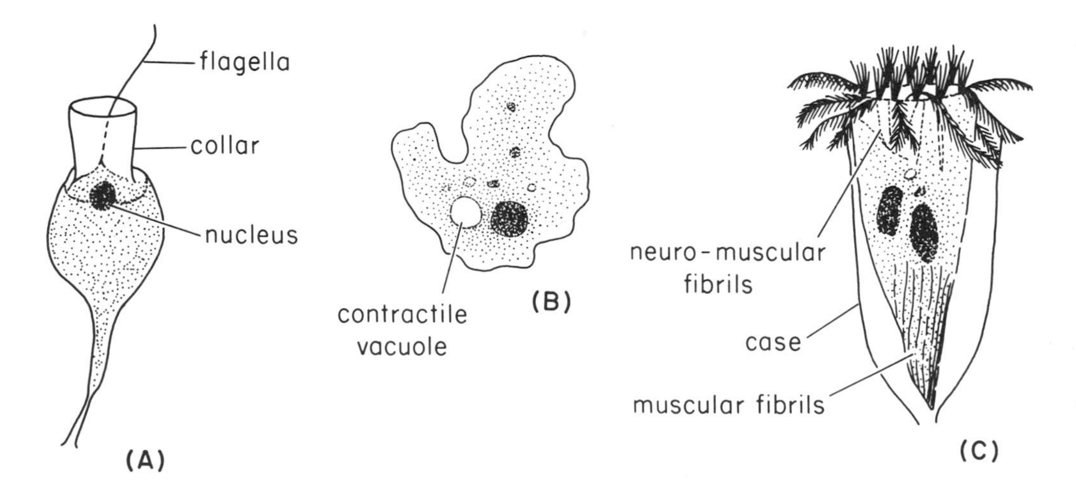

Fig. 9-2. VARIETIES OF PROTOZOA. **(A)** Subphylum Flagellata. Attached flagellate. **(B)** Subphylum Sarcodina, Class Rhizopodea. Amoeboid cell; contractile vacuole for water excretion. **(C)** Subphylum Ciliophora. Attached form with cilia modified into bristles about free end. [**(A)** *after* Lapage; **(B)** *after* Hyman, *The Invertebrates,* Copyright, McGraw-Hill, New York, *used with permission;* **(C)** *after* Campbell.]

cellular sponges, and the "true" multi-celled animals. I group the single-celled "animals" as the Phylum Protozoa. Although conventional and convenient, this phylum is almost certainly not a single phyletic radiation, and the *Treatise* authors have employed the alternative of setting these forms, plus some single-celled plants, into a distinct kingdom, the Protista. Since this is basically a change in taxonomic rank rather than content, I have instead accepted the traditional grouping.

Considered from the protozoan viewpoint, the evolution of a multicellular organization is merely one of several possible advances in organizational complexity; the other possibilities are represented within the flagellates and in the remaining classes (Figures 9-1 and 9-2). The simplest ones have differentiated only nucleus and cytoplasm, but most have specialized *organelles,* some of which function as muscles and others as nerves. Indeed, some protozoans have an equivalent of an intestinal tract and a highly developed sensory system of tactile bristles and complex photoreceptors. Among the various adaptations have been resistant skeletons of calcium carbonate and silica. Groups bearing these skeletons have, alone among the multiplicity of protozoans, been preserved as fossils, and they tell what little is known directly of the history of this antique group.

The Foraminifera

Structural plan and variations

The commonest fossil protozoans, the Foraminifera, stand low in the organizational scale of the phylum—more differentiated than the simplest flagellates, less so than the ciliates and the advanced flagellates. They

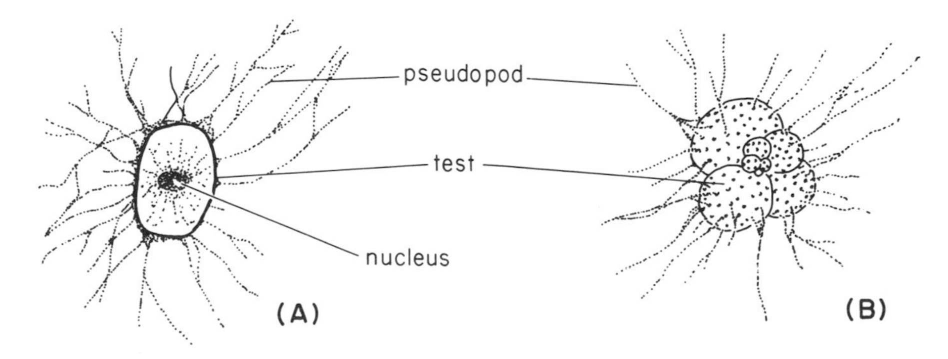

Fig. 9-3. MORPHOLOGY OF THE FORAMINIFERA. **(A)** *Gromia.* **(B)** *Globigerina.* [**(A)** *after* Hyman, *The Invertebrates,* Copyright, McGraw-Hill, New York, *used with permission;* **(B)** *after* Jepps.]

have, of course, the fundamental elements of nucleus (sometimes several nuclei) and cytoplasm but are distinguished by irregular body shape, by slender threadlike extensions, the *pseudopods,* that branch and anastomose, and by a skeleton or *test* (Figure 9-3). The bulk of the protoplasm lies within the chamber or chambers of the test, but some of it extends through openings to form a layer over the exterior.

On this basic plan of the Foraminifera a diversity of species has evolved, with variation in test composition, structure, and form—variation that can be correlated in part with variation in ecology. Tests are composed either 1) of *proteinaceous* (pseudochitinous) material (thought to be the most primitive type), 2) of foreign particles such as sand grains embedded in an organic matrix, i.e., *agglutinated,* or 3) of calcareous material. Calcareous tests are further characterized as *porcelaneous* (opaque, dull, porcelaneous appearance, without perforations, brown or amber in transmitted light), *granular* (tightly packed, very tiny calcite grains) or *hyaline* (glassy, perforate). Aragonite forms the test in a few hyaline types; the majority of hyaline tests and all porcelaneous and granular ones are calcareous. The test may also include considerable percentages of "trace elements," e.g., 0.3 to 16% $MgCO_3$, 1 to 5% Sr, 0.5 to 7% Na, 1 to 5% Si. It varies significantly 1) in microstructure, 2) in size and overall shape, 3) in number, shape, and arrangement of the internal *chambers,* 4) in openings, the *apertures* and *pores,* and 5) in characteristics of the test surface (Figures 9-4, 9-5, 9-6).

Form and function

The foraminifer uses its pseudopods for attachment, for locomotion, to capture other microorganisms for food, and in construction of the test. The food particles are engulfed and digested within the cytoplasm.

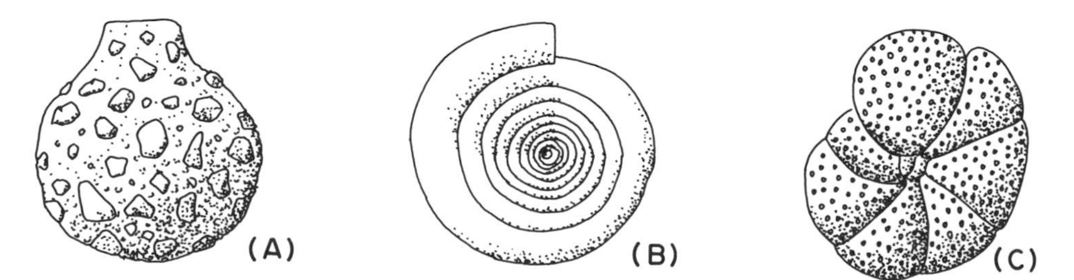

Fig. 9-4. CHARACTERISTICS OF THE FORAMINIFER TEST. **(A)** *Saccammina sphaerica;* Recent; magnification about 8 times. Test comprises grains of sand cemented together (agglutinated) and a chitinous base. Form globular; single chamber. **(B)** *Cornuspira involvens;* Recent; magnification about 20 times. Test of calcium carbonate, imperforate, spirally coiled; single chamber. **(C)** *Anomalina grosserugosa;* Recent; magnification about 25 times. Test of calcium carbonate, perforate, spirally coiled; many chambers. (*After* Brady.)

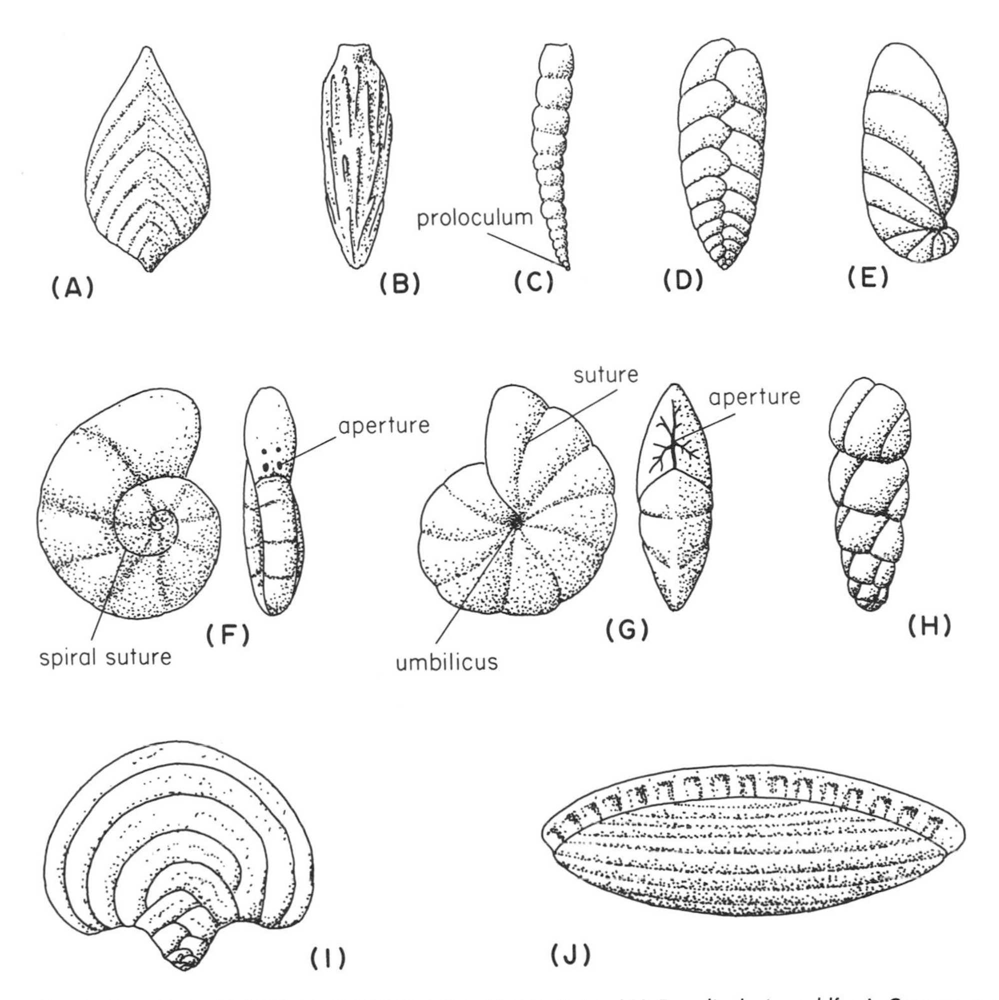

Fig. 9-5. CHARACTERISTICS OF THE FORAMINIFER TEST. **(A)** *Frondicularia goldfussi;* Cretaceous; magnified 8 times. Calcareous test; single row of chambers (uniserial), each of inverted V-shape. **(B)** *Technitella legumen;* Recent; magnified 50 times. Agglutinated test (of sponge spicules and sand); single chamber. **(C)** *Nodosinella glennensis;* Pennsylvanian; magnified 40 times. Agglutinated (of fine sand); uniserial. **(D)** *Bolivina incrassata;* Cretaceous; magnified 50 times. Two rows of chambers (biserial); test calcareous and finely perforate. **(E)** *Polymorphinoides spiralis;* Pleistocene; magnified 35 times. Test calcareous; coiled in immature portion but straight in mature. **(F)** *Entzia tetrastomella;* Recent. Coiled on surface of cone (trochoid) but cone very flat; chitinous test. **(G)** *Dendritina arbuscula;* Miocene. Tightly coiled in a single plane (planispiral) so that outer coil covers inner ones (involute); test calcareous, imperforate; aperture dendritic. **(H)** *Turrilina andreaei;* Oligocene; magnified 40 times. Coiled on elongate cone; test calcareous, finely perforate. **(I)** *Pavonina flabelliformis;* Recent; magnified 40 times. Biserial in early stages, uniserial in later; fan-shaped calcareous test. **(J)** *Schwagerina huecoensis;* Permian; magnified 5 times. Coiled, fusiform; septa between chambers folded. [**(A)** *after* Cushman, *Foraminifera.* Copyright, Harvard University Press, Cambridge, *used with permission.* **(B)** *after* Norman. **(C)** *after* Cushman *and* Waters. **(D)** *after* Cushman. **(E)** *after* Cushman *and* Hanzawa. **(F)** *after* Daday. **(G)** *after* d'Orbigny. **(H)** *after* Andreae. **(I)** *after* Parr. **(J)** *after* Dunbar *and* Skinner.]

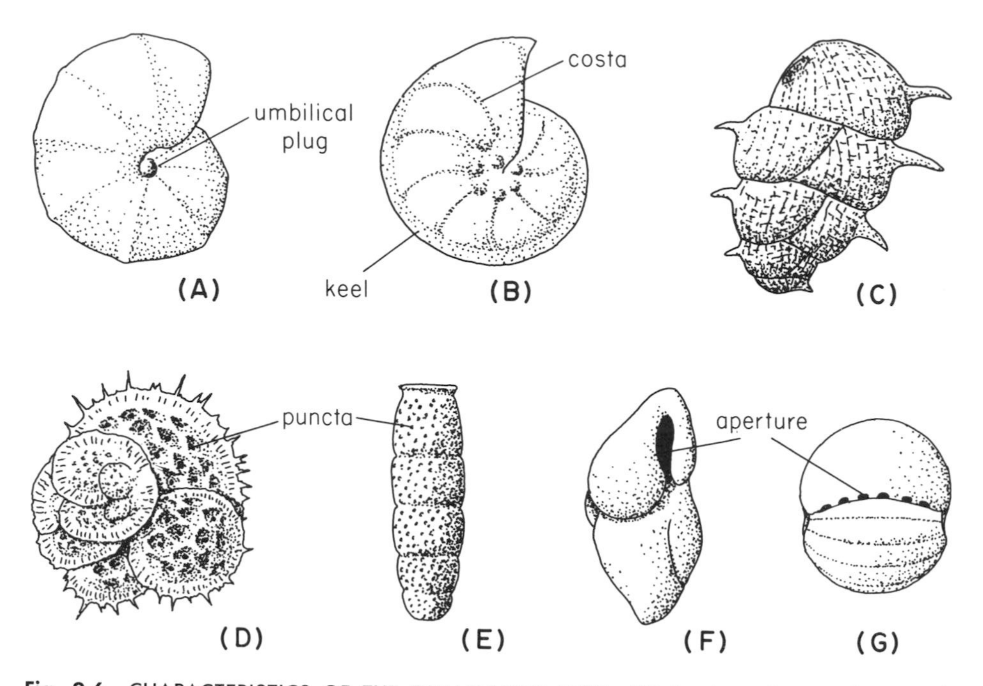

Fig. 9-6. CHARACTERISTICS OF THE FORAMINIFER TEST. **(A)** Involute planispiral test with "plug" of calcareous material in umbilicus. **(B)** Involute planispiral test with raised radial ribs or costae, a narrow keel on the periphery, and bosses at inner ends of the costae. **(C)** *Mimosina;* Recent; magnified 50 times. Later stages biserial; each chamber bears a spine; a calcareous, vesicular test. **(D)** *Globigerina triloba;* Recent. Test calcareous; surface cancellated, coarsely perforate, bears long slender spines; loose trochoid coiling. **(E)** Uniserial, coarsely perforate (punctate) test. **(F)** Trochoid spines; aperture elongate vertical slit. **(G)** Planispiral test; series of small apertures along inner border of apertural face. [**(A)**, **(B)**, **(E)**, **(F)**, and **(G)**, *after* Cushman, *Foraminifera.* Copyright, Harvard University Press, Cambridge. *Used with permission.* **(C)** *after* Millett. **(D)** *after* Rumbler.]

Oxygen for metabolism diffuses directly through the cell periphery, and carbon dioxide and other metabolic wastes diffuse back into the surrounding water. Some organelles are present; their significance is little understood; but they may function in digestion, food storage, or excretion. The animal-as-a-whole perceives changes in light intensity and in the physiochemistry of the water and responds with a limited behavior pattern. Apparently, there are no specialized parts for perception and coordination.

The foraminifers show complex reproductive patterns based on alternation of sexual and asexual generations (Figure 9-7). In some, the individuals formed by asexual division are relatively large initially, and construct *megalospheric* tests with a large initial chamber (the *proloculus*), but grow into relatively small adults; the sexually-formed zygotes are relatively small, and construct *microspheric* tests with a small proloculus, but grow to a large adult size. In others, however, dimorphism of the test is obscure. Some dimorphs have been classified as different species, so careful comparison of developmental series and of occurrence (p. 81) is necessary.

Ecology and paleoecology

Most foraminifer species seem sensitive to minor environmental differences, particularly of salinity, temperature, oxygen, food, and substrate. A few recent types occur in fresh, brackish, or hypersaline waters. An abundant and varied foraminifer fossil assemblage presumably indicates marine water of normal salinity—at least no exceptions have yet been

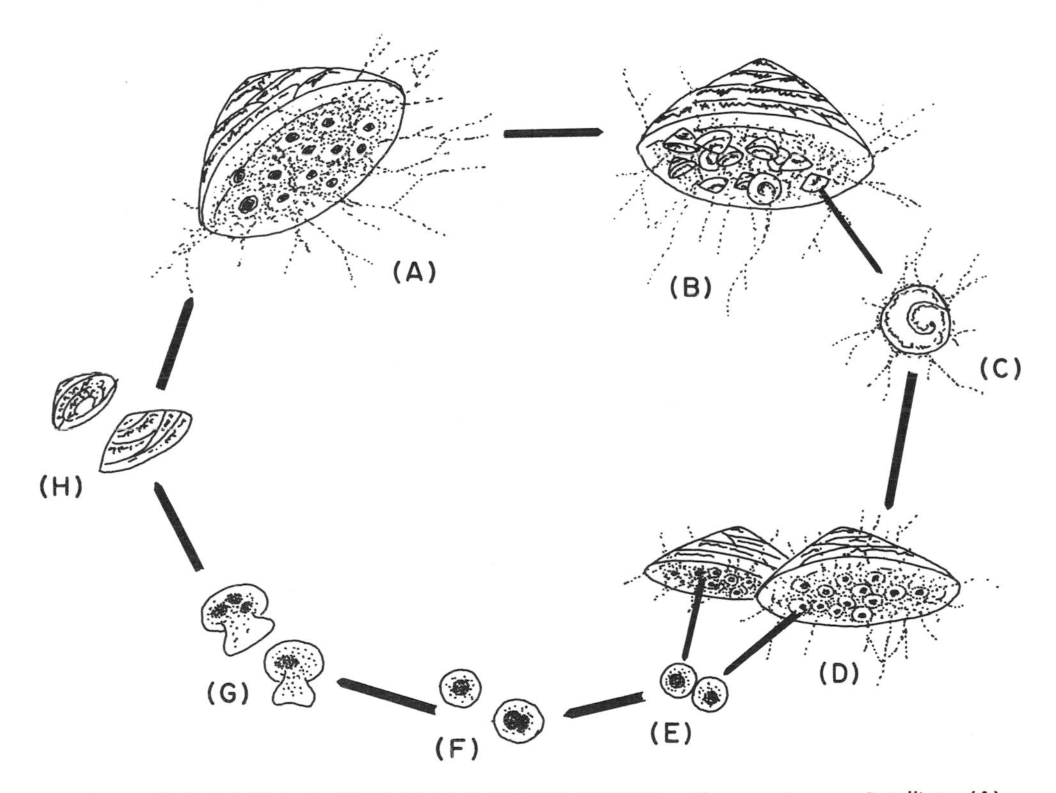

Fig. 9-7. ALTERNATION OF GENERATIONS IN FORAMINIFERA. The recent genus *Patellina*. **(A)** A microspheric individual in process of reproduction. Nucleus has divided repeatedly. **(B)** Each nucleus serves as a center for the development of new foraminifers, each forming its own shell. **(C)** The juvenile foraminifer formed by asexual division of parent. The test has a large initial chamber, is therefore megalospheric. **(D)** The adult megalospheric individual produces a large number of gametes by division of nucleus and cytoplasm. **(E)** Gametes escape and cells from different adults pair. **(F)** Paired gametes fuse to produce new individuals. **(G)** New foraminifers begin to form test with small initial chamber (microspheric). **(H)** Individuals continue growth to adult size. (*After* Myers.)

shown. Some assemblages of rare genera have been interpreted as brackish water faunas.

Foraminifer species with calcareous tests are most common at present in warmer waters, e.g., near the ocean's surface or in the tropics. This phenomenon is probably related to the lesser solubility of $CaCO_3$ in warm water, and fossil faunas with a high percentage of calcareous forms were probably warm-water faunas. Modern tropical faunas also include a large proportion of large Foraminifera with diameters between five and twenty millimeters. Cold-water faunas contain a high proportion of species that build a test from siliceous particles (quartz grains, sponge spicules, and so on).

Recent foraminifer faunas show high correlation between generic distribution and temperature. For example, on the Pacific coast several different generic associations can be distinguished, each related to a distinct range of temperature. Because temperature of the water varies inversely with depth, the generic associations are also characteristic of depth zones, although these change with latitude. Cenozoic faunas which contain a high proportion of living genera can be analyzed in terms of these associations and an estimate made of temperature and depth (Table 9-1). One would expect the accuracy of these estimates to decrease in progressively older faunas.

TABLE 9-1

Control of foraminifer assemblages by temperature, Cenozoic, west coast of North America *(information from Natland, 1957)*

Association I. Lagoonal. 0-20 feet; 24°-5°C.
Rotalia beccarii, Buliminella elegantissima, Trochammina inflata.

Association II. Inner sublittoral. 0-125 feet; 21°-13°C.
Elphidium poeyanum, E. articulatum, E. hannai, Eponides frigidus.

Association III. Inner sublittoral to upper bathyal. 125-900 feet; 13°-8.5°C.
Cassidulina limbata, C. tortuosa, C. californica, Eponides repandus, Polymorphina charlottensis.

Association IV. Upper bathyal. 900-2000 feet; 8.5°-5°C.
Uvigerina peregrina, Epistominella pacifica, Bolivina argentea, B. interjuncta, B. spissa.

Association V. Lower bathyal. 2000-4000 feet; 5°-3°C.
Buliminella subacuminata, Uvigerina pygmea.

Association VI. Lower bathyal to upper abyssal. 4000-7500 feet; 3°-2°C.
Buliminella rostrata, Nonion pompilioides, Pullenia bulloides, Uvigerina senticosa.

The Foraminifera are both benthonic and planktonic—the latter usually have globose, inflated tests (Figure 9-8). Many benthic species live in the interstitial water within the substrate. Fossil foraminiferal assemblages show associations with sedimentary types, but this may be

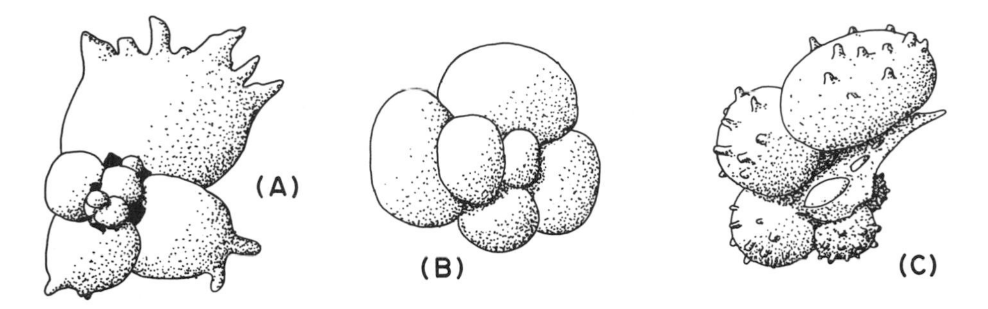

Fig. 9-8. PELAGIC FORAMINIFERA. **(A)** *Globigerinoides sacculifera;* Recent. **(B)** *Globigerina bulloides;* Recent. **(C)** *Hastigerina pelagica;* Recent. All magnified about 40 times. [**(A)** *and* **(C)** *after* Brady, **(B)** *after* d'Orbigny.]

either a direct response to the sediment composing the substrate, or a result of a common response of animals and sediment to variation in another factor such as depth.

Adaptation, evolution, and classification

The adaptive significance of most foraminifer test characteristics is unknown. Because of the difficulty of distinguishing adaptive similarities from features inherited from a common ancestry, specialists on the Foraminifera disagree on the phylogenetic relationships of the various lineages (some fifty families are recognized). The consensus holds that the proteinaceous forms are the most primitive, that the agglutinated tests represent the next evolutionary step, and that the calcareous tests evolved from the agglutinated by an increase in the proportion of calcareous cement to the proportion of foreign particles. Most of the earliest foraminifers had agglutinated tests, and calcareous types did not become common until the Devonian.

Phylogenetic inferences (and resulting classifications) based on test composition and microstructure contradict those derived from gross form. The inferred phylogeny (and classification) depends, therefore, on the relative weight given composition as against form. The interpretation of form itself is heavily weighted by ontogeny. In some phyletic lines, individuals of the later species retain in their ontogeny the adult characteristics of the ancestral species. Development was accelerated so that these adult features appear in the immature foraminifer, and new characteristics were added in subsequent stages. In other lines, however, development was retarded, and the characteristics of the immature were preserved in the adult. In these, the adults of the later species resemble the immature of the earlier. The interpretation of phylogeny from development is made with the recognition that morphologic features may shift back and forth

TABLE 9-2

The fossil Protozoa *(after Loeblich and Tappen, 1964)*

Precambrian(?) to Recent. Unicellular animals. Some colonial types. Marine, freshwater, parasitic. Benthonic and pelagic.

SUBPHYLUM FLAGELLATA

Late Cambrian(?) to Recent. Moderate to high complexity. Locomotion by long whiplike threads, the flagella. Marine, freshwater, parasitic. Benthonic and pelagic.

Order *Chrysomonadina*

Late Cambrian to Recent. Two groups recognized as fossils: Silicoflagella, Cretaceous to Recent forms, with a skeleton of siliceous rings and spines, and Coccolithida, late Cambrian(?) to Recent, with skeleton of calcareous disks. Both marine, typically planktonic.

Order *Dinoflagellata*

Late Jurassic to Recent. Marine plankton with cellulose skeletons.

Other orders unknown as fossils.

SUBPHYLUM SARCODINA

Cambrian to Recent. Locomotion and feeding by extensions, pseudopods, of amoeboid body.

CLASS RHIZOPODEA

Mississippian to Recent. Lobose pseudopods. Very rare as fossils.

CLASS RETICULAREA

Cambrian to Recent. Pseudopods filamentous.

Order *Gromida*

Eocene to Recent. Chitinous test with silicious plates or with agglutinate particles. Pseudopods not anastomosing. Freshwater.

Order *Foraminiferida*

Cambrian to Recent. Pseudopods anastomosing. Tests pseudochitinous, agglutinate and/or calcareous. Principally marine. Pelagic and benthonic, both sessile and vagrant.

Suborder *Allogromiina*

Late Cambrian to Recent. Membraneous and pseudochitinous tests.

Suborder *Textulariina*

Early Cambrian to Recent. Agglutinate test.

Suborder *Fusulinina*

Ordovician to Triassic. Granular calcareous test.

Suborder *Miliolina*

Carboniferous to Recent. Porcelaneous test.

Suborder *Rotaliina*

Permian to Recent. Perforate, hyaline test.

Order *Radiolaria*

Cambrian to Recent. Skeleton is latticework of silica, typically with spherical symmetry. Marine, largely planktonic.

SUBPHYLUM CILIOPHORA

Jurassic to Recent. Locomotion by hairlike cilia. One suborder, the Tintinnia, has gelatinous or chitinoid skeleton with agglutinated particles. Tintinnia typically planktonic, marine.

---

in time of development; that the early stages are adaptive as well as the later, and that new features may appear by alteration of developmental patterns which obscure developmental similarities (see p. 78).

The arrangement adopted here (Table 9-2) is that of Loeblich and

Tappen in the *Treatise on Invertebrate Paleontology*. Variation in composition and microstructure defines five suborders. They consider these characteristics, controlled by protoplasmic structure and physiology, more "basic" than shell form, which varies markedly with habitat and niche. The Foraminifera would seem an ideal group for the application of numerical taxonomy.

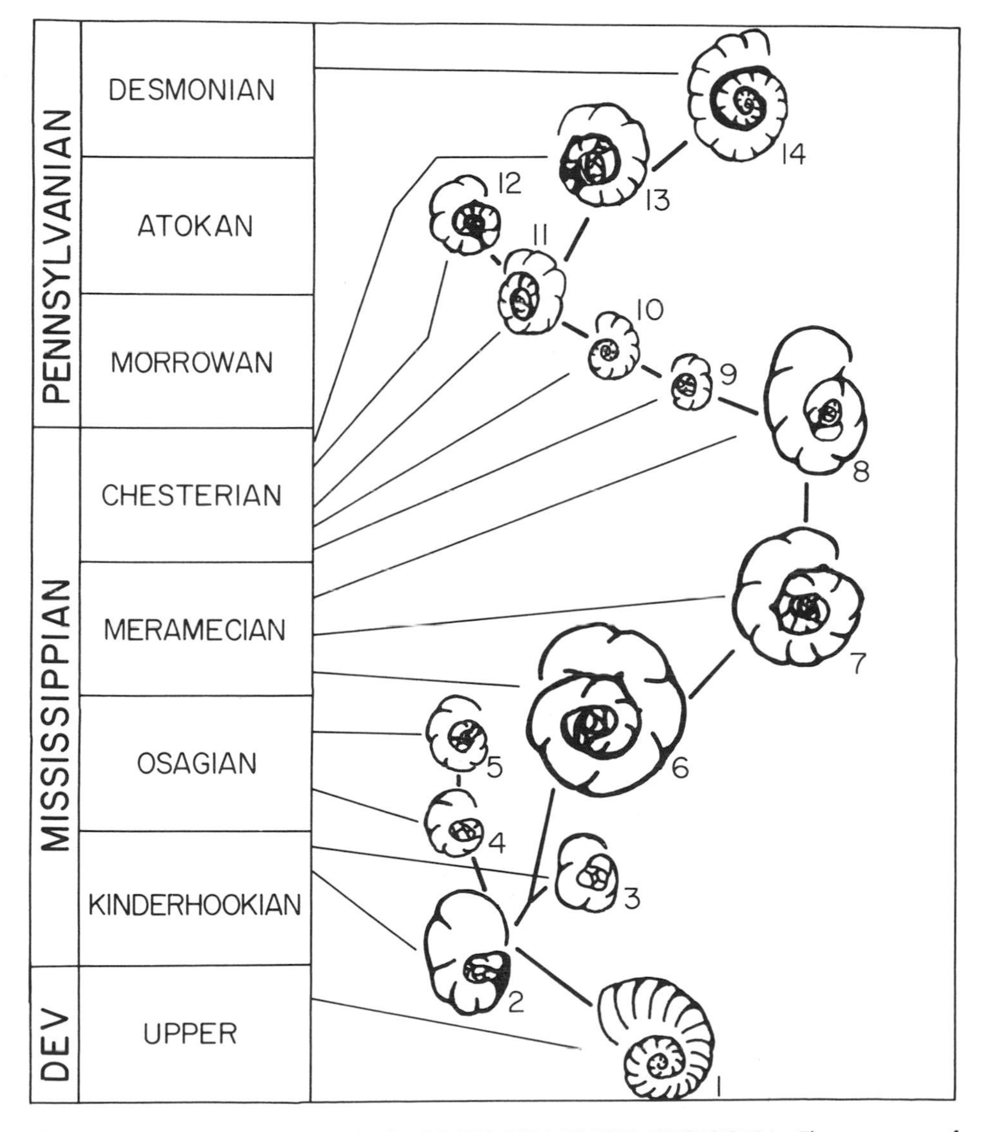

Fig. 9-9. STRATIGRAPHIC SEQUENCE OF THE FORAMINIFER *PLECTOGYRA*. The sequence of *Plectogyra* from the ancestral, late Devonian *Endothyra* (**1**), through the Mississippian, to the middle Pennsylvanian (**14**) as shown by more or less typical specimen for each stratigraphic level. (*After* Zeller, 1950.)

TABLE 9-3

Glossary for the Foraminifera ***(numbers in parentheses refer to pertinent illustration).***

Agglutinated. Test composed of grains of foreign material—sand, sponge spicules, etc.—cemented together. (9-4A; 9-5B.)

Apertural face. Flattened surface of chamber adjacent to aperture. (9-5F, G; 9-6F, G.)

Aperture. Relatively large opening to exterior in last-formed chamber. (9-5F, G; 9-6F, G.)

Biserial. Test with two rows of chambers. Chambers alternate on either side of the plane between the rows. (9-5D; 9-6C.)

Cell membrane (cm). Indistinct layer forming periphery of an animal cell. (9-1, 9-2.)

Chamber (ch). The space within the test as well as the enclosing walls. Test may be divided into several chambers separated by partitions. (9-5G, *et al.*)

Cilia. Numerous short hairlike processes on surface of cell. Occur in many animals besides protozoans. (9-2D.)

Convolute. Coiled test in which inner portion of the last whorl extends to the center of the spiral and covers the inner whorls. (9-4C; 9-5G; 9-6A, B, G.)

Costae. Ridges on external surface of test. They may run along sutures or be transverse to them. (9-6B.)

Cytoplasm. Semi-fluid, living portion of cell surrounding the nucleus. (9-1, 9-2, 9-3.)

Flagellum. Long, slender, whiplike process extending from cell. (9-1, 9-2A.)

Foramina. Opening connecting adjacent chambers in test. Typically formed as aperture(s) and enclosed by development of additional chambers.

Fusiform. Spindle-shaped. (9-5J.)

Involute. Coiled test in which inner portion of last whorl extends to cover part of the adjacent inner whorl. (9-4B.)

Keel. Keel-like ridge along outer margin of the test. (9-6B.)

Megalospheric. Test with relatively large initial chamber. Constructed by individual formed from asexual division of parent. (9-7.)

Microspheric. Test with relatively small initial chamber. Constructed by individual formed from sexual union of two cells. (9-7.)

Nucleus. Dense body suspended in interior of living cell. Bears most of hereditary material and controls most of cellular activity. (9-1, 9-2, 9-3.)

Perforate. Test with many small openings in chamber walls. (9-6D, E.)

Periphery. Outer margin of coiled test. (9-5G, *et al.*)

Planispiral. Coiled test with whorls of coil in single plane. (9-5G; 9-6A, B, G.)

Proloculum. Initial chamber of test. Typically at small end of series or at center of coil. (9-5C, *et al.*)

Pseudopod. Lobate or threadlike extension of cell periphery that changes in shape, character, and position with activity of cell. (9-3A, B.)

Punctae. Small to large holes in external walls of chambers. (9-6D, E.)

TABLE 9-3 (continued)

Spiral suture. Line of contact between whorls in coiled test. (9-5F.)

Suture. Line of contact between adjacent chambers of test. (9-5G, *et al.*)

Tentacle. Slender arm-like extension of cell periphery, of variable length and shape like a pseudopod but of fixed position. (9-2E.)

Test. Skeleton of protozoan. Also refers to skeleton of some other kinds of animals. (9-3A, B, *et al.*)

Trochoid. Coiled test in which whorls form a spiral on the surface of a cone. (9-5F, H; 9-6D, F.)

Uniserial. Test in which chambers form a single linear or curved series. (9-5C; 9-6E.)

Umbilical plug. Deposit of skeletal material in axis of coiled test. (9-6A, B.)

Umbilicus. Depression in axis of coiled test. (9-5G.)

Stratigraphic significance

Although the phylogeny of the Foraminifera has not been firmly established, the group is of great importance in stratigraphy. Many genera and species have short stratigraphic ranges, and the sequence of species in many evolutionary lineages is established. On the other hand, the sensitivity of foraminifers to environment changes and the knowledge of the ecology of recent genera provide a tool for the interpretation of past environments. Since foraminifers are one of the few groups of fossils common and identifiable in well cuttings, they have attracted intensive study, perhaps more than any other fossils. In some sedimentary sequences, such as the late Paleozoic of the Midcontinent area, foraminifers are the principal zonal fossils, and correlation of both surface outcrops and subsurface units rests on analysis of extensive foraminifer assemblages (Figure 9-9).

Geologic record of the Protozoa

The remaining groups of protozoans are less significant to the geologist, but some, the Radiolaria, the Silicoflagellata, the Coccolithophoridae, the Dinoflagellata, and the Tintinnidae, are sufficiently common as fossils to require illustration here (Figure 9-10).

An organization of cells: The phylum Porifera

Among the recent flagellate protozoans are species which form a gelatinous mass including numerous individuals (Figure 9-11). Typically,

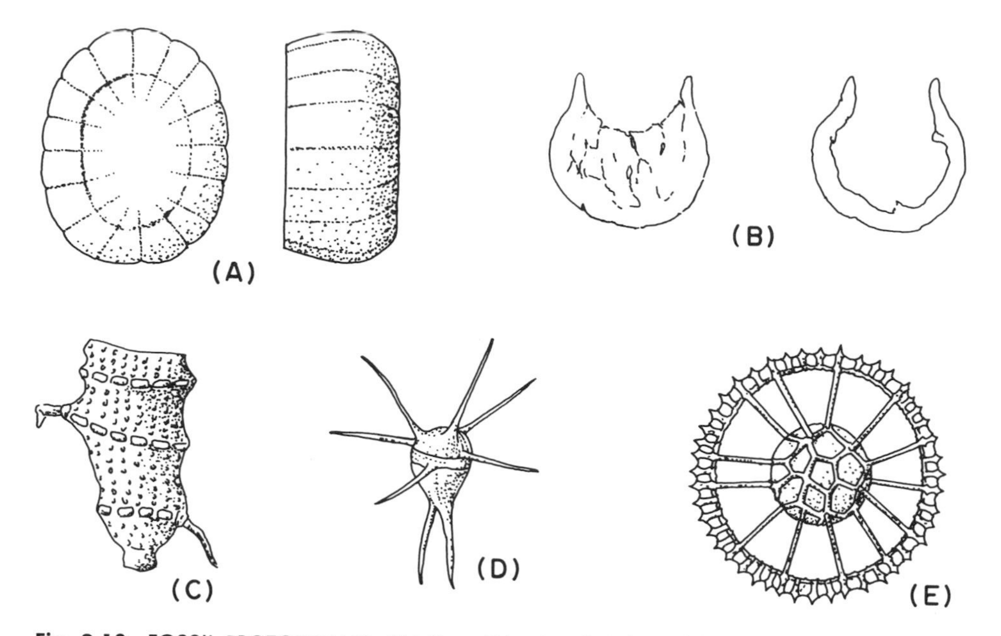

Fig. 9-10. FOSSIL PROTOZOANS. **(A)** Coccolithophorid *Calyptrolithus;* Miocene. **(B)** Tintinnid *Calpionella;* Jurassic; sections; enlarged approximately 300 times. **(C)** Dinoflagellate *Wetzelodinium;* Oligocene. **(D)** Dinoflagellate *Raphidodinium;* Cretaceous. **(E)** Radiolarian *Rhodosphaera;* Devonian. [**(A)** *after* Kamptner; **(B)**, **(C)**, *and* **(D)** *after* Deflandre; **(E)** *after* Rust.]

these masses are spheroidal, and the individual cells "face" outward with their flagella projecting into the water. In many species, the cells in a mass are alike and form, therefore, a colony of nearly independent protozoans. Among other species, however, the cells are differentiated; those near the "anterior" pole serve vegetative functions, i.e., locomotion, food-getting, respiration, and so forth, and those near the "posterior"

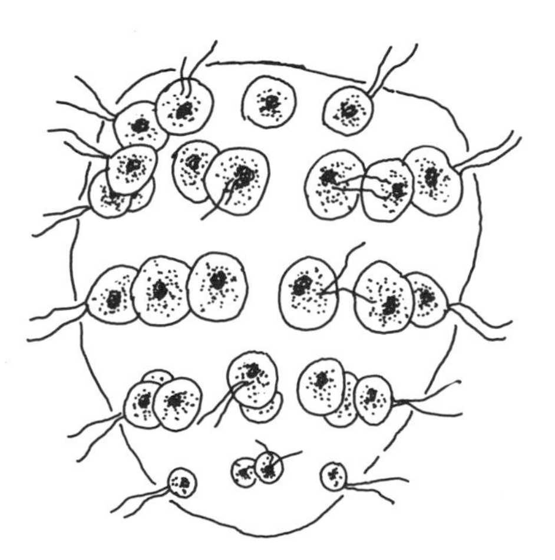

Fig. 9-11. COLONIAL PROTOZOAN. The colonial flagellate *Pleodorin.* Note difference in cell size at upper and lower poles. (*After* Merton.)

pole are specialized for reproduction. In addition, the beat of flagella is coordinated throughout the colony.

The organization of these colonies exemplifies quite a different evolutionary trend than that in most protozoans. Rather than including a variety of specialized structures, different cells are adapted to different functions and are organized with respect to one another in a definite way. This trend should result in a colony of many different kinds of cells, and an ever more complete integration to take advantage of their different specializations. One could, in fact, predict the existence of the phylum Porifera from characteristics of the colonial Protozoa.

General character and mode of life

Any discussion of the Porifera, that is, the sponges, begins with negatives: No other animal group evolved from this phylum, and the individual sponge is neither a real individual nor a colony. The body is roughly globular or sac-like, and most have an internal skeleton. Some sponges show radial symmetry—usually quite imperfect, but many have

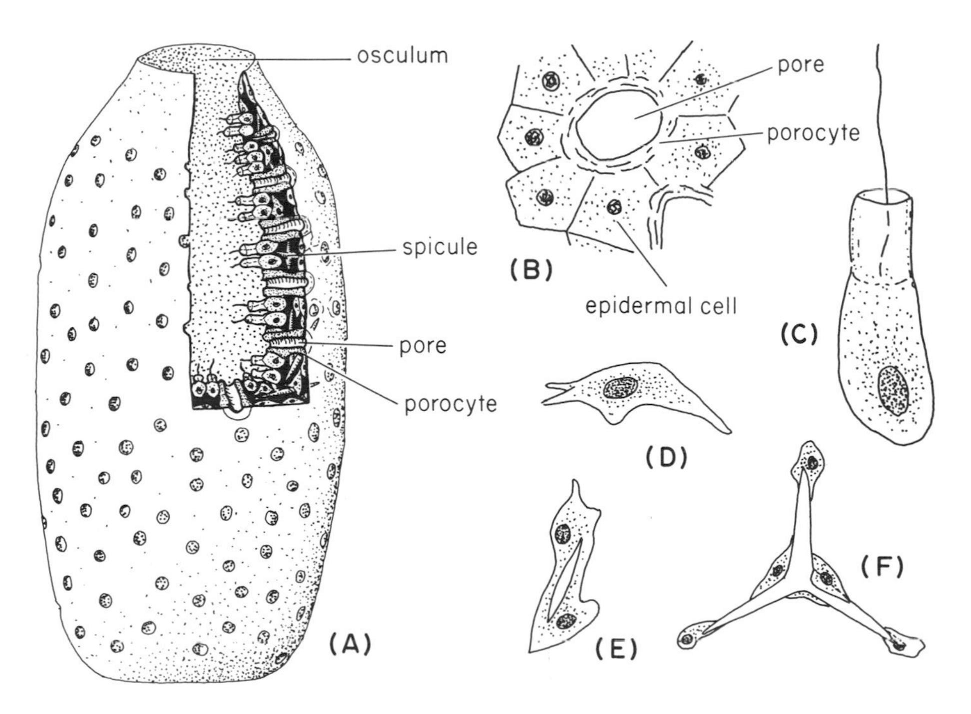

Fig. 9-12. MORPHOLOGY OF THE PORIFERA. **(A)** General body form, diagrammatic with section cut out of body wall. **(B)** View of body surface much enlarged. **(C)** Collar cell. **(D)** Amoeboid cell. **(E)** and **(F)** Amoeboid cells constructing spicules. [**(E)** *and* **(F)** *after* Woodland.]

irregular form that varies greatly among individuals of the same species. Three types of cells are recognized in the sponges. These are organized about a network of canals, the flattened *epithelial cells* covering the external surface of the individual (or the colony as you please), the flagellate *collar cells* lining the canals, and *amoeboid cells* wandering through the gelatinous material between canals (Figure 9-12).

The beat of the flagella of the collar cells maintains a current through canals where food particles are removed by adhesion to the collar. The amoeboid cells serve as jacks-of-all-trades in digesting food, in passing it from cell to cell, in constructing the skeleton, and in forming the epithelial and sex cells. Each cell is responsible for respiration and excretion much as are the single protozoan cells. The skeleton itself is composed either of *spicules* (Figure 9-12), consisting of radial spines or rods which have an organic axis and inorganic walls (calcite or opaline silica), or of organic fibers of *spongin.* The fertilized eggs develop into free-swimming ovoid larvae; after a brief period, they settle to the bottom.* Some attach to rocks, plants, and shells; others rest directly on the substrate, anchored by their flattened shape or rootlike spicules. Once attached or anchored, they remain fixed as sessile benthos. Small pieces broken from the individual will regenerate a new individual, and some form irregular colonies by asexual splitting.

Ancestry

Almost certainly the sponges evolved from flagellate protozoans. The larva has simple flagellate cells and lacks collar cells. This may be a recapitulation of evolutionary history, or a larval specialization. If the former interpretation is correct, the sponges developed from some unknown primitive flagellate; if the latter, their ancestry lies in the flagellates characterized by collar cells.

The Porifera represent one of the two (or more) shoots from protozoan stock that lead to multicellular animals—or, if I can switch metaphors, they are one of the railway lines leading from the protozoan terminus. They failed, however, to be more than a narrow-gauge branch line along which ran an evolutionary local. Why was this the local? The slick answer would be that the structural plan and its realized and potential variation lacked the necessary stuff to go further than an organization of canals and spicules. But that isn't very satisfactory as an answer. Fossils probably will never be found to supply the solution—still, speculation is interesting.

* The development of the larval sponge parallels, in a general way, that of the coral described on p. 26, although differing in important details (Hyman, 1940). The mode of development is a primary reason for considering a sponge an individual and not a colony.

Trends in adaptation

Canal system. Regardless of ancestry, subsequent evolution was an elaboration of the structural plan of canals, collar cells, epithelial cells, and amoeboid cells. The simplest structure consists of a simple tube, closed and attached at the base; open at the other end, and pierced by many minute pores (Figure 9-13). The inner surface of the tube is

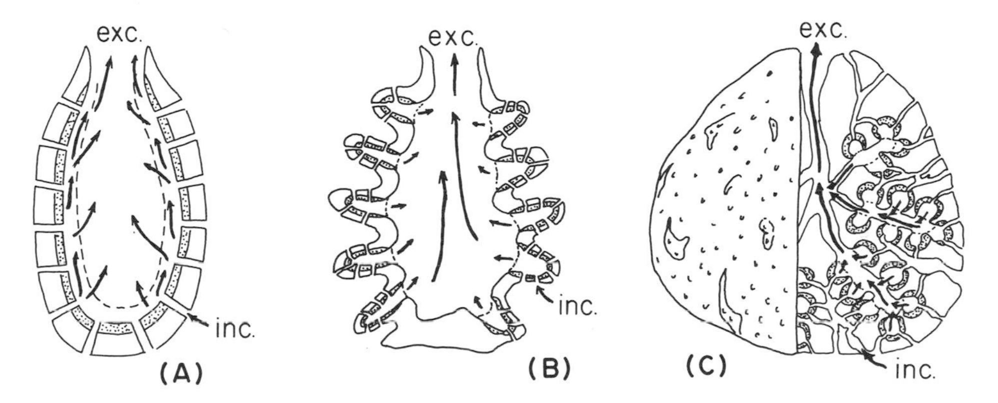

Fig. 9-13. TYPES OF CANAL SYSTEMS IN THE PORIFERA. Location of collar cells indicated by stippling. **(A)** Simple type; collar cells line entire surface of cloaca. **(B)** More complex system. Pores open into side canals and these in turn into the cloaca. Collar cells limited to lining of canals. **(C)** Highly complex system. Pores open through small canals into circular chambers lined by collar cells. Chambers empty into canals leading ultimately to osculum. Abbreviations: exc., excurrent; inc., incurrent.

lined by collar cells; the outer by epithelial cells. The beat of the flagella forces water out the open end of the tube, and fresh water flows in through the pores. The food-getting, respiratory, and excretory functions depend on the flow of water through this system. The rate of flow is a product of the current flow about the sponge and the current induced by the flagella. Any structural modification which increases the efficiency of the internal "pump" would increase the ability of the animal to survive in different external currents. For reasons related to the size of the flagella and to their beat, they are most efficient in small chambers; evolution, therefore, consists of subdivision of the tube into small bore canals (Figure 9-13B), and the development of an elaborate system of chambers in these canals (Figure 9-13C).

General form. As one might expect, the secondary evolutionary trends are adjustments to the external current and to the sediment carried by that current. Overall form varies with the current intensity—species with low, encrusting form grow in turbulent water; species with erect form in calm water. Within a species, individuals also vary in response to environmental differences. If the current is constant in direction, the incurrent openings are on the current side and the excurrent openings on the lee side of the sponge. The sponge form is also adapted to sedimentation—which clogs the pores and openings—by erect or elevated form that carries the body above the zone of maximum sedimentary movement and by reduction of sediment collecting surfaces.

Skeleton. Sponges are also adapted to turbulence and to support of the canal system through elaborate skeletal structures. The spicules take various forms as *monaxons, tetraxons, triaxons,* and *polyaxons,* and these, in turn, are woven into rigid skeletons (Figure 9-14). The siliceous (spicules of silica) and calcareous (calcite spicules) sponges are more common in cold waters; those with spongin skeletons predominate in tropical and semitropical seas. The adaptive significance of spicule composition is not clear, although siliceous skeletons of deep-water sponges are less soluble at low temperatures than calcite.

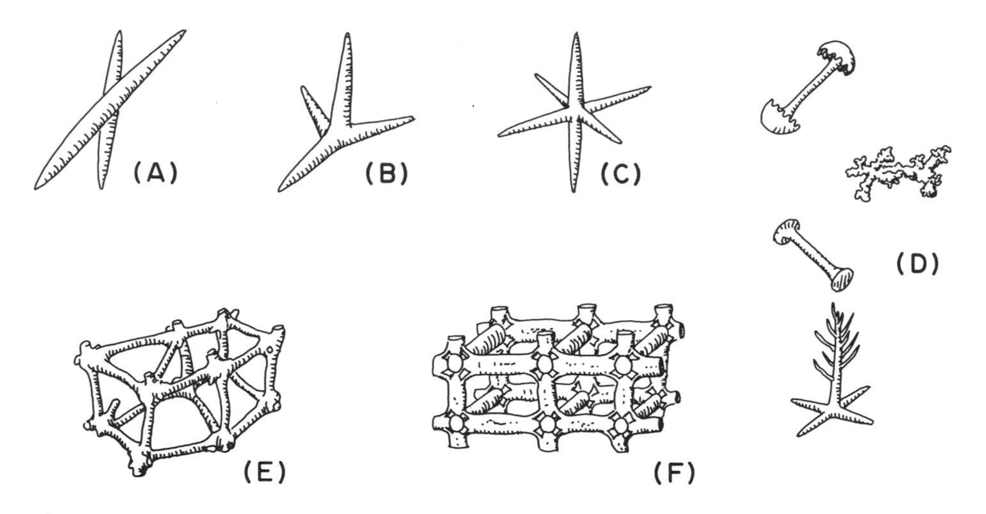

Fig. 9-14. SPONGE SPICULES. **(A)** Monaxons. **(B)** Tetraxon. **(C)** Triaxon. **(D)** Variety of modified spicules. **(E)** and **(F)** Spicules fused in tight latticework.

Paleoecology

The sensitivity of sponges to current and sediment makes individual species and genera fairly good environmental indicators. They are most abundant in shallow waters today, although some are characteristically in deep water. The phylum is predominately marine, but some genera occur in fresh waters. Most modern sponges inhabit relatively clear water, but some Paleozoic types apparently thrived in muddy environments. A few environmental analyses based on sponge adaptations and some efforts to associate particular species with particular environments of deposition have been made (Table 9-4, also Okulitch and Nelson, 1957; de Laubenfels, 1957). As I will report in most of the succeeding chapters, much remains to be done in this field.

TABLE 9-4

Paleoecology of sponges

Taxonomic group	*Paleozoic occurrences (Okulitch and Nelson, 1957)*	*Post-Paleozoic occurrences (Laubenfels, 1957 a, b)*
Calcispongea	Permian reef associations—warm, turbulent, low turbidity environments.	Low turbidity, normal marine salinity, depths less than 100 m.
Hyalospongea		Recent species deep, cold waters, 1000 + m.
Demospongia	Inhabited wide range of environments—turbid to clear. Devonian *Hindia* particularly associated with shaly beds.	Most recent genera occur at moderate depth, 10-300 m., require water of normal salinity, low turbidity.
Archaeocyatha	Reef formers. Probably limited to low turbidity, shallow water environments.	

Classification and distribution in time

Since most sponges that build rigid skeletons (the kind preserved as identifiable fossils) are relatively advanced types, the study of fossil sponges helps little in determining the relationships of the major groups of sponges; the study of recent sponges helps no more. Three classes (Table 9-5) are recognized on the basis of spicule composition and structure, the Calcispongea with calcareous spicules, the Hyalospongea with siliceous spicules of triaxon type, and the Demospongia with siliceous or spongin spicules not triaxons. Sponge spicules have been reported (doubtfully) from the Precambrian. Cambrian, Ordovician, and Silurian sponges had siliceous skeletons, but in Devonian time, some species with calcareous spicules appeared (Figure 9-15). Since both siliceous and calcareous sponges are locally abundant, some species and genera with short geologic ranges serve as guide fossils. Although isolated spicules are common microfossils, they have been little used because of the variety of types that can be found in a single sponge.

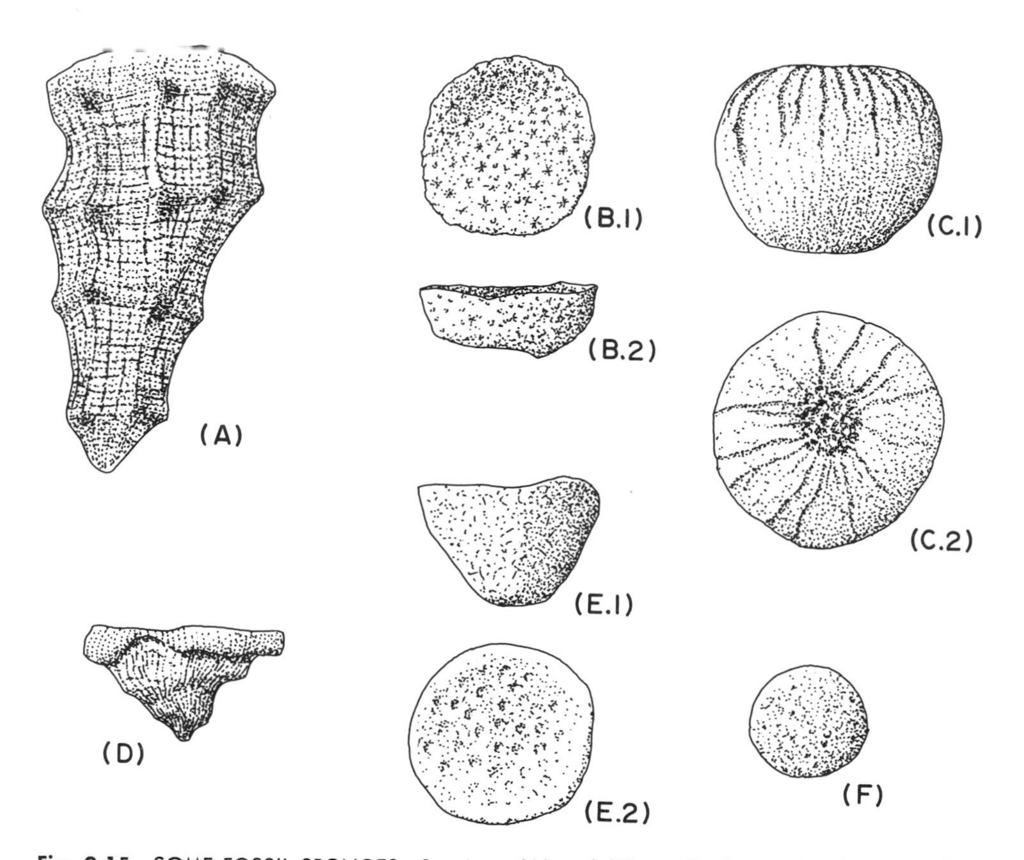

Fig. 9-15. SOME FOSSIL SPONGES. Specimen **(A)** and **(B)** are Hyalospongia; the remainder are Demospongae. **(A)** *Hydnoceras;* Devonian to Mississippian; length 13.5 cm. **(B)** *Astraeospongia;* Silurian to Devonian; diameter 5.9 cm. **(B.1)** top, **(B.2)** side view. **(C)** *Astylospongia;* Ordovician to Silurian; diameter 3.0 cm. **(C.1)** side view; **(C.2)** top. **(D)** *Hyalotragos;* Jurassic; side view; height 3.2 cm. **(E)** *Palaeomanon;* Silurian; diameter 3.1 cm. **(E.1)** side view; **(E.2)** top. **(F)** *Microspongia;* Ordovician to Permian; diameter 1.1 cm.

Cousins or in-laws?

Several groups of fossils show no clear affinities to any of the recognized phyla. Some have very simple skeletons; others have relatively complex ones, but their general and/or detailed form differs sharply from that known in other groups. In none of these fossil scraps-and-tagends does the skeleton give sufficient information about the organization of soft parts to help much in identification. Some may represent distinct and now extinct phyla, and others may be atypical branches from the known phyla. Paleontologists customarily arrange these as appendices to the recognized phyla on the basis of real or fancied resemblances or

TABLE 9-5

Classification of the Porifera

Precambrian to Recent. Multicellular units. Cells not organized into tissues, but three distinct types of cell present. Canal system. Skeleton of calcite, silica, or spongin. Radial symmetry or asymmetric. Marine or freshwater; sessile benthonic.

CLASS CALCISPONGEA

Devonian to Recent. Skeleton formed of spicules of calcium carbonate. Canal system simple to complex.

CLASS HYALOSPONGEA

Cambrian to Recent. Skeleton formed of siliceous spicules of triaxon type.

CLASS DEMOSPONGIA

(?) Precambrian, Cambrian to Recent. Skeleton of siliceous spicules or fibers of spongin. Spicules monaxial or tetraxial.

Simple fossil types with some spongelike features. May represent distinct phyla.

CLASS ARCHAEOCYATHA

Early and middle Cambrian. Skeleton of closely united calcareous spicules. Cup or cone shaped with outer and inner wall. Walls perforate. Sessile benthos, probably all marine.

CLASS RECEPTACULITIDA

Ordovician to Devonian. Skeleton discoid, of closely joined calcareous spicules arranged in double spiral. Sessile benthos, probably all marine.

---

on their apparent level of organization. Those with the simplest form are usually placed among the sponges or coelenterates.

The Archaeocyatha

One of these, the Archaeocyatha, comprises about sixty-five genera from lower Cambrian rocks. They are somewhat like a coral in form (Figure 9-16), rather cone or cup shaped, and are calcareous. Most consist of two cones, one inside of the other, with various types of plates between. Typically, the walls of the cones and the plate are perforated. The skeleton is radially symmetrical about the axis of the cones. The soft anatomy is entirely unknown. Though they resemble corals superficially, they are quite different in detail. They have a few sponge characteristics, but lack the typical spicules, and the arrangement of cones and plates is unlike the skeleton of the typical sponges. Okulitch, in the *Treatise on Invertebrate Paleontology* (1955), classifies them as a separate phylum; others have placed them among the coelenterates and even as algae or protozoa. If their assignment to a separate phylum

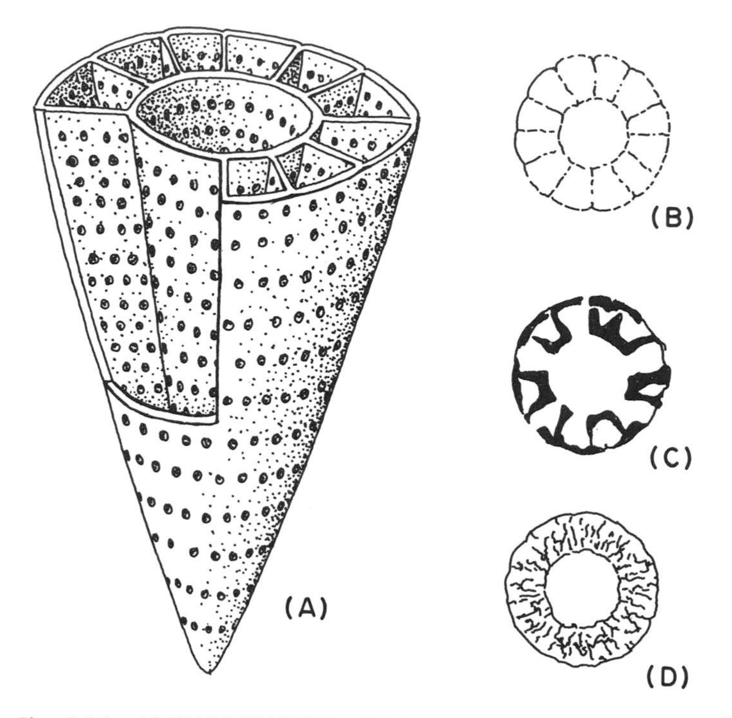

Fig. 9-16. ARCHAEOCYATHIDS. **(A)** Diagrammatic view of archaeocyathid. Note double wall with radial plates (parieties) in space (intervallum) between inner and outer walls. Note also pores in walls and parieties. **(B)** *Archaeocyathus;* early Cambrian; cross section; diameter 8 mm. **(C)** *Ajacicyathus;* early Cambrian; cross section; diameter 5 mm. **(D)** *Archaeocyathus;* early Cambrian; cross section; diameter 2.5 cm. [**(B)** *after* Bedford and Bedford; **(C)** and **(D)** *after* Okulith.]

is correct, they presumably represent an independent shoot from the Protozoa or from one of the primitive metazoan (many-celled animals) stocks.

TABLE 9-6

Glossary for Porifera *(numbers in parentheses refer to pertinent illustration).*

Amoeboid cell. Cell with irregular and changing shape. Characterized by lobose pseudopods which may be extended from body surface. Lies between epidermal and collar cells in body well. (9-12A, D, E, F.)

Canal. Tube leading from external pore to cloaca and serving for water flow. May lead into small chambers. (9-12, 9-13)

Cloaca. Central cavity of individual sponge into which pores and/or canals empty and which opens externally through the osculum. (9-12A; 9-13A, B, C.)

TABLE 9-6 (continued)

Collar cell. Cell with a distinct tubular collar about the base of a long, slender whip-like extension, the flagellum. Collar cells line inner surface of canals and/or cloaca. (9-12A, C.)

Desma. Spicule of irregular form bearing knotty growths. (9-14D.)

Epidermal cell. Brick-like cell on external surface of sponge. (9-12A, B.)

Monaxon. Spicule with single axis of growth. May be curved or straight and may bear expansions at one or both ends. (9-12E; 9-14A, D.)

Osculum. Large opening from internal cavity of sponge to exterior. Serves for outward (excurrent) flow of water. (9-12A; 9-13A, B, C.)

Polyaxon. Spicule which has several rays diverging from a point.

Pores. Numerous small openings from exterior of sponge into canal or cloaca. Serve for inward (incurrent) flow of water. (9-12A, B; 9-13A, B, C.)

Porocyte. Cell which surrounds each pore. Expansion or contraction serves to open or close pore. (9-12A, B.)

Spicule. Skeletal element of sponge. Typically small needlelike rod or fused cluster of such rods. (9-12E, F; 9-14.)

Tetraxon. Spicule having four axes of development. (9-14B.)

Triaxon. Spicule having three axes of development. (9-14C.)

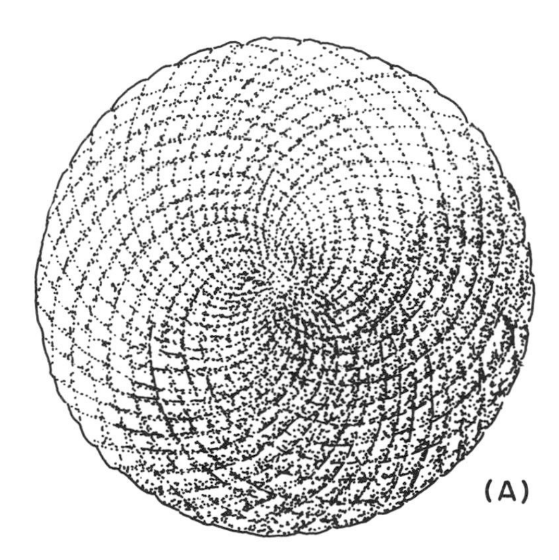

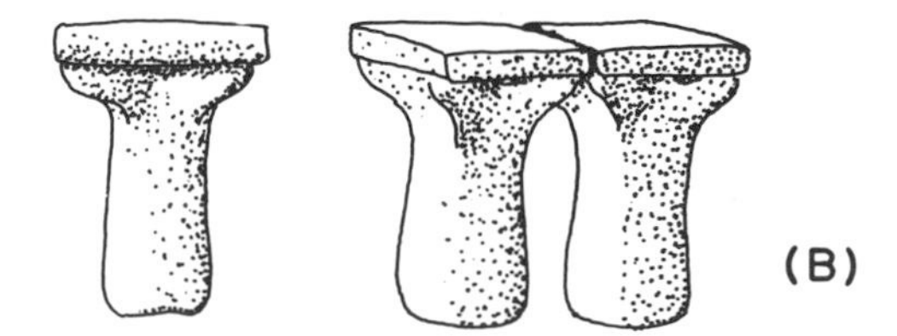

Fig. 9-17. THE RECEPTACULIDAE. **(A)** *Receptaculites,* Middle Ordovician to Devonian, view from above, diameter 4.5 cm. **(B)** Individual "spicules" of *Receptaculites;* lateral views. The rectangular plates which form the external surface of the fossil rest on circular pillars.

The archaeocyathids are useful guide fossils in Cambrian rocks. Apparently they were worldwide in the early Cambrian but restricted to Eurasia by the middle Cambrian. They occur in abundance in some calcareous beds.

The receptaculids

The fossils placed in the class Receptaculitida are somewhat more sponge-like. They are ovoid or discoid and formed of closely joined calcareous spicules. The spicules are perpendicular to the external surface and are typically arranged in a pattern of double spirals (Figure 9-17). The wall formed by the spicules encloses a central cavity. No pores or canals have been surely identified. The absence of these and the peculiar structure and arrangement of the spicules distinguishes the receptaculids from the typical sponges.

Only ten genera are known at present, all from Ordovician, Silurian, and Devonian rocks, but several, particularly *Receptaculites* and *Ischadites* are common fossils.

REFERENCES

Bandy, O. L. 1964. "General Correlation of Foraminiferal Structure with Environment," in Imbrie, J. and N. D. Newell (Eds.), *Approaches to Paleoecology*. New York: John Wiley.

Be, A. W. H. and L. Lott. 1964. "Shell Growth and Structure of Planktonic Foraminifera," *Science*, vol. 145, pp. 823-824.

Campbell, A. S. and R. C. Moore. 1954. "Protista 3 (Chiefly Radiolarians and Tintinnines) in Moore, R. C. (Ed.), *Treatise on Invertebrate Paleontology, Part D.* New York: Geological Society of America.

de Laubenfels, M. W. 1955. "Porifera," in Moore, R. C. (Ed.), *Treatise on Invertebrate Paleontology, Part E*, pp. 22-112. New York: Geol. Soc. of America.

———. 1957a. "Sponges of the Post-Paleozoic," in Ladd, H. S. (Ed.), "Treatise on Marine Ecology and Paleoecology," vol. 2, *Geol. Soc. of America, Memoir 67*, pp. 771-772.

———. 1957b. "Marine Sponges," in Hedgpeth, J. (Ed.), "Treatise on Marine Ecology and Paleoecology," vol. 1, *Geol. Soc. of America, Memoir 67*, pp. 1083-1086.

Hill, D. 1964. "The Phylum Archaeocyatha," *Biol. Review*, vol. 39, pp. 232-258.

Hyman, L. 1940. *The Invertebrates*, vol. 1, *Protozoa through Ctenophora*. New York: McGraw-Hill.

Jones, D. J. 1956. *Introduction to Microfossils*. New York: Harper.

Loeblich, A., Jr. and H. Tappen. 1964a. "Foraminiferal Facts, Fallacies, and Frontiers," *Geol. Soc. of America Bull.*, vol. 75, pp. 367-392.

———. 1964b. "Protista 2. Sarcodina, Chiefly 'Thecamoebians' and Foraminiferida," in Moore, R. C. (Ed.), *Treatise on Invertebrate Paleontology, Part C.* New York: *Geol. Soc. of America.*

Okulitch, V. J. 1955. "Archaeocyatha," in Moore, R. C. (Ed.), *Treatise on Invertebrate Paleontology, Part E.*, pp. 1-20. New York: Geol. Soc. of America.

——— and S. J. Nelson. 1957. "Sponges of the Paleozoic," in Ladd, H. S. (Ed.), "Treatise on Marine Ecology and Paleoecology," vol. 2, *Geol. Soc of America, Memoir 67,* pp. 763-770.

Phleger, F. B. 1964. "Foraminiferal Ecology and Marine Geology," *Marine Geology,* vol. 1, pp. 16-43.

Walton, W. R. 1964. "Recent Foraminiferal Ecology and Paleoecology," in Imbrie, J. and N. D. Newell, (Eds.), *Approaches to Paleoecology.* New York: John Wiley.

Zeller, E. J. 1950. "Stratigraphic Significance of Mississippian Endothyroid Foraminifera," *Univ. Kansas Paleon. Contrib.,* Art. 4, pp. 1-23.

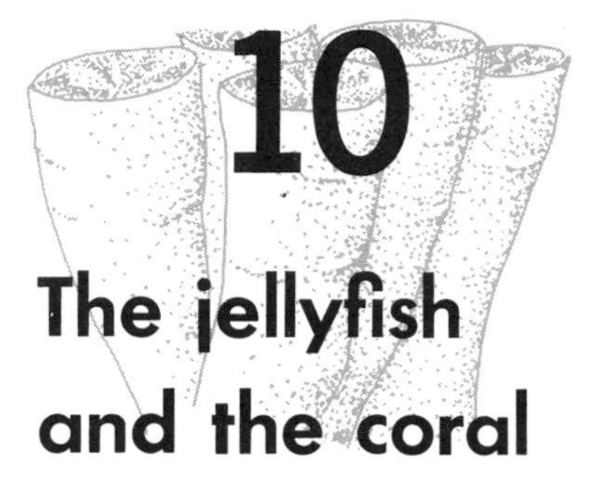

10 The jellyfish and the coral

The origins of multicellular animals (the Metazoa) lie somewhere in the murk of the Precambrian, lighted only by the study of the comparative morphology of modern organisms. The organization of the sponges, their ontogeny and the characteristics of their cells, demonstrate that the poriferan lineage either evolved independently from the flagellate protozoans or diverged from the other multicellular line very early. The sponges, then, give no information on the evolutionary "mainline."

With a single exception, the phylum Mesozoa, the remaining multicellar animals are more highly organized than the sponges and more difficult to connect with a primeval protozoan. The Mesozoa, which consist of an aggregate of slightly differentiated cells and thus are structurally intermediate between unicellular and multicellular animals, may have regressed from a more complex organization as a result of parasitic adaptations. The fossil data is, at present, of essentially no assistance—except in suggesting that the story is even more complicated than it appears from modern organisms. One observes a number of groups, e.g.,

Archaeocyatha, which apparently represent independent "attempts" at metazoan organization. In consequence, the suggestions made below are —at the very best—good guesses.

In a preceding chapter (pp. 147-157), I outlined the methodology of phylogenetic interpretation. In essence it involves the analysis of variant states of a set of common characters. The characters themselves are outside the field of inquiry. In comparing the high ranking taxa, i.e., superphyla, phyla, and subphyla, however, the characters are the subject of study and must be regarded as variant adaptations, e.g., polar organization and bilateral symmetry to active locomotion on the substrate, metamerism to swimming or burrowing, etc. Organization is so disparate that legitimate comparison becomes very difficult.

Two methods are brought to bear on this difficulty. One seeks structural unity in early ontogeny, and the other in subcellular structure, e.g., organelle form, fine structure of organelles, and molecular structure. Dillon (1962) finds similarities in fine structure which lead him to unite red and brown seaweeds with the metazoans; Hyman (1940) relies on ontogenetic pattern to reconstruct metazoan origins from a colonial flagellate protozoan; Hadzi (1963) derives the metazoans by cellular partitioning of a ciliate protozoan, because of gross similarity in organelles. There is a tendency—though not universal—to neglect the possibilities of adaptive parallelism and convergence.

The second leg of the phylogenetic stool is an evolutionary model that explains in reasonable fashion the inferred transformation. Thus Hyman argues for adaptation in the primitive metazoa for a semiplanktonic, pelagic life; Hadzi for adaptation to the requirements of the sessile benthos. The comparison of "reasonableness" for the various inferences is subjective.

The third leg, the empiric rules of linearity, irreversibility, simplicity, etc., are relatively ineffectual in this situation. For example, in a linear, irreversible phyletic model, the Mesozoa would lie in the protozoan-metazoan lineage, but there are very convincing arguments against this model. More to the point, a linear model demands separate origins for the sessile, radially symmetrical coelenterates and the vagrant, bilaterally symmetrical flatworms. But radial symmetry appears in sessile animals clearly derived from active, bilateral ancestors, e.g., the echinoderms and, among the chordates, the sea squirts. Thus the coelenterates could derive from early, simple flatworms (Hadzi).

In general, of the three legs necessary to phylogenetic models, one is sawed off altogether, one is loose in its socket, and the other is badly cracked. Of the various postulated sequences of metazoan evolution, the following one, hybridized principally from Hyman and Hadzi, seems at least as probable as any other and serves to organize the available information effectively:

a). Development of a colonial, vagrant, benthonic "protozoan" that moved by ciliary gliding on the substrate and fed on very small benthic organisms and on organic detritus.

b). Organization of *a*) a ventral feeding and locomotion layer, *b*) a dorsal protective layer, *c*) an interior cell mass for circulation and reproduction, and *d*) an anterior—posterior polarity with respect to unidirectional locomotion.

c). Differentiation and specialization of cells in various layers and at different poles.

d). Invagination of a part (or all) of the ventral layer to form a blind gut.*

Of these, stage **a).** is widely observed among recent protozoans and occurs in the ontogeny of sponges and of the metazoans. The sponges, however, adopted at this grade of development a sessile habit with a dorsal feeding surface and a protective ventral one. Polarity in sponges derived from the feeding polarity, and the dorsal layer invaginated to form a gut equivalent, the cloaca. Stage **b).** is unknown either among adult or larval metazoans; its loss in ontogeny is reasonable, however, since the common pelagic larval stage must move in three rather than two dimensions. Stage **c).** occurs in colonial protozoans, in larval sponges (though not in adults), and in all the metazoans. Stage **d).** is characteristic of metazoan ontogeny and provides the basic organization for the flatworms (Platyhelminthes) and Coelenterata.

In this phylogenetic scheme, the coral-jellyfish lineage then adopted a sessile, benthonic habit. The evolutionary steps are not obvious but may have started with development of tentacles around the mouth—with initial sensory functions and subsequent modification for grasping "larger" food—living organisms or debris—on the substrate. Collection of swimming or floating organisms could follow; if these became a principal food source, a sessile adaptive niche would become available. Adaptation to this habit would result, as in other sessile organisms, in reorientation of the body with the mouth and tentacles upward and in the development of a supportive "column" or "stem." Since the gut lacked a terminal anus, the posterior end of the body could be utilized for this function. With sensory input, food, and attacks arriving from above and all sides, any bilateral symmetry was suppressed by radial development. The ontogeny of the internal partitions, in coelenterates the pattern of muscles on these partitions (see below), and elongation of the mouth defines a

* Hadzi envisages stages **b).**, **c).**, and **d).** in a noncolonial multinucleate protozoan. Metazoans would have evolved directly from highly organized protozoans, the ciliates, by partitioning. No multinucleate ciliates are, however, known, and the complex organization and organelle differentiation of ciliates seems an opposite trend to cellular subdvision.

plane of bilateral symmetry that may be a residuum from this early stage —or a subsequent specialization.*

The coelenterate organization

The coelenterate radiation has a basic structural plan which appears in both adult organization and larval development (Figure 10-1) of animals as diverse as jellyfish and corals. The plan encompasses *radial symmetry* about an axis between the *mouth* (oral pole) and *base* (aboral pole). The body possesses a central cavity, the *coelenteron,* that develops either within the larva or by invagination of the "hollow ball" stage (Figure 2-9 and p. 26). The cavity opens to the surface at the mouth but

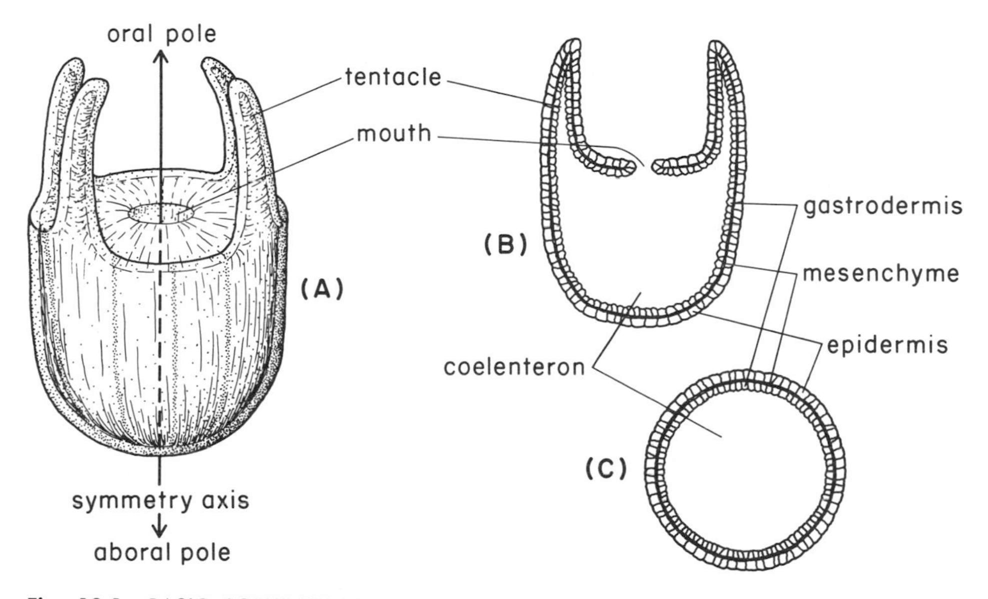

Fig. 10-1. BASIC COELENTERATE PLAN. **(A)** Generalized form in perspective drawing. **(B)** Longitudinal section. **(C)** Transverse section.

* In the first edition of this text (1960, p. 238), I suggested an origin for the Coelenterata in a bell-shaped pelagic protozoan colony. Subsequent evolution at the metazoan level produced differentiated cell layers, oral-aboral polarity, oral tentacles, and radial symmetry as in the jellyfish. The most telling arguments against origin of radial symmetry in this fashion are 1) the limitation of radial symmetry in protozoans, sponges and metazoans other than coelenterates to sessile benthonic forms and to vagrant forms clearly derived from sessile ancestry, 2) the universal presence of functional bilateral form in all free-living (nonsessile) animals except the coelenterates and a few crinoids.

ends blindly at the aboral end. The mouth is surrounded by *tentacles* (possibly the primitive number is four). Food is grasped by the tentacles, passed through the mouth into the coelenteron, and partly digested there before absorption. Respiration and excretion occur by diffusion from the individual cells.

The body wall consists of two distinct and specialized layers, the *epidermis* and *gastrodermis,* conjoined by a middle layer, the *mesenchyme,* with a mixture of jellylike matrix and undifferentiated cells (Fig. 10-2).

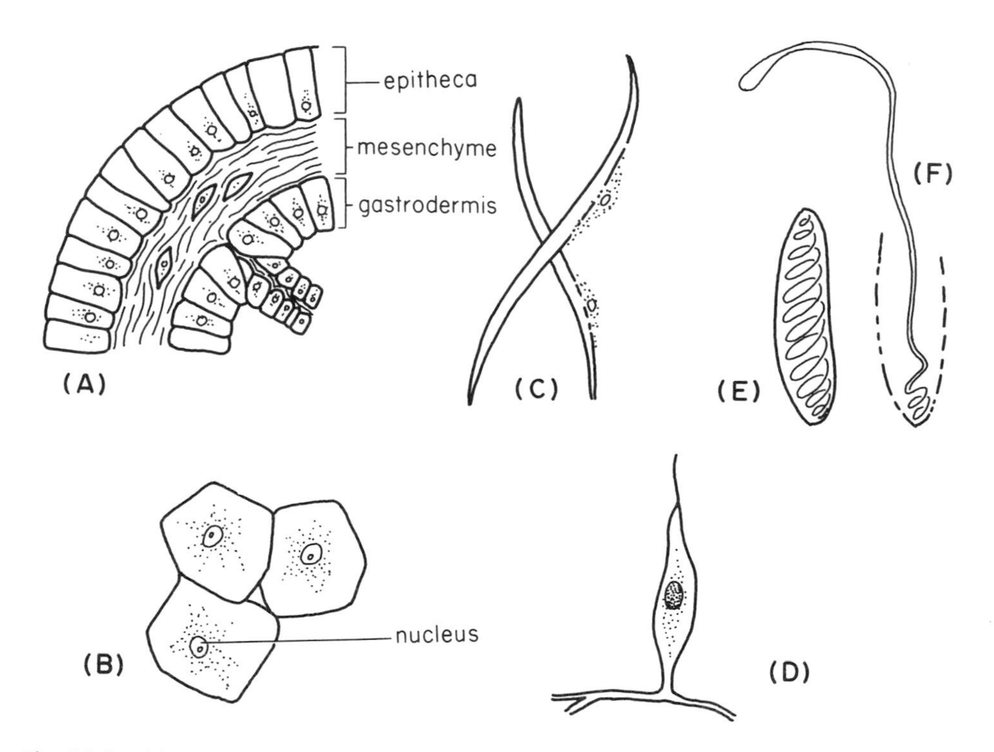

Fig. 10-2. CORAL TISSUES AND CELLS. **(A)** Section of body wall. **(B)** Epithecal cells, section parallel to body surface. **(C)** Muscular cells. **(D)** Nerve cell. **(E)** Nematocyst, unexploded. **(F)** Nematocyst, exploded.

The epidermis includes cells specialized for contraction (muscular strands) and for perception, intracellular communication, and coordination (nerve cells in a network), as well as the brick-like *epithelial* cells. Also interspersed in the epidermis, particularly on the tentacles and about the mouth, are the stinging cells, the *nematocysts,* which serve in food-getting and protection, and the gland cells, which secrete mucous-like substances. The gastrodermis consists mainly of epithelial cells, some of which are specialized for ingestion and intracellular digestion of food. These nutritive cells bear flagella that maintain circulation within the coelenteron.

Muscular, nervous, and mucous gland cells also occur in the gastrodermis. Other gland cells produce digestive enzymes.

The adaptive radiation of the coelenterates

The abundance and variety of coelenterates from the early and middle Cambrian indicate radial evolution, extending down into the Precambrian.

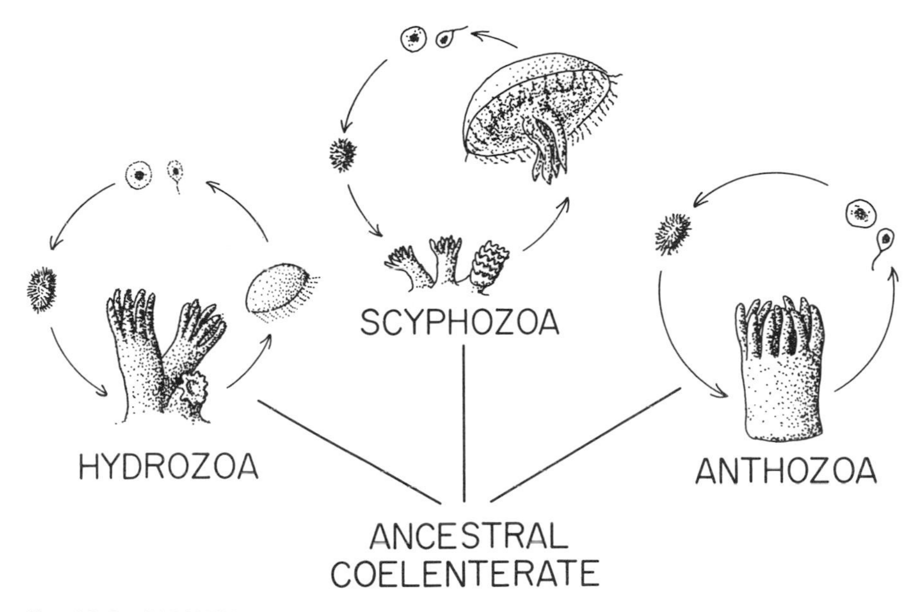

Fig. 10-3. RADIATION OF THE COELENTERATES. The relative importance of the attached polypoid stage and the free-swimming medusoid stage are shown by their size in the drawing. The hydrozoans have a large polyp (often colonial) that produces small medusoids, which in turn produce eggs and sperm. The scyphozoans have an insignificant polyp and a large medusoid stage. The anthozoans have no medusoid stage, and the polyp itself produces egg and sperm.

The phylum displays a fundamental trichotomy* based on reproductive and developmental adaptations (Table 10-1, Figure 10-3). The primitive

* Two obscure groups are considered by some paleontologists to constitute two additional classes. One of these has assigned to it a jellyfish-like object from the Precambrian Grand Canyon Series. Since little is known of them, and they are very rare, I have omitted them from this discussion.

Some authorities classify the "comb-jellies," the *ctenophores,* in the Phylum Coelenterata; others place them in a phylum of their own. When ranged in the coelenterates they are recognized as a separate subphylum, and the three classes described here are grouped in the subphylum Cnidaria. No fossil ctenophores are certainly known.

TABLE 10-1

Classification of the Coelenterata *(based on the* Treatise on Invertebrate Paleontology, Part F.*)*

Precambrian to Recent. Cells organized into tissues but not into distinct organs. Central digestive cavity with single opening. No head. Radial or biradial symmetry. Many have calcareous or horny skeleton. All aquatic; marine and freshwater; sessile benthos and pelagic, either planktonic or nektonic.

? CLASS PROTOMEDUSAE

Precambrian to Ordovician and possibly Silurian and Pennsylvanian. Subellipsoidal bodies with lobate radial segments. Some appear to have tentacles. Probably marine pelagic.

CLASS HYDROZOA

Cambrian to Recent. Radial symmetry. Most species have both medusoid and polypoid stages. Coelenteron not partitioned; mesoglea gelatinous, noncellular. Solitary or colonial; mostly marine; pelagic and sessile benthonic.

Order *Stromatoporoidea*

Cambrian to Cretaceous. Colonial; colony comprising calcareous lamellae and vertical pillars. Sessile benthonic; reef builders.

Order *Trachylinida*

Late Jurassic to Recent, possible Cambrian fossils. Medusoid; polypoid stage reduced or absent. Predominately pelagic.

Order *Hydroida*

Cambrian to Recent. Polypoid well developed, colonial or solitary. Marine and freshwater; pelagic and sessile benthonic.

Order *Spongiomorphida*

Triassic to Jurassic. Colonial. Massive calcareous skeleton with horizontal plates and vertical pillars; pillars may be grouped to form vertical tubules. Probably marine; sessile benthonic.

Order *Milleporina*

Late Cretaceous to Recent. Polypoid colony with massive to encrusting calcareous skeleton with pores through which polyps protude. Medusoid generation reduced. Marine, predominately sessile benthos.

Order *Stylasterina*

Late Cretaceous to Recent. Massive to encrusting skeletons with pores for polyps. Non-medusoid. Marine, sessile benthos.

Order *Siphonophorida*

Ordovician to Recent. Colonial. Polyps and medusae attached to stem or disk. Supported by swimming bell or float. Marine; pelagic. Rare as fossils.

CLASS SCYPHOZOA

Cambrian to Recent. Solitary. Coelenteron divided in some by mesentaries. Radial symmetry. Polyp much reduced, highly modified or lost. Mesoglea cellular. Entirely marine, predominately pelagic.

Subclass Scyphomedusae

Late Jurassic to Recent. True jellyfish. Marine, predominately pelagic. Rare as fossils.

Subclass Conulata

Middle Cambrian to early Triassic. Skeleton of chitin and calcium phosphate; form, elongate pyramidal. Attached by apex of pyramid or free. Traces of tentacles preserved as well as evidence of tentacular muscle attachments. Probably marine, sessile or planktonic.

TABLE 10-1 (continued)

CLASS ANTHOZOA

Middle Ordovician to Recent. Biradial or bilateral symmetry. Medusoid stage never present. Coelenteron partitioned by mesentaries. Marine; sessile or sluggish vagrant benthonic.

Subclass Ceriantipatharia
Miocene to Recent. Colonial or solitary. Mesentaries unpaired with weak musculature; new mesentaries develop in dorsal intermesenteric space. May have skeleton of horny material. Very rare as fossils.

Subclass Octocoralla
Permian to Recent. Eight tentacles and mesentaries. Skeleton of horny or calcareous spicules, separate or closely united. Position of polyp in colony marked by pit, but not divided by radial septa.

Subclass Zoantharia
Ordovician to Recent. Mesentaries paired and coupled, typically in cycles of six, six pairs in initial set.

Order *Rugosa*
Ordovician to Permian. Solitary or colonial. Calcareous skeleton with calcareous wall, the epitheca, septa, and typically, tabulae and dissepiments. Six primary septa, others develop in ventrolateral and lateral spaces but none in dorsolateral. Sessile benthos. Some, particularly colonial types, important as reef builders.

Order *Heterocorallia*
Carboniferous. Solitary. Four primary septa split in two near periphery, and additional septa develop between the split portions. Epitheca may be reduced or absent. Edges of tabulae may be inflected to replace epitheca.

Order *Scleractinia*
Middle Triassic to Recent. Solitary or colonial. Calcareous skeleton with epitheca, septa, tabulae, and dissepiments. Six primary septa; others develop in all interseptal spaces. Edge zone of trunk deposits skeletal material on outer surface of skeleton. Some important as reef builders.

Order *Tabulata*
Middle Ordovician to Permian. Colonial. Calcareous skeleton with epitheca and tabulae; septa typically small or absent, most common number, 12. Order of development unknown. Some important reef formers.
(Three other orders not recognized as fossils.)

---

coelenterates may have been solitary polyps with a flattened *base,* a columnar *trunk,* and a circlet of *tentacles* around the mouth. One or more phyletic lineages developed an adult stage* that floated and/or swam freely in the water. From these lines, with elaboration and modification of the trunk into a gelatinous float or bell, were derived *medusoid* (or jellyfish) forms.

One class, the *Hydrozoa,* emphasizes the elaboration and adaptation of the *polypoid* stage so that the medusoid stage is almost an afterthought designed primarily for reproduction. The distinctive polyp develops and reaches maturity (of form, not sexual maturity) with quite different morphology and adaptations than the medusa. The budding of the

* Or remotely, a medusoid larva which was "forced back" in ontogeny into the preceding generation.

medusoid individuals from the polyp (or from specialized polyps in a colony) ordinarily does not end the existence of the polyp, and the larval stage continues to live beside the adult into which it developed. For this reason, the polyp and medusa are considered separate individuals and the cycle, fertilized egg → free-swimming larva → polyp → larval medusoids → adult medusoids → fertilized eggs, an alternation of sexual and asexual generations.

A second class of coelenterates, the *Scyphozoa,* evolved elaborate medusoid adults and a polyp stage highly adapted for asexual budding but with little individuality otherwise. The third class, the *Anthozoa,* lacks a medusoid stage, and the individual animal retains the polypoid form to sexual maturity.*

Upon this fundamental division are imposed a series of trends, adaptively parallel in all three classes, but, in some cases, morphologically divergent. These trends include greater differentiation of cells to perform specialized duties, modification of tentacles and the area about the mouth to assist in food-getting, and subdivision of the coelenteron to increase surface area and to improve circulation of nutrients to all parts of an enlarging body.

Among the Hydrozoa and the Anthozoa, the sessile polyps have evolved a hard skeleton for protection and support in several independent lines. The hydrozoan skeletons, perhaps because of the small size of the individual polyp and the lack of internal structures that need support, are relatively simple, laminated, vesicular or tubular networks with supporting pillars or spines (Figure 10-6). The anthozoan polyps, which are larger and possess an internally divided coelenteron, characteristically have distinct, well-developed skeletal walls, and internal septa (Figures 10-9, 10-13).

The Scyphozoa

The true jellyfishes, the *scyphozoans,* are primarily animals of the nekton or plankton, where they float or swim by rhythmic contractions of their body. Those incapable of active swimming attach to seaweed for support. They are important and diversified members of the modern marine faunas but, as one would expect, rare as fossils. Four of the five recent orders are known as fossils, plus some genera not easily assigned to those orders. This is a relatively high percentage of preservation for a group lacking skeletons, but is based on a very few specimens preserved

* Some zoologists, e.g., Hyman, 1940, argue an alternative phylogeny in which the medusoid is regarded as the primitive form and the polyp as a specialized developmental stage that evolved later.

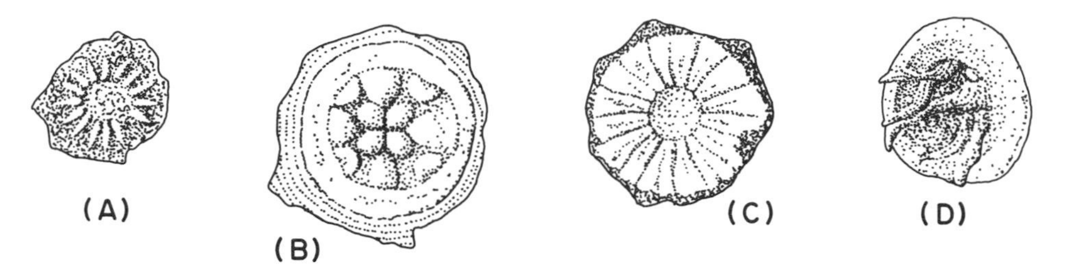

Fig. 10-4. EXAMPLES OF FOSSIL JELLYFISH. **(A)** *Lorenzinia apenninica;* Eocene; aboral view; 25 mm diameter. **(B)** *Rhizostomites admirandus;* late Jurassic; oral view; diameter 70 mm. **(C)** *Rhizostomites admirandus;* late Jurassic; aboral view; diameter 25 cm. **(D)** *Leptobrachites trigonobrachium;* late Jurassic; oral view; diameter 16 cm. Specimen **(B)**, **(C)**, and **(D)** are from the Solenhofen lithographic limestone. [**(A)** *after* Gortani; **(B)** *after* Brandt; **(C)** *after* Von Ammon; **(D)** *after* Brandt.]

as carbonized films or as molds and casts of a relatively resistant mesoglea (Figure 10-4).

A small group of Paleozoic and early Mesozoic fossils, the *conularids,* long driven from pillar to post in classification, are currently (Moore and Harrington, in Moore, 1956) set among the Scyphozoa as a distinct subclass. Typically, they have an elongate pyramidal skeleton composed of chitin, a chitinophosphatic compound, and calcium phosphate (Figure 10-5). A few specimens show traces of tentacles at the open lower end of the pyramid, and internal ridges and thickened portions of the shell margin seem to mark attachment of longitudinal and tentacular muscula-

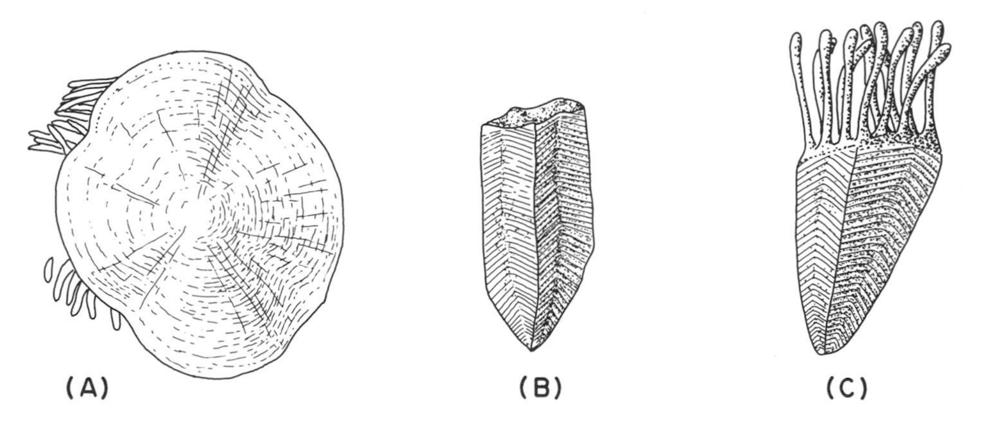

Fig. 10-5. CONULATA. **(A)** *Conchopeltis alternata;* middle Ordovician; aboral view; diameter 50 mm. The ends of the tentacles project from beneath the skeleton along one edge. *Conchopeltis* may have been a free-swimming form. Note biradial symmetry. **(B)** Conularid sp; lateral view; 10 cm high. **(C)** Reconstruction of conularid showing tentacles. Presumably this type was attached by its base. [**(A)** *after photograph by* Wells. *From Treatise on Invertebrate Paleontology.* Courtesy of Geological Society of America and University of Kansas Press. **(C)** *Based on restorations by* Kiderlen.]

ture. Only twenty genera have been described, though a few are fairly common fossils.

The Hydrozoa

All five modern hydrozoan orders are known as fossils, and two extinct orders, the *Spongiomorphida* and the *Stromatoporida,* are commonly assigned to this class. The latter order comprises genera with laminated and vesicular skeletons (Figure 10-6) that might well have been produced by either algae, sponges, bryozoa, or protozoans.

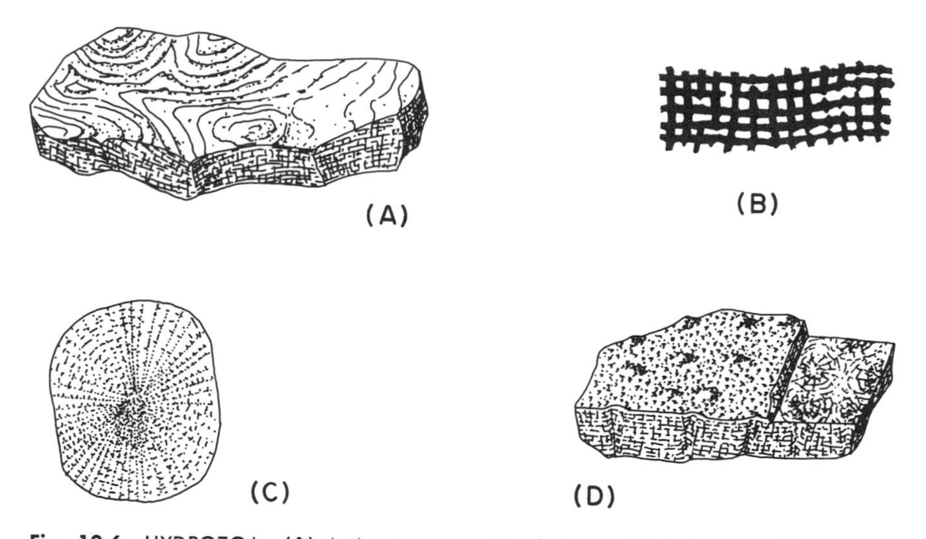

Fig. 10-6. HYDROZOA. **(A)** *Actinostroma* sp; Cambrian to Mississippian; oblique view, length 8.5 cm. **(B)** *Actinostroma* sp; thin section; 2.0 mm × 6.9 mm. Vertical elements are *pillars;* the horizontal elements are *laminae.* **(C)** *Discophyllum;* middle Ordovician; aboral view; about 7.5 cm in length. **(D)** *Stromatopora;* Ordovician to Permian; oblique view, 5.8 cm long. [**(C)** *after* Walcott.]

Species of the orders Milleporida and Stylasterida are major components of recent and Cenozoic coral reefs; the stromatoporoid species, of Paleozoic reefs. As such they are important in interpreting the ecology of the reef community. Some fossils of hydrozoan genera are widely distributed and have short geologic ranges; these, of course, serve as guide fossils. Identification is based primarily on variation in microscopic detail as shown in polished or thin section. Because of the relatively simple organization and low grade of developmental integration, the form of hydrozoans (and of the other coelenterates) is quite plastic under different environments of development. The "ecospecies" produced by this plasticity prove valuable environmental indicators—if they are properly distinguished from genetic species.

The Anthozoa

Generalities

The anthozoans are dominant elements of fossil marine assemblages, as they are of modern sea-floor communities. Of the fifteen or so orders of anthozoans, eleven are known as fossils, and three of these are known only as fossils. The form, the color, the actual existence of many coral reefs depend in large part upon anthozoans, and, although most recent members of the class are warm-water forms, the sea anemones, some corals, and some other anthozoans occur far from the tropics. The variety of anthozoans (and their conspicuous appearance) is demonstrated by their common names, organ pipe coral, soft corals, sea whips, sea fans, blue coral, sea pens, stony corals, thorny corals.

The basic anthozoan plan is much the same as that given above for the primitive polyp. The body is a hollow cylinder, closed below by the *base,* above by the *oral disk* through which the mouth opens, and divided internally by vertical partitions, the *mesentaries.* The oral disk bears one or more circles of tentacles. The body wall consists of two distinct layers, the epidermis and gastrodermis, separated by a mesenchyme containing many undifferentiated cells. A layer of soft tissue connects the polyps in colonial forms. The anthozoans have diverged during evolution in the number and arrangement of tentacles and mesentaries, in the details of mouth structure, in the form and mode of formation of the skeleton, and in symmetry. Unfortunately, paleontologists know little of the adaptive significance of these divergences.

Ceriantipatharia

Of the three anthozoan subclasses, the Ceriantipatharia are least important to paleontologists, for they have so far identified only a single fossil genus, and that Miocene. The skeleton, in those which possess one, consists of an axial rod of horny material bearing short spines (Figure 10-7). A fleshy layer surrounds the skeleton; from this arise small polyps. Some ceriantipatharians have six mesentaries, apparently a primitive character (see below and Figure 10-8).

Octocoralla

Octocorallians possess eight mesentaries and eight tentacles. Their skeleton consists of calcareous spicules variously arranged in the polps and in the fleshy layer that connects the polyps in a colony. Some also have a calcified axial structure. The spicules may be closely packed to produce a rigid (and preservable) structure or may be scattered in the body wall (Figure 10-7).

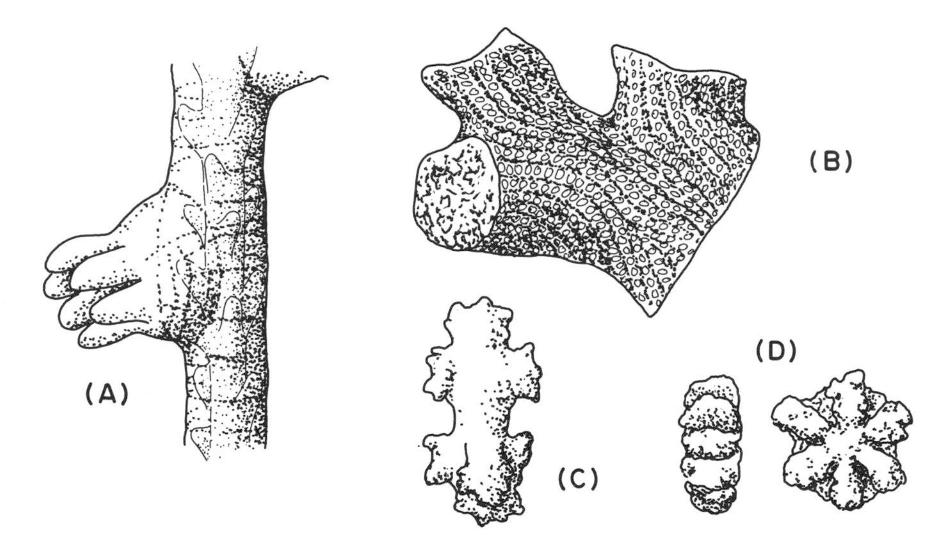

Fig. 10-7. CERIANTIPATHARIA AND OCTOCORALLA. **(A)** Polyp of ceriantipatharian, *Antipathella*. Central horny axis shown. **(B)** Octocorallan *Trachypsammia*, fragment of colony; Permian; length 2 cm. **(C)** Octocorallan *Corallium borneense*, spicule; Recent; length 0.6 mm. [**(B)**, **(C)**, *and* **(D)** *after Treatise on Invertebrate Paleontology*. Courtesy of Geological Society of America and University of Kansas Press.]

Fossil octocorals are relatively rare and include only a few genera. They have not been certainly identified before the Cretaceous, though some doubtful ones are known from rocks as old as very late Precambrian. They are important members of recent coral reef associations and presumably played a similar role in the formation of Cenozoic reefs.

Zoantharia

The subclass Zoantharia includes the greatest number of recent and fossil anthozoans—Wells and Hill (1956) record a total of 831 fossil genera. They are distinguished by the number and arrangement of the mesentaries; in addition to the primitive six, they have two or more additional

TABLE 10-2

Glossary for Coelenterata

Alar septum. One of a pair of the initial septa (protosepta) of rugosan. Located about midway between cardinal and counter septa. Secondary septa may insert pinnately on side away from cardinal. (10-8H; 10-9F, G.)

Axial vortex. Longitudinal structure in axis of corallite formed by twisting together of inner ends of septa. (10-10A.)

Basal disk. Fleshy wall closing off lower (aboral) end of the polyp. (10-13A.)

Basal plate. Skeletal plate formed initially beneath basal disk of polyp. The septa and walls extend up and out from the basal plate. (10-13A.)

Calice. Upper (oral) end of corallite on which basal disk of polyp rests. Typically bowl-shaped. (10-9A, B, C; 10-10A, B, C.)

Cardinal septum. One of initial septa. Lies in plane of bilateral symmetry of corallite. Presumably formed between "ventral" pair of mesentaries (10-8H). Distinguished by pinnate insertion of secondary septa on either side. (10-9F, G.)

Carinae. Longitudinal or oblique flanges on sides of septa. (10-9B, D; 10-11E.)

Ceratoid. Corallite with angle of about 20° between sides expanding from apex. (10-14E.)

Cerioid. Type of colonial coral skeleton (corallum) in which walls of individual corallites are closely united. (10-10J; 10-11E, G.)

Coelenteron. Spacious cavity, enclosed by body wall of coelenterate and opening externally through mouth. (B; 10-1A, B, C; 10-13A.)

Coenosarc. Layer of soft tissue connecting polyps in a colonial coral.

Coenosteum. Skeletal tissue connecting corallites of a colonial coral. Deposited by coenosarc.

Columella. Longitudinal rod in axis of corallite, formed by inner ends of septa and typically projecting up into calice. (10-10B; 10-11G, I.)

Corallite. Skeleton formed by individual polyp and consisting of walls, septa, and accessory structures such as tabulae and dissepiments. (10-9; 10-11; 10-12.)

Corallite wall. Skeletal wall forming sides of corallite and comprising various elements such as edges of septa, epitheca, or accessory deposits.

Corallum. Skeleton of colony or solitary polyp—in the latter, the corallum and corallite are identical. (10-9; 10-10; 10-11.)

Counter lateral septum. One of pair of the initial septa next to counter septum. No secondary septa are developed between counter and counter lateral septa. (10-8H; 10-9F.)

Counter septum. One of initial septa. Directly opposite cardinal septum (which see) in plane of bilateral symmetry. Distinguished by position of cardinal when that can be determined. (10-8H; 10-9F.)

Cupolate. Corallite with flat base and convex oral surface. Rather like a button in shape. (10-14A.)

Cylindrical. Nearly straight corallite which has essentially parallel sides except near base. (10-10G; 10-11D; 10-14F.)

Discoid. Corallite with very flat, button-like form. (10-10D; 10-14B.)

TABLE 10-2 (continued)

Dissepiment. Small curved plate forming a vesicle. Typically occur between septa near periphery of corallite with convex surface facing inward and upward. (10-9C, E; 10-11A, B, E, H.)

Edge zone. Fold of body wall of polyp that extends laterally and/or downward over sides of corallite. (10-13A.)

Epidermis. External layer of cells in body wall. Typically the cells are brick-like in form if they are not interconnected to form continuous multinucleat sheet. (10-2A, B; 10-1A, B, C.)

Epitheca. Sheath of skeletal material that forms the wall of the corallite. (10-9A, B, C; *et al.*)

Fasiculate. Type of colonial coral skeleton (corallum) in which corallites stand separate though they may be connected by tubules. (10-10I.)

Fossula. Unusually wide space between septa caused by failure of one or more septa to develop as rapidly as others. Most commonly due to abortion of cardinal septum and therefore a cardinal fossula. (10-10C; 10-11F.)

Gastrodermis. Layer of cells lining body cavity. Forms the inner layer of the body wall. (10-2A; 10-1A, B, C.)

Interseptal ridge. Longitudinal ridge on outer surface of corallite wall. Occurs between position of septa on inner surface. (10-9A.)

Major septum. One of initial or secondary septa. Typically the major septa are of subequal length and extend most of distance from wall to axis.

Medusoid. Type of coelenterate of free-living, jellyfish form. Inverted bowl-like form with mouth and tentacles downward. (10-4.)

Mesentary. One of several radial sheets of soft tissue that partition the internal body cavity. (10-13A.)

Mesoglea, mesenchyme, or *mesoderm.* Layer of cells and gelatinous connective material between inner and outer layers of body wall. (10-2A, B; 10-1B, C.)

Minor septum. One of a third cycle of septa formed between the initial and secondary septa and much shorter than they.

Mouth. External opening of body cavity. Serves in coelenterates for discharge of indigestible material as well as intake of food. (10-1A, C; 10-13A.)

Mural pore. Circular or oval hole in wall between corallites. (10-16A, B.)

Oral disk. Fleshy wall closing off upper end of the cylindrical column that forms the polyp's sides. (10-13A.)

Patellate. Corallite with angle of about 120° between sides expanding from apex. (10-14C.)

Polyp. Type of coelenterate of hydra-like form. Columnar body with base attached to bottom and with tentacles and mouth directed upward. (10-1; 10-13A.)

Septum. One of several longitudinal plates arranged radially between axis and wall of corallite. Presumably alternated in position with mesentaries and supported basal disk and lower wall of polyp. (10-9A, I; 10-13A; *et al.*)

Septal groove. Longitudinal groove on outer surface of corallite wall. Corresponds to position of septum on inner surface. (10-9A.)

TABLE 10-2 (continued)

Stereozone. Zone of dense skeletal deposits—typically along or near wall of corallite. (10-9E; 10-11A, B, I.)

Tabellae. Small horizontal plates near axis of corallite. In essence, incomplete tabulae.

Tabulae. Transverse partitions in corallite. Either flat or convex upward. May extend from axis to wall or be limited to area near axis. (10-9B, D, E; *et al.*)

Tabularium. Axial portion of corallite in which tabulae occur (10-9E.)

Tentacles. Arm-like extensions around mouth that serve primarily for food getting. (10-1A, B; 10-14A.)

Theca. Skeletal deposit enclosing corallite and, presumably, sides of polyp. (10-13A.)

Trabeculae. Rod of radiating calcite fibers that forms an element of the coral skeleton. (10-9H, I; 10-13B.)

Trochoid. Corallite with angle of about 40° between sides expanding from apex. (10-10F.)

Turbinate. Corallite with angle of about 70° between sides expanding from apex. (10-10E.)

pairs. The sea anemones (three orders, Zoanthiniaria, Corallimorpharia, and Actiniaria) lack skeletons, and only a few fossil genera of doubtful affinities have been collected. The other four orders (the extinct Rugosa, Tabulata, and Heterocorallia, the still extant Scleractinia) have calcareous skeletons, and, of course, were more often fossilized. They are important components of modern and fossil coral reefs, but many also occur in non-reef habitats.

Origins. The arrangement and ontogeny of the mesentaries in recent zoantharians indicate a common ancestry in a species with twelve mesentaries (Figure 10-8). These would include the primitive six found in the ceriantipatharians and in the early developmental stages of the zoantharians. In the next stage of evolution, a pair of mesentaries occurs in the space between two of the original six. This is the general octocoral plan and occurs also in zoantharian development. For the sake of description, these two new mesentaries are termed the *dorsal;* the pair on the opposite side, the *ventral;* and the pairs between the dorsal and ventral, the *lateral.* In the third stage, two additional mesentaries develop on each side, one between the lateral mesentaries, one between the ventral and the adjacent lateral. This produced six pairs of mesentaries, a *dorsal* pair, a *ventral* pair, two *dorso-lateral* pairs, and two *ventro-lateral* pairs.

This sequence occurs in the development of recent zoantharians and provides a phylogeny that seems to be in general agreement with other morphologic features. In theory, it should be possible to find fossils to test this hypothesis, since skeletal plates, the *septa,* are located between

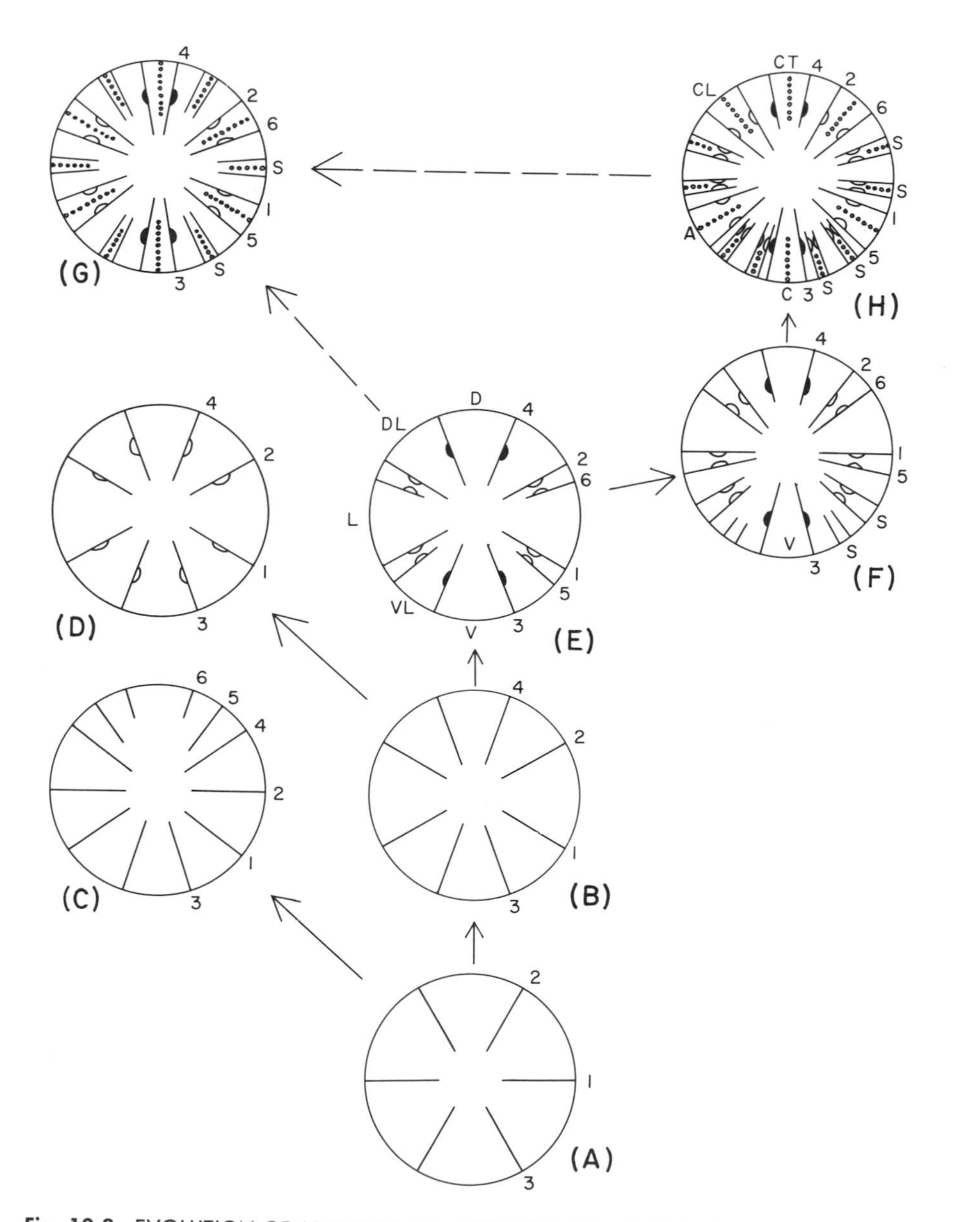

Fig. 10-8. EVOLUTION OF ANTHOZOAN MESENTARY ARRANGEMENT. Cross section of body; body wall and mesentaries shown by solid lines, septa by dotted lines. **(A)** Anthozoan stem, hypothetical. **(B)** Common ancestor of octocorallans and zoantharians, hypothetical. **(C)** Ceriantipatharian. **(D)** Octocorallan. **(E)** Zoantharian stem, hypothetical. **(F)** Zoanthinarian. A sea anemone. **(G)** Scleractinarian. **(H)** Rugosan, reconstructed from septal pattern. Code and abbreviations: 1, 2, 3, 4, 5, and 6, primary mesentaries; A, alar septum; C, cardinal septum; CL, counterlateral septum; CT, counterseptum; D, dorsal; DL, dorso-lateral; L, lateral; S, secondary; V, ventral; VL, ventro-lateral. (*After Treatise on Invertebrate Paleontology*, courtesy of Geological Society of America and University of Kansas Press.)

the pairs of mesentaries. Unfortunately, the oldest fossil zoantharians are not much help. The Rugosa apparently had already acquired the basic pattern before they evolved a skeleton in the middle Ordovician. The Tabulata, which appeared at about the same time, have twelve rudimentary septa and very little is known of their pattern of development. The Scleractinia which appear rather abruptly in the Triassic may have derived from either the Rugosa or a soft-bodied zoanthiniarian, in either case with the basic pattern already established.

The Rugosa. In the middle Ordovician Black River strata appear

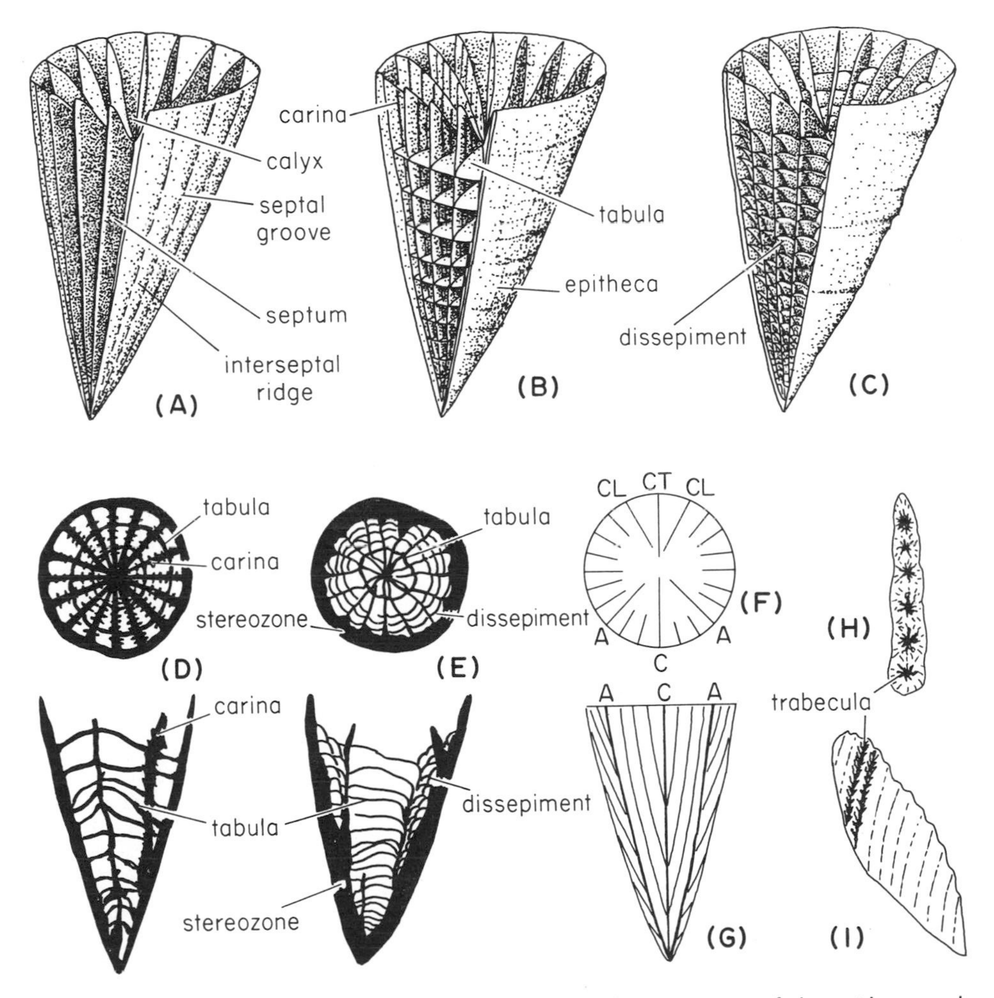

Fig. 10-9. MORPHOLOGY OF THE RUGOSA. (A) Lateral view, a part of the epitheca and the distal ends of several septa removed. (B) Lateral view, tabulae shown in open section. (C) Lateral view, dissepiments shown in open section. (D) and (E) Transverse and longitudinal thin sections. (F) Diagrammatic view looking down on calice. (G) Diagrammatic view from side showing relation of grooves and ridges of cardinal, alar, and secondary septa. (H) Transverse section of septum. (I) Longitudinal section of septum. See abbreviations in 10-8.

several genera of corals characterized by a conical calcareous skeleton, a *corallite* (Figure 10-9), comprising a wall, the *epitheca,* septa, and horizontal partitions (*tabulae* and *dissepiments*). Thin sections of the initial part of the skeleton reveal six primary septa that presumably correspond to six pairs of mesentaries. Additional septa developed between the ventral septum, the *cardinal,* and the ventral lateral septa, the *alars;* and between the alar septa and the dorsal lateral, the *counter lateral,* septa. No septa appear between the counter laterals and the dorsal (*counter*). The development of secondary septa in just four of the six interseptal

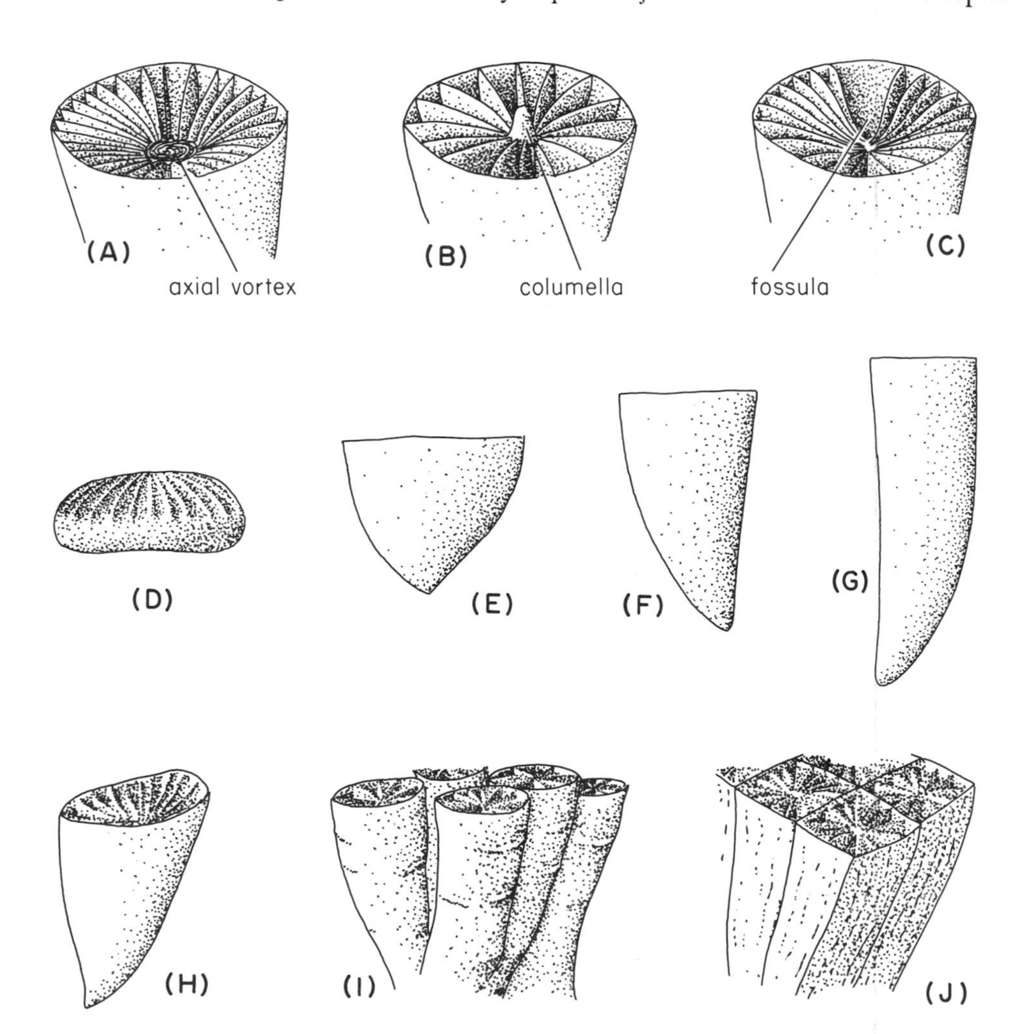

Fig. 10-10. MODIFICATIONS OF RUGOSAN MORPHOLOGY. **(A)** View of calice with axial vortex. **(B)** View of calice with columella. **(C)** View of calice with fossula. **(D)** Discoidal corallite. **(E)** Turbinate corallite. **(F)** Trochoidal corallite. **(G)** Cylindrical corallite. **(H)** Solitary coral, corallite and corallum identical. **(I)** Colonial coral, corallum formed by loosely attached coralites (fasiculate). **(J)** Colonial coral, corallum formed by tightly joined corallites (ceroid).

spaces produces a general fourfold (biradial) symmetry and accounts for a name often used for the order, Tetracoralla. Presumably, the base of the polyp rested on the open end, the *calyx,* of the corallite.

Evolutionary trends (Figures 10-9 and 10-10) within the order include: variation of general shape; modification of the tabulae and dissepiments; reduction of septa so they fail to reach the axis of the coral; complete reduction of one or more of the primary septa, usually the cardinal; appearance of axial rods of one sort or another; development of a *marginal zone* of thickened septa or closely spaced dissepiments; and the derivation of colonial species from solitary types. Unfortunately, little is known of the adaptive significance of these changes.

Rugosans are not common in Ordovician rocks but become increasingly abundant in the Silurian and Devonian. Toward the end of the Paleozoic, they decrease in variety and numbers and are not known above the lower Triassic. Many genera are important guide fossils; Figure 10-11 gives a few examples.

The ecology of the rugosans has had some attention, but, unhappily, not enough (Wells, 1957). They were, of course, benthonic. The solitary corals had a very small area of attachment and commonly toppled over or sank into the bottom sediments. The growth of a curved or twisted skeleton reflects these minor disasters. The colonial rugosans were supported by their undersurface; very few were attached or encrusting. Few rugosans, either solitary or colonial, occurred on the upper part of reefs in the zone of wave action, but they thrived on the reef flanks. Most seem to have lived on shallow bottoms (below wave base) in warm, well oxygenated waters of normal marine salinity. They apparently did best in environments of slow deposition, and, therefore, are more typical of limestone beds than of shales and sandstones.

The Rugosa-Scleractinia problem. The last known rugosans are apparently early Triassic (Il'ina, 1963); the oldest scleractinians, which replace them ecologically, are middle Triassic. As evidenced by skeletal similarities, the two shared many adaptive trends, and occupied many of the same niches. Because of this similarity and because of the succession of the two in time, paleontologists have been extremely interested in their relationships. The causes for the extinction of the Rugosa seem to be beyond our reach at present—there is no evidence that competitive Scleractinia appeared until after that extinction. Conversely, though many scleractinians occupied the same ecological position as rugosans, there seems no good reason why some Scleractinia could not have evolved before the extinction of the older order—unless the scleractinian radiation (in middle Triassic and later) was based on a persistent, evolving, rugosan line.

Some authorities have argued for independent origin of the Scleractinia from some soft-bodied anthozoan. The common ancestry of the two would then be in the Cambrian or earlier. This view is supported by the

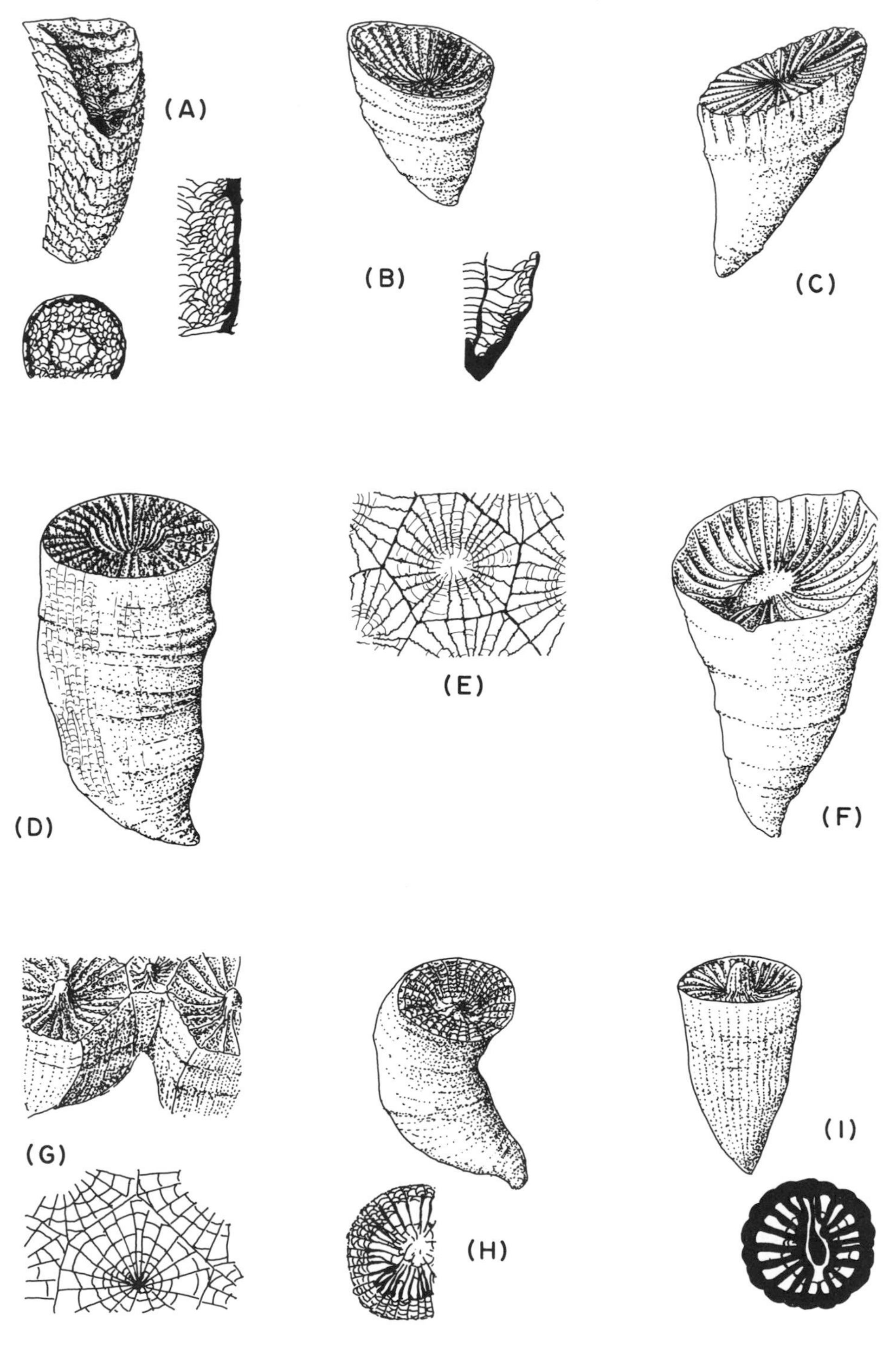

Fig. 10-11.

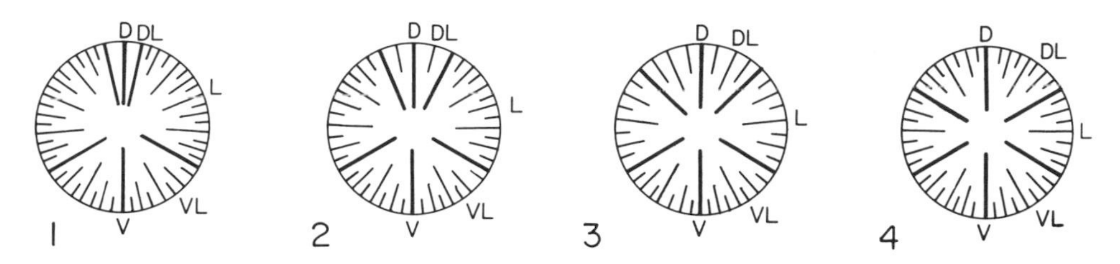

Fig. 10-12. EVOLUTION OF THE SCLERACTINIA. The evolutionary sequence postulated by Schindewolf (1950). **(1)** Rugosan septal plan—no septa in dorso-lateral sextants. **(2)** Septal arrangement in some late rugosans—short septa in dorso-lateral sextants. **(3)** Septal arrangement in some early scleractinians—fewer septa in dorso-lateral sextants than in others. **(4)** Scleractinian septal plan, full hexagonal symmetry. Abbreviations: D, dorsal; L, lateral; DL, dorso-lateral; V, ventral; VL, ventro-lateral.

existence of skeletonless anthozoans in recent faunas that would be suitable structurally to evolve into Scleractinia types, and by the occurrence among the earliest scleractinians of some that are very different from the rugosans.

The evaluation of coral phylogeny depends largely on the pattern of development of the septa (Figure 10-12). In Rugosa, the first six septa develop in pairs, cardinal and counter, alar and counter lateral. The remaining major septa develop in cycles of four, one adjacent to the alar in each alar-counter lateral quadrant on either side, and one adjacent to the cardinal in each cardinal-alar quadrant. No major septa appear between the counter lateral and counter septa. The septa of the Scleractinia, on the other hand, develop in cycles of six and are added in the spaces between all six original septa. Some late Paleozoic rugosan corals have long secondary septa between the counter and counter lateral septa, and a few Triassic scleractinian genera develop fewer septa in two adjoining sextants

---

Fig. 10-11. RUGOSAN CORALS. **(A)** *Cystiphyllum;* Silurian; lateral view and transverse and longitudinal sections; length 7.8 cm. Septa reduced to low, incomplete ridges on surface of dissepiments and tabulae. Dissepimentarium wide. **(B)** *Zaphrenthis;* Devonian; oblique view of corallite and longitudinal section; length 2.3 cm. Narrow dissepimentarium appears in calical rim; septa long, in contact in axis, bear carinae near rim. **(C)** *Aulacophylum;* Early and Middle Devonian; oblique view; length 8 cm. Deep cardinal fossula; septa between cardinal and alar directed pinnately toward fossula; others radial; marginal dissepimentarium. **(D)** *Heliophyllum;* Early to Middle Devonian; oblique view; length 7.6 cm. Fossula indistinct; wide dissepimentarium; carinae. **(E)** *Hexagonaria;* Devonian, transverse section; width of corallite 1.1 cm. Ceroid; thin carinate septa; tabulae, dissepiments. **(F)** *Amplexizaphrentis;* Mississippian-Pennsylvanian; oblique view; length 3.0 cm. Fossula; tabulae; no dissepiments. **(G)** *Lithiostrotion;* Mississippian; oblique view and transverse section; width of corallite 1.8 cm. Ceroid; columella, tabulae, dissepiments. **(H)** *Bothrophyllum;* Mississippian-Pennsylvanian; oblique view and transverse section; length 6.8 cm. Wide dissepimentarium; narrow fossula; major septa join to produce weak axial structure; septa dilate in tabularium. **(I)** *Lophophyllidium;* Pennsylvanian-Permian; oblique view and transverse section; length 2.7 cm. Columella large; tabulae but no dissepiments. [**(A)** *sections after* Lang *and* Smith; **(B)** *section after* Schindewolf.]

of the original six. These variants from the more typical rugosan and scleractinian pattern bridge part of the gap but do not explain the difference in order of insertion, i.e., in only the spaces adjacent to the alars and cardinals of rugosans but in all interseptal spaces of the scleractinians.

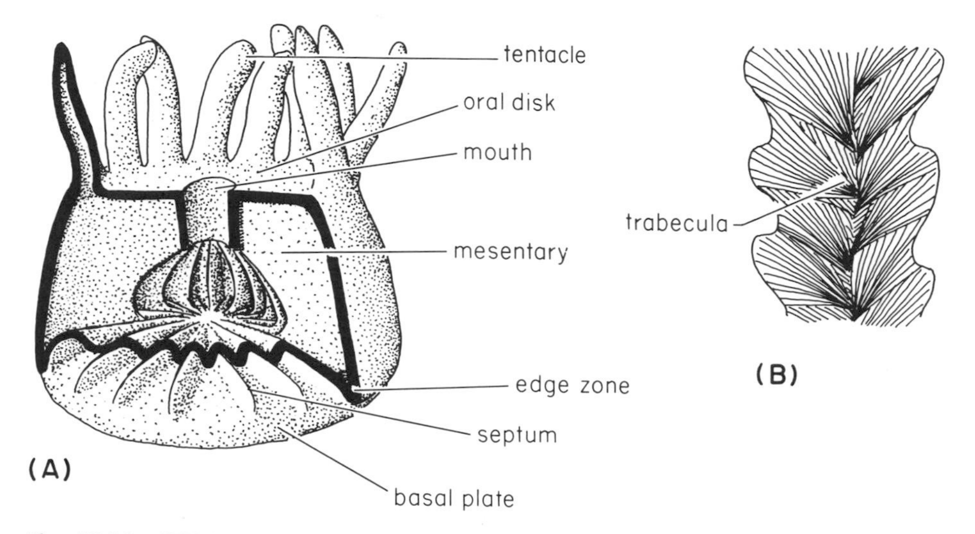

Fig. 10-13. STRUCTURAL PLAN OF SCLERACTINIA. **(A)** Diagram of polyp and skeleton. Portion of body wall and mesentaries cut away. **(B)** Transverse section of septum.

The Scleractinia. The general plan (Figure 10-13) of the scleractinians, like that of the rugosans, comprises a calcareous cup formed by the epitheca and divided internally by septa and dissepiments. The septa differ from those of the Rugosa in microstructure as well as in arrangement. The shape of the skeletal cup, the *corallum,* differs considerably among different scleractinians (Figure 10-14) and is also strongly modified by environmental factors during growth. Most families of the order have evolved colonial genera.

An important feature of scleractinia evolution was the development of the *edge zone*. This consists of the part of the polyp that lies outside of the wall of the corallum. In some, this is merely a bulge of tissue around the end of the cup, but in many, it folds down over the side of the skeleton. In the latter, the epitheca, which is formed by the lower body wall within the cup, is suppressed and replaced by elements deposited by the edge zone. Since the edge zone extends down and laterally, these deposits broaden the base of the coral and strengthen its attachment to the substrate.

The edge zone, and its equivalent in the colonial types, permits construction of massive encrusting skeletons. This may have been a preadaptation to reef building or an adaptation for survival in the upper wave zone of reefs. Regardless of its origin, it permitted expansion of the scleractinians into environments unavailable to the Rugosa. Recent reef-building scleractinians also contain within their soft tissues algae which apparently supply nutrients for the polyps in return for support and protection. Goreau and Goreau (1960) educe evidence that the nutrient is a critical vitamin. Whatever the relationship it seems necessary for vigorous growth of the corals, and, consequently, of the reef.

The dependence of the modern reef-building corals on algae limits reef formation to shallow, well lighted waters, the zone of photosynthesis.

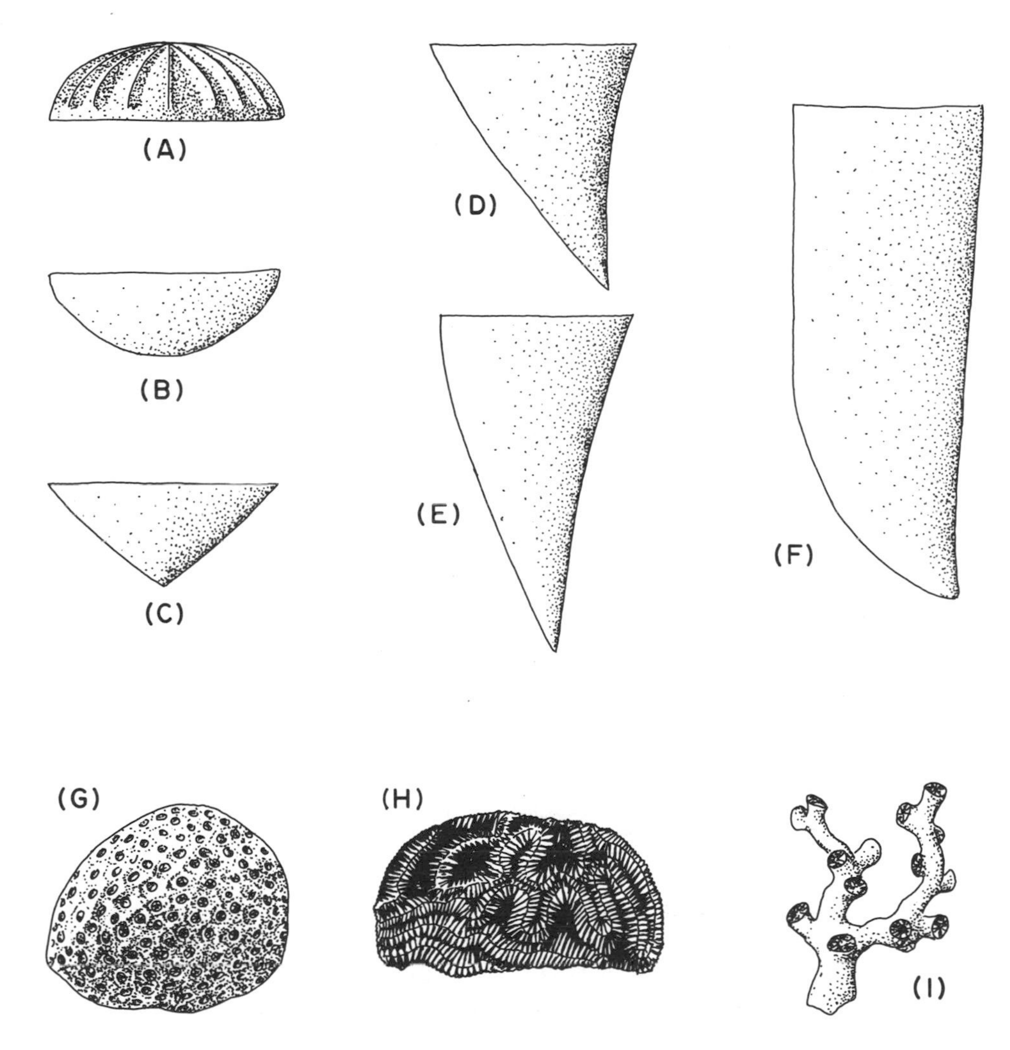

Fig. 10-14. VARIATION IN SCLERACTINIAN FORM. **(A)** Cupolate. **(B)** Discoid. **(C)** Patellate. **(D)** Trochoid. **(E)** Ceratoid. **(F)** Cylindrical. **(G)** Colonial coral, massive, distinct corallites. **(H)** Colonial coral, massive, corallites in linear series with no walls between. **(I)** Colonial coral, branching, distinct corallites.

Reef corals are further limited by temperature—they flourish only between 25° and 29°C—and by circulation of the water to supply oxygen and food. For these reasons, reef growth is now limited to the shallows of tropical and subtropical seas. Presumably they have had similar habitats since the Mesozoic, and, therefore, are used as an index to Mesozoic and Cenozoic climatic zones (Figure 10-15). Teichert, however, has pointed

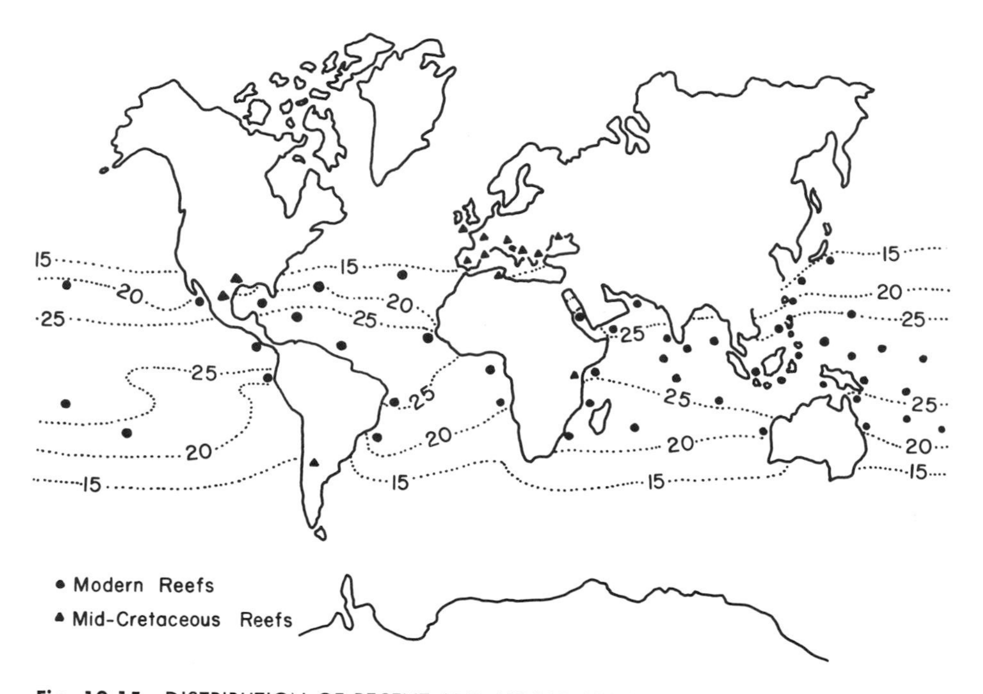

Fig. 10-15. DISTRIBUTION OF RECENT AND MIDDLE CRETACEOUS CORAL REEFS. The present mid-winter isotherms are shown by dotted lines. Temperatures are given in degrees centigrade. The distribution of Cretaceous reefs indicates that the 15° and probably the 20° isotherms were further north and south than at present.

out (1958) that an abundance of colonial corals in itself does not demonstrate the existence of a true shallow-water reef but that a variety of frame-building types and the presence of calcareous algae help to distinguish them from cold-water bank deposits.

The nonreef-builders are less limited in distribution, and some survive at temperatures as low as −1.1°C and at depths of 6000 meters. Particular species, of course, have a more limited tolerance of these factors and, in addition, are controlled in their distribution by salinity and particularly by the character of the bottom.

The Tabulata. The tabulates, a large group of Paleozoic corals, have been assigned various positions in coelenterate taxonomy. Their location here as an order among the Zoantharia follows the *Treatise on Inverte-*

brate Paleontology. The tabulate skeleton (Figure 10-16), consists of a simple tube partitioned by horizontal tabulae. Septa are rudimentary or, in many genera, absent. They are all colonial and the individual tubes are fused or connected by short tubules. In some, *mural pores* perforate the walls and connect adjacent corallites.

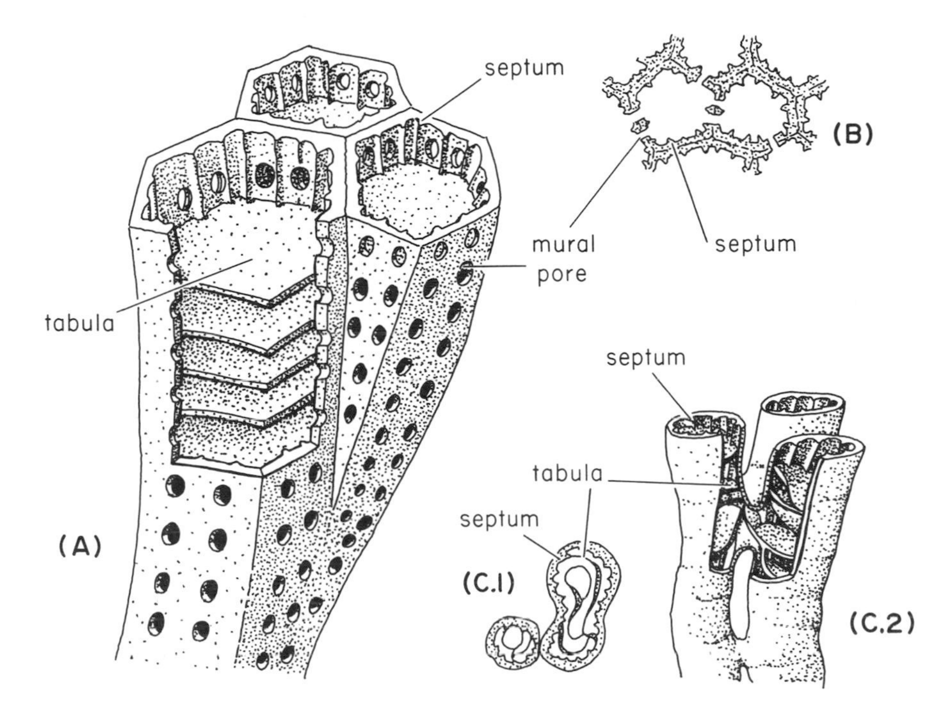

Fig. 10-16. STRUCTURAL PLAN OF THE TABULATA. **(A)** Oblique view of *Favosites,* portion of one corallite wall removed. **(B)** Transverse section of *Favosites* corallite. **(C)** *Syringopora:* **(C.1)** Transverse section. **(C.2)** Oblique view with portion of corallite wall removed.

The tabulates underwent a considerable radiation (over a hundred described genera) and abundant in late Ordovician, Silurian, and Devonian rocks (Figure 10-17). Like the Rugosa, the Tabulata had a sinking spell in the late Paleozoic, and only a few genera of doubtful affinities occur in post-Permian beds.

Because of their massive colonial habit, tabulates were important framework builders in some Paleozoic reefs. They also occurred in non-reef habitats, though they apparently thrived only in low turbidity environments. The form of the colony and the characteristics of the individual coral both seem responsive to environmental differences during growth and some "species" appear to be growth variants.

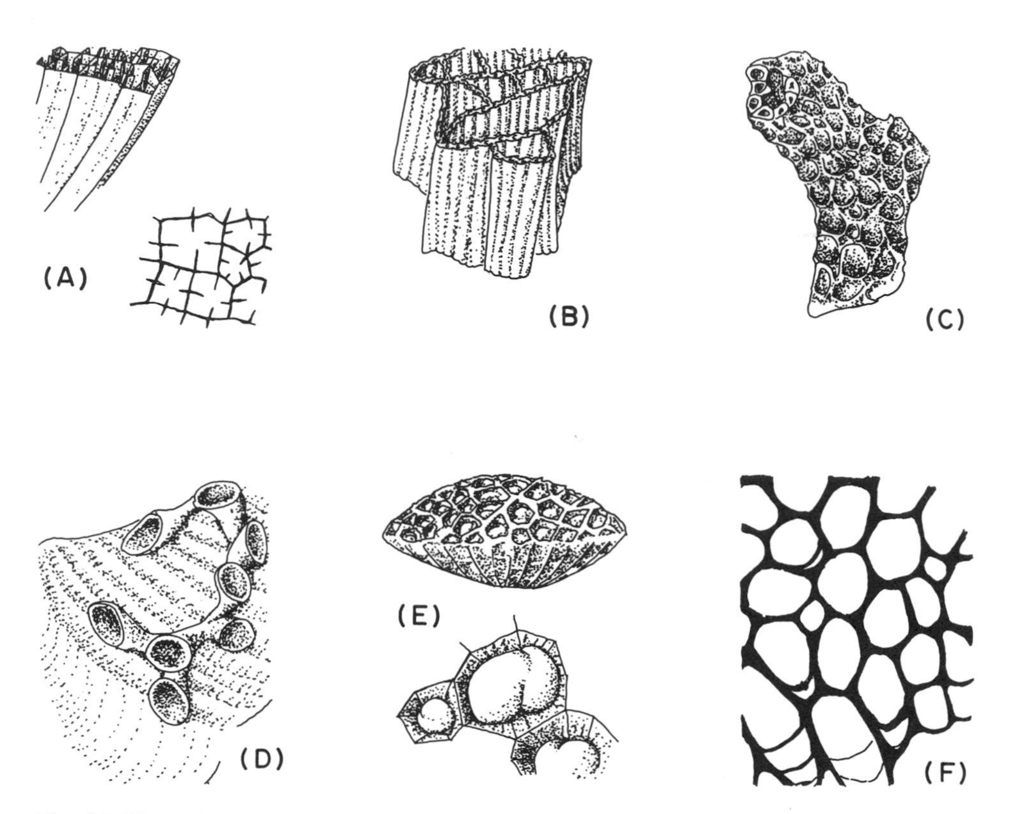

Fig. 10-17. VARIATION IN THE TABULATA. **(A)** *Tetradium;* middle-late Ordovician; oblique view and transverse section; width of corallite 2 mm. Prismatic, ceroid; four septa in corallite. **(B)** *Halysites;* Ordovician-Silurian; oblique view; length of corallites 3.7 cm. Corallites united along two sides; long slender tubes joined to form palisade-like structure; palisades interconnected to form network; chain-like in transverse section. **(C)** *Striatopora;* Silurian-Permian; lateral view; length of fragment 3.4 cm. **(D)** *Aulocaulis;* Devonian; lateral view; diameter of calice 2.5 mm. Fasiculate; basal portions of corallites recumbent and cemented to substrate—in this case a brachiopod shell. **(E)** *Pleurodictum;* Early Devonian; lateral view of corallum and vertical view of several corallites; length of corallum 7.5 cm. and diameter of largest corallite 7.5 mm. Corallum discoidal; corallites polygonal; tabula, thin blister-like. **(F)** *Chaetes;* Ordovician to Permian; transverse section; diameter of corallite 0.5 mm. [**(F)** *from Treatise on Invertebrate Paleontology.* Courtesy of Geological Society of America and University of Kansas Press.]

REFERENCES

Dillon, L. S. 1962. "Comparative Cytology and the Evolution of Life," *Evolution,* vol. 16, pp. 102-117.

Dougherty, E. C., *et al.* (Eds.) 1963. *The Lower Metazoa: Comparative Biology and Phylogeny.* Berkeley, California: University of California Press.

Goreau, T. F. and N. I. Goreau. 1960. "Distribution of Labeled Carbon in Reef-building Corals with and without Zooxanthellae," *Science,* vol. 131, pp. 668-669.

Hadzi, J. 1963. *The Evolution of the Metazoa.* New York: Macmillan.

Hyman, L. H. 1940. *The Invertebrates. Vol. I, Protozoa through Ctenophora.* New York: McGraw-Hill. See also vol. II published in 1951 and vol. V published in 1959.

Il'ina, T. G. 1963. "Some New Data on the Origin of the Hexacorals," *Dokl. Acad. Sci. U.S.S.R.*, vol. 148, pp. 156-158.

Kato, M. 1963. "Fine Skeletal Structures in Rugosa," *Hokkaido Univ., Faculty of Sci., Journ., Ser. IV, Geol. and Miner.*, vol. 11, pp. 571-630.

Ladd, H. S. 1961. "Reef Building," *Science*, vol. 134, pp. 703-715.

Moore, R. C. (Ed.) 1956. "Coelenterata," *Treatise on Invertebrate Paleontology, Pt. F.* New York: Geol. Soc. of America.

Schindewolf, O. H. 1950. *Grundfragen der Paleontologie.* Stuttgart: Schweizerbart.

Scrutten, C. T. 1964. "Periodicity in Devonian Coral Growth," *Palaeont.*, vol. 7, pp. 552-558.

Teichert, C. 1958. "Cold- and Deep-Water Coral Banks," *Bull. of American Assoc. of Petrol. Geologists,* vol. 42, pp. 1064-1082.

Wells, J. W. 1957. "Corals," in Ladd, H. S. (Ed.) "Treatise on Marine Ecology and Paleoecology," vol. 2. *Geol. Soc. of America, Memoir 67,* pp. 773-782.

——— and D. Hill. 1956. "Zoantharia-General Features," in Moore, R. C. (Ed.) Pp. 231-232. See above for ref.

11

A step upward: Bryozoa

Somewhere in the Precambrian rocks, a fine dark shale bears the carbonized impressions of inhabitants of an ancient lagoon. The geologist who has the good fortune to see the glistening films of carbon will immortalize himself and help solve some of the profound problems of animal evolution. Biologists can predict, mostly from a study of the simpler living metazoans and their embryology, some of what they will find. Simple plants, algae, swung with the current and drifted with the tide. These plants furnished the primary energy for the animal community. There must have been a variety of protozoans, some simple types, others with the beginnings of complex organelles. Some shared with algal cousins the ability to photosynthesize, but others fed on the algae or preyed on other protozoans. Colonial protozoans swarmed to browse on the microscopic pastures or crept along the bottom to collect their food. Anchored to the bottom, simple sponges and primitive polypoid coelenterates netted passing microorganisms from the water. Other metazoans foraged along the bottom. The former occupied a world in which impressions of danger and of food impinged from all sides, and in which the only constant direction was that given by the animal's movement. The coelenterate was adapted with strong polarity and radial symmetry.* You saw some consequences of these adaptations in Chapter 10.

The bottom-creepers inhabited a world with intimations of up and down as well as front and back. The feeding surface was, of course, "down," and the animals flattened out along the axis of polarity. The dominant direction of perception was "forward" rather than radial, and, presumably, selection acted to concentrate organs of perception and coordination in the "forward" position. A gut developed as a blind internal tube surrounded by the nutritive cells that originally filled the interior of the animal.

The potentialities of this structural plan have yet to be fully exploited, but it has yielded already such complex creatures as squids, praying mantises, and men. The flatworms, the phylum Platyhelminthes, have developed the immediate adaptive features of this plan without progressing in general efficiency. They are the most primitive of this great metazoan stock and retain the primitive gut which serves both in nutrition and in circulation. The nervous system, derived from the epidermal cell layer, is more complex and more highly integrated than the diffuse nerve net of the coelenterates, and a fairly complex system of visual, chemical, and tactile receptors feeds data into the coordinating centers. The mesenchymal tissue, between gut (the gastrodermal cell layer) and body wall, is more abundant and important than in the coelenterates, for it furnishes material for muscles, excretory organs, and nutritive glands as well as for reproductive organs.

Many of the platyhelminthes now exist as poor relatives at the table of the more complexly organized metazoans. Some, such as the free-living planarians, compete successfully with their more highly endowed cousins. The remainder continue as parasites, a habit for which they are preadapted by their simple body structure. The evolution of these parasitic types, which is unrecorded from fossils, has consisted primarily of a regression in complexity, of the appearance of some peculiar mechanisms connected with parasitic life, of the biochemical adjustment to the host, and of the complex reproductive cycles.

Size and activity

The increased complexity and integration in the early metazoans created evolutionary problems and started correlated evolutionary trends. Integration means a higher level of activity; activity demands heightened

* Again, I urge skepticism about these phylogenetic reconstructions—a large, if not overwhelming, element of dramatic fiction underlies this one as well as the others now available.

efficiency and further structural modification, and greater activity results in increased resistance to and exploitation of the environment. Larger animals are, in general, better able to resist and exploit, but larger animals need superior integrative mechanisms and greater efficiency. The surface area for respiration, nutrition, and excretion increases approximately as the square of linear size increases, but the bulk increases as the cube. More cells are buried away from these functional surfaces, and mechanisms evolve to supply food and oxygen and remove wastes from the buried cells. Such mechanisms, in turn, control the effects of environmental factors on the bulk of the cells. The propulsion provided by the beat of microscopic cilia or by peristaltic waves in a thin sheet of muscle cannot provide rapid, efficient locomotion in large animals; distinct groups of antagonistic muscles operating against some sort of rigid supportive elements are essential and, once evolved, permit new modes of burrowing and swimming. Animals, therefore, resist and exploit the environment to a greater extent.

Evolutionary trends

The major trends of metazoan evolution are increased size, increased complexity of nutritive, circulatory, excretory, and integrative structures, and improved locomotor and perceptor adaptations. Specific adaptive requirements for smaller size have sometimes reversed the first trend; the adoption of parasitic or sessile habits, the second and third. These trends appear in all the metazoan phyla above the platyhelminthian level and may have evolved independently in several lineages.

In place of the diffuse nerve net of the coelenterates and flatworms, most of the other Metazoa have a concentrated system. The flatworms illustrate the first steps in this change, for they possess a small number of longitudinal nerve cords that collect impulses from all parts of the body. Transverse cords connect the longitudinal trunks. The cell bodies of many of the nerve cells are concentrated in masses, the ganglia, and the anterior ganglia in more highly organized metazoans form a definite brain.

The circulation of nutrients, gases, and wastes in the coelenterates and platyhelminthes depends largely on branches from the gut which ramify through the body and, in the mesoderm, on diffusion from cell to cell. Although this works very well in small animals which have all their cells close either to the exterior surface or the gut wall, it would hardly suffice for an elephant or even a snail. The most obvious solution is the development of fluid-filled spaces within the body—obvious at least to all organisms above the coelenterate-platyhelminthes level (Figure 11-1). The various materials to be circulated diffuse through the fluid, and currents increase the rate of circulation. Some of these spaces, those intimately related to the excretory organs, may have evolved primarily to collect the metabolic wastes. Fluid-filled spaces also form a hydrostatic "skeleton" on which the muscles may operate (Clark, 1964). Spaces between the gut and the body walls assume other functions as well. Since neither paleontologists nor zoologists have direct knowledge of the evolution of such spaces, they cannot be sure of the primary adaptation that produced them. Among most of the metazoan phyla, part or most of their functions are taken over by the blood vessels, sinuses, and rigid skeletons.

Cells specialized for separation of metabolic wastes from body fluids occur in the flatworms. In evolution, they tend to increase in number and in efficiency, and, among the metazoans with well developed circulatory systems, they are typically grouped to form "kidneys."

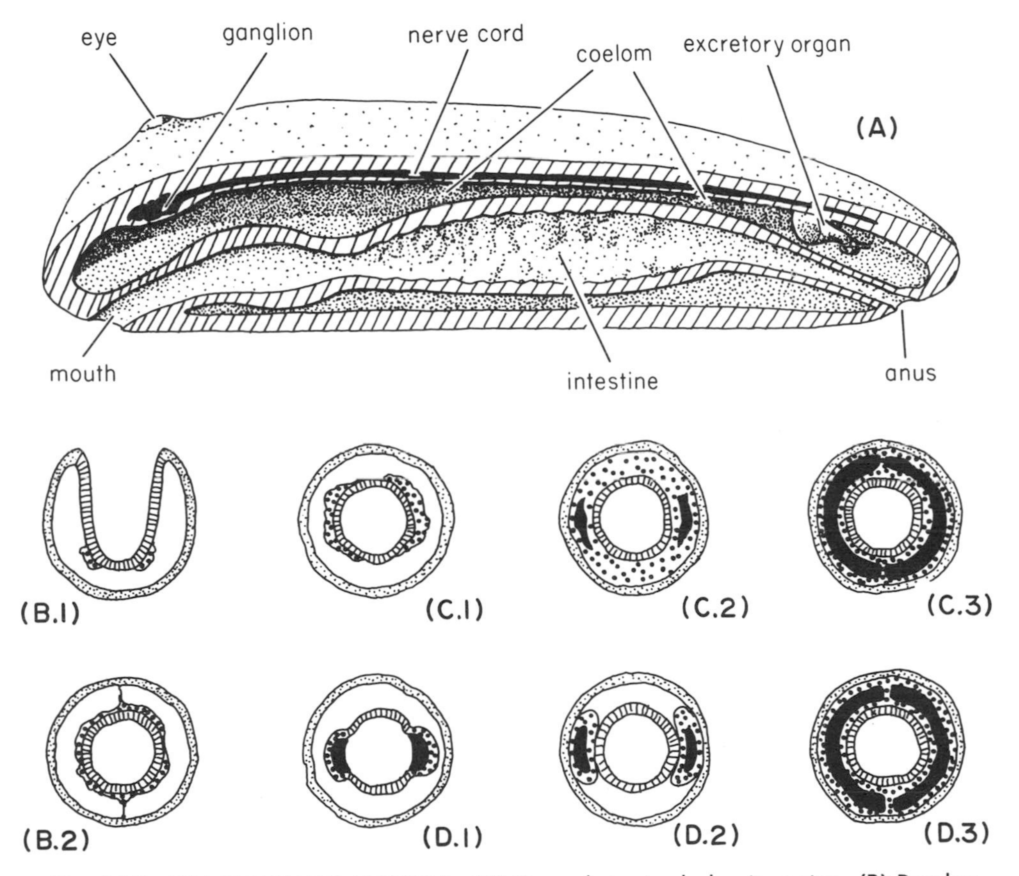

Fig. 11-1. THE COELOMATE METAZOA. **(A)** General structural plan in section. **(B)** Development of body cavity in pseudo-coelomates—cavity is remnant of internal space in "hollow ball" stage of development. Mesoderm, heavy stipple; endoderm, lines; ectoderm, light stipple. **(C)** Development of body cavity as it occurs in molluscs, annelids, and arthropods. **(C.1)** Mesoderm buds from sides of endoderm lining gut. **(C.2)** and **(C.3)** Coelomic cavities (solid) develop as split within mesoderm. **(D)** Development of body cavity as it occurs in echinoderms and chordates. **(D.1)** Pouches develop on sides of gut. **(D.2)** and **(D.3)** Pouches close off from gut to form coelomic spaces; lining of pouches becomes mesoderm.

The coelenterates and platyhelminthes take food particles in through the mouth and expel indigestible wastes through the same opening. This obviously is inefficient, and animals with separate incoming and outgoing currents possess a selective advantage. In some coelenterates, the mouth is modified so that incoming material enters in a partly separate opening, the siphonoglyph. Among the other metazoans, the blind end of the gut breaks to the surface. In these, food enters the anterior mouth, and the undigested materials are discharged through the posterior anus. Because the direction of movement is constant, the gut is differentiated along its length for digestive and absorptive functions (Figure 11-1).

Significance of a space

Many zoologists accept the form and mode of development of the body spaces as prime criteria in inference of metazoan phylogeny. Developmental mode, however, is primarily an ontogenetic adaptation; thus some of the "simpler" modes may represent "short-cut" approaches rather than retention of a primitively simple plan. Phylogenetic schemes based on other fundamental adaptations agree in part with this one, but some disconcerting contradictions appear. Each of the two major types of body spaces, the *pseudocoel* and the *coelom,* probably evolved independently in several different lines* as an adaptation to similar selective pressures. The pseudocoel plan encompasses animals in which the body cavities are remnants of the "hollow" in the early "hollow ball" stage of development. The body cavity lies between the ectoderm and entoderm and is partly filled by the mesoderm or organs developed from it. The "pseudocoels" are less diversified and possess a simpler organization than the coelomates. Many are now parasitic; most are quite small; their organization is comparatively simple, and they comprise a relatively small number of species. They all lack a rigid skeleton, and none of the pseudocoelomate phyla are definitely known from fossil specimens.

The coelomate plan includes animals in which the body cavities lie entirely within the mesoderm, i.e., the cavities are fully surrounded by tissue of mesodermal origin. In many animals, among them the echinoderms, the bryozoans, some brachiopods, the protochordates, and the more primitive chordates, the coelom forms as pouches from the cavity of the gut. The connections to the gut close over; the pouches expand to fill the space between the gut and body wall, and their lining becomes the mesoderm. In other animals, including the molluscs, the annelids, and the arthropods, the mesoderm buds off the gut wall and fills the space between gut and body wall. The coelom then develops as cavities within the mesoderm.

* A convenient and noncommittal escape gambit for this phylogeneticist.

All of the animals common as fossils—except the Protozoa, Porifera, and Coelenterata—are coelomates, and they predominate in numbers and variety in modern faunas. The bryozoans and the brachiopods display the simplest organization among the coelomates, are closely related to each other, and are difficult to relate to the remaining phyla. For these reasons I'll take up these two next in this survey of animal evolution.

The origin of phyla

On a gross level, the bryozoans and brachiopods are strikingly different: the former with colonial habitus and a basically tubular skeleton fused into the colony; the latter with solitary habitus and a bivalve shell. On a finer scale, however, they display a number of similarities that suggest a common phyletic origin:

a). Possession of ring of ciliated tentacles about the mouth, i.e., the *lophophore*.

b). Similarity of larval form, an inverted, flattened cone marked by a horizontal band of cilia, an apical sense organ, and a bivalved shell—in a dorso-ventral position in the brachiopods and right-left in the bryozoans, however.

Their relationship is obscured by the peculiar metamorphosis of the bryozoans which involves complete destruction of the larval structures and the generation of the adult pattern from the outer cell layer, the ectoderm. This ontogenetic shortcut obscures the primitive developmental pattern.

Granting this uncertainty, the ancestral bryozoan-brachiopod was a simple, sessile coelomate probably attaching to the bottom and collecting its food with the ciliated tentacles of the lophophore. It possibly resembled the living acoelomate Entoprocta, though more complexly organized. The evolutionary choices for such creatures are relatively clear-cut—either evolve into moderately large, shelled individuals or form colonies of small ones. Some species may have reproduced asexually and formed weak aggregates of individuals. These foreshadowed the Bryozoa. A tubular skeleton, a standard adaptation for the sessile benthos, replaced a bivalved shell (if present). Emphasis on asexual reproduction would result in an extensive colony, a bryozoan. The species without asexual reproduction increased in size, developed a bivalve shell, and became a brachiopod.

How about the evidence for this reconstruction? It's rather slim. Some of the simplest (therefore possibly most primitive) Bryozoa occur in Ordovician rocks—along with more complex genera. These primitive species have skeletons composed of simple tubes, and the individuals are widely separated and relatively weakly connected (Figure 11-3A). The most primitive brachiopods occur in lower Cambrian rocks but are hardly bryozoan-like. The time relation suggests the brachiopods as the basic stock from which a colonial species diverged with the benefit of asexual reproduction; however, their evolution could hardly be this simple.

The divergence of brachiopods and bryozoans, as well as many other major categories, appears to result from a transformation extending well back into the embryology. One paleontologist, Schindewolf, has argued that such transformations were abrupt (see p. 173). Since fossil evidence is lacking in many such cases, proof or disproof is difficult. The brachiopods and bryozoa could have diverged in a single major step, but, as this reconstruction indicates, the changes could have been adaptive and can be explained as gradual evolution in response to low or moderate selective pressure. The inadaptive condition (or barrier) suggested in some cases does not appear here.

The bryozoans

The development of the colony (the *zoarium*) is the critical feature of the bryozoan organization (Figure 11-2). As described in the preceding section, the initial individual degenerates after attachment to the bottom, and a new individual (the *ancestrula*) regenerates from parts of the larval tissue. One or more buds appear and develop into distinct individuals (called *zooids*). These in turn bud until a large colony is formed. Each zooid initiates a skeletal tube, the *zooecium*. As the zooid grows, it lengthens this tube. Individuals in some species form successive horizontal partitions (*diaphragms*) across the tube; each of these may reflect a regeneration of the zooid similar to that which occurred after the initial individual metamorphosed. Wide spacing of the diaphragms and thin walls in the lower part of the tube indicate rapid initial growth; close spacing and thick walls in the upper portion result from slower growth. This difference defines the *immature* and *mature* regions of the zooecium. The colony itself may show ontogenetic changes as younger individuals develop in a different fashion from the older ones. Some at least of these changes during development of the individual and of the colony are adaptations to changing environmental conditions, for the ecology of a relatively large, erect colony may be quite different from that of its initial, small, encrusting stage.

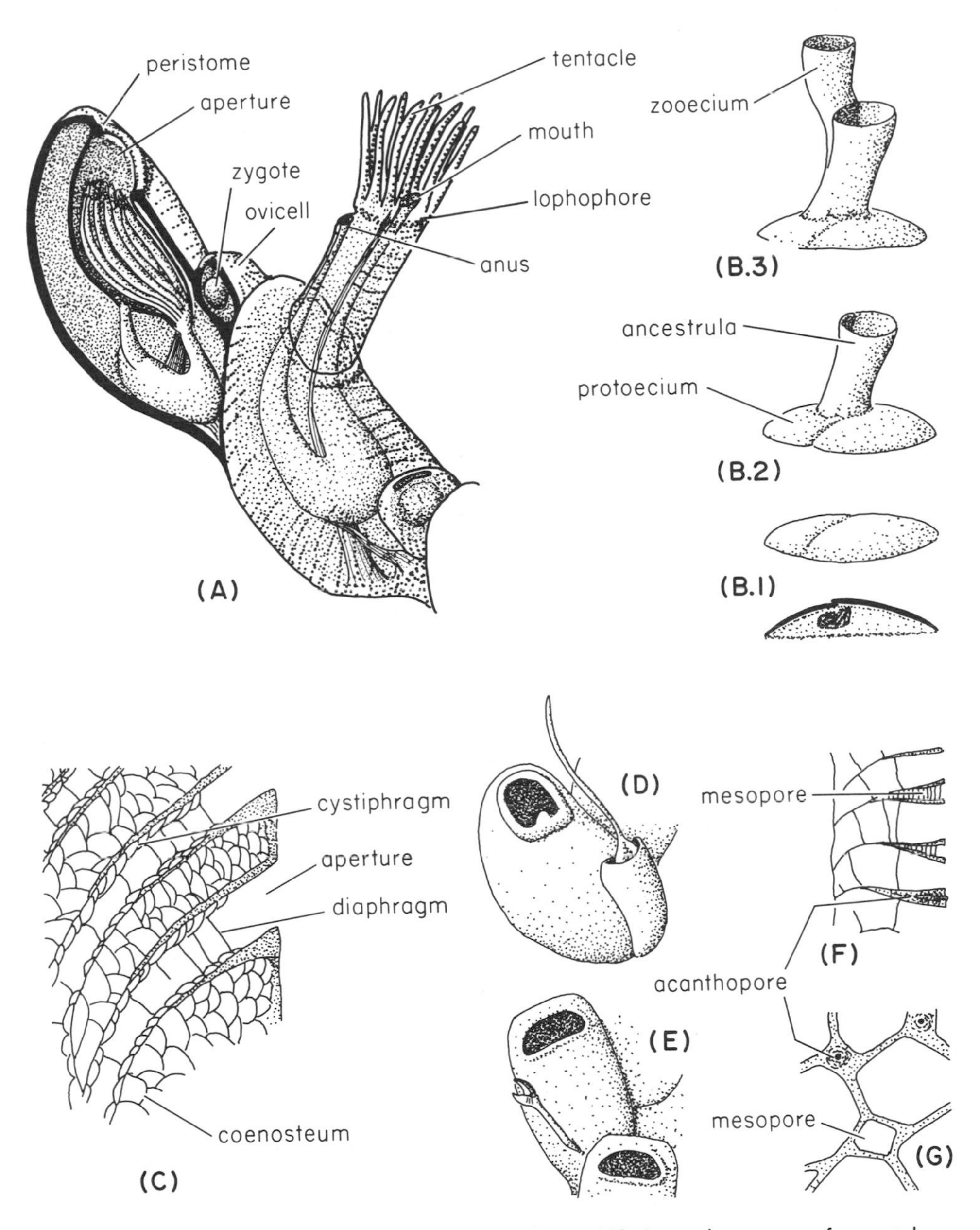

Fig. 11-2. ESSENTIALS OF BRYOZOAN MORPHOLOGY. **(A)** General anatomy of recent bryozoans; zooecial wall of lower zooid cleared; wall of upper zooid sectioned. Lower polypide extended; upper retracted. ×50. **(B.1)** to **(B.3)** Stages in formation of zoarium. **(B.1)** Skeleton formed by larva after attaching—the protoecium. Cross-section and oblique view. **(B.2)** Protoecium and ancestrula, oblique view. The latter is the zooecial tube formed by the original polypide after metamorphosis of larva. **(B.3)** First zooecium formed. **(C)** Longitudinal section of zoarium. ×20. **(D)** Specialized zooid, a vibraculum, attached to another zooecium. **(E)** Avicularium. **(F)** Longitudinal section of Paleozoic bryozoan. Acanthopores and mesopores shown. **(G)** Tangential section of same. [**(B)** *after* Bord; **(D)** *after* Hincks; **(E)** *after* Busk.]

Morphology and adaptations

The primitive bryozoan zooecium consists of a simple tube of chitin (Hyman, 1958) or calcite (the former probably more primitive) open at one end. The calcareous well displays a variety of distinctive microstructures useful in classification. The zooid, while feeding, extends its tentacles and part of the body from the zooecium (Figure 11-2) but retracts rapidly when disturbed. Among more advanced Bryozoa, the zooecium forms a distinct, rather box-like chamber. The opening, the *aperture,* is restricted to one end or one side of the chamber. The chamber and aperture vary in shape in different species; the aperture in some is recessed, and internal supporting structures occur in many. Most of the morphologic variation is adaptive to hydrostatic differences arising in extension and retraction of the lophophore. As a consequence, the paleontologist, in the study of individual zooecia, deals principally with adaptations in development and in the extension-retraction function.

The structure of the zoarium gives considerable information about the natural history of the species. Attachment is adaptive to the substrate and to the overall shape and bulk of the colony. The shape, in turn, must be adapted to current velocity (or wave action), to substrate support, to sedimentation rates, and to feeding and respiratory functions. Because the tentacles of the zooids form a sort of net to filter microorganisms from the water, the distribution of individuals, as well as colony form, modifies the food-getting mechanisms. Some of the different types of zoaria are shown in Figure 11-3.

Remarkable, and, therefore, important identifying features of the colonies are specialized zooids. Some individuals assume protective roles as *avicularia* or *vibracula.* The former resemble a bird's head and consist of two "jaws" with muscles to open and close them. The latter are rather whiplike and likewise have a set of muscles that keep the "whip" in motion. These seem to be responsible for destroying or removing sedimentary particles and encrusting organisms from the colony. Some species also have individuals specialized for reproduction, typically by the formation of an *ovicell* in which one egg or a number of eggs develop. Since such specialized individuals lose all or part of their usual functions, they must be supported in this share-the-wealth scheme by other zooids.

The specialized zooids, of course, have modified zooecia and can be recognized in fossil bryozoans. In these fossil groups with close modern relatives, the function of these particular zooecia are easily determined; the functions of the *mesopores, acanthopores,* and so on, found only in fossil bryozoans, are unknown.

The circulatory, respiratory, and excretory functions do not have, so far as is known, any association with skeletal modifications. The Bryozoa lack any definite organs for these functions, and, presumably, diffusion from the water into the organism, from cell to cell, and from cell to

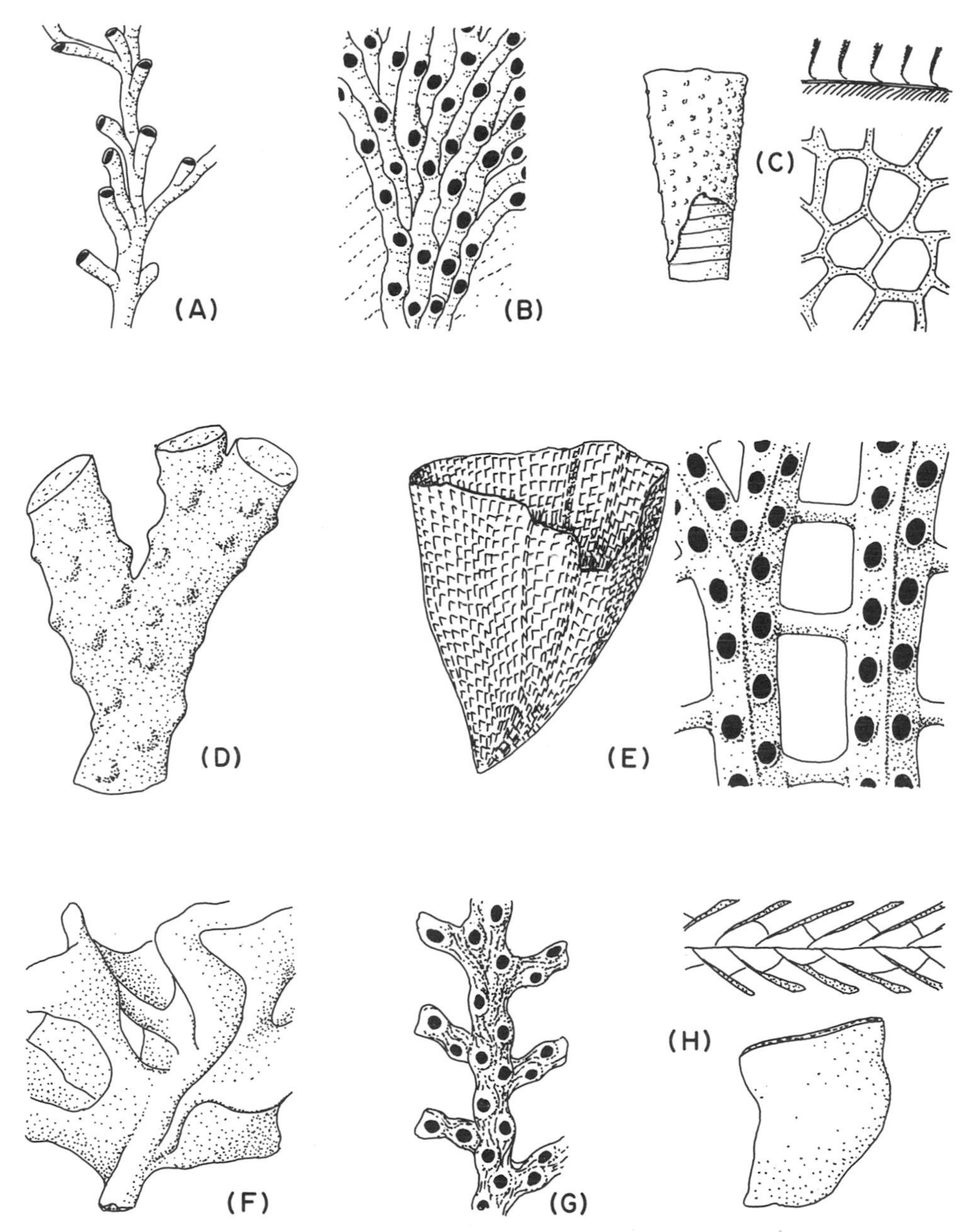

Fig. 11-3. VARIATION IN ZOARIA. **(A)** A repant form. *Hederella;* Silurian to Pennsylvanian; lateral view; (×5). **(B)** An encrusting form. *Sagenella;* Silurian lateral view (×8). **(C)** An encrusting form; *Leptotrypa;* Ordovician; lateral view (×1); longitudinal and tangential sections (×25). Zooecia polygonal in tangential section. **(D)** Ramose zoarium. *Hallopora;* Ordovician to Devonian; lateral view. **(E)** Reticulate, funnel-shaped form. *Fenestella;* Ordovician to Permian; lateral view of entire zoarium (×0.5) and enlarged view (×20) of a small part of colony. **(F)** Frondose zoarium. A recent species. **(G)** Ramose form—short, regularly-spaced side branches. *Penniretepora;* Devonian to Permian; lateral view (×15). **(H)** Bifoliate frond. *Phyllodictya;* Ordovician; longitudinal section (×10) and lateral view (×1). [(**C**) *section after* Ulrich; (**H**) *after* Ulrich.]

TABLE 11-1

Glossary for the Bryozoa *(numbers at end of definition indicate figure that illustrates structure).*

Acanthopore. Small tube adjacent and parallel to zooecial walls. Formed of cone-in-cone layers with minute central tubule—the latter may contain transverse partitions. Typically marked on surface by projecting spines. (11-2F, G; 11-5D.)

Ancestrula. Initial individual of colony. Derived from metamorphosis of larva. (11-2B.)

Anus. Terminal opening of digestive tract. Serves primarily for discharge of indigestible materials. (11-2A.)

Aperture. Opening in wall of skeleton (zooecium) through which the living animal extends lophophore and portions of body (11-2A, *et al.*).

Avicularium. Specialized individual bearing a beak-like structure worked by strong muscles. (11-2E.)

Bifoliate. Colony consisting of two layers of zooecia growing back-to-back. (11-3H.)

Coenosteum (coen). Vesicular or dense skeletal material between zooecia. (11-2C; 11-4B; 11-5B, D.)

Cystiphragm. Calcareous plate extending from zooecial wall part-way across tube. Surface domed, convex upward and inward. (11-2C; 11-5A.)

Diaphragm. Calcareous plate extending transversely across width of zooecial tube. Surface flat or gently curved. (11-2C.)

Encrusting. Colony which forms broad sheet over substrate to which it is attached. (11-3B, C.)

Esophagus. Portion of digestive tract leading from mouth to stomach. (11-2A.)

Frondose. Erect colony consisting of broad, flat branches. (11-3F.)

Interzooecial space. Portion of zoarium between zooecia. (11-2C.)

Longitudinal section. Section of zoarium cut parallel to zooecial tubes. (11-2C, F; *et al.*)

Lophophore. Circular or horseshoe-shaped ridge around mouth which bears a circlet of tentacles. (11-2A.)

Macula. Cluster of small zooecia. Marked superficially by shallow depression which typically is surrounded by unusually large zooecia. (11-5B.)

Massive. Colony form consisting of thick heavy zoarium, generally hemispherical or subglobular in shape. (11-5A.)

Mesopore. Zooecium of unusually small size set between larger zooecia and characterized by numerous transverse partitions (diaphragms). (10-2F, G.)

Monticule. Cluster of modified zooecia—typically of relatively small size—that project as elevation on zoarial surface. (11-3C, D.)

Mouth. Opening of digestive tract for intake of food.

Ovicell. Specialized skeletal structure—usually a chamber that houses the bryozoan larva during development. (11-2A; 11-5E, G.)

Peristome. Elevated rim surrounding aperture. (11-2A, D.)

Polypide. The living portion—soft parts—of the individual bryozoan (zooid).

Protoecium. Skeleton of larva formed when it attached to the substrate. Consists of the two chitinous valves of the free-living larva cemented to the substrate. (11-2B.)

TABLE 11-1 (continued)

Ramose. Colony form consisting of erect, round, or moderately flattened branches. (11-3D; 11-4B, C; 11-5B.)

Reptant. Colony form consisting of largely separate zooecial tubes which lie attached to substrate. (11-3A.)

Stomach. Pouch in anterior portion of digestive tract in which part or most of digestion occurs. (11-2A.)

Tangential section. Section of zoarium cut at right angles to zooecial tubes. (11-2G; 11-3C; *et al.*)

Tentacle. Flexible "arm" borne on ridge (lophophore) that surrounds mouth. Ciliated and functioning primarily in food getting. (11-2A.)

Unifoliate. Colony form consisting of single layer of zooecia—all opening onto one surface. (11-3C.)

Vibracula. Specialized individual in colony, consisting largely of whiplike process. (11-2D.)

Zoarium. Skeleton of entire bryozoan colony. Composed of calcite and/or chitin. (Portions of zoaria in 11-2, 11-3, 11-4, and 11-5.)

Zooecial wall. Sides of skeleton (zooecium) of individual bryozoan. (11-2A, C; *et al.*)

Zooecium. The skeleton, either chitinous or calcareous, of the individual bryozoan. Consists of tubular walls and various internal structures. (11-A, B, C; *et al.*)

Zooid. The individual bryozoan—including the soft parts (polypide) and the skeleton (zooecium). (11-2A.)

Zygote. Cell formed by union of egg and sperm cells.

coelom suffices to meet the moderate requirements of these small, inactive animals. Some experimental studies indicate that circulation within the coelom is accelerated by extension and retraction of the lophophore (Magum and Schopf, 1967).

Geologists and Bryozoa

The abundance of bryozoans in post-Cambrian rocks and their relatively rapid evolution make them valuable in rock-time correlation. Since individual zooecia are microscopic, they can be identified from fragments in well cuttings—a very desirable characteristic. In spite of difficulties in the preparation of identifiable specimens from the matrix and an unwieldy morphologic terminology, they contribute much to stratigraphy and could contribute more.

Most genera and species occur in a very limited series of rock types and were apparently sensitive to small environment changes. Unfortunately, these limiting environmental factors have hardly been studied. Duncan, in her review (1957), cites approximately 80 papers that touch on bryozoan paleoecology; most of these describe the more obvious features of bryozoan distribution:

a). They are most abundant in impure calcareous clastics, less abundant in shales, and rare in sandstones.

b). Species with fragile skeletons must have lived in fairly quiet waters.

c). Some robust groups contributed to reef formation.

d). Particular species tolerated a limited range of depth and temperature. For species with modern representatives, this range can be determined and used in interpretation.

e). Agitated water and low turbidity are favorable to the growth of Bryozoa.

Only a few paleontologists have examined the finer details of morphology in terms of environmental significance. Stach (1936) found that certain types of zoaria occur in only a limited number of habitats and are controlled primarily by the nature of the substrate. Others appear in a variety of habitats and modify their growth and form to suit the local environment. Variation in depth and in turbulence determines the development of different forms in this group of species. Similarly, Ross (1964) describes the ecological zonation of bryozoans in middle Ordovician carbonates of New York and Vermont, notes the absence of bryozoans from calcareous muds, and distinguishes bryozoan associations characteristic of a) calcareous sand substrates, b) small bioherms, c) large bioherms.

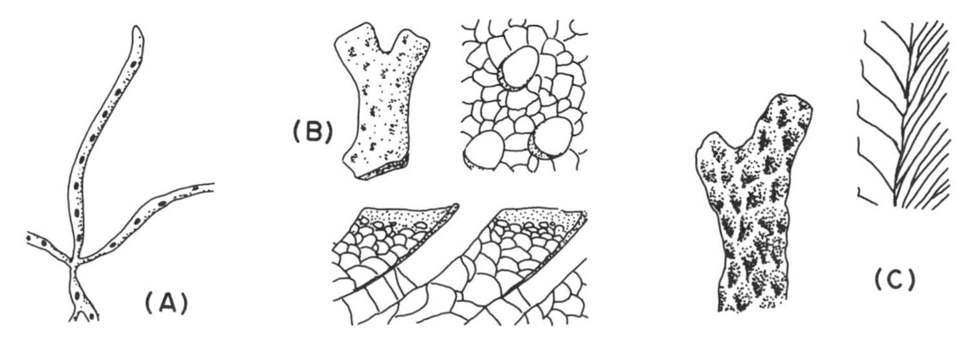

Fig. 11-4. REPRESENTATIVE BRYOZOA: CTENOSTOMATA, AND CYCLOSTOMATA. **(A)** Ctenostome *Vinella;* Ordovician to Cretaceous; repant; lateral view (×12). **(B)** Cyclostome *Meekopora;* Silurian to Permian; ramose; lateral view of zoarium (×1) and tangential and longitudinal sections (×10). **(C)** Cyclostome *Pleuronea;* Eocene to Pliocene; lateral view (×12) and longitudinal section (×12). [**(A)** *after* Ulrich; **(B)** *sections after* Ulrich; **(C)** *after* Canu *and* Bassler.]

Geologic occurrence

Figures 11-4 and 11-5 show some of the most useful guide fossils in this group. The order Ctenostomata, which ranges from the early Ordovician to the Recent, comprises zoaria with membraneous or calcified zooecia. The zooecia of most are simple tubes with terminal apertures and arise as isolated individuals from a threadlike tube, the *stolon.* In most features, the Ctenostomata seem the most primitive bryozoans. Since most recent species lack calcified skeletons, their comparative rarity as fossils probably does not measure their actual abundance or diversity.

The Cyclostomata are more advanced, for they have closely-grouped calcareous zooecia (though thin-walled and porous) and ovicells. The zooecia, however, remain simple undivided tubes. The cyclostomes, too, appeared in the early Ordovician and survived to the Recent. In the

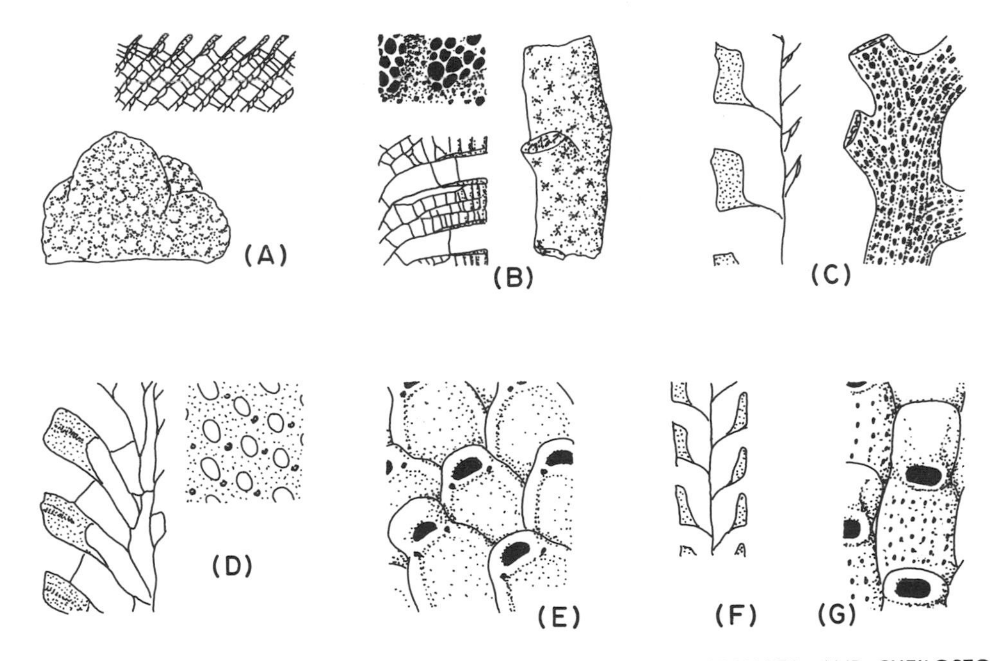

Fig. 11-5. REPRESENTATIVE BRYOZOA: CRYPTOSTOMATA, TREPOSTOMATA, AND CHEILOSTOMATA. **(A)** A massive trepostome, *Monticulipora;* Ordovician; lateral view (×0.5) and longitudinal section (×10). **(B)** A ramose trepostome, *Constellaria;* Ordovician; lateral view (×0.5) as well as tangential and longitudinal sections (×20). **(C)** Cryptostome *Sulcoretopora;* Devonian to Permian; ramose; longitudinal section (×20) and fragment (×5). **(D)** Cryptostome *Rhombopora;* Devonian to Permian; mesopores; longitudinal (×20) and tangential (×10) sections. **(E)** *Micropora,* a cheilostome; Cretaceous to Recent; zooecium at left with ovicell; ×25. **(F)** Another cheilostome, *Trigonopora* shown in longitudinal section; Eocene to Recent; bifoliate. **(G)** A final cheilostome, *Metracolposa;* Eocene; detail of zoarium; (×15). Large ovicell above aperture. [**(A)** *section after* Ulrich *and* Bassler; **(B)** *sections after* Ulrich; **(C)** *section after* Ulrich; **(D)** *after* Ulrich; **(E)** and **(F)** *after* Canu *and* Bassler; **(G)** *after* Bassler.]

TABLE 11-2

Classification of the Bryozoa *(based on Bassler, 1953).*

Ordovician to Recent. Colonial. Coelomate. Gut U-shaped; mouth opens in center of group of tentacles, the lophophore. Skeleton membraneous, chitinous, or calcareous. Aquatic, predominately marine, sessile benthonic.

CLASS GYMNOLAEMATA

Ordovician to Recent. Lophophore circular without a "lip" overhanging the mouth. Body wall not muscular. All marine.

Order *Ctenostomata*
Ordovician to Recent. Zooecia simple tubes with terminal apertures. Zooids develop by budding from a threadlike stolon. Skeleton membraneous, but calcified in some.

Order *Cyclostomata*
Ordovician to Recent. Closely grouped calcareous zooecia. Zooecia simple tubes; simple aperture.

Order *Trepostomata*
Ordovician to Permian. Massive, lamellate, or stem-like zoaria. Zooecia calcareous, long and slender; diaphragms; mature and immature regions. Some specialized zooids.

Order *Cryptostomata*
Ordovician to Permian. Frondlike or branching zoaria. Calcareous zooecia with very distinct mature and immature regions and with recessed apertures.

Order *Cheilostomata*
Jurassic to Recent. Calcareous and membraneous skeletons. Zooecia short; aperture small and may be surrounded by elevated peristome.

CLASS PHYLACTOLAEMATA

Cretaceous to Recent. Lophophore horseshoe shaped; "lip" overhangs mouth. Skeleton chitinous or gelatinous. Fresh water. One fossil genus recognized.

Paleozoic they were overshadowed in numbers and diversity by the trepostomes and cryptostomes, but they outlasted these to become the most abundant bryozoans in the early and middle Mesozoic.

The order Trepostomata includes species with massive, lamellate, or stem-like zoaria. Their zooecia consist of long calcareous tubes partitioned by diaphragms and divided into immature and mature regions. Specialized zooids occur between the normal type in the mature region. Their form is fairly robust, and some species helped, to a moderate degree, in building Paleozoic "coral" reefs. Trepostomes are among the oldest known bryozoans; they occur in lower Ordovician rocks, and a middle Cambrian genus has been described—though the identification is very suspect. They underwent a rapid radiation in the middle Ordovician, declined in numbers and variety after the Silurian, and apparently failed to survive the Paleozoic.

The Cryptostomata have delicate frondlike or branching zoaria. The zooecia are calcareous and resemble those of the Trepostomata. The boundary between the mature and immature regions is more distinct than in the latter, and the apertures are recessed below the zooecial surface. They are another product of the early Paleozoic radiation of the Bryozoa and had a history similar to that of the trepostomates except that they did rather better in the Devonian and Carboniferous.

No Triassic cryptostomes are known, but apparently one line, as yet undetected in the fossil record, survived and gave rise to the Cheilostomata in the Jurassic. By Cretaceous, the cheilostomes had diverged into a variety of species, genera, and families, and they predominate in Cenozoic faunas. The Cheilostomata include species with either calcareous or membraneous skeletons. The zooecia are short and rounded or angular. The aperture is most commonly small, and, in some, is surrounded by an elevated rim, the *peristome*. Ovicells of different types are developed, as are specialized zooids, the avicularia and vibracula.

A final group, regarded as a different class, are the freshwater bryozoans. They are known as fossils only from a single genus. The zoarium is membraneous or gelatinous. This class, the Phylactolaemata, is characterized by a horseshoe shaped lophophore and a "lip" overhanging the mouth. The other five orders compose the class Gymnolaemata, which has a circular lophophore and lacks a "lip."

REFERENCES

Bassler, R. C. 1953. "Bryozoa," in Moore, R. C. (Ed.) *Treatise on Invertebrate Paleontology, Pt. G.* New York: Geol. Soc. of America.

Boardman, R. S. and K. M. Towe. 1966. "Crystal Growth and Lamellar Development in Some Recent Cyclostome Bryozoa," abstract in *Program 1966 Annual Meeting, Geol. Soc. of America.*

Clark, R. B. 1964. *Dynamics in Metazoan Evolution.* Oxford: Clarendon Press.

Duncan, H. 1957. "Bryozoans," in Ladd, H. S. (Ed.) "Treatise on Marine Ecology and Paleoecology," vol. 2. *Geol. Soc. of America, Memoir 67,* pp. 783-800.

Hyman, L. 1959. *The Invertebrates, Vol. V, Smaller Coelomate Groups.* New York: McGraw-Hill.

Magum, C. P. and T. J. M. Schopf. 1967. "Is an Ectoproct Possible," Nature, vol. 213, pp. 264-266.

Ross, J. P. 1964. "Morphology and Phylogeny of Early Ectoprocta (Bryozoa)," *Geol. Soc. of America, Bull.,* vol. 75, pp. 927-948.

Stach, L. W. 1936. "Correlation of Zoarial Form with Habitat," *Jour. of Geology,* vol. 44, pp. 60-65.

12

Individualism reconsidered: The Brachiopoda

The brachiopods, with the same general structural plan as the Bryozoa and a somewhat similar mode of life, evolved in a different direction. They abandoned, or never attained, asexual "budding" and colony formation. The two chitinous plates of the larva become the two valves of the adult shell. The individual animal grows much larger than the individual bryozoan and approximates in size an entire bryozoan colony.

In spite of, or perhaps because of, these differences, the brachiopods were as important, as abundant, and as diversified in the Paleozoic as the Bryozoa and were among the dominant members of sea-bottom societies throughout that era. After the end of the Paleozoic they declined in numbers and variety, but a few genera survived to the Recent.

Success in the sessile benthos

Animals fixed on the sea floor face quite different environmental challenges than do the active terrestrial ones with which we are more familiar. They cannot stalk their prey nor range widely to graze on algae, but must either filter small drifting organisms from the water or seize larger prey that swim into tentacle reach. Waste products are expelled into the water to be carried off by currents or simply diffuse. On shallow sea bottoms, turbulence removes wastes, supplies ample oxygen, and constantly renews the supply of plants and animals for the filter feeders. In quiet waters, however, the oxygen concentration is reduced, organic wastes accumulate, and noxious compounds like hydrogen sulphide appear. Turbidity also modifies feeding and respiration, particularly for the filter feeders, by clogging gills, ciliary tracts, and so on, with mud or sand. If deposition is rapid, sessile animals are likely to be buried; if turbulence is great, they may be broken or otherwise injured. Some require solid objects for attachment, such as coarse detritus or shells of other animals. Others live unsupported on the bottom; these may need special mechanisms to prevent their sinking into the sediment. The sessile benthos cannot flee their predators and parasites nor avoid the vagaries of the physio-chemical environment. Since copulation is impossible, the sperm must swim or float some distance to fertilize the egg. The larvae must be motile and, typically, pelagic to ensure dispersal. High larval mortality in the plankton requires production of numerous eggs.

On the other hand, requirements of circulation, of perception, of internal regulation, and of coordination are less stringent than among active animals. As a consequence, these systems in the bryozoans, brachiopods, bivalves, and other sessile groups are relatively simple—either primitive or "degenerate." The brachiopod, for example, has only a pair of ganglia above and below the anterior portion of the gut, with nerves to the arms and to the muscles of the shell; it lacks distinct sense organs; it possesses only a few simple blood vessels with a slightly differentiated muscular portion as a heart; and it has small and simple excretory organs.

The difficulties of being a brachiopod

Brachiopods are animals of the marine sessile benthos. They construct a shell with two valves, one dorsal, the other ventral.* The hinge between valves is posterior; they open anteriorly. Since each valve has somewhat different functions, they differ in form; since the plane of bilateral symmetry is perpendicular to the surface of the valves, each is bilaterally symmetrical—has a "right" and "left." They feed by an elaborate filter mechanism. Evolution has impressed on their form a variety of adaptations to the peculiarities of their environment. The fossil shell

* Relative to organ position—not necessarily with respect to orientation on the bottom.

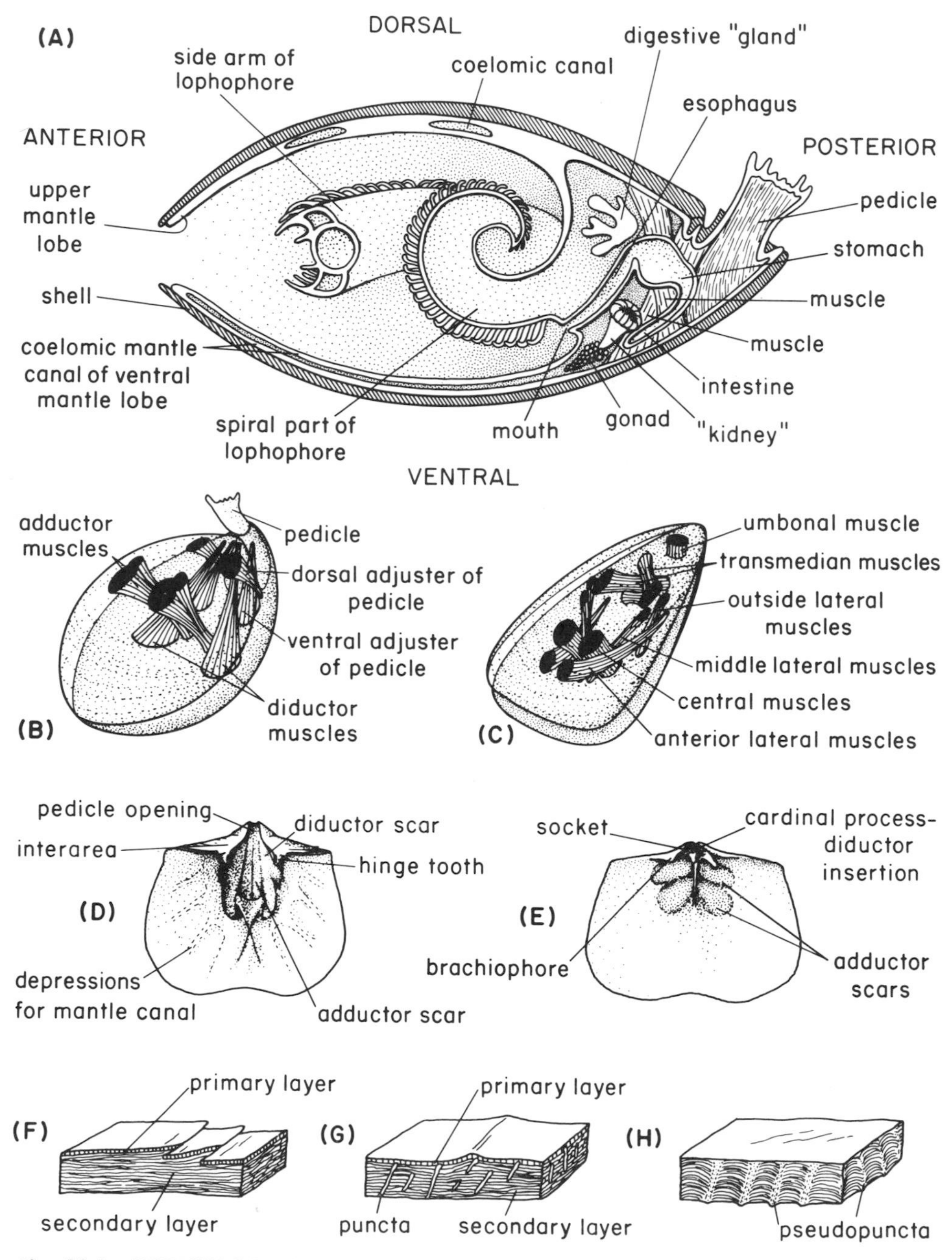

Fig. 12-1. MORPHOLOGY OF THE BRACHIOPODA. **(A)** Diagrammatic view of brachiopod sectioned in the median plane. **(B)** Muscle system of articulate brachiopod. **(C)** Muscle system of inarticulate brachiopod. **(D)** Interior of pedicle valve. **(E)** Interior of brachial valve. **(F)** Cross-section of shell, nonpunctate. **(G)** Cross-section of shell, punctate. **(H)** Cross-section of shell, pseudopunctate.

TABLE 12-1

Glossary for the Brachiopoda ***(figures in parentheses refer to pertinent illustration).***

Adductor muscles. Muscles that close and/or hold valves together. May leave attachment scars on interior surface of valve. (12-1B, C, D, E.)

Adjuster muscles. Two pairs of muscles in articulate brachiopods that insert on the brachial and pedicle valves and have their origin on the pedicle base. They adjust the position of the shell on the pedicle. (12-1B, D, E.)

Alate. Shell form in which the valves are drawn out at the lateral ends of hinge line to form wing-like extensions. (12-5E; 12-10A.)

Anterior. Direction of shell margin where the valves separate when open. Opposite the position of the hinge line. (12-1A.)

Beak. Pointed extremity of valve adjacent to or posterior to the hinge line and in midline of valve.

Brachial valve. Valve to which brachidium is attached. In most—but not all—brachiopods the smaller valve, typically with a small or indistinguishable beak and bearing only a small part of the pedicle opening. (12-1A, E, *et al.*)

Brachidium. Calcareous support for the lophophore. (12-2.)

Brachiophore. Short, typically stout, processes that project from hinge line of brachial valve into the interior of the valve. Serve for attachment of lophophore. (12-1E; 12-2A.)

Cardinal margin. Curved posterior margin along which the valves are hinged. Equivalent to *hinge line* in shells with straight margin.

Cardinal process. Ridge or boss on inner surface of brachial valve between (or part of) the "scars" marking the insertion of the diductor muscles. (12-1E.)

Chilidial plate. Plate at side of the opening (notothyrium) in the brachial valve for pedicle. (12-3G.)

Commissure. Junction between edges of valve.

Costae. Ridges on external surface of valve that extend radially from beak. Costae do not involve any folding of inner surface of shell.

Crura. Basal portions of calcareous support (brachidium) of lophophore. (12-2B, C, D.)

Delthyrium. Opening in pedicle valve adjacent to hinge line. Serves for passage of pedicle. (12-1D; 12-3A, B.)

Deltidial plate. Plate on either side of pedicle opening (delthyrium) in pedicle valve that constricts opening or even, with its mate from opposite side, closes it off completely. (12-3E, F.)

Dental plate. Plate extending up from floor of pedicle valve to hinge line that serves to support tooth. (12-3A.)

Diductor muscles. Muscles that open valves. Insert on floor of pedicle valve and along or adjacent to hinge line on brachial valve. Attachments may show as scars on inner surface of valves. (12-1B.)

Fold. Elevation (up-arch) of a valve along the midline. Accompanied by complementary depression (sulcus) of other valve. (12-4E; 12-8C, D; *et al.*)

TABLE 12-1 (continued)

Foramen. Circular opening adjacent to beak of pedicle valve. Serves for passage of pedicle. (12-3F; 12-10B.)

Growth lines. Series of fine to coarse ridges or breaks on outer surface of shell. Subparallel to edges of valve and concentric about beak. (12-5C, D; *et al.*)

Hinge line. Edge of shell where two valves are permanently articulated and which serves as hinge when shell is opened or closed. (12-3A; *et al.*)

Hinge plate. Plate, simple or divided, that lies along the hinge line in the interior of the brachial valve. Typically nearly parallel to plane between valves, it bears hinge sockets and is joined to bases of crura. (12-3A.)

Interarea. Plane or curved surface between beak and hinge line on either valve. Generally distinguished by a sharp break in angle from the remainder of the valve and by the absence of costae, plications or coarse growth lines. (12-1D; 12-3A, H through L; *et al.*)

Jugum. Simple or complex skeletal connection between the right and left halves of the brachidium. (12-2B, C, D.)

Loop. Brachidium consisting of a pair of simply curved or doubly bent longitudinal "arms" and a relatively simple jugum connecting the anterior ends of these arms. (12-2B.)

Lophophore. Appendage extending anteriorly from mouth and consisting of a lobed disk or a pair of coiled arms (brachia). The arms or lobes attach on either side of the mouth and bear on their edges slender, ciliated threads (cirri). (12-1A.)

Mantle. Two folds of body wall that lie respectively above and below viscera and that line the inner surface of each valve. (12-1A.)

Mantle canal. Canal within mantle that connects to coelomic cavity of body. (12-1A, D.)

Median septum. Calcareous ridge built along midline of interior of valve. (12-2A, B.)

Notothyrium. Opening in brachial valve adjacent to and outside the hinge line. Forms part of the opening for the pedicle. (12-3A, G.)

Pedicle. Muscular and/or fibrous stalk which is attached to the inner surface of the pedicle valve and which passes out posteriorly to attach to substrate. (12-1A.)

Pedicle opening. Opening adjacent to or along hinge that serves for passage of pedicle. Opening may be in pedicle valve only or in both pedicle and brachial valves. (12-1A, D; 12-3A through G; *et al.*)

Pedicle valve. Valve to which the pedicle is attached. By convention ventral in position. (12-1A, D.)

Plica. Radial ridges and depressions that involve entire thickness of shell and thus appear as corrugations on inner as well as outer surfaces. Distinguished from fold and sulcus by small amplitude and by occurrence to sides of midline. (12-8C, D; *et al.*)

Posterior. Direction defined by position of hinge line and/or pedicle opening. (12-1A.)

Pseudodeltidium. Single plate covering all or part of pedicle opening in pedicle valve. (12-3C, D.)

Pseudopunctate. Shell microstructure characterized by structureless rods of calcite in prismatic layer perpendicular to shell surface. May weather out in fossil shells, leaving tiny openings like those in punctate shells. (12-1H.)

TABLE 12-1 (continued)

Punctate. Shell microstructure characterized by small canals extending perpendicularly from inner to outer surface of shell. Typically appear under hand lens as closely spaced pores. (12-1G.)

Socket. Depression along hinge line of brachial valve which receives the hinge tooth of the pedicle valve. (12-1E; 12-2A, B; 12-3A.)

Spiralium. One of a pair of spirally coiled calcareous ribbons that form the brachidium in some brachiopods. (12-2C, D, E.)

Spondylium. Curved plate in midline of beak of pedicle valve. Formed by union of dental plates from either side of midline and serving for muscle attachment. (12-7C.)

Sulcus. Major longitudinal depression, down-arch, along midline of valve. Typically associated with up-arch, the fold. (12-4E; 12-10A, B; *et al.*)

Tooth. Projection along hinge line of pedicle valve that fits into socket on opposing valve. (12-1D; 12-3A.)

Umbo. Relatively convex portion of valve next to (anterior to) beak.

Valve. One of the two curved, chitino-phosphatic or calcareous plates that form the brachiopod shell and that surround and lie, respectively, above and below the soft parts.

and skeletal structures within the shell reproduce some part of these adaptations—a greater part possibly than in any other group of sessile animals.

Feeding

Tracts of fine hairlike processes (*cilia*) on the lophophore and about the mouth form the brachiopod filter mechanisms. The lophophore consists of a pair of coiled arms (*brachia*) bearing ciliated, tentacle-like *cirri.* The beat of these cilia maintains currents into and out of the shell. In some, water is drawn in through the center of the gap between the valves and expelled laterally; others, with a different lophophore structure, have a median excurrent path and lateral incurrents. In the burrowing lingulids, stout chitinous hairs or *setae* that fringe the edge of the brachiopod shell are grouped to form incurrent and excurrent tubes. The brachia and cirri lie before the mouth in the anterior part of the shell cavity and in many brachiopods occupy the greater part of the space within the shell. They are supported by attachment to one valve, named for this attachment, the *brachial valve.* A calcified *brachidium* supports the lophophore in some. The brachidia are quite simple in some brachiopods, even ones with complex lophophore structure, but where complex they seem to parallel that structure closely. Such complex brachidia, as well as impressions of the lophophore on the interior of some shells, give information

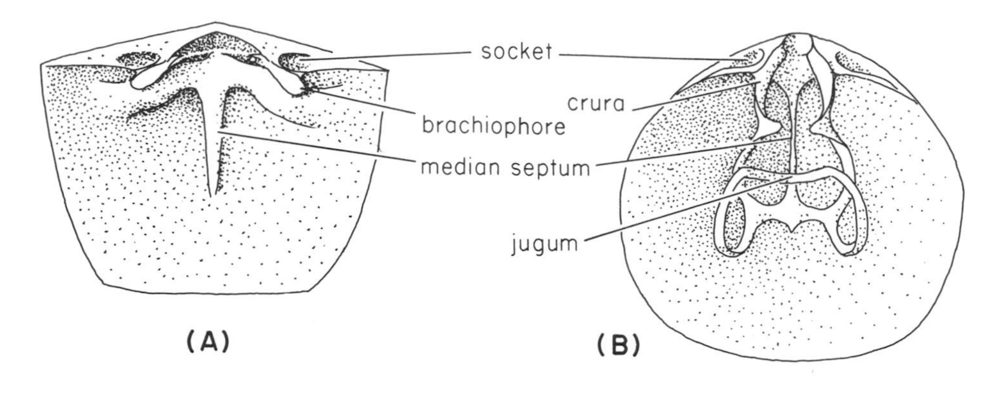

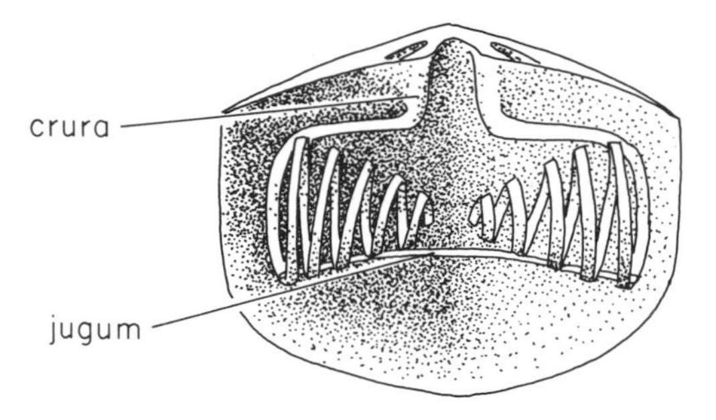

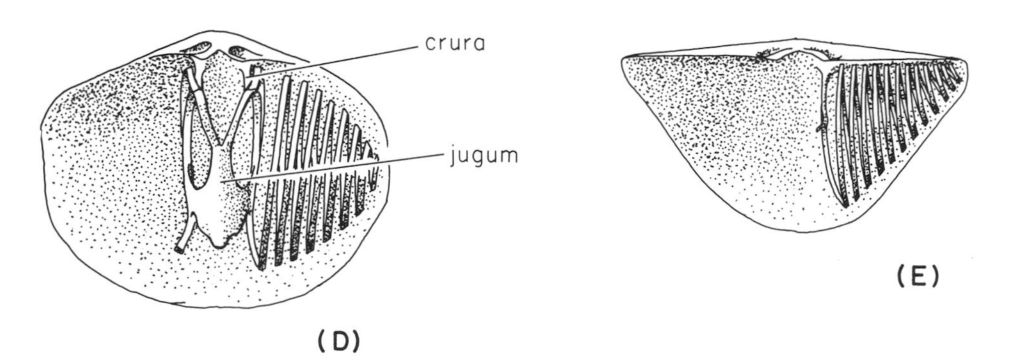

Fig. 12-2. VARIETIES OF BRACHIDIA. **(A)** Brachiophore. **(B)** Loop. **(C)** Spiralium, atrypoid. **(D)** Spiralium, athyroid. **(E)** Spiralium, spiriferoid. [**(A)** *after* Schuchert and Cooper; **(B)** *after* Davidson; **(C)**, **(D)**, and **(E)** *after* Beecher.]

about feeding mechanisms (Figure 12-2). The size and structure of the lophophore probably affects valve shape and modifications of shell shape, such as central folds, and must influence circulation of water in and out of the shell. Some of the elaborate plication of shells may reflect feeding adaptations (Rudwick, 1964): They maintain a large opening for water flow but restrict the width of the opening so that detritus and small predators are excluded.

Respiration and circulation

Respiration is accomplished primarily by the mantle lobes that line the inside of the valves, but the thin-walled cirri and their cilia may also function in exchange of gases. The adaptation may be expressed in general shell shape, possibly in *plications* of the shell if these correspond to folds in the mantle, and possibly in the *pallial markings* on the inner surface of the shell, which record the position of branches on the coelomic space in the mantle (Figure 12-1). Very little is known of the relation of structural variation to environmental variation.

Support and protection

Some brachiopods attach by a fleshy "stem," the *pedicle;* others, by a reduced rootlike or threadlike pedicle, and still others lie free on the

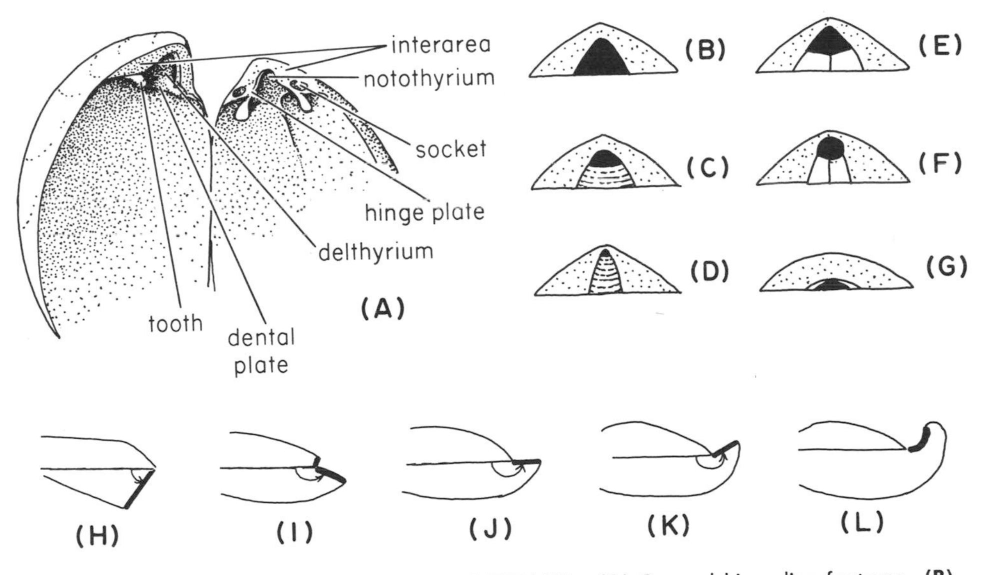

Fig. 12-3. BRACHIOPOD HINGE LINE AND INTERAREA. **(A)** General hinge line features. **(B)** through **(G)** Posterior view of valve. **(B)** Pedicle valve; delthyrium completely open. **(C)** Pedicle valve; delthyrium partly closed by pseudodeltidium. **(D)** Pedicle valve; delthyrium nearly closed by pseudodeltidium; pedicle vestigial. **(E)** and **(F)** Delthyrium partly covered by deltidial plates. **(G)** Brachial valve; notothyrium partly covered by chilidial plates. **(H)** through **(L)** Lateral view of valves showing attitude of interarea by heavy line. **(H)** Interarea on pedicle valve at acute angle to plane of commissure. **(I)** Interareas on both valves, at obtuse angle to commissure. **(J)** Interarea on pedicle valve in plane of commissure (straight). **(K)** Interarea on pedicle valve, at angle greater than 180° (reflex). **(L)** Interarea on pedicle valve changing with growth.

bottom, or else cement one valve to solid objects. Some fossil species have spines that apparently served for support and, in some immature individuals, for attachment. The pedicle attaches to the valve (the *pedicle* one) opposite the one bearing the brachia. It passes posteriorly through the hinge area through an opening in one or both valves. The size of the pedicle opening and its location (Figure 12-3) reveal something of pedicle structure and function in fossil brachiopods, as do the scars left in the shell by the attachment of the pedicle and its auxiliary muscles (Figure 12-1). Robust pedicles support the shell above the substrate or in a burrow; small ones anchor it as it rests on one valve or the other.

Species that lack a muscular pedicle risk settling into soft sediments or burial by continued deposition. The strongly convex lower valves of some and the high *folds* (Figure 12-4) of others functioned, perhaps, to keep the valve margin from being covered. Elongate, round, and winged shapes (Figure 12-5) must also be or have been adaptive either to bottom conditions or to turbidity and turbulence. Since the substrate character, turbidity, and turbulence are determinable in some fossil occurrences, paleontologists have investigated this aspect of brachiopod paleoecology (for examples, see Cooper, 1957, and the papers he cites) more than any other.

A calcareous or chitinophosphatic shell protects the individual brachiopod against its predators and parasites and against some inclemencies of the physical environment. Some hold the valves in articulation entirely with an elaborate set of muscles; others have a hinge with teeth and sockets. Muscle structure and operation is indicated by attachment plat-

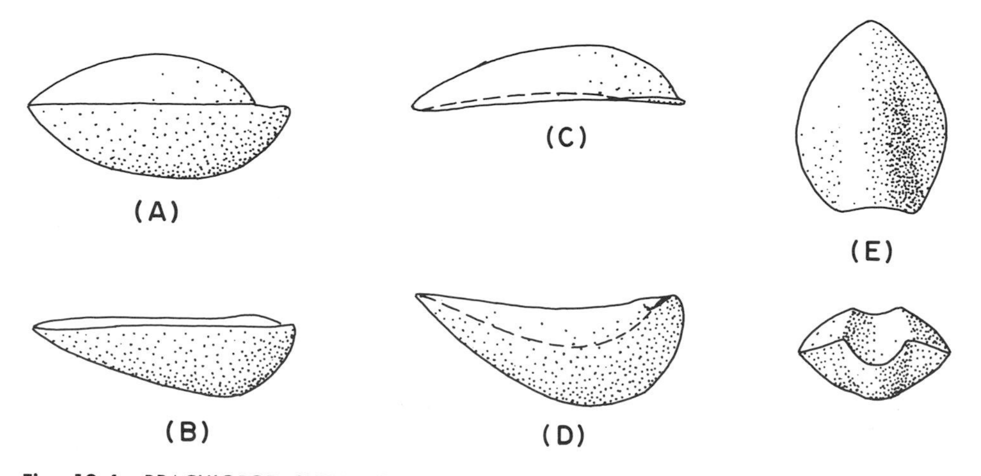

Fig. 12-4. BRACHIOPOD SHELL FORM. **(A)** through **(D)** Lateral views of shell form. **(A)** Both valves convex (biconvex). **(B)** Brachial valve plane, pedicle convex (plano-convex). **(C)** Brachial valve convex, pedicle plane (convexi-plane). **(D)** Brachial valve concave, pedicle convex (concavo-convex). **(E)** Pedicle valve and anterior view of both valves. Sulcus in brachial valve and fold in pedicle.

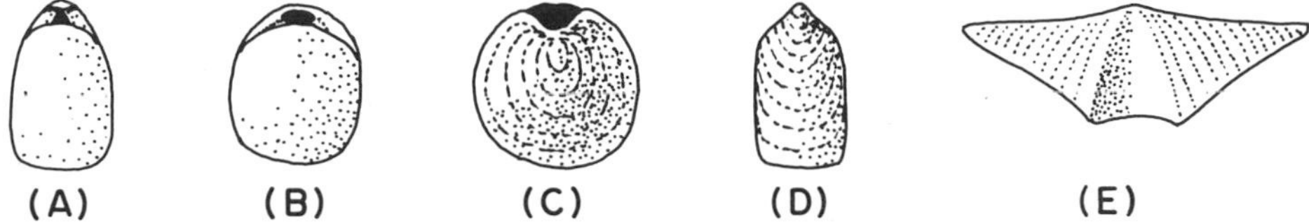

Fig. 12-5. VARIATION IN BRACHIOPOD SHELL FORM. **(A)** Elongate ellipsoidal. **(B)** Ellipsoid. **(C)** Subcircular. **(D)** Linguliform. **(E)** Winged or alate. **(A)** through **(D)** are brachial views; **(E)** a pedicle view.

forms and/or scars (Figure 12-1). Are variations in the muscular apparatus that opens and closes the valves or in the articulation of valves adaptively significant? Does the thickness and composition of the shell influence predation or resistance to turbulence, or is the chemistry of the environment more important? To what extent are shape or plications (Westermann, 1964) related to strengthening of the shell? The variations in fine structure of the shell such as small openings (*punctae*) and calcite rods (*pseudopunctae*) have at present no functional explanation; they may be results of special adaptations in the mantle for shell formation.

The difficulties of being a paleontologist

Questions about adaptive significance loom large in interpreting brachiopod phylogeny, as well as in paleoecology. Many of the recognized taxonomic units in many groups of fossil animals probably are artificial, i.e., they do not correspond to phyletic units. Since paleontologists do not know and probably never will know the adaptive significance of all fossil characteristics, they can not distinguish characteristics inherited from common ancestors from those evolved to fit similar environments. In classification, they must simply do the best they can. In only a few kinds of animals, like the vertebrates, do they understand adaptive trends so thoroughly that there can be very few artificial taxonomic assemblages.

Brachiopods fall tantalizingly between these two extremes. Their characteristics, obviously, have adaptive significance, but that significance is unknown for many of them. For this reason, the interpretation of brachiopod evolution has changed periodically as additional information has accumulated. Paleontologists specializing in brachiopod phylogeny and taxonomy have sought key, conservative character states that appeared early in the divergence of the various stocks and remained stable through their subsequent evolution. Since, however, the character states cannot be evaluated except by empirical usefulness in ordering genera,

TABLE 12-2

Classification of the phylum Brachiopoda *(after Williams and Rowell, 1965)*

Cambrian to Recent. Bivalve, valves dorsal and ventral. Lophophore consists of brachia bearing ciliated cirri. Suspension feeders; marine; sessile benthos, attach by pedicle, lie free, or cement valve to substrate.

CLASS INARTICULATA

Cambrian to Recent. Valves never articulated by teeth and sockets. Shell chitinophosphatic or calcareous.

Order *Lingulida*
Cambrian to Recent. Calcium phosphate, rarely calcareous, shell; biconvex; beak typically terminal on both valves; pedicle typically emerging between valves posteriorly.

Order *Acrotretida*
Cambrian to Recent. Subcircular, phosphatic or punctate calcareous shell; pedicle opening confined to pedicle valve; beak marginal to subcentral.

Order *Obolellida*
Cambrian. Calcareous, biconvex, subcircular shell; pseudointerarea on pedicle valve; pedicle may emerge between valves or through pedicle valve; beak of brachial valve marginal.

Order *Paterinida*
Cambrian to Ordovician. Phosphatic, rounded shell; pedicle valve convex to hemiconical; pseudointerarea divided by delthyrium; brachial valve similar.

CLASS ARTICULATA

Cambrian to Recent. Valves calcareous, articulated by hinge teeth and sockets.

Order *Orthida*
Cambrian to Permian. Typically unequally biconvex shells with relatively wide, straight hinge lines and with interareas on both valves. Shell impunctate, rarely punctate or pseudopunctate.

Order *Strophomenida*
Ordovician to Jurassic. Plano- to concavo-convex, less commonly biconvex. Interareas highly variable; hinge line typically long. Pseudodeltidium rarely absent; pedicle opening much reduced or absent. Typically pseudopunctate.

Order *Pentamerida*
Cambrian to Ordovician. Biconvex. Spondylium in pedicle valve. Delthyrium open or partly closed by deltidium. Impunctate shell. Interareas commonly small; hinge line short or moderately long.

Order *Rhynchonellida*
Ordovician to Recent. Typically biconvex, shell with strong beaks on one or both valves, short hinge line, functional pedicle. Impunctate, rarely punctate.

Order *Spiriferida*
Ordovician to Recent. Typically biconvex; interareas, pedicle, and hinge length highly variable; punctate or impunctate. Spiral brachidium.

Order *Terebratulida*
Devonian to Recent. Typically biconvex, short hinge line, interarea on pedicle valve only. Punctate; functional pedicle; looped brachidium.

---

the result is a key rather than a valid phyletic model. The classification and the phyletic scheme used in the *Treatise on Invertebrate Paleontology* employs a different approach (Williams and Rowell, 1965, pp. 223, 227), related in rationale to that of numerical taxonomy (see pp. 159-160):

genera are assembled into larger taxa on their similarity in a very large number of characters. No character state is exclusive to a taxon, but the set of character states of each taxon is distinct (polythetic system). In turn the superfamilies are marshaled into orders (Table 12-2) largely connected by families or subfamilies which overlap two or more superfamilies. The definitions of each taxon are filled with restrictive terms such as "rarely," "commonly," "typically," "variably," "may be." In consequence, identification of a specimen as a member of an order, e.g., Strophomenida, is only a probability, i.e., it has an assemblage of character states found commonly in this order but only rarely in other orders. Positive assignment to an order depends on generic identification, so it may be best to refer to it as a strophomenid-like shell until that critical identification is possible.

A provisional phylogeny

In spite of difficulties in phylogenetic interpretation, certain broad trends and patterns appear if one examines brachiopod evolution broadly and with reference to sets of characteristics rather than individual structures.

The proto-brachiopod. When the brachiopods appeared in the early Cambrian, they were already diversified, representing six distinct orders and both recognized classes. Most of these early Cambrian genera had chitinophosphatic shells, lacked a definite hinge between the valves, and had an elaborate set of muscles to hold them in articulation (Figure 12-1). Some also had a large pedicle that extended through notches developed on both valves. Since these are most similar to larval brachio-

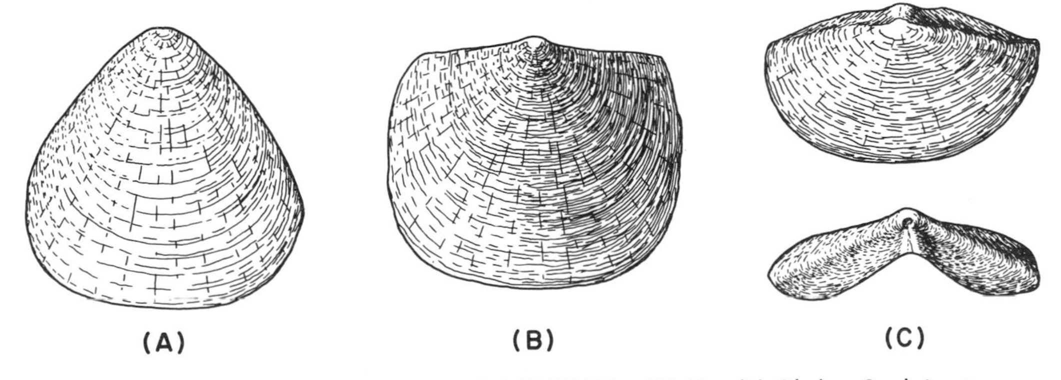

Fig. 12-6. THE BRACHIOPOD RADIATION: INARTICULATA. **(A)** Lingulid *Obolus;* Cambrian to Ordovician; pedicle view; length 1.6 cm. **(B)** Acrotretid *Acrotreta;* Cambrian to Ordovician; brachial view of brachial valve; length 7.5 cm. **(C)** Paterinid *Paterina;* Cambrian; pedicle and posterior views; length 2.0 mm.

pods, and since recent genera with these characters possess what seems a generally primitive anatomy, they are considered closest to the hypothetical ancestral species.

Initial radiation. This primitive stock of hingeless brachiopods forms a class, Inarticulata. When they first appear in the record, some had the pedicle opening shifted to the pedicle valve alone or had lost the pedicle altogether; some cemented themselves to the substrate by one valve; some formed calcareous shells in place of the chitinophosphatic. Four orders are recognized, the Lingulida, the Acrotretida, the Obolellida, and the Paterinida (Figure 12-6).

Among the early Cambrian brachiopods that evolved calcareous shells are some of special interest as forerunners of the class Articulata. They have a definite hinge between valves, typically marked by a tooth-and-socket articulation and flattened areas, *interareas,* next to the hinge line on one or both valves. These are grouped in the order Orthida (Figure 12-7), but even at this time level two distinct phyletic groups had diverged, one, the superfamily Billingsellacea, the other, the superfamily Orthacea, distinguished primarily by different origins of the pedicle.

The second round. The diversification of the brachiopods proceeded

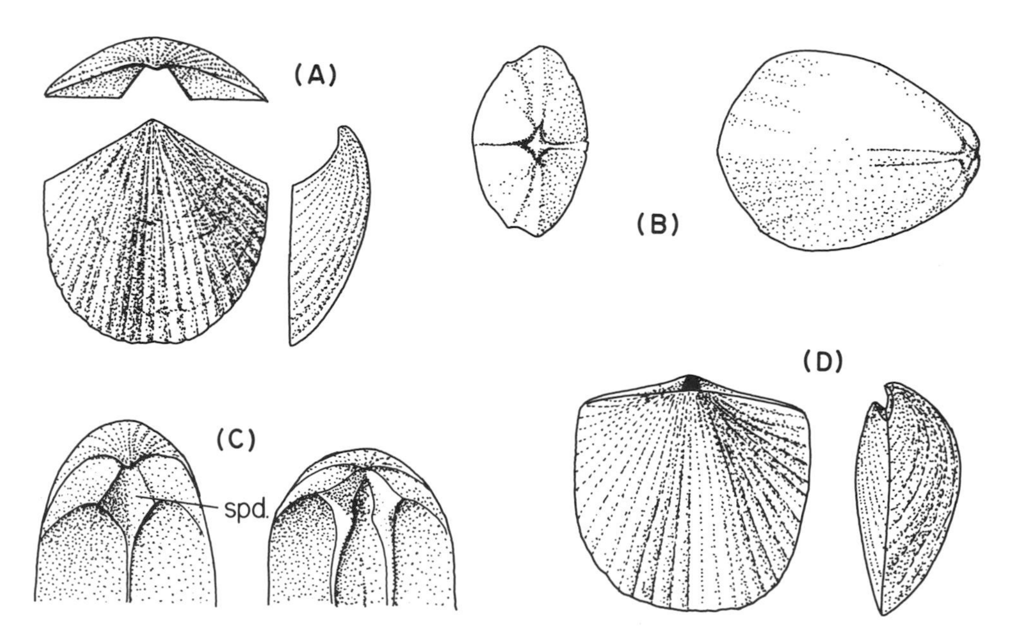

Fig. 12-7. THE BRACHIOPOD RADIATION: ORTHIDA AND PENTAMERIDA. **(A)** Orthid *Eoorthis;* middle Cambrian to early Ordovician; pedicle valve, posterior, pedicle, and lateral views; length 1.8 cm. **(B)** Pentamerid *Pentamerus;* middle Silurian; posterior and pedicle views; length 8.0 cm. **(C)** Pentamerid *Conchidum;* middle and late Silurian; internal views of posterior end of pedicle and brachial valves. Note specialized hinge structures; abbreviation: spd., spondylium. **(D)** Orthid *Dalmanella;* early Silurian; brachial and lateral views; length 1.2 cm. [**(A)** *after* Walcott; **(C)** *after* Schuchert *and* Cooper.]

through the middle and late Cambrian, but the only real novelty was the appearance of the articulate order, Pentamerida. This line continued the elaboration of hinge structure with the formation of an internal platform for muscle attachment (Figure 12-7) and also developed simple lophophore supports, the *crura,* and a persistent *primary* shell layer.

This progressive and moderate increase in brachiopod variety was interrupted by a burst of evolution in the Ordovician. The first new group to appear was the order Strophomenida (Figure 12-8), presumably derived from unknown orthid ancestors. Their hinge line, like that of orthids, was typically very wide, but commonly one valve was flattened or concave, and the shell had vertical rods of structureless calcite in the primitive laminated layer, a *pseudopunctate shell* (Figure 12-1). Two additional orders appeared in the middle Ordovician. The Rhynchonellida characteristically have biconvex shells that are strongly plicate (Figure 12-8). The posterior portion of one or both valves is commonly drawn out into a prominent beak. The hinge line is very short and the interareas

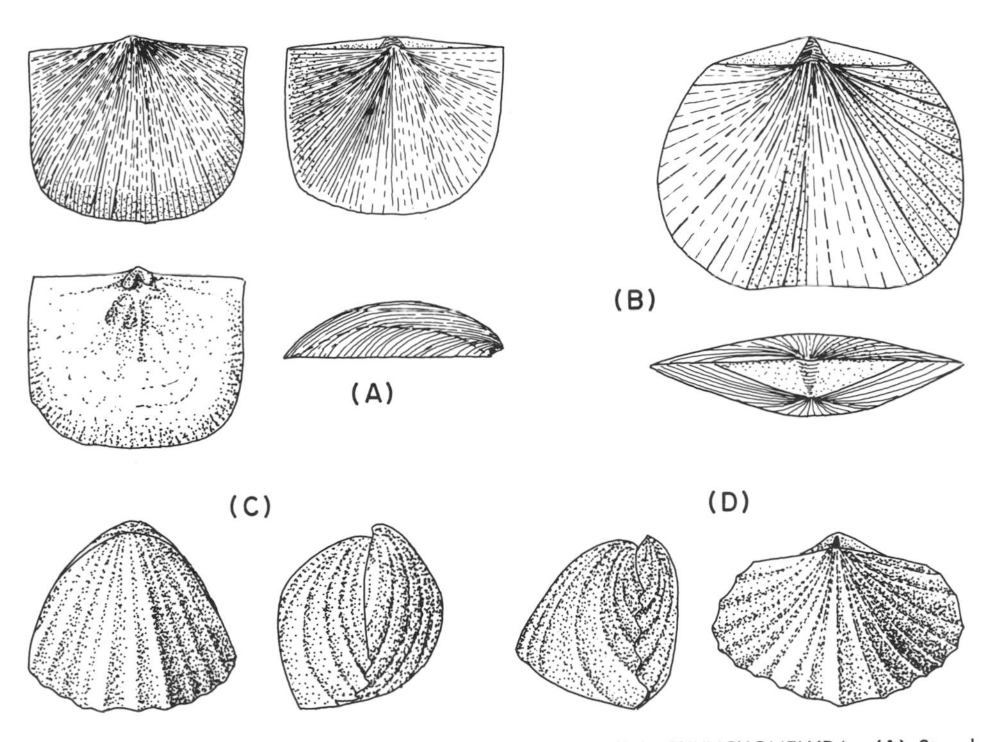

Fig. 12-8. THE BRACHIOPOD RADIATION: STROPHOMENIDA, RHYNCHONELLIDA. **(A)** Strophomenid *Rafinesquina;* late Ordovician; upper views pedicle and brachial, lower views, interior of brachial valve and lateral; length 3.1 cm. **(B)** Strophomenid *Orthetes;* Devonian to Pennsylvanian; brachial and posterior views; length 1.7 cm. **(C)** Rhynchonellid *Rhynchotrema;* Ordovician; brachial and lateral views; length 1.9 cm. **(D)** Rhynchonellid *Camarotoechia;* Silurian to Permian; lateral and brachial views; length 1.4 cm. [**(B)** *after* Piveteau, ed., *Traité de Paleontologie.* Copyright Masson et Cie, Paris. Used with permission.]

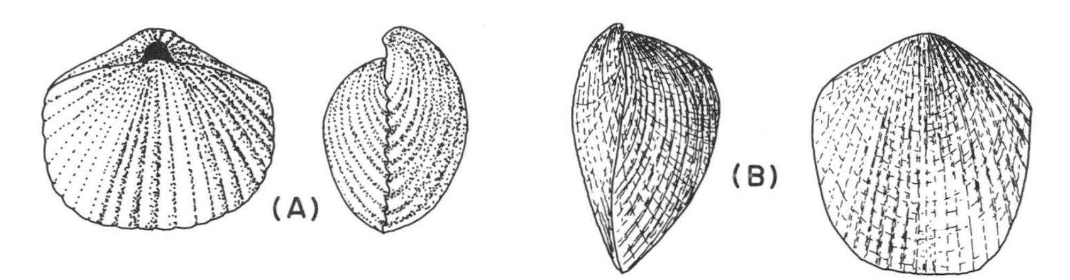

Fig. 12-9. THE BRACHIOPOD RADIATION: SPIRIFERIDA. **(A)** *Zygospira;* Ordovician to Silurian; brachial and lateral views; length 1.5 cm. **(B)** *Atrypa;* Ordovician to Devonian; lateral and brachial views; length 3.8 cm.

rudimentary or absent, with the pedicle opening restricted by some sort of deltidial plates. The rhynchonellids probably derived from early pentamerids.

The Ordovician Spiriferida are somewhat similar to the Rhynchonellida in external features, with a rounded biconvex shell, a short hinge line, and a small interarea limited to the pedicle valve (Figure 12-9). Internally, the simple cural bases for lophophore support are replaced by spiral brachidia, which first appear as a simple loop but are soon elaborated (Figure 12-2). Their similarity to the rhynchonellids suggests a common origin.

In addition to these new groups, the older orders produced a series of advanced forms. Among the inarticulates, genera with calcareous shells appeared in the Acrotretida and Lingulida. This change seemingly occurred by parallel evolution, and it also paralleled an earlier event in the Articulata. The orthids now included a wide variety of forms, some with one valve flat or concave, a group with punctate shells, and a number of genera with short basal supports, the *brachiophores,* for the lophophore.

Boom. By the end of the Silurian, all of the orders of brachiopods—and all but three of the different suborders—had joined the benthonic associations. Some lines became extinct; two orders of the inarticulates failed to survive the end of the Ordovician. Two inarticulate orders survived and, though never prosperous, hung on to the Recent. Some of these inarticulate genera have extremely long stratigraphic ranges (*Lingula,* Ordovician to Recent; *Orbiculoidea,* Ordovician to Permian; *Discinisca,* Jurassic to Recent; *Crania,* Cretaceous to Recent) and have demonstrated remarkable stability of form.

The orthids with nonpunctate shells began to fall off in variety and numbers before the close of the Ordovician, but the ranks were filled out by punctate types. The pentamerids lasted out the Devonian after a maximum of diversity in the Ordovician. The strophomenids and rhynchonellids continued to be abundant.

The spiriferids diversified in a remarkable fashion in the later Silurian

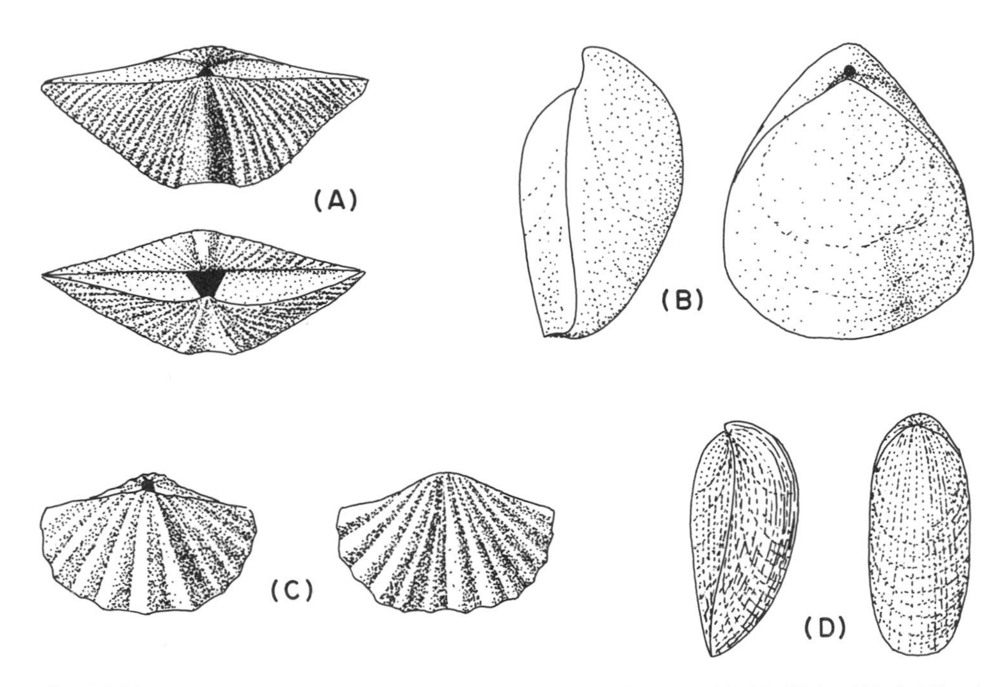

Fig. 12-10. THE BRACHIOPOD RADIATION: SPIRIFERIDA AND TEREBRATULIDA. **(A)** Spiriferid *Mucrospirifer;* middle and upper Devonian; brachial and posterior views; length 2.0 cm. **(B)** Spiriferid *Meristella;* middle Silurian; lateral and brachial views; length 2.6 cm. **(C)** Spiriferid *Trematospira;* middle Silurian to middle Devonian; brachial and pedicle views; length 1.2 cm. **(D)** Terebratulid *Etymothyris;* lower and middle Devonian; lateral and brachial views, length 4.5 cm.

and in the Devonian (Figure 12-10). The initial group (the Atrypidina) expanded considerably, and three additional suborders appeared. One of these lines (the Spiriferidina) typically has a wide hinge line, a large interarea on the pedicle valve, and costate or plicate shells. Another (the Athyrididina) retain the short hinge and small interarea of the Ordovician genera but have smooth shells and strong beaks. The final group includes a rather small group of genera in the Retziidina.

A distinctly new order, the Terebratulida, also showed up in the late Silurian, possibly as a late deviant from the orthid main line (Figure 12-10). They, like the spiriferids, have brachidia, but rather than a spire a more or less complex loop. Characteristically, the terebratulids have a short hinge line and a small interarea on the pedicle valve. The portion of the pedicle notch next to the hinge line is closed over so that a round opening is left at the posterior edge of the interarea. Typically, the beak on the pedicle valve is strongly developed.

And bust. The brachiopods reached their zenith in the late Silurian and early and middle Devonian. The punctate orthids survived in moderate numbers to the end of the Paleozoic, but their nonpunctate relatives

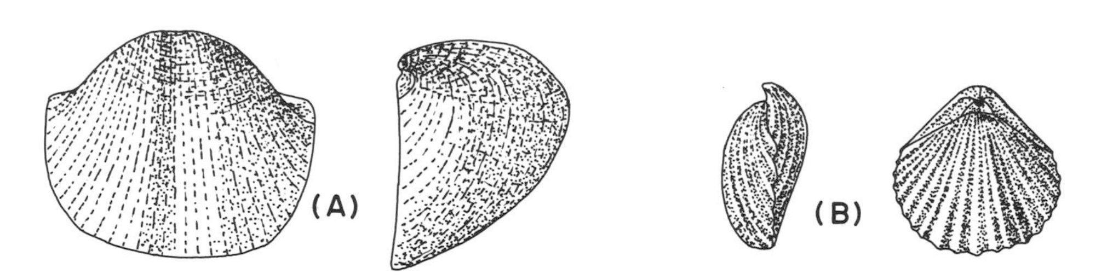

Fig. 12-11. THE BRACHIOPOD RADIATION STROPHOMENIDA AND RHYNCHONELLIDA. **(A)** Strophomenid *Dictyoclostus;* Mississippian to Permian; pedicle valve, pedicle and lateral views; length 3.4 cm. **(B)** Rhynchonellid *Kallirhynchia;* Jurassic; lateral and brachial views; length 2.8 cm.

are unknown later than the middle Devonian. The pentamerids lasted only to the end of that period. All the spiriferid suborders declined in variety; the Atrypidina became extinct in the late Devonian; the other suborders, though they persisted until Jurassic time, are represented by only a few genera after the early Pennsylvanian.

The typical strophomenids also declined in the later Paleozoic, even though they managed to survive to the early Jurassic, but two suborders, the Chonetindina, which appeared in the late Ordovician, and the Productidina, which are first known from the Devonian, did flower in the late Paleozoic (Figure 12-11). Marked typically by rudimentary interareas, deeply concave brachial valves and strongly convex pedicle valves, they are among the most common and characteristic fossils through the Mississippian, Pennsylvanian, and Permian. The chonetids survived until Jurassic times, but the productids are unknown beyond the Permian. Some late members of the productidinid group are very coral-like in form and, presumably, in ecology.

The rhynchonellids outlasted most of their brachiopod contemporaries, underwent a considerable radiation in the Mesozoic, and survived to the Recent. Even so, they dwindled in numbers in the late Paleozoic. The latecomers in brachiopod evolution, the terebratulids, hung on in small numbers after a truncated early Devonian radiation. They, like the rhynchonellids, diversified in the Mesozoic, presumably into habitats vacated by the Paleozoic and early Mesozoic brachiopod extinctions, and they are the most diversified of recent brachiopods.

A stratigrapher's-eye view

As you should expect from the preceding description of brachiopod evolution, many genera had short stratigraphic ranges. Although most also had limited environmental ranges and geographic distribution, they are of great importance as guide fossils. Many lower and middle Paleozoic sequences are zoned by brachiopod assemblages, and productid and terebratulid genera and species are useful stratigraphic markers in the late Paleozoic and in the Mesozoic respectively. Since the various orders and suborders have a fairly short period of abundance, the brachiopod faunas have a characteristic appearance in the major time units, even though particular genera are not identified.

As sessile benthos, brachiopods show a particular sensitivity to the environment of sedimentary deposition, and a particular type of sedimentary rock is likely to bear a particular brachiopod assemblage. The significance of this to paleoecology and paleogeography is obvious. From a very general viewpoint, the occurrence of abundant brachiopods indicates water of moderate to shallow depth and of approximately normal salinity. The modern inarticulate, *Lingula,* can tolerate brackish waters, and fossil lingulids commonly occur in beds that lack a normal marine faunal assemblage and that are intimately associated with beds bearing freshwater and terrestrial plants and animals. Lingulids also occur in normal marine environments.

REFERENCES

Cooper, G. A. 1957. "Brachiopods," in Ladd, H. S. (Ed.), "Treatise on Marine Ecology and Paleoecology," vol. 2. *Geol. Soc. of America, Memoir 67,* pp. 801-804.

Ferguson, L. 1963. "The Paleoecology of *Lingula squamiformis* Phillips during a Scottish Mississippian Marine Transgression," *Jour. of Paleontology,* vol. 37, pp. 669-681.

Grant, R. E. 1965. "A Permian Productoid Brachiopod: Life History," *Science,* vol. 152: 660-662.

Greiner, H. 1957. "*Spirifer disjunctus:* Its Evolution and Paleoecology in the Catskill Delta," *Peabody Mus. of Nat. Hist., Yale Univ., Bull. 11.*

Hyman, L. 1959. See ref. p. 279.

Paine, R. T. 1963. "The ecology of the brachiopod *Glottidia pyramidata,*" *Ecological Monog.,* vol. 33: pp. 187-213. The most complete study of modern brachiopod ecology yet published.

Rudwick, M. J. S. 1960. "The Feeding Mechanisms of Spire-Bearing Fossil Brachiopods," *Geological Mag.,* vol. 97, pp. 369-383 (also pp. 514-518 for discussion).

——— 1964. "The Function of the Zigzag Deflexions in the Commissures of Fossil Brachiopods," *Palaeontology,* vol. 7, pp. 135-171.

Sass, D. B. et al. 1965. "Shell Structure of Recent Articulate Brachiopoda," *Science,* vol. 149, pp. 181-182.

Westermann, G. E. G. 1964. "Possible Mechanical Function of Shell Plication in a Triassic Brachiopod," *Canadian Jour. of Earth Sci.,* vol. 1, pp. 99-120.

Williams, A. 1966. "Growth and Structure of the Shell of Living Articulate Brachiopods," *Nature,* vol. 211, pp. 1146-1148.

——— and A. J. Rowell. 1965. "Classification," in Moore, R. C. (Ed.), *Treatise on Invertebrate Paleontology, Pt. H, Brachiopoda.* New York: Geological Society of America.

13 Segments: The Annelida

The bryozoans and brachiopods explored some of the evolutionary possibilities of the coelomate plan. What of those coelomates whose ancestors were active, rather than sessile? The recent and fossil bryozoans and brachiopods number perhaps 2000 genera. Are the active coelomates more or less diversified? These questions are obviously loaded, for we, insects, crustaceans, clams, snails, and many other groups are coelomates.

Within this variety of coelomates, biologists distinguish two groups of phyla, which may have evolved independently from acoelomate ancestors (p. 268). One of these includes the vertebrates and echinoderms; the other comprises the annelid worms, the arthropods, the molluscs, and several "minor" phyla. The phyla within each group share a similar mode of development from the egg and a similar method of coelom and mesoderm formation. The features shared by both groups include only a few fundamental metazoan characters and a variety of things like brains, circulatory systems, and body segmentation that probably evolved independently to fill similar adaptive needs. When did they separate? In what radiation of primitive metazoans and under what circumstances? The fossil evidence, as usual, helps little to answer these questions. Rather advanced representatives of the annelid-arthropod-mollusc stock and rather primitive members of the echinoderm-chordate stock occur in lower Cambrian rocks, but they show no convergence toward a common ancestry. Here again, paleontologists can only hope for critical Precambrian fossil finds.

Another dichotomy

The brachiopod-bryozoan line may have branched from the annelid-arthropod-mollusc stock early in its development. If it did, the chief feature conserved from their common ancestry was a microscopic larva characterized by an apical tuft of cilia, a girdle of cilia about its midriff, and a mouth and anus opening below the ciliary girdle. Other than this, there is little to unite the two groups of phyla—each went its own way before the beginning of the Cambrian.

The annelid-arthropod-mollusc stock itself did not continue unbranched for long after it differentiated from the acoelomate stock. The annelids and the arthropods are surely very close, and, as I will detail shortly, the latter evolved from the annelids, possibly quite late in the Precambrian. The origin of the molluscs is not nearly so clear. The annelids have their bodies divided into large numbers of similar segments. Some molluscs show rudiments of similar segmentation, and one recently discovered genus displays quite distinct segmentation. These may be vestiges retained from a primitive segmented annelid ancestor, or they may have evolved independently in the mollusc line. If the latter is correct, then the common ancestor of molluscs and annelids was a nonsegmented coelomate "worm" with several longitudinal nerve cords, a slightly differentiated head, and a simple circulatory system—or at least these are common features in the annelid and molluscan structural plan.

The origins of segmentation

Repetition of organs along the symmetry axis (*metamerism*) is widespread among animals—extending even to acoelomates. This repetition is limited in the acoelomates, however, to specific functional systems, typically the nutritive, reproductive, and/or excretory organs. Among the simpler coelomates, the coelom is divided into three segments, which have different functions related to hydrostatic requirements, e.g., the protrusion of tentacles or proboscis. Among the annelids, arthropods, and vertebrates, the body is separated into numerous similar segments by sheets of connective tissue. Each segment repeats organ systems, as in acoelomate metamerism, but also divides the body musculature and the skeleton (a hydrostatic one in the annelids) into autonomous functional blocks.

Clark (1964) has explained annelid and vertebrate segmentation in terms of locomotor adaptation. He recognizes in aquatic animals three locomotor modes: ciliary creep, peristaltic waves, undulatory waves. Ciliary creep is inefficient except for very small animals, and even acoelomates replace it with peristaltic or undulatory mechanisms in larger forms. In general, peristaltic and undulatory waves alter body shape and require some sort of skeleton to oppose shape changes and to permit antagonistic muscle action. For example, contraction of muscles along one side of the body produces a bowed shape, but, unless other body muscles operating on the "skeleton" reverse the "bow," a traveling wave is impossible. A fluid-filled coelom functions as a skeleton through its hydrostatic properties, but such skeletons are relatively inefficient. Because a hydrostatic skeleton cannot oppose bending movements, lever systems are impossible and muscle lengths must be great. Individual muscle contractions affect the entire hydrostatic skeleton and therefore operate against the entire body musculature. Shape changes slowly as a consequence of the great contraction necessary in long muscles and of the complete deformation of the body. However, division of the body musculature and the skeleton into autonomous blocks overcomes most of these deficiencies.

Clark argues that annelid segmentation evolved for peristaltic rather than for undulatory locomotion. Since the coelomic chambers and muscle blocks coincide exactly, and a fluid offers no resistance to bending, forces are limited to the segment rather than being transmitted along the skeleton. The most efficient locomotion would be by peristaltic waves; alternate contractions of the longitudinal and circular muscles in each segment would induce concomitant shortening and lengthening. Annelids that employ undulatory waves have reduced segmentation 1) by perforation of the intersegmental partitions and interconnection of the coelomic chambers, 2) by reduction of the circular muscles, and 3) by development of intersegmental muscle *groups.**

In Clark's opinion the primitive annelid was an active burrower in which segmentation arose gradually by shortening and grouping of the primitive longitudinal muscles. The longitudinal nervous system would necessarily reflect this development, and it might induce segmental repetition of other organ systems. Later adoption of a free-swimming existence subordinated peristaltic to undulatory locomotion and reduced the need for complete segmentation.†

* In vertebrates, the segmental muscles transmit bending forces through septa to an axial skeleton. The vertebrae are intersegmental, so that the forces are transmitted to adjacent blocks, fore and aft. The consequence is generation of undulatory waves.

† As Clark himself recognizes, this sequence could work the other way, but this leaves unexplained the nonfunctional segmentation in the free-swimming annelids.

Rewards of virtue

But—returning now from the realm of semimythologic ancestors—the segmentation creates important evolutionary possibilities. Successive segments, first of all, acquire different functions and specialize for those functions. Thus, an anterior segment or segments form a distinct head. If, as in annelids, portions of the body wall fold outward to form metameric paired appendages, these may adapt to special functions. Differentiation of segments subdivides the body into functional areas. These areas expand

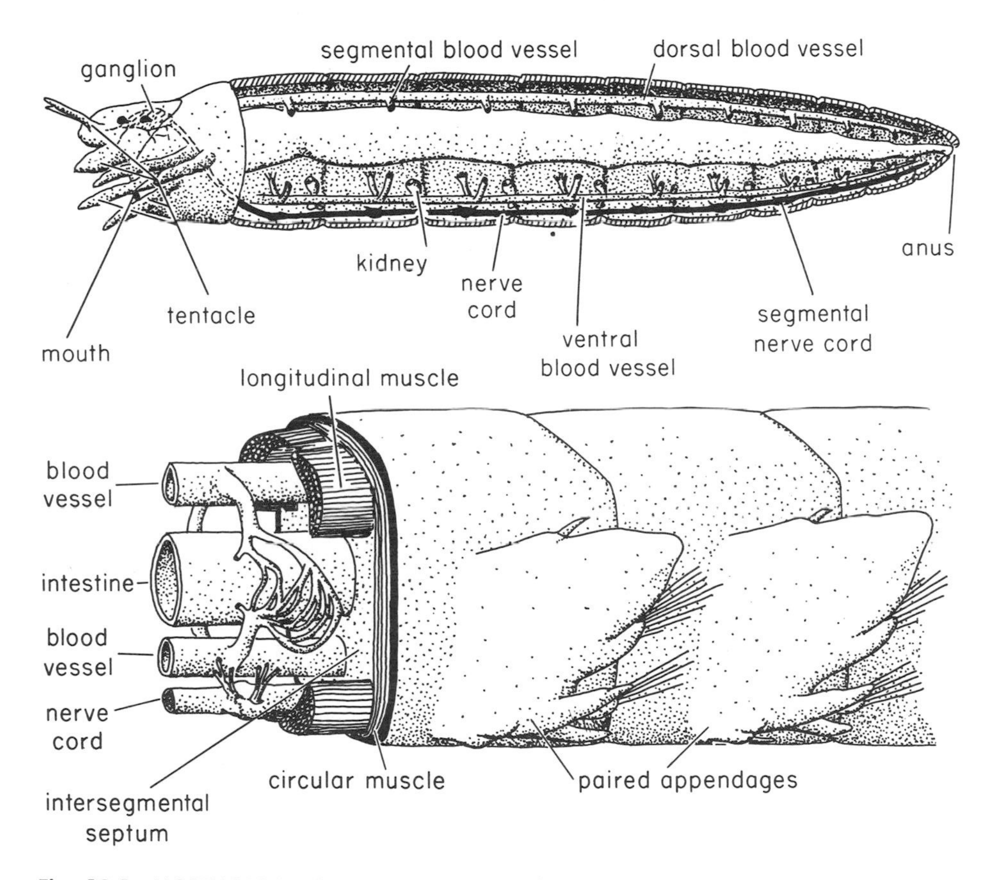

Fig. 13-1. MORPHOLOGY OF ANNELID. Upper drawing is a diagrammatic view of a generalized annelid. Head consists of a *prostomium* bearing antenna and palps and of several fused segments bearing modified tentaculate appendages. Segments behind head are shown in section with muscles, digestive glands, and reproductive organs omitted. Dorsal and ventral blood vessels give off lateral vessels in each segment. The ventral nerve cord terminates anteriorly in a cerebral ganglion and also gives off segmental nerves. The gut opens anteriorly through the mouth, leads back into an intestine and opens posteriorly through the anus. A pair of "kidneys" occur in each segment behind the head. The lower drawing shows several enlarged segments, one with cuticle and portions of the muscles removed. The longitudinal muscles are partly segmented. The segmental blood vessels give rise to a capillary network. Blood is pumped from the dorsal vessel through the capillaries into the ventral vessel.

or contract as adaptations change. The head region can, in this fashion, spread backward to include additional segments. Since ontogeny is related to localized biochemical activity, the extent of the head may increase simply through extension of the existing developmental fields. An important morphological change may arise from a minor biochemical change, and, if advantageous, may initiate a large evolutionary change. The annelids, however, have not gone very far in exploiting these possibilities (Figure 13-1). Typically, the three anterior segments differentiate from the body segments by modifications of the appendages to form sensory antennae and feeding palps and tentacles and by the lack of excretory ducts.

To sustain an active life, the annelids possess a more complex organization than the brachiopods and bryozoans. They respire through the body surface, particularly through parts of the paired appendages that have a fine net of blood capillaries. Their circulatory system consists of a set of longitudinal vessels that feed into and drain capillaries. Portions of one of the longitudinal vessels are enlarged to form one or more "hearts." The excretory system consists of paired tubules in each segment (with a few exceptions). A pair of longitudinal ventral nerve cords connects the sensory and motor nerves of the individual segments, and, in turn, are connected to a pair of ganglia in front of the mouth. In some, eyes are present as well as antennae.

Geologic occurrence

Since the group lacks skeletal structures, except for the chitinous jaws (scolecodonts) or the calcareous tubes of a few genera, the fossil record of the annelids is scant. Some of the "worm" burrows and trails known from Precambrian, as well as later rocks, probably are of annelid origin. The famous middle Cambrian Burgess shale locality in British Columbia has yielded some ten genera of annelids (Figure 13-2), preserved as carbonized films in the dark, fine-grained shale. Burrows are common, though no one can be certain of the zoologic affinities of the burrowers. Some burrow "species," *Scolithus, Arthrophycus,* are guide fossils. Certain distinctive burrow types are associated with particular depositional environments; calcareous tubes also occur in a restricted range of environments, and may be associated with particular kinds of shells on which they build their tubes. Some borings in fossil pelecypod and brachiopod valves may be the work of predaceous annelids.

The paucity of fossil annelids contrasts with their biologic importance in modern marine faunas. They are among the most abundant and diversified benthonic animals; they comprise free-swimming, crawling, sessile, and burrowing forms; they include some of the most common and voracious predators, many of the scavengers, and a large percentage of the fil-

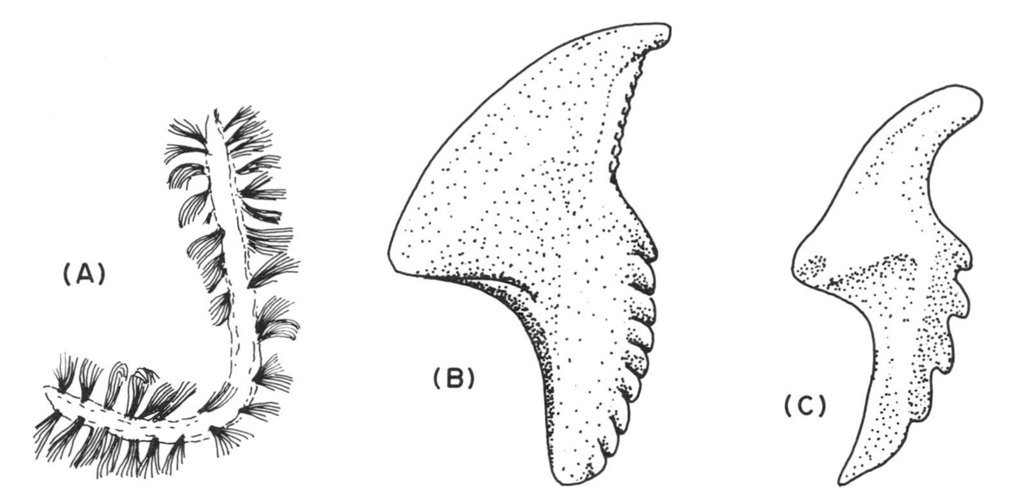

Fig. 13-2. FOSSIL ANNELIDA. **(A)** *Canadia;* middle Cambrian; ×2; paired appendages bear long clusters of bristles. **(B)** and **(C)** Elements of annelid jaws. **(B)** *Arbellites;* Silurian; upper face of left maxillae; ×30. **(C)** *Ildraites;* upper face of left maxillae; ×50. [**(A)** *after* Walcott; **(B)** *after* Piveteau, ed. *Traité de Paléontologie;* Copyright, Masson et Cie, Paris; used with permission; **(C)** *after* Eller.]

ter feeders. Even if they have just now reached their climax of diversification—in itself an improbable coincidence—the most significant gap in paleoecology is the lack of the annelid components of ancient faunas.

Homeless waifs

Among the frustrations of the paleontologists are the various *"problematica."* Several of these, e.g., the archeocyathids and conularids, I have already discussed. Among these orphans are a series of jaw-like or tooth-like fossils (Figure 13-3) that occur in Ordovician to Triassic rocks. These fossils, the conodonts, are composed of calcium phosphate, and,

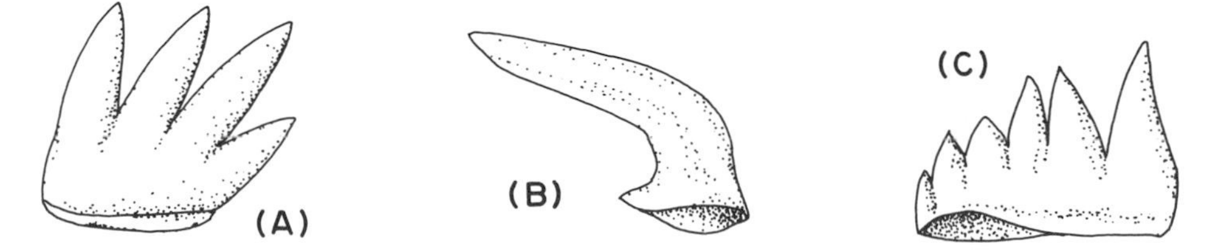

Fig. 13-3. CONODONTS. Three conodonts selected to show the general character of the group. **(A)** *Leptochirognathus;* Ordovician; lateral view; ×20. **(B)** *Stereoconus;* Ordovician; lateral view; ×20. **(C)** *Neocoleodus;* Ordovician; lateral view; ×30. (*After* Branson and Mehl.)

in this, resemble the vertebrates. Several types commonly occur together and form an "assemblage;" assemblages probably derive from single individuals with several types of these structures. They show no evidence of wear (unlike teeth). Some possess healed fractures indicating they were internal structures. Most invertebrate paleontologists, anxious perhaps to be rid of the problem, call them vertebrates. Vertebrate paleontologists prefer to regard them as invertebrates. Since they are of major importance as guide fossils, they cannot be disregarded completely.

All other things equal, unidentifiable fossil groups gravitate in the direction of the "worms." Conodonts resemble, at least vaguely, the chitinous jaws of annelids. In consequence, I place them in this chapter without judgment on their true affinities.

Missing links

One of the most interesting members of the Burgess shale fauna is *Aysheaia pedunculata* (Figure 13-4). These fossils resemble a rather plump worm, show indistinct segmentation, and possess a series of short,

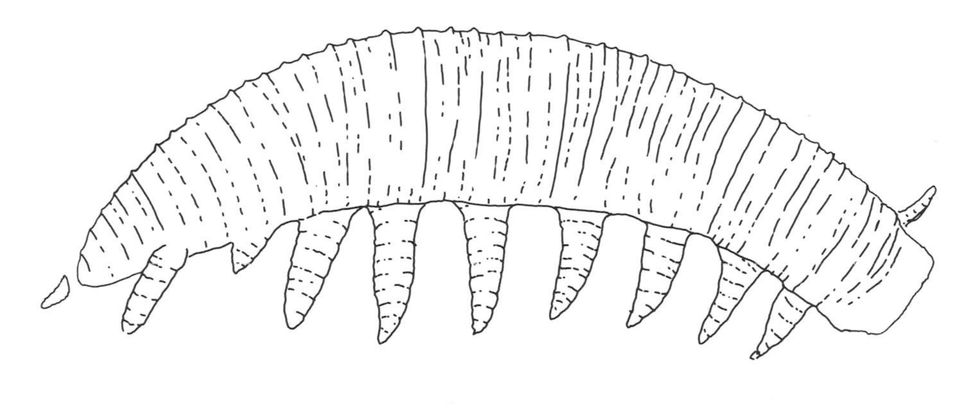

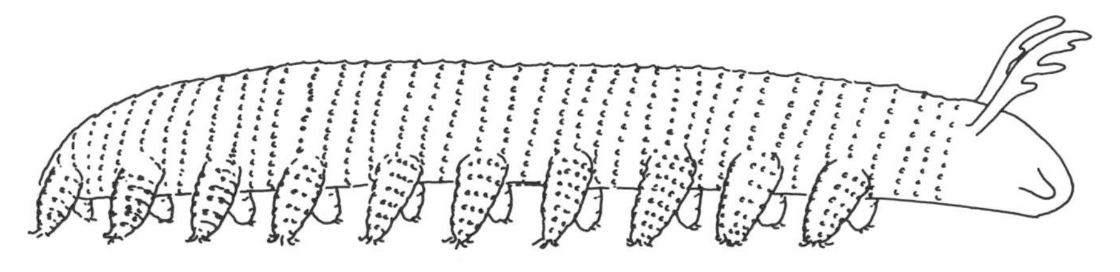

Fig. 13-4. FOSSIL ONYCHOPHORAN. The upper drawing is of *Aysheaia pedunculata;* middle Cambrian of British Columbia; length 3.6 cm. The lower drawing is a restoration of Aysheaia. Even superficially, the resemblance to annelids and arthropods is obvious. (*After* Hutchinson.)

massive, paired appendages. Details of the head and of the appendages are obscured, for, like other Burgess fossils, they are crushed and carbonized. What can be seen agrees rather closely with the features of a modern genus, *Peripatus,* and they presumably belong to the same class, the Onychophora. As I implied above, the onychophorans are rather annelid-like. Like some annelids, they are covered by a thin impervious waxy layer, the cuticle, and have specialized appendages in the head region that function as antennae and jaw parts. The nervous and excretory systems also resemble those of the annelids. But *Peripatus* diverges from the annelids in several important characteristics. First, the three head segments tend to fuse and lose their distinctness. Second, respiration is carried on within the body, through the walls of minute branching tubes (the tracheae) which open in a pit on the animal's surface. Third, the circulatory system consists of a tube-like dorsal heart that pumps blood into the cavities between the organs. Both the heart and the circulatory cavities (which are called the *haemocoele*) form within the mesoderm, and the haemocoele replaces the coelomic cavity except in the internal cavities of the gonads and excretory tubules. In these three characteristics, the onychophores resemble the arthropods.

This group of animals, then, is intermediate between the phyla Annelida and Arthropoda. Because they possess a haemocoele and tracheae, they are often classified as arthropods. The morphologic similarities of annelids and arthropods, such as segmentation and the paired appendages, suggest a common ancestor with annelid structure. If this interpretation is correct, a linking form, with a mixture of annelid and arthropod features, should have appeared during the evolution of the Arthropoda. The Onychophora show this mixture of features and thus may be said to be a "non-missing" link.

The classification of the onychophorans is difficult. They show the same level of organization as the annelids and many structural similarities. They have diverged no further from the primitive annelid stock than many "good" annelids. If one prepares a horizontal classification based on this radiation from the ancestral annelid, the Onychophora constitute a separate subphylum or class within the phylum Annelida. On the other hand, an onychophoran lineage (though certainly neither *Peripatus* or *Aysheaia*) probably gave rise to the arthropods. Ergo, the Onychophora form a subphylum of the phylum Arthropoda. Since the *Treatise on Invertebrate Paleontology,* Part O (Moore, 1959, pp. 16-20) includes the onychophorans in the Arthropoda, the same arrangement will be followed here (Table 14-1).*

* The hideous category name "supersubphylum" is applied in the *Treatise.* The use of "subphylum" for "supersubphylum" and the insertion of "superclass" for "subphylum" would seem a desirable improvement.

REFERENCES

Clark, R. B. 1964. See ref. p. 279.

Fahlbusch, K. 1964. "Die Stellung der Conodontida im Biologische System," *Palaeontographica, Abt. A,* Bd. 123, s. 137-201.

Moore, R. C. (Ed.) 1959. "Arthropoda 1," *Treatise on Invertebrate Paleontology, Pt. O.* New York: Geological Society of America.

———. 1962. "Miscellanea," *Treatise on Invertebrate Paleontology, Pt. W.* New York: Geological Society of America. Deals with variety of body and trace fossils commonly called "worms," including conodonts.

Schwab, K. W. 1966. "Microstructure of Some Fossil and Recent Scolecodonts," *Jour. of Paleontology,* vol. 40, pp. 416-423.

Seilacher, A. 1953. "Studien zur Palichnologie," *Neues Jahrb. Geol. und Paläont.,* Abh., Bd. 96, s. 455-480, Bd. 98, s. 87-124.

14

Jointed limbs: The Arthropoda

Arthropods—trilobites—characterize the basal Cambrian rocks. Arthropods—insects, crustaceans, and arachnids—by number of species and of individuals, dominate nearly all modern natural societies. The fossil record preserves only fragments of this long and successful history, but even these fragments are impressive.

This evolutionary success rests on a substrate of annelid characteristics, among them segmentation and paired appendages on each segment. These characteristics are modified (Figure 14-1): 1) by specialization of segments and appendages; 2) by the division of the body in distinct functional regions in which a number of segments may be fused; 3) by occurrence of a rigid chitinous exoskeleton beneath the cuticle; 4) by the formation of the blood cavities, the haemocoele, and 5) by a large number of detailed differences in the nervous, respiratory, and excretory systems. Division of the exoskeleton into jointed elements permits independent movement of segments and appendages. "Arthropoda" (jointed + legs) refers to this latter characteristic.

The diversification of the arthropods obscures the phylogeny of the

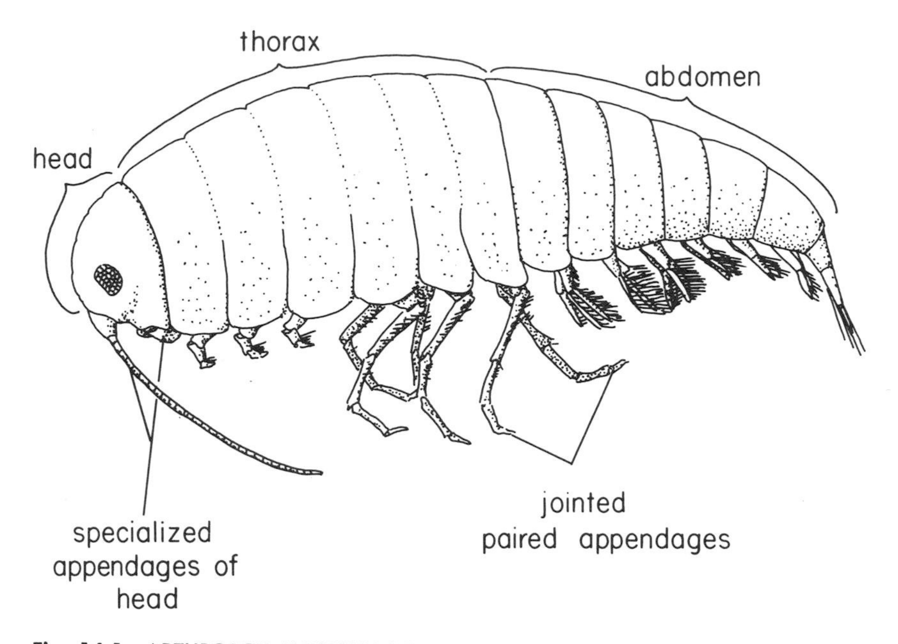

Fig. 14-1. ARTHROPOD MORPHOLOGY. Note segmentation, fusion of segments (six in head), differentiation of several regions (three, head, thorax, and abdomen, in this case), jointed paired appendages, and differentiation of appendages. The number of segments, the number and character of the body regions, and the form of the appendages vary widely within the Arthropoda; the general pattern does not.

major groups. The fossil record helps little, for most of the important lines appear suddenly. The arrangement of subphyla and classes varies from one paleontologist to another, between paleontologists and biologists, and among biologists. At present, the structure and arrangement of the appendages are the primary criteria in classification. Some groups, like the insects, seem homogeneous and represent descendants from a single lineage. Others appear heterogeneous and may represent several parallel or convergent phyletic lines. Special problem children are a variety of Paleozoic forms, among them the trilobites, that exhibit a bewildering mixture of primitive characteristics, of unique specializations, and of features otherwise limited to recognized homogeneous classes.

The classification of Table 14-1 follows the arrangement in the *Treatise on Invertebrate Paleontology,* Part O, except for the substitution of the "subphylum" and "superclass" categories for "supersubphylum" and "subphylum" respectively. The early, "difficult" types are grouped as superclass Trilobitomorpha, united principally because of a common array of primitive characters in antennae and the other paired appendages. Thus defined, Trilobitomorpha represents the earliest arthropod radiation and includes the ancestors of more "advanced" classes, though these have not yet been spotted. The superclass Chelicerata includes spiders, scorpions,

TABLE 14-1

Classification of the phylum Arthropoda (*after Treatise on Invertebrate Paleontology, Pt. O*)

Cambrian to Recent. Segmented; paired appendages on segments. Three or more segments fused into head. Open circulatory system with cavities around organs. Typically with jointed, chitinous exoskeleton. Marine, freshwater, terrestrial. Aquatic forms vagrant and sessile benthonic, more rarely pelagic.

SUBPHYLUM PROTOARTHROPODA

Cambrian to Recent. Exoskeleton not rigid nor jointed.

SUBPHYLUM EUARTHROPODA

Cambrian to Recent. Rigid, jointed exoskeleton.

SUPERCLASS TRILOBITOMORPHA

Cambrian to Permian. Primitive arthropods with antennae, simple two-branched appendages, and no appendages specialized as mouth parts.

CLASS TRILOBITOIDEA

Cambrian to Devonian. Assemblage of forms with various specializations of body regions and appendages not present in trilobites.

CLASS TRILOBITA

Order *Agnostida*
Cambrian to Ordovician. Small trilobites with subequal cephalon and pygidium. Possess only two or three thoracic segments.

Order *Redlichiida*
Cambrian. Large semicircular cephalon; typically, large genal spines; numerous thoracic segments; diminutive pygidium; facial sutures opisthoparian or fused. Glabellar segments typically distinct; eyes commonly elongate crescent.

Order *Corynexochida*
Cambrian. Subelliptical, typically with large pygidium. Cephalon semicircular—commonly with large genal spines; glabella distinct, expands forward; eyes elongate and narrow. Opisthoparian sutures; rostral plate fused with hypostoma or rudimentary. Thorax with 5 to 11 segments with spinose pleurae.

Order *Ptychopariida*
Cambrian to Permian. Typically opisthoparian, more rarely proparian or with marginal sutures. Pygidium small (early) to large. Eyes more distinct from eye ridges than in Redlichiida. Glabella typically tapers forward.

Order *Phacopida*
Ordovician to Devonian. Typically proparian or gonatoparian, rarely opisthoparian. Glabella either expanding or tapering forward. Pygidium typically medium to large.

Order *Lichida*
Ordovician to Devonian. Glabella broad, extending to anterior border; glabellar furrows elongated longitudinally; occipital ring tends to fuse with glabellar lobe. Opisthoparian. Pygidium large, includes leaflike or spinose pleurae.

Order *Odontopleurida*
Cambrian to Devonian. Strongly convex cephalon. Glabella widest at occipital ring; ring elongated posteriorly and commonly bearing tubercles or spines. Opisthoparian. Large genal spines typical; small spines on anterior border of cephalon. Pleurae each bear pair of spines, posterior one long. Paired spines on pygidium.

TABLE 14-1 (continued)

SUPERCLASS CHELICERATA

Cambrian to Recent. Body divided into prosoma and opisthosoma. No antennae; anterior pair of appendages bear claws; appendages about mouth only slightly modified. Aquatic, terrestrial; carnivores or scavengers.

CLASS MEROSTOMA

Cambrian to Recent. Gills on paired appendages of opisthosoma. Terminal segment bears a spine, the telson. Marine, ?freshwater.

Subclass Euripterida

Ordovician to Permian. Opisthosoma segmented. Sixth pair of appendages oar-like. Freshwater and marine.

Subclass Xiphosura

Cambrian to Recent. Opisthosomal segments typically fused. Sixth pair of appendages unspecialized walking legs. Terrestrial, ?freshwater.

CLASS ARACHNIDA

Silurian to Recent. Book lungs or tracheae; possibly gills in early scorpions. No appendages on opisthosoma; four pairs of walking legs. Terrestrial. Includes scorpions and spiders.

SUPERCLASS MANDIBULATA

Cambrian to Recent. Paired appendages of head specialized as antennae and as mouth parts. Marine, freshwater, and terrestrial.

CLASS CRUSTACEA

Cambrian to Recent. Two pairs of antennae and, typically, some two-branched appendages. Body divided into two parts; anterior covered by carapace. Primarily aquatic, marine and freshwater. Includes ostracods.

CLASS MYRIAPODA

Silurian to Recent. Elongate; numerous undifferentiated segments behind head, each bearing pair of walking legs. Single pair of antennae. Tracheae.

CLASS INSECTA

Devonian to Recent. Body consists of three regions, head, thorax, and abdomen. Three thoracic segments bear walking legs. Single pair of antennae. Tracheae. Most have wings on 2nd and 3rd thoracic segment. Terrestrial and freshwater.

---

king crabs, and euripterids, distinguished by the specialization of the single pair of appendages in front of the mouth as pincers, by the absence of antennae, and by the division of the body into two parts, the head (*prosoma* or *cephalothorax*) and abdomen (*opisthosoma*). The superclass Mandibulata apparently does not map phylogeny, for the crustaceans and the myriapods probably arose independently from the trilobitomorphs and/or the protoarthropods, and the insects evolved in turn from a myriapod stem. Thus interpreted, the principal criteria for this superclass—possession of one or two pairs of antennae and of appendages modified for jaws—result from parallel evolution.

Est omnis divisa in partes tres: The Trilobita

The trilobites appear in the earliest Cambrian rocks—in fact define the lower limit of the period and era. What appear to be trilobite trails occur even earlier. They diversified moderately in the early Paleozoic but then fell into a decline and disappeared into the evolutionary graveyard at the end of the Permian. Fortunately for their posthumous fame, the trilobites deposited calcium carbonate in their exoskeletons and abounded in shallow marine environments. The first assured their preservation; the second guaranteed their value to stratigraphers, who must deal largely with rocks formed in those environments.

The trilobite adaptation

The trilobites exhibited most of the general evolutionary tendencies of the arthropods, with particular modifications for their mode of life (Figure 14-2). The flattened body (often accentuated by compaction of sedi-

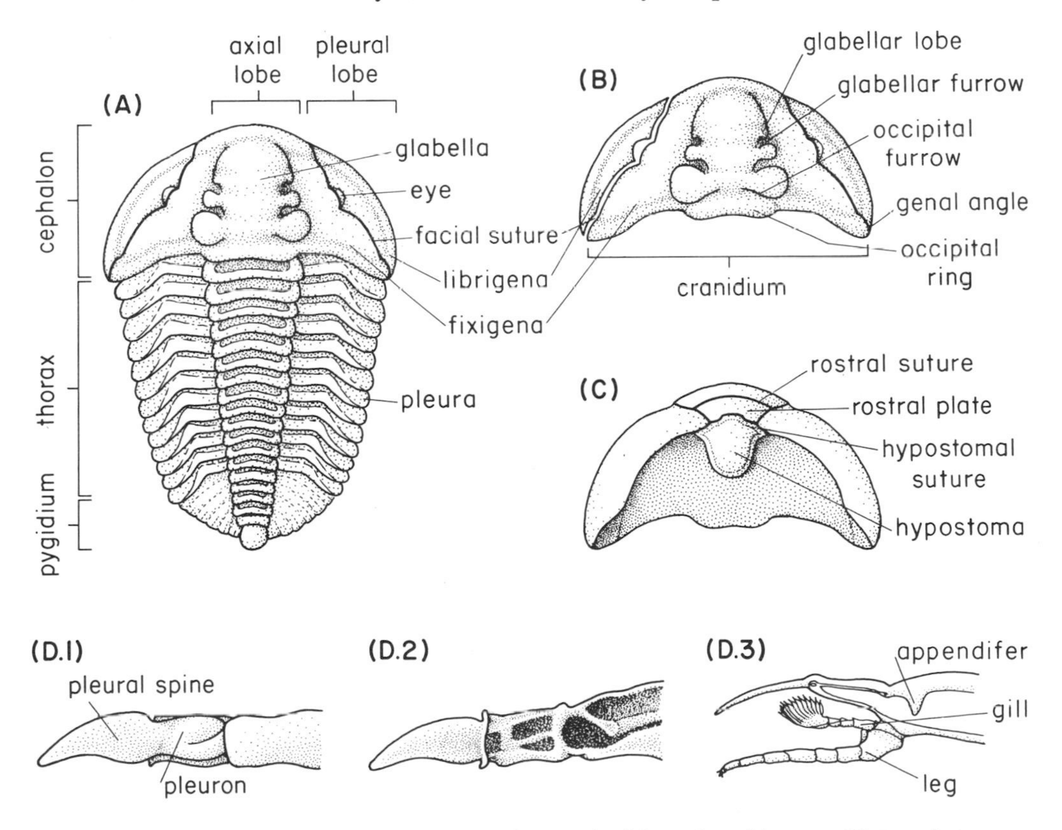

Fig. 14-2. TRILOBITE MORPHOLOGY. **(A)** Dorsal view of trilobite about life size. **(B)** Dorsal view of cephalon with left librigena separated from cranidium. **(C)** Ventral view of cephalon. **(D.1)**, **(D.2)**, dorsal and ventral views of pleura. **(D.3)** Pleura cut in transverse section with leg restored.

TABLE 14-2

Glossary for Trilobita ***(illustration numbers cited at end of definition)***

Antenna. One of a pair of slender many-segmented appendages attached to the ventral surface of the head in front of the mouth.

Apodeme. Downward projection from dorsal interior of thoracic segment which served for attachment of muscles. (14-2D3.)

Axial furrow. Groove that bounds axial lobe of trilobite. (14-2A; *et al.*)

Cephalon. Anterior portion of trilobite—the head—consisting of several fused segments and bearing the eyes and mouth. (14-2A, B; *et al.*)

Cheek. Portion of dorsal surface of cephalon lateral and anterior to glabella. Typically much lower and flatter than glabella and separated from it by a furrow.

Cranidium. Central part of cephalon including axial lobe (glabella) and bounded by the facial suture. (14-2B; *et al.*)

Doublure. Portion of dorsal exoskeleton that is bent under to form a border about the ventral surface of the animal. (14-2C, 14-6.)

Facial suture. Line along which exoskeleton of head split when trilobite molted. May be limited to margin of cephalon or may pass as fine line across dorsal surface of cheek. (14-2A, B; 14-6.)

Genal angle. Posterior lateral corner of cephalon. Typically terminates in spine but may be rounded. (14-2A, D; 14-7.)

Genal spine. Spine extending posteriorly from posterior-lateral corner (genal angle) of cephalon. (14-2B; 14-7.)

Glabella. Elevated axial portion of cephalon. Represents anterior part of axial lobe. (14-2A; 14-4.)

Hypostoma. Plate on undersurface of cephalon directly in front of mouth. (14-2C; 14-6.)

Hypostomal suture. Line of juncture between front of hypostoma and the rostral plate or librigenal doublure. (14-2C; 14-6.)

Lateral glabellar furrows. Groove extending transversely across glabella. May be complete or consist of short grooves extending part way from lateral border of glabella toward medial line. (14-2B; 14-4.)

Lateral glabellar lobes. Transverse lobe on glabella bounded by complete or partial transverse furrows—remnant of original segments fused in cephalon. (14-2B; 14-4.)

Leg. Long, rather slender, jointed appendage that forms the lower branch of the paired appendages. (14-2D3.)

Librigena. Portion of cheek lateral to facial suture and freed from cranidium during molt. (14-2B; *et al.*)

Mouth. Opening of digestive tract for ingress of food. In trilobites it lies in the ventral midline of the head a short distance behind the anterior margin.

Occipital furrow. Transverse groove that marks off the posterior segment of the glabella from the anterior portion of that structure. (14-2B.)

Ocular platform (oc. plt). Elevated portion of trilobite cheek that extends laterally from eye. (14-4B.)

TABLE 14-2 (continued)

Opisthoparous. Type of facial suture which crosses cheek, passes along the medial border of the eye and intersects posterior margin of cephalon medial to genal angle. (14-6B, C, D.)

Palpebral furrow. Furrow along medial border of palpebral lobe. (14-5A.)

Palpebral lobe. Elevated portion of cheek along medial border of eye. (14-2A, B; 14-5A; *et al.*)

Pleura. Portion of thoracic segment lateral to axial lobe. (14-2A, D; 14-8.)

Pleural spine. Lateral extremity of pleura. Pointed or sharply rounded; narrower than the medial portion of the pleura. (14-2D; 14-8.)

Proglabellar field. Portion of cranidium between front of glabella and anterior border of cephalon.

Proparous. Type of facial suture which crosses the dorsal surface of the cephalon, passes along medial edge of eye, and intersects lateral border of cephalon in front of, or at, the genal angle. (14-6E.)

Pygidium. Posterior portion of trilobite consisting of one or more fused segments. (14-2A; 14-8.)

Rostral plate. Median, ventral plate in cephalon between doublure of cranidium and hypostoma. (14-2C; 14-6.)

Rostral suture. Line of juncture between rostral plate and cranidial doublure.

Telson. Spine mounted on terminal or one of near-terminal segments and directed posteriorly along midline.

Thorax. Portion of body between head (cephalon) and tail (pygidium). Consists of a series of separate articulated segments. (14-2A.)

ments), the location of the mouth on the ventral surface and of the eyes on the dorsal, and the structure and arrangement of the appendages indicate they were, typically, benthonic animals that crept or swam along the bottom and fed on small organisms and organic debris in the bottom muds. The body is differentiated into three regions, the head, the *cephalon,* consisting of fused segments (probably six with the anterior one rudimentary or absent in the adult), the *thorax*, with a variable number of distinct articulated segments, and the tail, the *pygidium,* which includes one or more segments fused into a rigid plate. A pair of longitudinal furrows divide the body into three lobes—the origin of "trilobite." The *axial lobe* presumably contained the internal organs; the lateral (*pleural*) lobes may have served to protect the paired appendages and/or as a confining hydrofoil for generation of feeding and locomotor currents by the appendages. The anterior pair of appendages form antennae; the remaining four on the cephalon consist of two-branched limbs, one branch leg-like, the other gill-like. Störmer (1939) concluded that these limbs served primarily in locomotion and little if at all in food-getting. In this respect, the trilobites are more primitive than other arthropods, which have the cephalic limbs modified for manipulation, cutting, and crushing. If this interpretation is correct, trilobites were predominately detritus feeders.

The thoracic limbs resemble the last four on the cephalon, though longer and with more slender proximal segments. The leg-like branch presumably functioned as a walking leg, the gill-like branch as a respiratory structure. The anterior limbs are the largest, and the size diminishes toward the pygidium. The pygidium bears still smaller but similar legs.

The evolutionary trends

Within the trilobite stock, several distinct evolutionary trends are observed. These were adaptive, but paleontologists can't agree what all of them were adaptive to. Early workers generalized—and their successors have demolished their generalizations. The most I can do here is to report some of the trends and suggestions which tentatively determine their significance.

The glabella. The axial portion of the cephalon forms a distinct lobe, the *glabella* (Figure 14-3). In the early Cambrian trilobites, the glabella is low and divided by *glabellar furrows* into *glabellar lobes,* the remnants of the fused segments. This primitive type of glabella is widest at the back of the cephalon and tapers toward the front. Later trilobites tended to lose the furrows and to modify the shape of the glabella in one of several ways. The adaptive significance of these changes is not clear; some

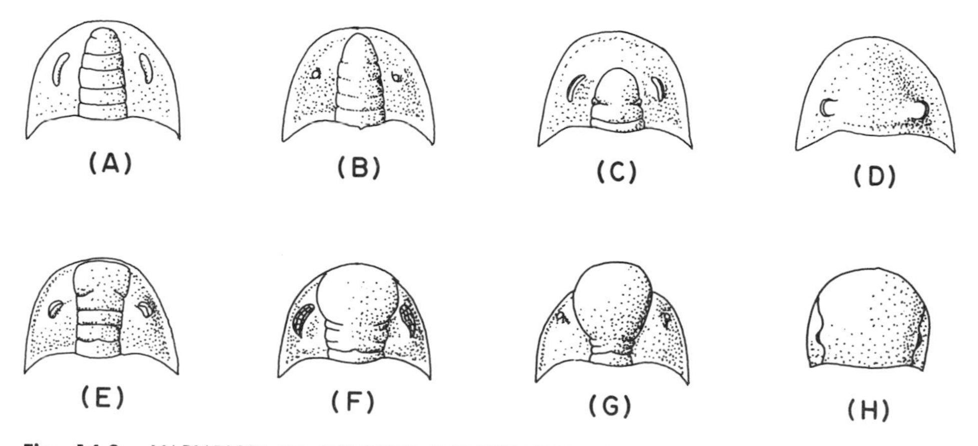

Fig. 14-3. VARIATION IN TRILOBITE MORPHOLOGY: GLABELLA CHARACTERISTICS. **(A)** Primitive; glabella segmented and tapering to front. **(B)** Anterior segments less distinct; furrows do not cross glabella. **(C)** Glabella relatively short; two furrows missing; two coalesce and fail to cross glabella. **(D)** Glabella broad, flat, and unsegmented; not distinct from cheek region. **(E)** Anterior lobe of glabella expanded slightly; anterior pair of furrows incomplete. **(F)** Anterior lobe of glabella expanded and reaching anterior edge of cephalon; occipital only complete furrow. **(G)** Glabella inflated, extended beyond anterior margin; furrows incomplete. **(H)** Glabella inflated, unsegmented, not distinct from cheek region.

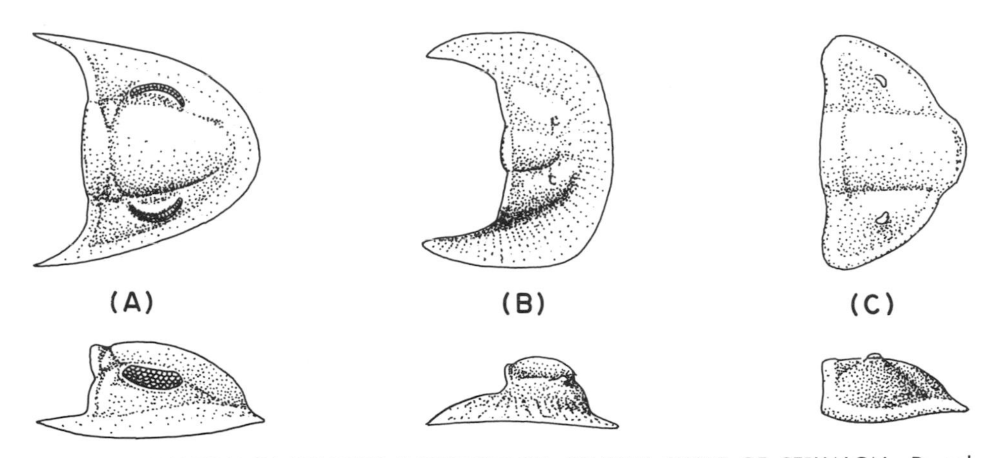

Fig. 14-4. VARIATION IN TRILOBITE MORPHOLOGY: GENERAL SHAPE OF CEPHALON. Dorsal views at top; lateral views beneath. **(A)** *Proteus.* Glabella distinct and high; ocular platform moderately elevated; border of cephalon relatively narrow and steeply sloping. **(B)** *Harpes.* Glabella small but moderately elevated; ocular patform very strongly elevated; edges broad and flattened. **(C)** *Trimerus.* Gabella not distinct from ocular platform nor platform from margin; margin narrow except for "lip" in front.

may relate to increased differentiation of the nervous system; some to specialization of the anterior part of the gut, and others to streamlining for swimming or burrowing. The configuration of the external surface may also reflect to some degree muscle attachments on the internal surface. The very much inflated glabellas of some species may have been a flotation mechanism—sort of built-in water wings for nektonic types.

The cheeks. The pleural lobes of the cephalon are the genal regions (or cheeks). The variation in width and convexity of the cheeks (Figure 14-4) may be adaptive to protection of the limbs that lie below them, to streamlining of the cephalon, or to development of a broad, stabilizing hydrofoil—or to causes unimagined.

The eyes. The trilobite's eyes resemble those of insects, for they are composed of a large number of separate lenses. In some lines, the eyes are

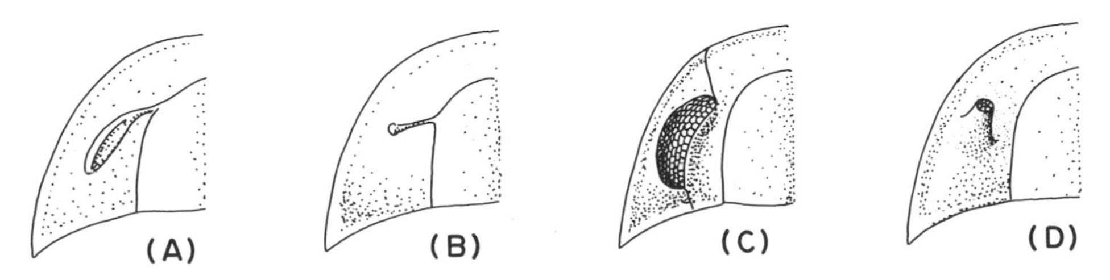

Fig. 14-5. VARIATION IN TRILOBITE MORPHOLOGY: THE EYES. **(A)** "Primitive"; eye narrow, flat lobe connected by ridge to glabella. **(B)** Eye rudimentary; ridge present. **(C)** Eye enlarged. **(D)** Eye mounted on end of stalk.

reduced or disappear altogether. In others, they are enlarged and may be raised on platforms or stalks (Figure 14-5).

There can be no question that these are adaptations to different requirements for sight. The problem would be simple if left at this vague generalization, but some researchers have not been satisfied. Large eyes have been assigned to:

a). Deep-water environments with low light intensities.

b). Shallow water environments with intense lighting.

Degenerate eyes have been stated to be adaptations to:

a). Deep-water environments with low light intensities.

b). A burrowing mode of life.

Possibly, each of these interpretations is correct for one or another trilobite species. Unfortunately, no one knows for certain which explanation goes with which species, although the skeletons of some appear to be adapted to burrowing habits. Detailed studies of the environments of deposition would help.

Elevated or stalked eyes increased the field of vision; they may also have permitted the animal to crawl or burrow in the bottom muds with only the eyes exposed. A few genera have extraordinarily large eyes that occupy most of the cheek and extend laterally and ventrally to form the side of the cephalon. These provided vision in all directions and may be adaptive to pelagic life in which dangers approach from below as well as above.

E. N. K. Clarkson has (1966) provided a more rigorous and satisfying analysis of trilobite eye function. Through study of the size, spatial arrangement, and bearing of the axes of the individual lenses, he defined the angular range of vision and the relative acuity of vision in different directions. He argues that in the genus *Acaste*, vision involved direction and movement perception but little or no form perception. The visual field lay predominately just above the plane of the body with vertical limits between approximately 0° and 30° but with complete lateral traverse (180°).

The sutures. Very narrow linear uncalcified tracts, the *sutures* divide the cephalic exoskeleton into distinct regions. Typically, they served as lines of weakness along which the exoskeleton split during molting. A longitudinal pair, the *facial sutures,* either cross the genal regions dividing them into lateral, *librigenal,* and medial, *fixigenal,* portions or else run along the lateral margins of the cephalon dividing the dorsal genal region from the *doublure,* i.e., the underturned rim of the cephalon. Typically the longitudinal sutures extend to the anterior edge of the cranidium and continue back across the doublure (Figure 14-6). In some, however, they may join

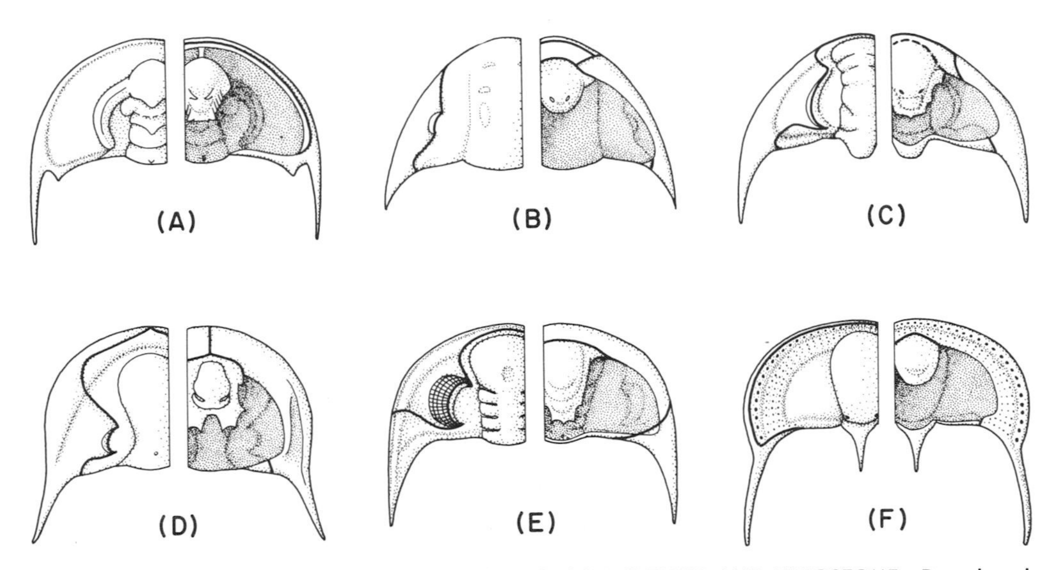

Fig. 14-6. VARIATION IN TRILOBITE MORPHOLOGY: SUTURES AND HYPOSTOME. Dorsal and ventral views of cephalon in trilobites. Heavy line indicates position of sutures. **(A)** *Paedeumias.* Only suture ventral—may be equivalent of rostral suture. **(B)** *Dysplanus.* Facial sutures opisthoparian. Rostral and hypostomal sutures well developed. **(C)** *Fieldaspis.* Facial sutures opisthoparian; rostral suture on anterior border; hypostomal suture partly fused—nonfunctional. **(D)** *Lachnostoma.* Facial sutures opisthoparian, coalesce anteriorly to form single median suture on doublure. No rostral suture or plate present. **(E)** *Odontochile.* Facial sutures proparian, coalesce anteriorly. No median or rostral sutures. **(F)** *Cryptolithus.* Facial sutures opisthoparian—near margin. They coalesce anteriorly. No median or rostral sutures.

at the midline and either terminate or cross the doublure as a single median suture. A transverse suture (the *rostral*) crosses the doublure anteriorly between the longitudinal sutures and separates the *rostral plate* from the fixigenae. A second transverse suture, the *hypostomal,* divides the rostral plate from the second ventral element, the hypostome.

The position and character of the cephalic sutures must have responded both to the size and character of the elements they bound and to the function of the sutures in molting. Since the trilobite shed its exoskeleton many times during growth, the manner in which it did so must have been important to survival. Two alternate styles of molting are now well known: the phacopid in which the division came between thorax and cephalon, and the animal, in effect, backed out of the cephalic exoskeleton; and the olenid mode in which the librigenae split off the front and sides of the cephalon and the animal crawled forward through the opening. As might be expected, the facial sutures are fused and nonfunctional in "phacopid" trilobites but fully functional in the "olenid" types.

If, as Hupé (1953) suggests, the facial sutures follow the boundary between an anterior, ocular, segment and a series of fused postocular segments, suture position and development reflect the development and adaptations of the segments—in particular the position of the eyes and the relative development of the marginal spines. In general, therefore, the facial suture passes inward and backward around the eye and thence outward and backward to the cephalic margin. In some trilobites in which the eye is reduced, the suture loses its connection with the eye lobe—possibly because of an absence of proper embryologic inductance by the rudimentary eye. Where the spine at the posterio-lateral corner (*genal angle*) of the cephalon derives from the ocular segment, the suture intersects the margin medial to the genal angle, i.e., is opisthoparian; in those in which a post-ocular segment supplies the spine, the sutural intersection is anterior to the angle, i.e., is proparian (Palmer, 1962).

Genal spines. The genal angles may be modified in several ways (Figure 14-7), but the significance of these variations is not clear. Immature trilobites have several pairs of spines on the genal border. In the typical adult only a single pair persists to form the "definitive" genal spines. This pair may derive from the ocular or from a post-ocular segment and therefore may be either librigenal or fixegenal. Closely related genera may differ in this character. Head spines are common among immature aquatic organisms, and function either as adaptations to floating (in the planktonic types) or for attachment to objects on the substrate (in benthonic types). They may also serve in the immature as in the adult for protection against predators. Some features of the genal spines may then be related to larval

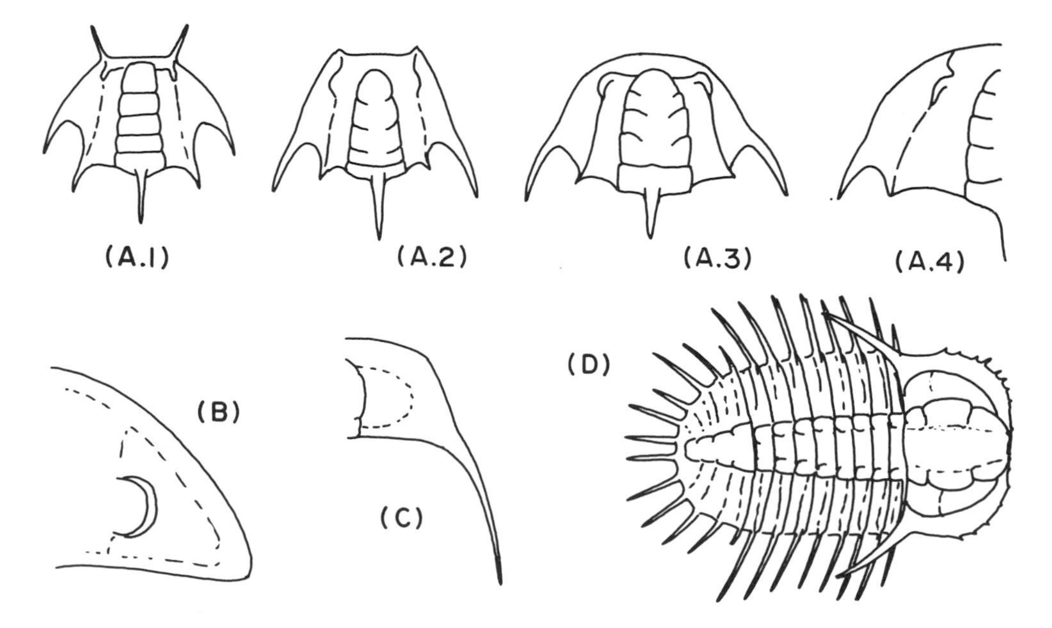

Fig. 14-7. VARIATION IN TRILOBITE MORPHOLOGY: GENAL SPINES AND SPINOSITY. **(A.1)** through **(A.4)**. Ontogeny of *Leptoplastus* illustrating change in size and position of cephalic spines from larva **(A.1)** to adult **(A.4)**. **(B)** Genal angle rounded, no spine present. **(C)** Genal spine slender and elongate. **(D)** Spinose form, *Acidaspis*. Length 2.4 cm. [**(A)** *after* Raw.]

adaptations. Many genera, however, show developmental changes—enlargement of the spines or their loss—quite late in growth, which culminate in some sort of typical arrangement in the adult. Since many bottom swimmers (sharks, king crabs) have posterior lateral "horns" or similar extensions of the head, the length, size, and angle of spines might have been important in stabilizing the animal on the substrate.

Some genera retain additional spines on the cephalon and develop elongate spines on the thoracic segments and on the pygidium. Such spines occur in recent animals of the plankton and nekton, and by their frictional drag prevent their bearers from sinking. They may have performed similarly in some small trilobites; in others they may have discouraged predators, served as camouflage, or supported the animal on particularly soft substrate.

The thorax. Since the arthropods evolved from annelids, the ancestral arthropod presumably had a large number of similar and relatively simple segments behind the head. Some Cambrian trilobites retain this primitive condition, but even they complicate the structure by the addition of the lateral processes, the *pleura* and the *pleural spines* (Figure 14-8). Since the pleura are related to the protection and function of the limbs, their variation may echo changes or improvement in the limbs or it could reflect changes in internal anatomy. Many trilobite lineages reduce progressively the number of thoracic segments by incorporation of some into the

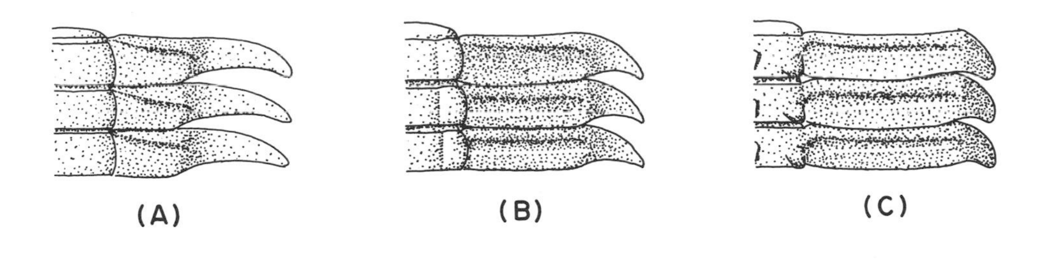

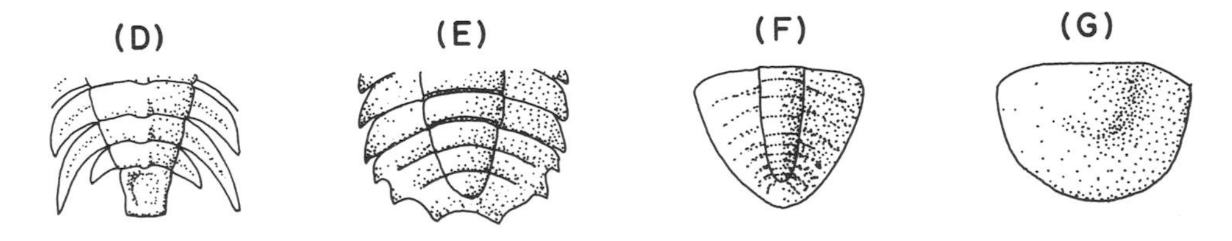

Fig. 14.8. VARIATION IN TRILOBITE MORPHOLOGY: PLEURA AND PYGIDIA. **(A)** Narrow pleura and long pleural spine. **(B)** Pleura wide and spines short. **(C)** Spines very short and blunt. **(D)** Pygidium consists only of terminal segment and is a small elongate plate. **(E)** Pygidium of three segments. Segments strongly marked and border spinose. **(F)** Pygidium of ten segments. Segments rather indistinct and those on the pleural portion do not correspond in position with the axial ones. **(G)** Axial lobe shortened and merged with pleural lobes. Segments obliterated by fusion.

pygidium and/or by a decrease in the number of segments formed. Two Cambrian orders, Eodiscida and Agnostida, have only two or three thoracic segments—presumably a reduction from an annelid condition. Some species have axial spines arising from the axial segments—the function is uncertain, though it may have been protective. Others have enormously extended pleural spines.

The pygidium. In some early Cambrian genera, the thorax continues to the terminal segment of the body, unmodified except by diminishing size; in others, the last segment was somewhat flattened and enlarged. This primitive condition is modified by incorporation of additional segments into a large rigid unit (Figure 14-8). The pygidium in some is as large as the cephalon. There is a general trend toward obliteration of segmentation, and the surface becomes quite smooth. In some, the boundary between axial and pleural lobes is lost.

The pygidium may have functioned as a propeller with rapid upward and downward movements, but many paleontologists question this interpretation. Some trilobites without pygidia have a long spine extending backward for one of the posterior segments; others have pygidia with similar spines. These may have been forced downward into the mud for support in locomotion—at least the modern king crab uses a similar structure in this way.

Whatever the locomotor adaptations of the pygidium, they are difficult to identify and difficult to separate from the protective functions. Presumably, a single rigid plate gave better protection than several small separate ones. Among those trilobites that rolled up like pill bugs to protect their delicate appendages and lightly armored undersurface, the pygidium covered the underside of the head.

In addition to all these possible or improbable functions, the development of the pygidium may correlate with changes in the soft anatomy, possibly in the reproductive organs. Since no close relatives survive, the character of the internal organs is largely unknown.

Appendages and ventral surface. Trilobite appendages were rarely fossilized because of their fragile attachment and structure; some apparently had only a chitinoid skeleton unsupported by calcareous deposits which was quickly destroyed after death or molting. For these reasons, the form of most genera is unknown and imperfectly known for many others. Störmer has distinguished some variations that may have had adaptive significance, but, inevitably, study of trilobite appendages lags.

For the same reasons, not much is known of modifications of the underside of the cephalon. The dorsal exoskeleton of the cephalon, pleura, and pygidium is folded onto the ventral surface (Figure 14-2) to form the *doublure.* A large plate, the *hypostoma,* lies in front of the mouth; a smaller one, the *metastome,* lies behind it. The walking legs arise along either side of the midline behind the mouth.

The hypostoma varies considerably in details of form and in attachment to the rostral plate and librigenal doublure (Figure 14-6). In the living crustaceans, a large labrum (the homologue or analogue of the hypostoma) is associated with secretions used in collecting fine food particles and, in some, with eddy currents useful in food gathering. On the other hand, filter feeders which employ sweeping hairs or gills on the paired limbs have small labra (Tiegs and Manton, 1958). Analysis of the form of trilobite hypostoma in the light of these relationships might prove interesting.

Correlated trends. Up to this point, I have discussed evolutionary trends in terms of isolated parts. This isolation simplifies the problem of analysis but does not provide a complete interpretation, for these "parts" did not function in isolation. Some trends are known to be correlated—the modification of eyes and facial suture already cited form one example. Unfortunately, less attention has been given to interpretation of the whole animal than to certain spectacular adaptations. Some trends, whose adaptive significance is now a mystery, may find explanation as additional studies of trilobites and their environments demonstrate the functional relationship between parts.

The material basis of evolution

In many fossil groups, the paleontologist deals primarily with the morphology of adult or near-adult animals. If he analyzes evolutionary trends, he does so from a series of adults and without knowledge of the changes in developmental pattern behind these trends. Evolutionary theorists and practical stratigraphers both suffer from this gap in knowledge. If a change involved many coordinated modifications in ontogeny, it required change in many genes and a considerable period of time. If the change was simple, it probably involved only a few genes and could have occurred in a moment of geologic time.

The extensive growth series known in some species of trilobites (Figures 14-9 and 14-10) permit the paleontologist to trace ontogenetic modifications. Some appear to result from relatively small changes in rate of growth in different parts of the animal. Sufficient information has accumulated for some tentative generalizations:

a). The shift of the eyes and facial suture onto the dorsal surface from the cephalon border must have resulted from an increase in rate of growth on the border as compared with the dorsal surface. These structures, therefore, didn't really shift dorsally, but rather the cephalic border moved away from them. The complexity of this change has not been determined—it may have involved simple acceleration of growth at the posterior lateral borders of the first segment, or complex coordinated changes in different parts of several segments.

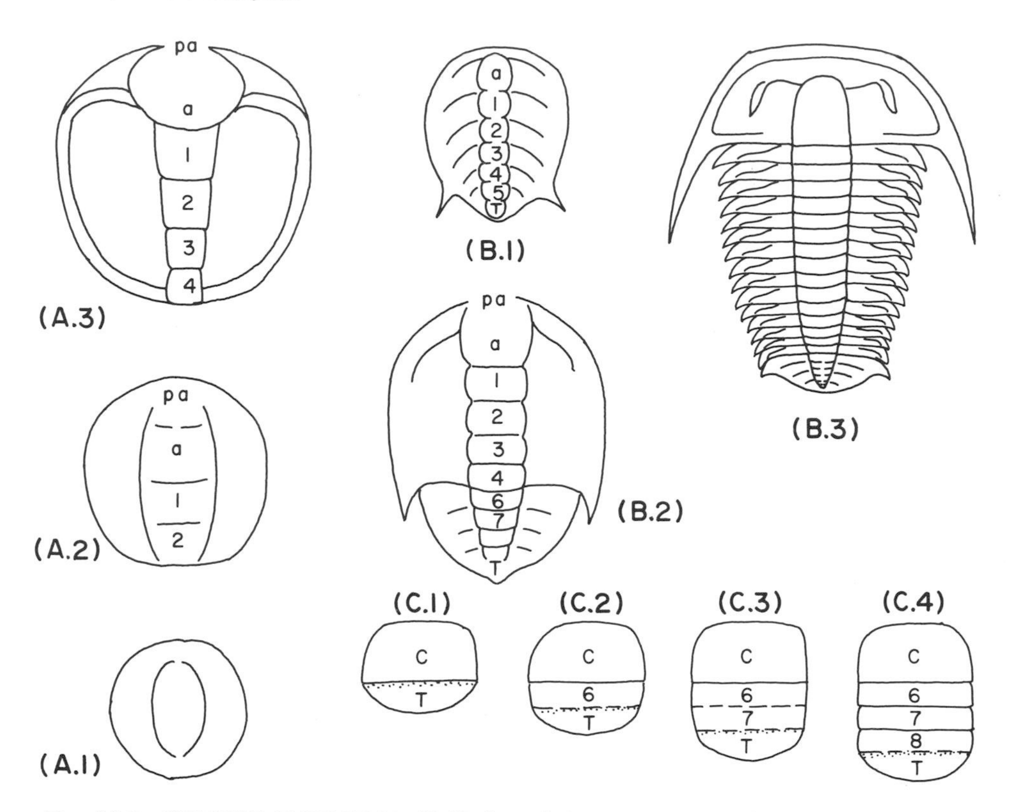

Fig. 14-9. TRILOBITE ONTOGENY. **(A.1)** through **(A.3).** Stages in the development of the cephalon in the species *Blackwelderia quadrata.* Note progressive differentiation of segments. **(B.1)** through **(B.3).** Stages in the development of *Olenus.* **(B.1)** shows individual just before pygidium (segments 5 and 6) separates from cephalon; the **(B.2)** stage consists of cephalon and a five-segment pygidium; **(B.3)** has a fully developed thorax and differs from adult only in size and minor details of proportion. **(C.1)** through **(C.4),** diagram of segmentation in pygidium and thorax. Segments develop at anterior border of terminal and form part of the pygidium **(C.2)** and **(C.3).** Later in development they separate successively along the anterior border of the pygidium **(C.4)** so that the anterior thoracic segment of the adult is the first thoracic to appear in the larva. Abbreviations: a, antennal segment; c, cephalon; pa, preantennal segment; T, terminal segment; numbers indicate other segments, the thoracic and pygidial segments are numbered in order of appearance. [**(A)** *after* Endo; **(B)** *after* Stormer.]

b). The post-cephalic segments appeared successively at the anterior border of a terminal generating zone. The first segment developed between the terminal zone and the back of the cephalon, the second between the terminal and the first, the third between the terminal and the second, and so on. These segments are added successively to the thorax from the front of the pygidium; retardation of growth without reduction in segmentation rate might increase the number of thoracic segments.

c). On the other hand, if the rate of segmentation were reduced, some segments would not be completely cut off from the terminal and would be incorporated in the adult pygidium.

A very small number of changes in genes might produce either or both of these morphologic changes. That these analyses are tentative is obvious; that they will ultimately yield interesting and valuable conclusions is, I hope, equally obvious.

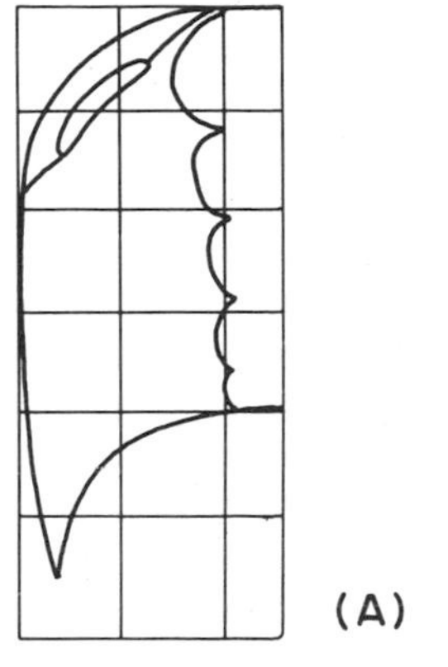

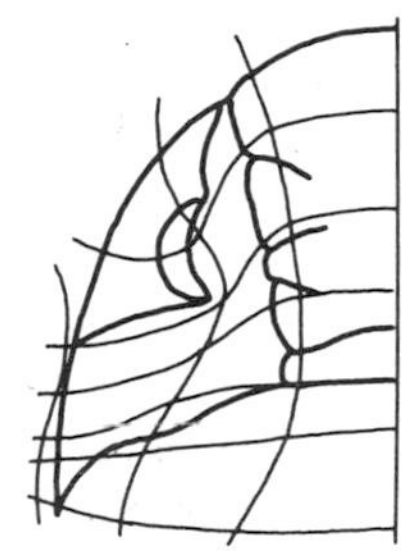

(A)

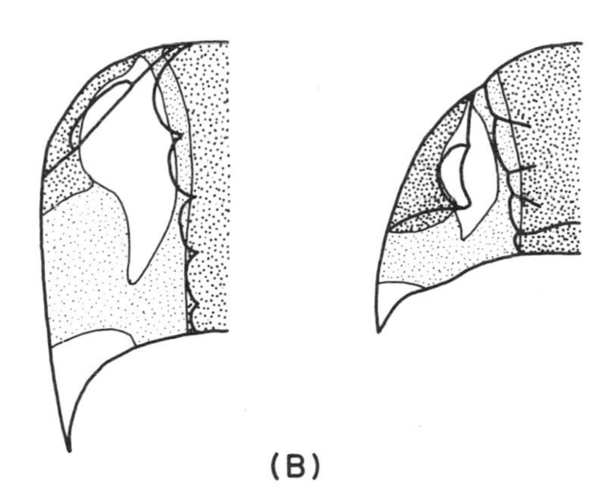

(B)

Fig. 14-10. GROWTH PATTERNS. **(A)** Transformation of trilobite cephalon during growth as shown by deformation of coordinate system (*also* see p. 31). **(B)** Regions of rapid growth (heavy stipple), moderate growth (light stipple), and slow growth (white), as indicated by relative deformation of coordinates.

Ontogeny and phylogeny

Interpretation of trilobite origins and phylogeny depends very largely on a priori assumptions. If one believes single-mindedly that ontogeny recapitulates phylogeny, one places full weight on features of ontogeny. If he believes particular changes in characteristics involve large numbers of genes (or, to use pre-genetic terminology, are "fundamental" changes of form) then he arranges the phylogeny on the basis of these characteristics. If he accepts neither, it will be difficult—even impossible—to arrange the various genera in a phylogenetic scheme. All this is true in any phylum or class, but it seems particularly striking in the trilobites.

Segmentation and phylogeny. Since the arthropods evolved from annelids, the ancestral arthropod presumably had a large number of segments. Was the proto-trilobite likewise multisegmented? The consensus holds that they were, and that trilobites with few segments are specialized.

Some students of the group have argued an opposite view. They point out that the early developmental stages of all trilobites have only a few segments and that the others are added later. If the rule "ontogeny recapitulates phylogeny" is taken as a dogma, the most primitive trilobites must have had no more than six segments. The trilobites of the orders Eodiscida and Agnostida, which possess two or three thoracic segments, would be most primitive. Members of these orders are either blind or have very small eyes, the cephalon is simple in structure, and segmentation of the glabella is distinct. All these are the common characteristics of trilobite larvae. Because of these presumed primitive characteristics, some paleontologists believe the trilobite ancestry lies in one of these orders.

All members of these orders, however, have a large pygidium. In most, the individual segments cannot be distinguished in the pygidium. These are surely not primitive characteristics if the annelid ancestor theory is accepted. The loss of distinct segmentation and specialization of the fused segments for a particular function occur late in the evolution of the arthropod lineages, not at the beginning.

Facial sutures and phylogeny. A major characteristic used in trilobite classification is the location of the genal angle relative to the facial suture. Four broadly defined states of this character are recognized:

a). *opisthoparian,* in which the genal angle is lateral to the posterior limb of the facial suture;

b). *proparian,* in which the angle is behind the posterior limb of the suture;

c). *gonatoparian,* in which the angle coincides with the suture;

d). *marginal,* in which the suture remains entirely along the lateral edge (or on the ventral surface) of the cephalon.

Harrington (1959) argues, from embryological evidence and from the variation in sutural type among trilobites otherwise quite similar, that the opisthoparian type is basic to the trilobite plan and that proparian and marginal types evolved independently in several different phyletic lineages from opisthoparian ancestors. A classification derived exclusively from suture state then cuts phylogeny rather than parallels it.

Glabellar segmentation and phylogeny. Hemingsmoen (1951) and others have emphasized the importance of reduction in glabellar segmentation and in the similarity of glabellar segments as a phyletic trend. Although this interpretation is consistent with derivation from the annelids and with apparent trends within compact taxa, strict and uniform application cannot be justified theoretically. In general, glabellar segmentation is more "primitive" in early developmental stages and this suggests ontogenetic recapitulation. On the other hand larval "segmentation" is continuously available by developmental retardation (p. 322) and may be utilized in descendant adults.

The overview and phylogeny. To paraphrase Harrington (1959), a completely satisfactory natural classification of the trilobites is still not possible. Many trilobites can be placed in compact families and these in turn associated on the superfamilial level. A considerable number of genera can not be so assigned, however, and some of the ordinal groupings are suspect. The application of multiple criteria is absolutely necessary; the weighting of these criteria is extremely difficult in the absence of well-established functional interpretation. The definition of morphologic sets through numerical taxonomy might be a desirable advance in this area. The arrangement here (Table 14-1) is taken without alteration from the *Treatise*.

A view from the rocks

A variety of trilobites including the blind, highly specialized (?) agnostids and eodiscids occur in lower Cambrian rocks. The eodiscids drop out above the middle Cambrian. Taken as a whole, the trilobites reached their maximum diversity during the Cambrian and Ordovician. The agnostids did not survive the end of the Ordovician, and, although some Silurian and Devonian trilobites are very common, the number of

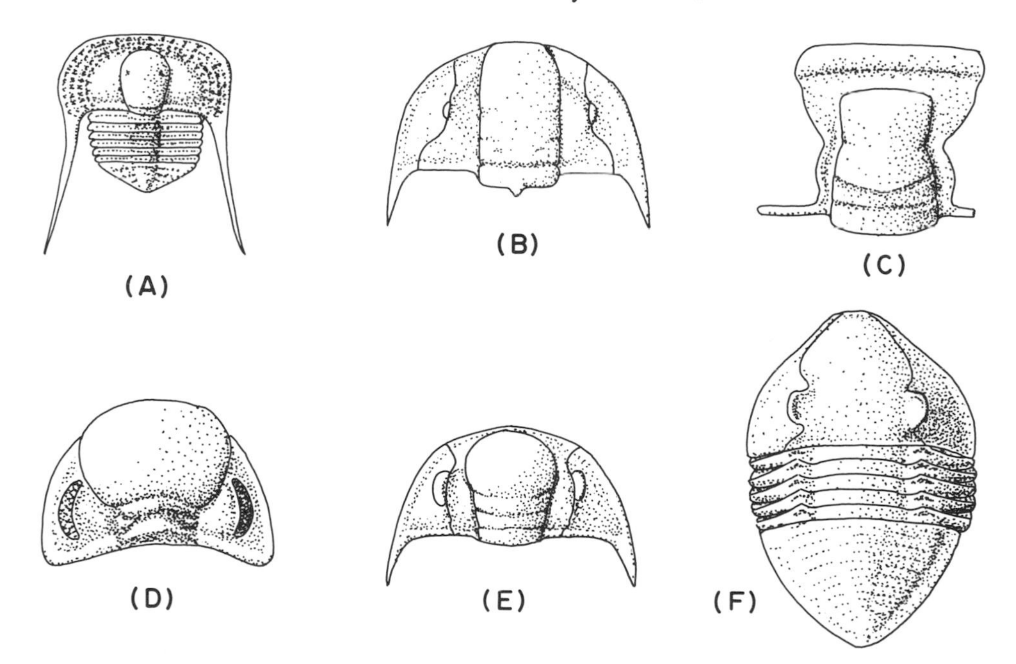

Fig. 14-11. A HODGEPODGE OF TRILOBITES. **(A)** *Cryptolithus;* middle to late Ordovician, dorsal view; length 2.7 cm. **(B)** *Olenoides;* middle Cambrian; dorsal view of cephalon; length 3.1 cm. **(C)** *Saukia;* late Cambrian; dorsal view of cephalon; free cheeks missing; length 1.6 cm. **(D)** *Phacops;* Silurian to Devonian; dorsal view of cephalon; length 1.5 cm. **(E)** *Paradoxides;* middle Cambrian; dorsal view of cephalon; length 2.3 cm. **(F)** *Isotelus;* middle to late Ordovician; dorsal view, thorax should show three more segments; length with added segments, 7.4 cm.

species declined progressively. Only a few genera have been described from Carboniferous beds in North America; this rarity is characteristic of other parts of the world, too. The class became extinct at the end of the Permian.

Trilobites are major zonal fossils for the Cambrian, and some genera and many species have short and thus useful stratigraphic ranges. Figure 14-11 shows a few of these. At any particular time, most trilobite genera were sharply restricted in geographic distribution, and, though some of these restrictions have been shown to be the result of local ecologic differences, others are of great value in defining ancient seas and lands.

Though much work has been done on the specific habits of trilobites, Brooks (1957) cites only eight papers that deal with interpretation of particular trilobite habitats. Although some other paleoecologic studies touch on the association of trilobite and environment, this number of publications would hardly suggest an oversupply of "trilobite paleoecologists." They were entirely marine animals and thrived in a variety of habitats; beyond that, little is known. The importance of the trilobites to stratigraphic analyses and evolutionary theory requires that much more be done in this field—one interesting area of investigation is the interpretation of trails and burrows associated with trilobites.

Crustaceans: the ostracods

As the trilobites slid toward extinction, the crustaceans—and others—slipped into their places. Their origin is obscure, though some of the trilobitomorphs found in the middle Cambrian foreshadow the crustacean form and may properly belong to the class. They include, in the Recent faunas, lobsters, crabs, crayfish, barnacles, shrimps, copepods, prawns, ostracods and water fleas; they inhabit freshwater as well as marine; and a few have adapted to terrestrial environments. They are among the commonest benthic animals and include, as well, a variety of free-swimming types.

In spite of their abundance and diversity, only one subclass, the Ostracoda, is particularly important to paleontologists. The difficulty lies partly in the composition of the skeleton, for many recent crustacea deposit little or no calcite in their chitinous exoskeletons. Typically, chitin is destroyed before burial or in the early stages of diagenesis. A few fossil genera are abundant; some are found in large numbers where the conditions for preservation were unusually good, but most are represented by only a few specimens.

The members of the Crustacea are distinguished by two pairs of preoral antennae and three pairs of appendages adapted as mouth parts. The anterior segments are fused and covered by a dorsal shield, the

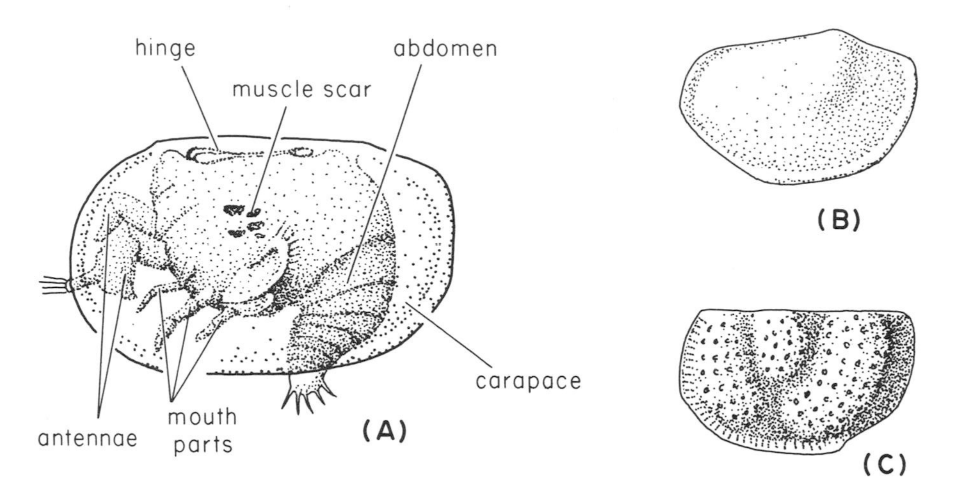

Fig. 14-12. OSTRACODA. **(A)** General morphology of the ostracods. Carapace forms bivalve shell covering head and abdomen. Appendages behind head are rudimentary. **(B)** *Lepeditia,* Ordovician to early Devonian; lateral view of left valve; length 4 mm. **(C)** *Beyrichia;* Silurian to Permian; lateral view of left valve; length 1.5 mm. [**(A)** *after* Sars; **(C)** *after* Bassler and Kellett.]

carapace. The posterior segments are distinct, and the more anterior bear paired legs adapted for walking or swimming. Some or all of the legs bear gills. The terminal segment typically is differentiated to form a spine-like or fin-like *telson.* As one expects in active animals, they have highly developed sense organs, a complex nervous system, and an efficient circulatory system.

In the ostracods (Figure 14-12), the carapace extends down and back to cover the appendages and the abdominal segments. The two sides of the carapace are joined in a hinge along the middle of the back to form a bivalve shell. A calcareous layer is deposited within the carapace. Ostracods are fairly abundant in many post-Cambrian beds. Evolution in form of the valves was rapid, and they are small enough to be preserved in well cuttings. Consequently, they supplement foraminifera and conodonts as guide fossils in subsurface stratigraphy.

To land: Myriapoda

The arthropods are preadapted to life on land. The exoskeleton prevents evaporation of fluids through the body surface; the walking legs can transport an animal on land as well as on the sea bottom, and the skeletal structure is strong enough to support the animal without the help of water. The only real problem was the replacement of the external gills by internal respiratory organs. This problem was solved by a group of primitive mandibulate arthropods by the Silurian if not earlier; the chelicerates managed it at about the same time.

The myriapods are the oldest and most primitive terrestrial mandibulates, the millipedes (Diplopoda) and centipedes (Chilopoda). The myriapod head, though it bears eyes, antennae, and paired mouth parts, is small and is followed by a large number of similar leg-bearing segments. Respiration depends on diffusion of gases through the walls of internal tubules, the *tracheae*. The tracheae open externally through one or two pairs of pores on each segment and branch extensively through the viscera. Although millipedes occur from the Silurian onward and the centipedes appeared in the early Cenozoic, a geologist is more likely to find one in his boots than in a rock.

Into the air: The Insecta

In one line of the myriapod radiation, an extraordinary modification occurred. The body wall of several segments (two at least) close behind the head developed lateral folds or flaps—for what purpose no one seems certain. Two pairs of these flaps enlarged to form wings, though the adaptive significance of the early steps in wing development are unknown. Among other flying animals, wings developed from legs in a series of changes which, whether hypothetical or observed, had adaptive significance.

The diagnostic features of insect form other than wings evolved before middle Devonian time. However wings evolved, they provided the key modification but did not "create" the insect form. This form (Figure 14-13) is a modification of the annelid plan of similar segments bearing similar paired appendages. The segments of the head are fused; one pair of appendages forms the antennae; the remainder, the mouth parts. The three segments behind the head are enlarged and partly fused into a *thorax*. Each thoracic segment bears a pair of walking legs, and the second and third develop wings in those insects that have them. The posterior pair of wings was lost in the order Diptera (flies, mosquitoes, and so on). The remaining segments, typically eleven, form the *abdomen*. The paired appendages of these segments are never more than rudiments, and are typically absent in the adult. Circulatory, nervous, and excretory systems are highly developed. The tracheal respiratory system is very complex and ends in tiny tubules that supply all parts of the body.

On this plan the insects have built a tremendous diversity of form. The wings as well as the limbs and body evolved to fit divergent modes of life. The larval stages are typically adapted to different environments than the adult and undergo a distinct change of form, *metamorphosis,* be-

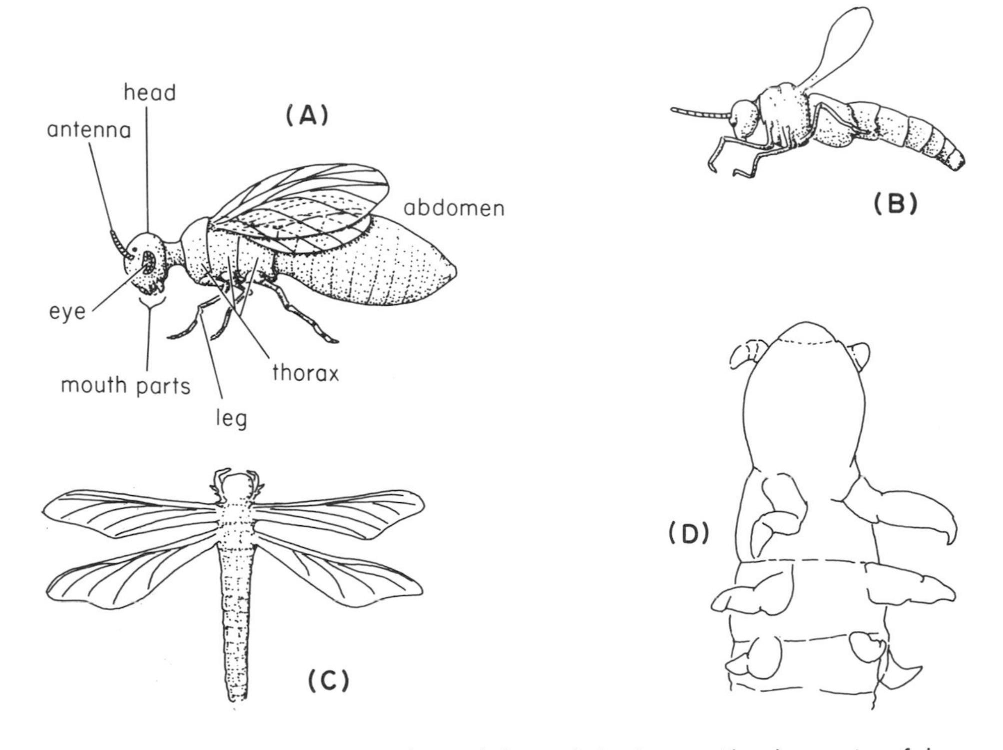

Fig. 14-13. INSECTA. **(A)** External morphology of the insects. The three pairs of legs, the wings, and the division of the body into head, thorax, and abdomen are distinctive. **(B)** *Dasyhelea;* Miocene to Recent, lateral view; length 3 mm. **(C)** *Mischoptera;* Pennsylvanian; dorsal view, wing span 17 cm. A representative of an extinct order. **(D)** *Rhyniella;* middle Devonian; ventral view. [(**B**) *after* Palmer; (**C**) *after* Carpenter; (**D**) *after* Scourfield.]

fore they reach maturity. In two orders social organization evolved, with caste systems and elaborate integrating behavior.

A paleontologist quails before the taxonomic complexity of the class. Modern faunas probably include over 1,000,000 insect species; about 13,000 fossil species have been described; approximately 25 recent orders are recognized; 44 extinct orders have been established (though Carpenter, 1953, reduces this number to 10). Fortunately for paleontologic sanity, insects are terrestrial and have chitinous skeletons. They are, therefore, rare fossils, preserved only in fine-grained lake, swamp, and lagoon sediments and in amber. The earliest insects identified are wingless types from the middle Devonian of Scotland. Since the Pennsylvanian beds in North America and Europe contain many lagoon and lake deposits, insects are relatively common and quite diversified. One modern order, that of the roaches, is present; five extinct orders have been recognized. The latter resemble several modern orders, e.g., dragonflies, mayflies, and so on, but are primitive in several important characteristics.

The early Permian faunas include representatives of nine extinct orders, mostly holdovers from the Pennsylvanian, and seven recent orders, among them dragonflies and "bugs." Beetles appeared in the late Permian; primitive flies and hymenopterans (wasps, bees, ants) in the Jurassic; and lepidopterans (butterflies and moths) and isopterans (termites) in the early Cenozoic. Many recent families are represented in the Jurassic insect faunas, and many recent genera in the early Cenozoic.

Chelicerates

Like the crustaceans, the chelicerates have structural predecessors if not actual ancestors in the middle Cambrian Burgess shale fauna. They diverge considerably from mandibulates (Figure 14-14), particularly in the

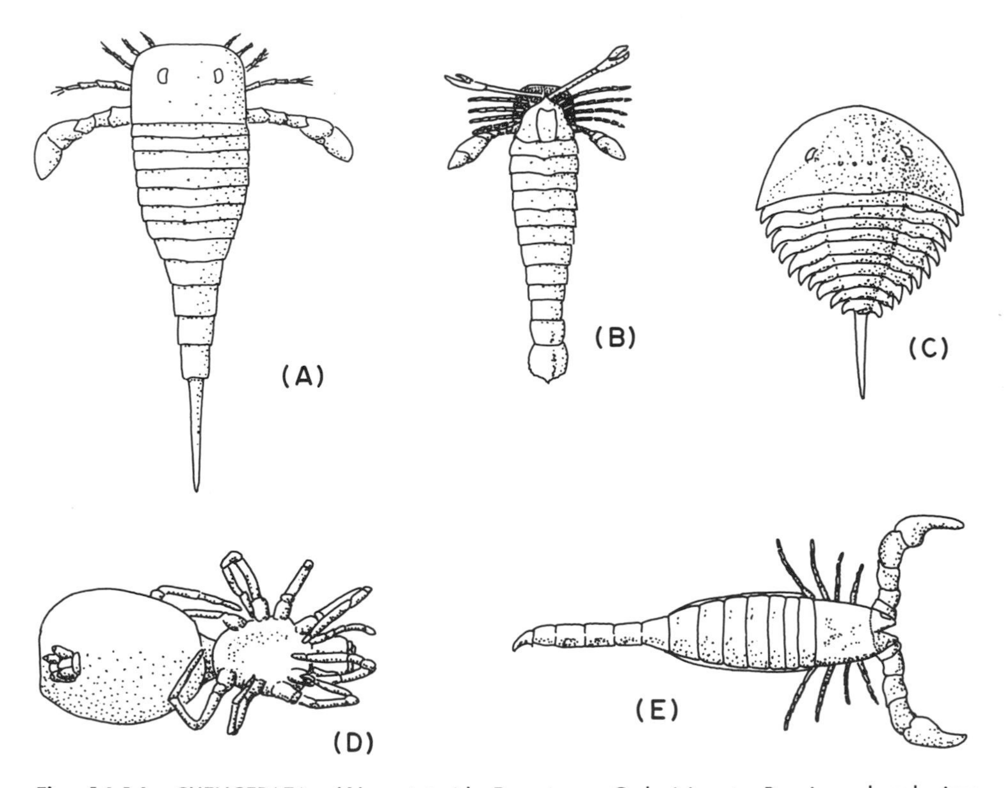

Fig. 14-14. CHELICERATA. **(A)** euripterid; *Eurypterus;* Ordovician to Permian; dorsal view; length 21 cm. Note prosoma, bearing legs, and appendageless opisthosoma. **(B)** A euripterid, *Pterygotus;* Ordovician to Devonian; ventral view; length 20 cm. Anterior pair of appendages possess claws (chelae). **(C)** A xiphosurian, *Neolimulus;* Silurian; length 2.2 cm. **(D)** A spider (Arachnida) *Argenna;* Miocene to Recent; ventral view; length 1.4 mm. **(E)** A scorpion (Arachnida), *Palaeophonus;* Silurian; dorsal view; length 4.8 cm. [**(A)** *and* **(B)** *after* Clark *and* Ruedemann; **(C)** *after* Woodward; **(D)** *after* Petrunkevitch; **(E)** *after* Pocock.]

absence of antennae, in the presence of claws on the single pair of preoral appendages (*chelicerae*), and in the fusion of extra segments into the head (the *cephalothorax* or *prosoma*). Some authorities have considered these divergences to imply that the chelicerates evolved independently from the annelids. The mixture of chelicerate and mandibulate characteristics in some of the Cambrian trilobitomorphs argues against this conclusion.

The gill-bearing aquatic chelicerates, the Merostoma, first appear in upper Cambrian rocks. The earliest merostomes all belong to the subclass Xiphosura, which is chiefly remarkable for its long range (Cambrian to Recent) and for the durability of the genus *Xiphosura* (Jurassic to Recent). The xiphosurans, so far as is known, have always been marine and primarily benthonic.

The other subclass, the eurypterids, scorpion-like in form, range from Ordovician to Permian. They are very rare in most faunal associations but form the major portion of a few. The absence of typical marine animals from these associations, and the character and location of the rocks indicate an unusual environment, either brackish or hypersaline. These eurypterid faunal associations are of unusual interest to human vertebrates, because they include the oldest known members of our group. For this reason, I'll mention them again in a subsequent chapter.

The class Arachnida now includes only air breathing forms, though the early scorpions were probably aquatic. This adaptation to terrestrial life began in the Silurian and was foreshadowed in the shift of the merostomes to brackish or freshwater habits. The key modification was in the respiratory system, either *"book lungs"* formed of many thin-walled leaves, or *tracheae*. The spiders and scorpions are of considerable importance in recent terrestrial communities, but they are rare as fossils. Several genera of scorpions have been described from Silurian rocks; the spiders do not appear until the Pennsylvanian.

REFERENCES

Brooks, H. K. 1957. "Chelicerata, Trilobitomorpha, Crustacea (Exclusive of Ostracoda) and Myriapoda," in Ladd, H. S. "Treatise on Marine Ecology and Paleoecology," vol. 2, *Geol. Soc. of America, Memoir 67*, pp. 895-930.

Carpenter, F. M. 1953. "The Evolution of Insects," *American Sci.*, vol. 41, pp. 256-270.

Clarkson, E. N. K. 1966. "Schizochroal Eyes and Vision of Some Silurian Acastid Trilobites," *Palaeontology*, vol. 9, pp. 1-29.

Harrington, J. J. 1959. See Moore, R. C. (Ed.) 1959, pp. 145-170.

Hemingsmoen, G. 1951. "Remarks on the Classification of Trilobites," *Norsk geol. tidsskr.*, vol. 29, pp. 174-217.

Hunt, A. S. 1967. "Growth, Variation, and Instar Development of an Agnostid Trilobite," *Jour. of Paleontology*, vol. 41, pp. 203-208.

Hupé, P. 1953. "Classification des Trilobites," *Ann. Paléont.*, t. 39, pp. 59-168.

Kjellesvig-Waering, E. N. 1966. "Silurian Scorpions of New York," *Jour. of Paleontology,* vol. 40, pp. 359-375. Analysis of adaptations and habitats of early scorpions.

Moore, R. C. (Ed.) 1955. "Arthropoda 2," *Treatise on Invertebrate Paleontology, Pt. P.* New York: Geological Society of America. Deals with the chelicerates.

———. 1959. "Arthropoda 1," *Ibid., Pt.* O. Covers the trilobitomorphs.

———. 1962. "Arthropoda 3," *Ibid., Pt.* Q. Deals with ostracods.

Palmer, A. R. 1962. "Comparative Ontogeny of Some Opisthoparian, Gonatoparian, and Proparian Upper Cambrian Trilobites," *Jour. of Paleontology,* vol. 36, pp. 87-96.

Robison, R. A. 1967. "Ontogeny of *Bathyuriscus fimbriatus* and its Bearing on Affinities of Corynerochid Trilobites," *Jour. of Paleontology,* vol. 41, pp. 213-221.

Tiegs, O. W. and S. M. Manton. 1958. "The Evolution of the Arthropoda," *Biol. Reviews,* vol. 33, pp. 255-337.

Whittington, H. B. 1963. "A Natural History of Trilobites," *Smithsonian Inst. Rept. for 1961,* pp. 405-415.

15

Some that crawled: Chitons, snails, and clams

The contrast between typical adult molluscs and annelids or arthropods is striking. The molluscs are commonly unsegmented—only one recent genus shows any indication of distinct segments; they lack paired appendages and possess instead a large muscular *foot;* and they have, in place of a flexible cuticle or jointed chitinous exoskeleton, a single rigid calcareous shell or pair of valves. Only the chitons have a segmented shell, and some molluscs lack a shell altogether. The earliest known molluscs are hardly more annelid-like than the modern forms. In spite of these differences, zoologists consider the two stocks to be closely related. Four major similarities, i.e., 1) in the form of the planktonic larva (p. 299); 2) in the early pattern of cellular division from the egg (spiral cleavage); 3) in the fixation of developmental regions at the earliest cell division (determinate development); and 4) in the formation of the coelomic pouches by splitting of the inner cell layer (schizocoelous origin), unite the molluscs with the annelids. In addition, they both possess a ventral nerve cord (or cords) and a major dorsal blood vessel. Some of these characteristics may have evolved independently in each group. If so, they appeared in response to similar selective pressures on animals with similar organizations and thus represent parallel evolution.

In spite of these indications of common origin, most biologists argued that the common ancestry was pre-annelid, i.e., in a nonsegmented coelomate worm. A few suggested that the absence of segmentation in the molluscs was secondary and not primitive and that their ancestry should be sought in the segmented worms, the annelids. The latter viewpoint is supported now by the discovery of a living mollusc, *Neopilina,* with several pairs of gills, of "kidneys," of "hearts," of sex organs, and of shell muscles. The general body musculature, the coelomic spaces such as they are, and the nervous system, however, show no clear evidence of segmentation.

The protomollusc

If the annelid segmentation evolved as an adaptation to burrowing, the primitive mollusc lost segmentation with the evolution of a different mode of locomotion. Since the oldest known molluscs and the simplest modern types crawled (or crawl) along the sea floor, the protomolluscs, presumably, were benthonic crawlers. The snails and chitons, which preserve this adaptation, progress by ripples of contraction in the ventral musculature like the peristaltic waves in annelids. Unlike the annelids, which force their way through the substrate, the gliding molluscs require little support from a hydrodynamic skeleton (p. 300). Suppression of the coelom and reduction of segmentation would not be surprising. A reconstruction (Figure 15-1) of the ancestral mollusc derives from the adult form and ontogeny of the various recent and fossil molluscans. They were elongate, coelomate, bilaterally symmetrical, and probably segmented.

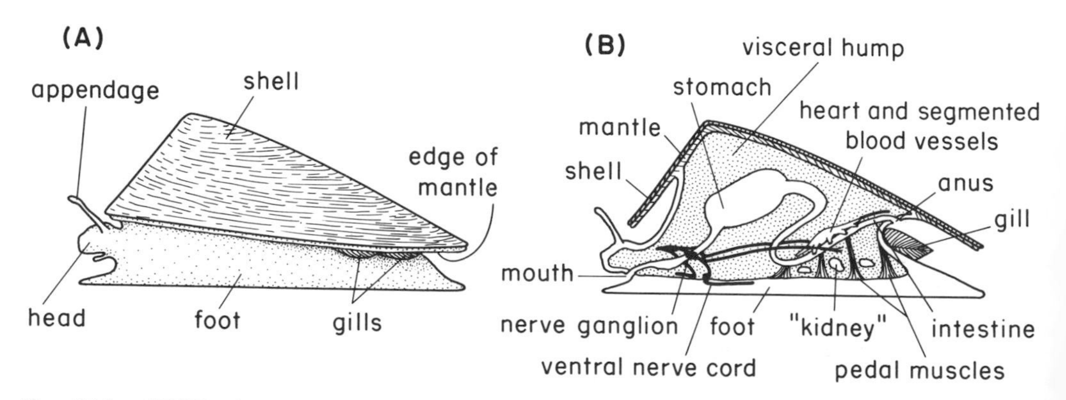

Fig. 15-1. GENERALIZED MOLLUSC. **(A)** Lateral view. **(B)** Same with shell, mantle, and foot sectioned to show viscera.

They moved on the substrate by action of the ventral body musculature, much thickened and concentrated as the *foot.* The dorsal surface bowed upward to form the *visceral hump* and was covered by a modified "skin," the *mantle,* which may have borne calcareous spicules or a complete shell. Several pairs of *pedal* muscles connected the foot with the mantle and either drew the shell down over the foot or pulled the foot up into the shell. The mantle folded laterally over the sides and back of the body enclosing an extensive *mantle cavity.* The intestinal tract, excretory system, and genital tubes all emptied into this cavity posteriorly. On either side of it was mounted a series of leaf-like gills. The *head* was partly differentiated from the body. Very possibly it bore sensory tentacles and eyes, and the mouth opened out at its anterior-ventral end. The animal fed with a *radula,* a slender horny strip that bore a row of small teeth and could be protruded from the mouth to work like a rasp. The major elements of the nervous system were a loop about the esophagus, a pair of ventral nerve cords innervating the musculature of the foot, and the nerves to the viscera and mantle. A tubular dorsal heart pumped blood through extensive cavities about the viscera. The excretory system consisted of several pairs of simple ducts.

The recent *Neopilina* approaches this postulated ancestral plan most closely. Fossil species belonging to this same group, the Monoplacophora, occur in lower Cambrian rocks, and, by their ancient stratigraphic occurrence, support the argument. The Amphineura (chitons) display somewhat similar features, though they lack metamerism except in shell and gills and are not known earlier than the Ordovician. The lower Cambrian faunas include a variety of straight, curved, and coiled shells (Fig. 16-3(A)) without clear affinities to recent molluscan classes. From the evidence of the fossil record, the ancestral molluscan line may have evolved in late Precambrian times and started to diversify shortly before the beginning of the Cambrian.

The Mollusca and locomotion

One of the primary features of molluscan evolution has been adaptation to different types of locomotion. The original "gliding" function is retained by the Amphineura, most Gastropoda, and a few genera of the Bivalvia. The "glide" is apparently satisfactory for sluggish bottom-living animals—snails have done quite well with it. The late Precambrian and Cambrian communities obviously contained suitable ecologic niches for these gliders, but could employ other types even to compete with the arthropods. These potential jobs included positions for burrowers and for sessile animals. The muscular foot is preadaptive to burrowing—if a hydrostatic skeleton is supplied. The small circulatory sinuses in the "primitive" creeping foot provide such a skeleton, and enlargement and specialization of these circulatory spaces (the haemocoel) mark the burrowing adaptation. The shape and location of these spaces and the control of blood flow by the circulatory vessels creates a subdivided hydrostatic skeleton analogous in function to the segmented annelid coelom. It is hardly astonishing that some members of the class Gastropoda, all of the Scaphopoda, and many of the Bivalvia adopted a burrowing habit. The latter line also evolved a heavy bivalve shell and an elaborate filter feeding system from the gills. Both features are preadaptive for sessile animals, and many bivalves have been and are sessile.

The step from a "glider" to a burrower is short and easy; the transformation to an active swimmer is more like a broad jump. A few gastropods and bivalves have accomplished it, but I, as a descendent of undulatory, segmented swimmers, will always be astonished at the success of the cephalopods. The modification is so little documented that a reconstruction must be largely speculative. One change was the development of *tentacles* from the head and/or part of the foot. They would serve to grasp and hold food (could the protocephalopod have been sessile?) but would also permit the animal to crawl along the bottom. The tentacles could be, and probably were, used as oars in swimming, as they are sometimes in recent squids. Another change was a complex and extraordinary evolutionary improvisation. The posterior end of the cephalopod body was folded under so that the anus lay beneath the head. The anterior and posterior portions of the mantle brought together in this change fused into a cone. The shell secreted by the mantle surface likewise became conical. Similar forms occur among sessile organisms and among some planktonic forms. Among the latter the bell or cone serves as a float. As the individual grew, the mantle pulled away from the small end of the shell and deposited a thin partition (a *septum*) across the interior. If, as is probable, the chambers between septae were partly gas filled, they reduced the specific gravity of the animal.

If the protocephalopods evolved an efficient swimming mechanism, they would be in business. But what could serve? Sculling about with threshing tentacles was possible—but not particularly efficient—at least for large animals without rigid skeletons.

One resource remained. In many molluscs, the muscles of the mantle pump water in and out of the mantle cavity. Distinct tubes evolve to direct these currents. The protocephalopod almost surely had some such mechanism. This pulsating jet system might not help a clam to burrow; it assuredly does help a cephalopod to swim. Given this toehold on an adaptation, selection acted on variant mantle cavity structures to produce a highly efficient jet jump.

The history of the molluscs seems to me sort of a Horatio Alger adventure. The annelid-arthropod stock was born with twin silver spoons, segmentation and a flexible or jointed exoskeleton. Molluscs began with

TABLE 15-1

Glossary for Mollusca (*figure numbers cited at end of definitions*)

Anus. Opens into posterior portion of mantle cavity in midline of body. (15-1B; *et al.*)

Cerebral ganglion. One of pair of large nerve centers in head at either side of digestive tube. Serves for reception of sensory perceptions and for coordination of activities. (15-1B.)

Dorsal nerve cord. One of pair of nerve bundles, passing from posterio-dorsal head. (15-1B.)

Foot. Ventral portion of body—forming a broad sole in some molluscs, a hatchet-shaped structure in some, and arms in others. Consisting chiefly of muscles used in locomotion. (15-1A, B.)

Gill. Leaf-shaped structure lying in posterior portion of mantle cavity and used in respiration. Typically occur in pairs, one mounted on either side of body.

Head. Anterior dorsal portion of body. Bears mouth, sensory organs, and major nerve ganglia. May be more or less distinct from foot and visceral hump. (15-1A, *et al.*)

Head appendages. See "tentacles."

Heart. Enlarged portion of medial dorsal tube that pumps blood through body. (15-1B.)

Intestine. Relatively narrow portion of digestive tube between stomach and anus. Typically consists of a number of coils occupying part of visceral hump. Serves for digestion and absorption. (15-1B.)

Kidney. Excretory organ—filters metabolic wastes, etc., from body fluids. Typically one or more pairs, one of pair on either side of body cavity. (15-1B.)

Mantle. Dorsal and lateral portions of body wall. Typically it covers the visceral hump and forms a fold extending around the sides of visceral hump and foot. (15-1A, B.)

Mantle cavity. Cavity between lateral and posterior folds of mantle and sides of visceral hump and foot. (15-1B.)

Pedal ganglion. One of a pair of large nerve centers which supplies nerve connections to muscles of foot. (15-1B.)

Pedal muscles. Paired muscles connecting foot to interior surface of shell. (15-1B.)

Radula. Strip of horny material bearing teeth like those of a file. It can be protruded out through mouth from its position in the floor of the digestive canal.

Segmental blood vessel. Lateral branch of dorsal blood vessel that extends to base of gill. (15-1B.)

Segmental nerve cord. Branch extending laterally from side of longitudinal nerve cord to base of gill.

Shell. Calcareous plate (or plates) deposited by cells in surficial layer of mantle and more or less covering body. (15-1.)

Stomach. Pouch in anterior portion of digestive tube that serves primarily for digestion. (15-1B.)

Tentacles. Used in two senses—for short, slender sensory head appendages (15-1A) and the grasping arms of the cephalopods. (16-1A.)

Ventral nerve cord. One of pair of nerve bundles passing posterio-ventrally from head to foot. (15-1B.)

Visceral hump. Portion of body behind head and above foot in which digestive and reproductive organs are concentrated. Not set off distinctly from foot though it may form a hump. (15-1B.)

a primitive method of locomotion and a rigid skeleton suitable only for sessile or sluggish bottom animals. On this base they have equaled the arthropods in the sea and have been reasonably common and diversified in fresh water and on the land.

Forethoughts and afterthoughts: Monoplacophora, Amphineura, and Scaphopoda

Paleontologists originally regarded the fossil monoplacophorans as a primitive order of snails. The conical shell is similar to that of some recent snails, and some genera show a slight degree of coiling (Figure 15-2). In particular, the juvenile part of the *Neopilina* shell coils quite strongly. The monoplacophoran shell differs from typical gastropods in that it bears

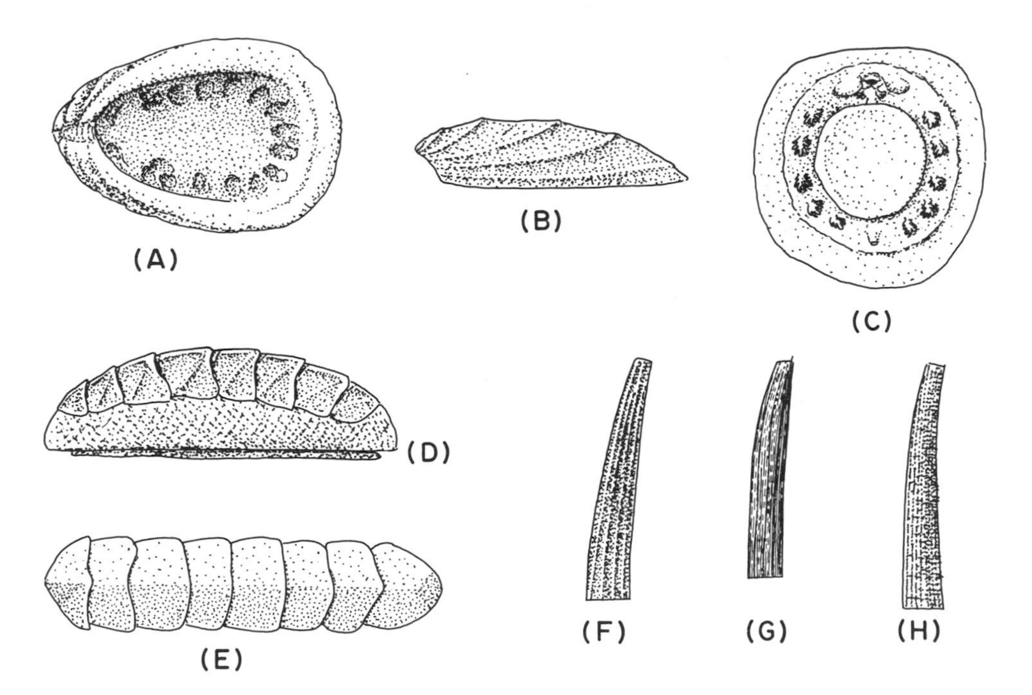

Fig. 15-2. MOLLUSCAN ODDMENTS. **(A)** Monoplacophoran *Tryblidium;* middle Silurian; ventral view of shell, anterior to left; length 3.2 cm. Note paired pedal muscle scars. **(B)** Lateral view of *Tryblidium.* **(C)** Monoplacophoran *Neopilina;* Recent; ventral view, anterior at top; length 3 cm. **(D)** Amphineuran *Chiton;* Recent; lateral view, anterior at left; length 3.5 cm. **(E)** Amphineuran *Helminthochiton,* Ordovician to Pennsylvanian; dorsal view of valves; length 6.5 cm. **(F)** through **(H)** are three different species of the scaphopod *Dentalium,* all fossil. Lateral views; length of **(F)**, 3.2 cm.; **(G)**, 3.8 cm.; **(H)**, 8.4 cm. [**(A)** and **(E)** *after* Piveteau, ed., *Traité de Paléontologie.* Copyright, Masson et Cie, Paris. Used with permission. **(C)** *after* Lemeche.]

on its inner surface the scars of several pairs of muscles. The discovery of the living *Neopilina* demonstrated that these paired scars reflect segmentation—not a gastropod characteristic—nor are the other characteristics particularly snail-like. I here follow Lemche (1957) and place them in a horizontal classification as a separate class, Monoplacophora.*

More than any other class of mollusc, the chitons, the Amphineura, retain the ancestral mode of life. They diverge from the founder of the phylum by reduction of the head, by increase in the number of gills, and by the development of a row of longitudinal calcareous plates along the mid-dorsal line (Figure 15-2). Some have a reduced foot, worm-like form, and calcareous spicules in the mantle. These may be primitive characteristics but more likely are specializations that simulate the primitive. The shell-less forms are unknown as fossils; the shelled are from several genera, the oldest of which is Ordovician.

A small group of burrowing molluscs, the scaphopods, may represent a distinct line from the primitive stock. The foot is conical. The mantle folds expand laterally and ventrally to fuse and form a cone. The foot extends from the wide end of the cone; the mantle cavity opens through a small hole at its apex. The mantle deposits a conical tusk-like shell. These shells appear in Devonian rocks, although some doubtful examples occur in Ordovician strata, but are neither varied in form nor more than locally abundant in the record.

Solution in a curve: The gastropods

Most snails, like the chitons, remain in an environment like that of the ancestral molluscs. Unlike the chitons, they evolved a marked modification of the general molluscan structural plan. The visceral hump coils forward and the whole body above the foot undergoes *torsion,* i.e., twists so that the mantle cavity lies above and alongside the head (Figure 15-3). Typically, the gills and the internal organs on the inside of the curve disappear so that bilateral symmetry is lost. The form of the shell follows the torsion of the visceral mass and forms, geometrically, a spiral coiled on a cone.

The origin of the gastropod coiling and torsion is obscure. As already

* Fossils that are clearly molluscan but not obviously members of the recent classes occur in lower and middle Paleozoic rocks. They include some forms suggestive of cephalopod shells—*Volborthella, Salterella, Wyattia,* etc.—some with vague resemblances to gastropods, *Hyolithes* for example, and some without any apparent affinities such as *Matthevia.* Several new classes have been proposed to receive these problem children (Fisher, 1962; Marek and Yochelson, 1964; Yochelson, 1966), but I can see no clear advantage of this procedure over their placement as "class" *incertae sedis* pending a more thorough knowledge of the basic radiation of the Mollusca.

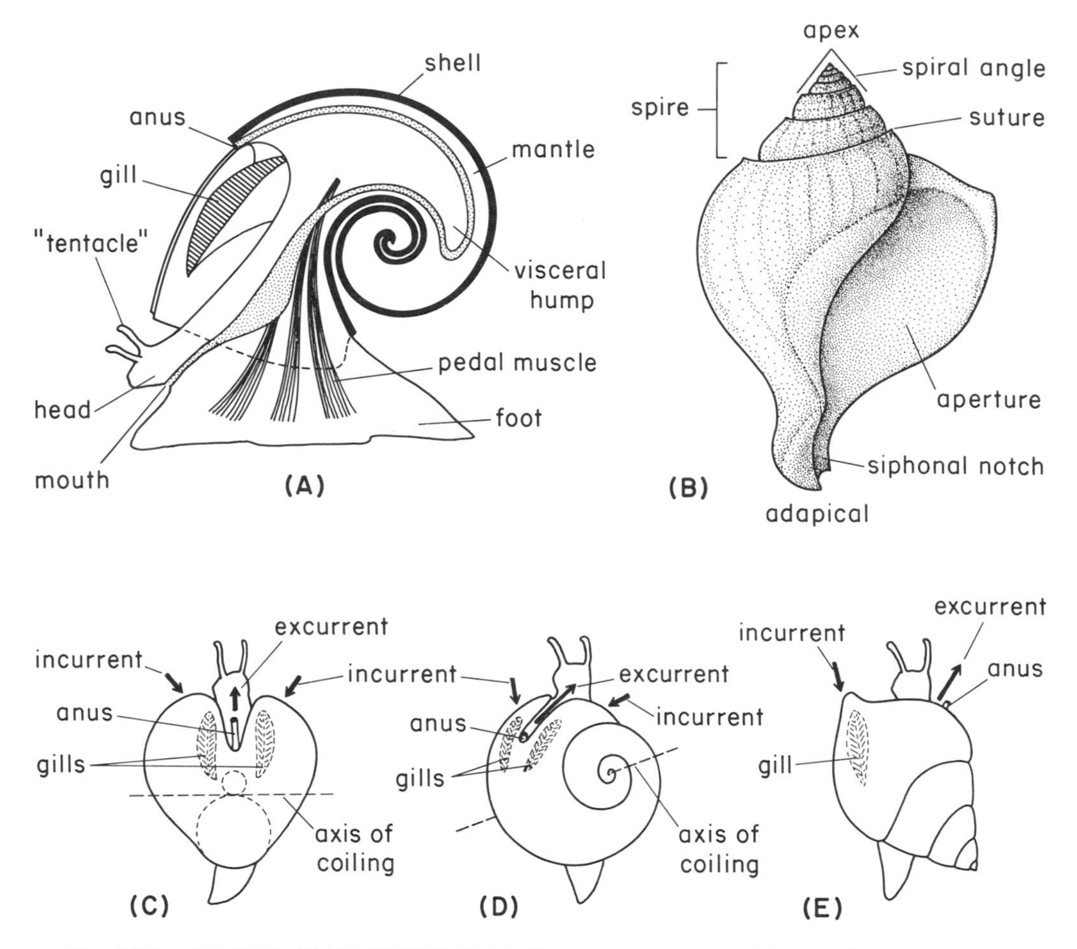

Fig. 15-3. GENERAL CHARACTERISTICS OF THE GASTROPODA. **(A)** Lateral view in section. Digestive tract stippled. **(B)** Lateral view of shell. **(C)** Primitive snail, dorsal view. Mantle cavity above head; anus in slit; gills bilaterally symmetrical; incurrent paths symmetrical. Shell planospiral. **(D)** Snail at intermediate evolutionary level, dorsal view. Mantle cavity to left of head; anus in slit; "right" gill and mantle cavity somewhat reduced. Shell conispiral. **(E)** Snail at advanced evolutionary grade. Mantle cavity to left of head with single functional gill and incurrent path. Anus to right of head at excurrent position.

described, the monoplacophoran *Neopilina* has the initial portion of its shell curved, but the significance of that coiling and its relation to larval and adult form is unknown. Some early Paleozoic snail shells coil spirally in the plane of bilateral symmetry, but most of these almost surely had undergone torsion. The coiling and torsion permit enlargement of the visceral mass above a foot small enough to be withdrawn into the shell. Coiling of viscera forward over the head creates a compact though high mass; torsion lowers it back toward the substrate and balances it over the

TABLE 15-2

Glossary for Gastropoda (*figures in parentheses refer to pertinent illustration*)

Abapical. Direction opposite that of apex along axis of spiral. (15-3B.)

Anterior. Direction along midline axis toward extremity of head. Generally defined in shell by anterior position of siphonal notch or canal and typically the abapical direction.

Anus. In recent snails it lies alongside—either to right or left—of head. (15-3A, C, D, E.)

Aperture. Opening of shell through which body is extended or withdrawn. (14-3B.)

Apex. In conical shells the small end of the cone or spire. (15-3B.)

Apical. In the direction of the apex or in its vicinity. (15-3B.)

Body whorl. The last (and typically the largest) complete loop in the spiral: The last-formed, it terminates in the aperture.

Columella. Medial pillar in spiral shell formed by coalescence of inner walls of whorls.

Conispiral. Shell type consisting of a spiral coiled on the surface of a cone. (15-3B, D, E; *et al.*)

Costa. Ridge on surface of shell—either parallel to axis of coiling or to border of spiral. (15-3B; 15-4A, E, G.)

Dextral. Direction of coiling. With axis of coil vertical, apex up (up in orthostropic shells, down in hyperstrophic) and aperture facing observer the aperture will be to the right of the axis. (15-3B, E, F; 15-4B, D, E, G.)

Eye. Photo-sensitive organ mounted on short stalk at side of head.

Foot. See Table 15-1. In gastropods a broad, flat muscular sole. (15-3A.)

Growth line. Low ridge or break on outer shell surface parallel to edge of aperture. Marks previous position of aperture. (15-4G.)

Head. See Table 15-1. (15-3A.)

Hyperstrophic. A rare shell type in which the whorls are coiled on an inverted cone so that the apex points forward rather than back. Not easily distinguished from orthostropic unless aperture shows siphon pointed in same direction as apex.

Inner lip. Inner border of aperture—portion adjacent to last whorl and to columella. (15-3B.)

Left gill. See Table 15-1. In all recent snails the left gill is rotated so it lies on the right anterior side of body. In dextral individuals it is typically smaller than the right or is absent. (15-3A.)

Left longitudinal nerve cord. See Table 15-1. In recent gastropods torsion of the viscera loops the left cord over to the right side of the body.

Orthostropic. The common shell type in which the whorls are coiled on an erect cone so that the apex points back rather than forward. If a notch for siphon is developed in aperture it will be opposite the direction of apex (abapical). (15-3B, D, E.)

Outer lip. Lateral border of aperture. (15-3B.)

Periphery. Portion of whorl most lateral to axis of coiling.

Planispiral. Shell type formed by spiral coiled in a single plane and symmetrical in that plane. Very rare in gastropods though many genera approach planispiral. (15-4A.)

TABLE 15-2 (continued)

Posterior. The direction opposite the head along the midline axis. In nearly all shells the apical direction—see "anterior" and "hyperstrophic." (15-3E.)

Pseudoplanispiral. Shell coiled in one plane, but whorls not symmetrical in that plane. (15-4B.)

Right gill. See Table 15-1. In all recent snails the right gill is rotated to a left anterior position. In dextral individuals it is the larger gill; in sinistral the smaller. (15-3C, D, E.)

Right longitudinal nerve cord. See Table 15-1. Torsion of the gastropod viscera loops the right cord to the left side of the body.

Selenozone. Band of closely-spaced crescentic growth lines on lateral surface of whorls. Typically it marks positions of apertural notch or slit during earlier stages of growth. (15-4G.)

Shell. External calcareous skeleton. Consists of a single plate—typically drawn into a spirally coiled tube—which encloses the viscera and into which the head and foot may be withdrawn. (15-3A; *et al.*)

Sinistral. Direction of coiling. In orthostropic shells with axis of coil held vertical, apex up, and aperture facing observer the aperture will be to the left of the axis. See dextral.

Sinus. Groove or reentrant in lateral margin of aperture (outer lip) distinguished from slit by nonparallel margins. (15-4G.)

Siphon. Tubular extension(s) of mantle border. Pierced by a canal that opens into the mantle cavity from the exterior. May be indicated in shell by groove or notch. (15-3E.)

Siphonal notch. Notch at anterior end of aperture occupied by siphon. Where present it lies between—virtually separates—the inner and outer lips. (15-3B; 15-4D.)

Slit. Parallel sided reentrant in lateral border of aperture. (15-3C, D; 15-4G.)

Spiral angle. Angle formed between two lines tangent to periphery of two or more whorls and on opposite sides of shell. Angle may change from initial whorls to later ones. (15-3B.)

Suture. Line of contact between two whorls. Typically a spiral on the outer surface—and also on the inner around the axis of coiling where that space (the umbilicus) is not closed. (15-3B; *et al.*)

Tentacle. See Table 15-1. (15-3A.)

Torsion. Rotation of viscera to right or left so that the anus and the mantle cavity are brought from a medial posterior to a lateral anterior position. (15-3.)

Umbilicus. Opening along central axis of spiral formed where inner walls of whorls fail to meet. Typically a conical opening, widest in the body whorl. (15-4B.)

Whorl. Single complete turn of spiral shell.

---

foot (Morton and Yonge, 1964). The evolution of torsion is less easily explained—some suggest it is the result of a larval adaptation.

The principal evolutionary grades among the more generalized snails, the prosobranchs, represent stages in adjustment of the viscera to torsion and in exploitation of its possibilities. In the early forms (the belleraphontids), torsion positioned the posterior mantle cavity above the head; water was drawn in latero-ventrally over the leaflike gills and flushed out anteriorly carrying with it fecal debris. The plano-spiral shell rose in the midline plane of the body, raising the visceral mass high above the foot. Presumably the gills and the kidneys were bilaterally symmetrical. A slit in the dorso-medial lip of the shell housed the excurrent opening of the mantle cavity and separated it partially from the head and the incurrent opening.

The evolution of conispiral coiling reduced the maximum hydrodynamic cross-section of the shell and also reduced the torque on the pedal musculature by lowering the center of the visceral mass. This modification shifted the mantle cavity to one side, crowded the conispiral shell into the right side of the mantle cavity, reduced the size of the right mantle cavity, and ultimately closed the right incurrent path. The excurrent slit shifts laterad—to the left of the head. This pattern is established in late Cambrian snails (the macluritines and pleurotomariines). Subsequent changes typically involve suppression of the gill and other organs of the right mantle cavity, the displacement of the anus and excurrent opening to the right, away from the incurrent opening, the modification of the left mantle margin into an incurrent tube, i.e., *siphon,* and the modification of the left gill to fit the asymmetry of the mantle cavity.

Of the more specialized gastropods, the opisthobranchs have reduced or eliminated the shell and typically the mantle cavity as well with consequent profound alteration of the gill and the other mantle cavity organs. In the other subclass, the pulmonates, the opening into the mantle cavity is closed except for a single small aperture. The gill is suppressed—in fully aquatic forms gill-like structures derive from the mantle. The anus and excretory ducts open alongside the mantle cavity rather than within it so that the cavity forms a lung, an analogue of the vertebrate structure, for air breathing. The developmental pattern—including the torsion—is very much modified.

The shell reflects these soft-part modifications—particularly in the form of the opening (*aperture*) through which the foot, head, and part of the visceral mass are protruded. For example, position of the excurrent opening is indicated in bellerophotines, macluritines, and pleurotomariines by a marginal slit. In others an anterior (*abapical*) notch marks the position of the incurrent siphon.

Many of the shell characteristics are simple properties of the spiral curve and are, to some extent, interrelated (Figure 15-4). Critical features include the type of coiling, the spiral angle, the cross profile of shell and of the individual whorl, the tightness of coiling, the ratio of whorl number to shell height, the contact of the inner surface of the whorls, and the characteristics of the aperture. The finer subdivisions in the classification of recent snails are based on the character of the radula. Unfortunately, this is very rarely fossilized. Knobs, ribs, spines, and so forth on

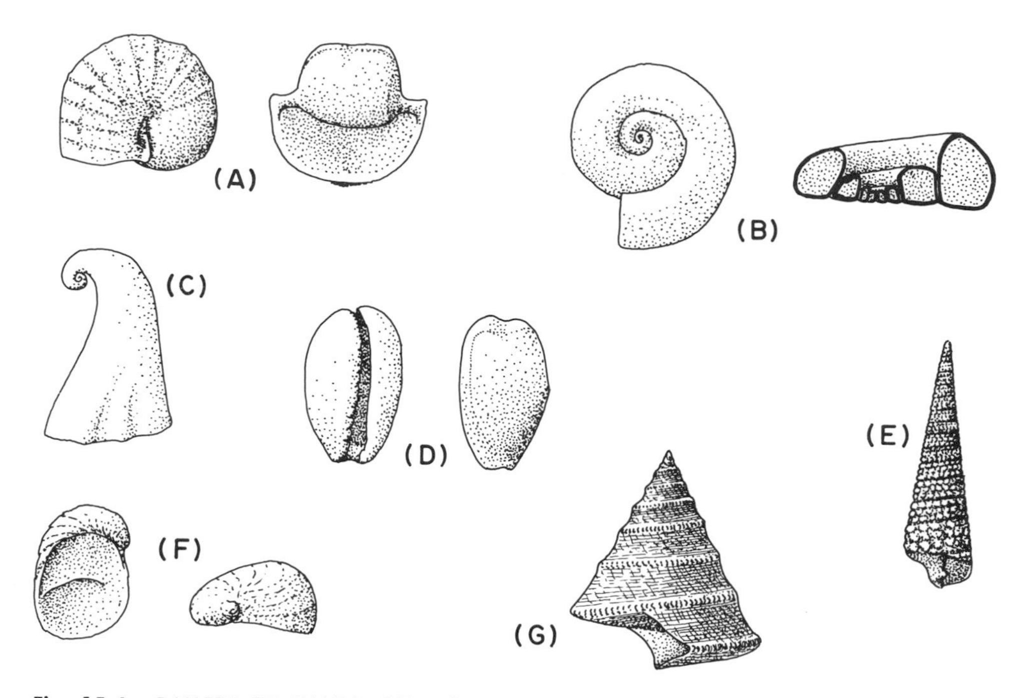

Fig. 15-4. GALLERY OF SNAILS. **(A)** *Bellerephon;* Ordovician to Triassic; lateral and apertural views, width 2.1 cm. **(B)** *Maclurites;* Late Cambrian to Ordovician; apical view and cross-section; maximum diameter 6.5 cm. **(C)** *Platyceras;* Silurian to Permian (?); apical view; length 7.3 cm. **(D)** *Cyprea;* Cenozoic to Recent; apertural and dorsal views; length 2.6 cm. **(E)** *Turritella;* Cretaceous to Recent; apertural view; length 4.1 cm. **(F)** *Crepidula;* late Jurassic to Recent; apertural and lateral views; length 2.6 cm. **(G)** *Pleurotomaria;* Jurassic; apertural view; length 2.7 cm.

the shell surface are also used, but have relatively little weight in classification.

The snails impose on these basic structural grades a spectrum of adaptation from burrowing benthos to seminektonic pelagia. They have, however, been predominantly benthic, and these are the commonest fossils. The pulmonates invade freshwater and terrestrial environments but retain the characteristic adaptations in feeding and in locomotion.

Snails are among the dominant predators and scavengers of the benthos and use their rasp-like radula to bore through shells and into the flesh of sluggish and sessile organisms. Others are herbivores and employ their radula to browse on aquatic plants. Presumably, a large part of the gastropod radiation corresponds to differences in food source as well as variation in substrate, turbulence, depth, and temperature. Unfortunately, little of this adaptive variation shows in the shell, and the information on gastropod paleoecology is not commensurate with their importance in marine communities.

The lack of paleoecologic control diminishes the value of gastropods in correlation. The long range of many genera and species diminishes it

TABLE 15-3

The Mollusca

Early Cambrian to Recent. Segmentation partial, rudimentary, or absent. Body comprises foot, head, visceral hump, and mantle. One or more pairs of gills in mantle cavity except in scaphopods and specialized gastropods. Highly organized circulatory, respiratory, and nervous system. External calcareous (calcite and/or aragonite) shell in most. Haemocoel largely replaces coelom.

CLASS MONOPLACOPHORA

Early Cambrian to Recent. Broad foot, low visceral hump, partly differentiated head. No distinct mantle cavity. Segmented. Conical shell, coiled in some genera. Benthonic, marine, ?herbivorous. Fossils from early Cambrian to Silurian.

CLASS AMPHINEURA

Ordovician to Recent. Elongate foot, low visceral hump, rudimentary head. Three or more pairs of gills. Shell absent or composed of eight pieces in series along mid-dorsal line. Benthonic, marine; typically herbivorous. Few fossils.

CLASS SCAPHOPODA

Silurian to Recent. Foot reduced, compressed, conical; visceral hump high; head simple; mantle forms cone about body. No gills. Shell tusk-shaped; foot protruding from large end; mantle cavity opening through apex. Marine, burrower. Feeds on small organisms encountered in bottom sediments, a mud grubber. Few fossils.

CLASS GASTROPODA

Early Cambrian to Recent. Foot large; head distinct with eyes and tentacles; visceral hump high and coiled forward. Two gills, one reduced or absent in some; some fossil genera probably had more gills. Shell typically coiled, covering visceral hump. Marine, freshwater, and terrestrial. Benthonic except for a few planktonic genera. Herbivores and carnivores. Includes several classes and a number of orders but not easily determined from shell characteristics. Some genera and species are guide fossils. Snails and slugs.

CLASS BIVALVIA

Ordovician to Recent. Foot laterally compressed, hatchet-shaped, may be greatly reduced; head rudimentary; visceral mass large; mantle folded over sides of body. Bivalve shell, one valve on either side of body, and hinged dorsally. Two gills, enlarged and elaborated. Benthonic; marine and freshwater. Vagrant, burrowing, and sessile. Primarily detritus or suspension feeders—currents through gill filaments.

Subclass Prionodesmacea

Ordovician to Recent. Prismatic shell structure. Mantle lobes separate; siphons poorly developed.

Order *Palaeoconcha*

Ordovician to Recent. Protobranch gills, no hinge teeth, subequal adductors. Predominantly burrowers.

Order *Taxodont*

Ordovician to Recent. Protobranch or filibranch gills. Taxodont hinge; subequal adductors. Vagrant and sessile benthonic.

Order *Schizodonta*

Ordovician to Recent. Filibranch; schizodont hinge; subequal adductors. Primarily vagrant benthonic.

Order *Dysodonta*

Ordovician to Recent. Filibranch and eulamellibranch; dysodont; anterior adductor reduced or absent; byssus. Chiefly sessile.

TABLE 15-3 (continued)

Order *Isodonta*
Triassic to Recent. Filibranch; isodont; no anterior adductor; some with byssus. Sessile.

Subclass Teleodesmacea
Ordovician to Recent. Shell laminate, nonprismatic. Mantle lobes fused. Siphons well developed.

Order *Heterodonta*
Silurian to Recent. Eulamellibranch; heterodont; subequal adductors. Predominantly vagrant benthonic.

Order *Pachyodonta*
Jurassic to Recent. Eulamellibranch; pachyodont; one valve very much enlarged, may be coral-like. Sessile benthonic.

Order *Desmodonta*
Ordovician to Recent. Eulamellibranch except for one suborder with septibranch gills; desmodont. Burrowers and borers. Razor clams, etc.

even more. The lot of a stratigrapher of gastropods is not a happy one. Some genera are, however, of particular value, and some, such as the freshwater and terrestrial pulmonates, occur in depositional environments otherwise devoid of fossils.

A study in shape and function: The Bivalvia

Although some bivalves retained the primitive mode of locomotion, the bivalve shell and a filter feeding mechanism created additional evolutionary potentialities. The valves of the shell (Figure 15-5) are formed by a mantle lobe on either side of the body, so that the plane of symmetry passes between the valves, and their articulation (*hinge line*) lies in the mid-dorsal line. The *beak* on the elevated area (*umbo*) above the hinge typically points forward and indicates the orientation of the shell, anterior and posterior, right and left. The valves are closed by the *adductor muscles* (primitively two, one anterior and the other posterior), and are opened by tension of a *ligament* above and/or within the hinge. Concentric growth lines record shell growth on the external surface.

A large mantle cavity extends forward on either side of the body to accommodate the enlarged gills. A few bivalves have a pair of simple

Fig. 15-5. GENERAL MORPHOLOGY OF A BIVALVE. **(A)** Lateral view with left valve and mantle fold removed. Anterior to the left. **(B)** Cross-section. Shell shown in black; mantle in closely spaced stipple. **(C)** Lateral view with valve, mantle fold, and body wall removed. Space around gut occupied by digestive glands and by gonad. Cavity of foot occupied by a blood sinus. **(D)** Cardinal area showing ligament grooves. **(E)** Cardinal area with resilifer. **(F)** Lateral view of interior of right valve. **(G)** Lateral view of exterior of left valve.

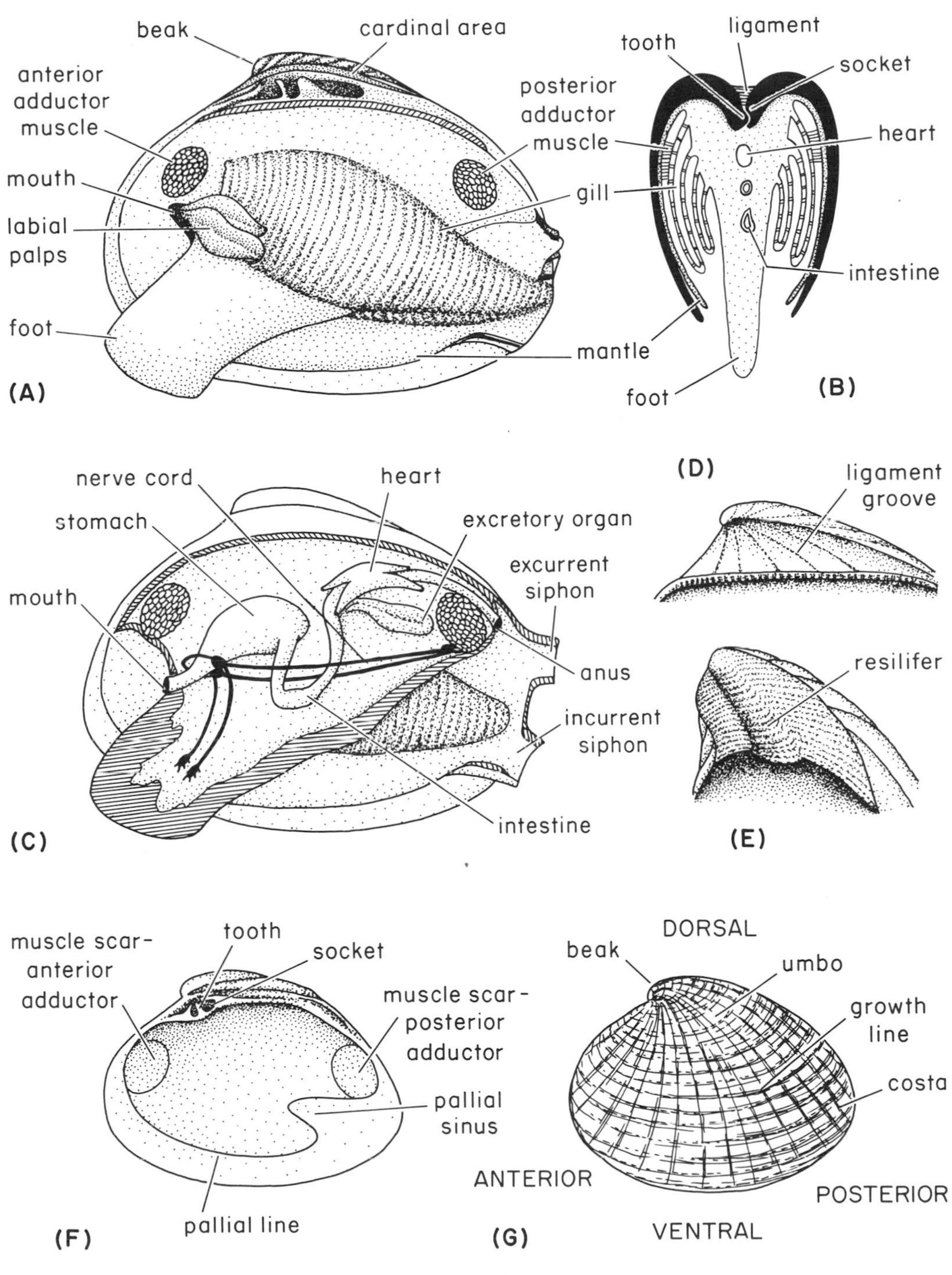
beak
cardinal area
anterior adductor muscle
posterior adductor muscle
mouth
gill
labial palps
foot
mantle
(A)
tooth
ligament
socket
heart
intestine
foot
(B)
nerve cord
heart
stomach
excretory organ
excurrent siphon
mouth
anus
incurrent siphon
intestine
(C)
(D)
ligament groove
resilifer
(E)
muscle scar-anterior adductor
tooth
socket
muscle scar-posterior adductor
pallial sinus
pallial line
(F)
DORSAL
beak
umbo
growth line
costa
ANTERIOR
POSTERIOR
VENTRAL
(G)

Fig. 15-5.

TABLE 15-4

Glossary for Bivalvia (*figures in parentheses refer to pertinent illustration*)

Adductor muscles. One or two large muscles that are attached to interior surface of both valves and serve to close them. See anterior and posterior adductor.

Anterior. Direction, in the plane of bilateral symmetry, of the mouth and rudiments of the head. Defined in shells by beaks which typically point anteriorly and/or by pallial sinus which occurs in posterior part of shell. (15-5G.)

Anterior adductor muscle. Adductor muscle (see which) in anterior position. Absent in some advanced groups. (15-5A, C, F.)

Anterior adductor muscle scar. Area of attachment of anterior adductor. Generally a semicircular area depressed below surface of valve interior and variously scarred. (15-5F.)

Beak. Projection of dorsal portion of valves above hinge line—the initial portion of the valve. Typically the beak shows strong curvature and points anteriorly. (15-5A, G; *et al.*)

Byssus. Threadlike process derived from anterior portion of foot and used to attach shell to substrate. (15-7A.)

Cardinal area. Flat or gently curved surface on valve between beak and hinge line and extending unbroken both anterior and posterior to beak. Set off from the remainder of shell by a sharp angle. (15-5A, D, E.)

Cardinal teeth. Projections along hinge line that fit into opposing sockets as part of valve articulation and that have a long axis perpendicular or oblique to hinge line. (15-5A, F; 15-10F.)

Commissure, plane of. Plane defined by line of juncture between valves. In pelecypods the plane of bilateral symmetry.

Costa. Ridge on external shell surface—one of several radiating from beak. (15-5G.)

Desmodont. Type of hinge line structure characterized by a large ligament inside the hinge line. Recognized in the shell by presence of large depression in hinge area (resilifer) in which ligament was located. (15-10H.)

Dorsal. Defined in pelecypods by the hinge line and by the beaks which form the dorsal margin. (15-5G.)

Dysodont. Type of hinge line structure characterized by the reduction of hinge teeth and development of a large internal ligament (resilium). Typically the hinge plate is broad and bears a deep depression (resilifer) marking the position of the ligament. (15-5E; 15-10D.)

Equilateral. Shell form in which the anterior and posterior portions of valves are subequal and nearly symmetrical. (15-5G; 15-6A, B.)

Equivalve. Shell form in which right and left valves are subequal and symmetrical about plane of commissure. (15-5B; 15-6A, B, C.)

Escutcheon. Flat or simply curved area between beak and hinge line and posterior to beak. Corresponds to posterior part of cardinal area and is separated from remainder of shell surface by sharp change in angle. (15-10F.)

Eulamellibranch. Type of gill structure. Each gill consists of two sheets of filaments which extend downward for a distance and then are folded back dorsally. Adjacent filaments are connected by interfilamentary junctions and the ascending and descending portions of each filament are also connected. (15-8C.)

TABLE 15-4 (continued)

Excurrent siphon. Posterior dorsal extension of mantle borders forming a tube which confines and directs outflowing current from mantle cavity. (15-5A, C.)

Filibranch. Gill type. Each gill consists of two sheets of filaments which extend some distance downward and then are folded up toward roof of mantle cavity. The adjacent filaments are not interconnected though the ascending and descending branches of each filament may be. (15-8B.)

Foot. See Table 15-1. May be rather broad though not flattened dorso-ventrally but more commonly is compressed laterally into a broad or narrow blade. (15-5A, B, C; 15-6; 15-7.)

Gill. See Table 15-1. Pelecypods possess a pair, one on either side of viscera. (15-5A, B; 15-8.)

Growth line. Slight ridge or break in shell surface approximately parallel to the borders of valve and representing shell margin at earlier growth stage. (15-5G.)

Heterodont. Type of hinge line structure characterized by a small number of hinge teeth of two types, the cardinals and the laterals. The former, directly below the beak, are perpendicular or oblique to the hinge line; the latter which lie ahead of and behind the cardinals parallel the hinge line. (15-5F; 15-10F.)

Hinge line. Line of articulation between the two lateral plates (valves) that form the pelecypod shell. (15-5A; *et al.*)

Hinge tooth. Projection along hinge line that fits into a pit (socket) on the opposing valve and assists in articulation of valves. (15-5A, B, F; *et al.*)

Incurrent siphon. Posterior-ventral extension of mantle borders forming a tube which confines the current flowing into the mantle cavity. (15-5A, C.)

Inequilateral. Shell form in which anterior portion of valves is much shorter than posterior. (15-6C; 15-7A, B, C.)

Inequivalve. Shell form in which one valve is flatter (and often smaller) than the other. (15-7C, D, E.)

Isodont. Type of hinge line structure characterized by two large subequal teeth in one valve and corresponding sockets in the other. (15-10E.)

Labial palps. Ciliated appendages on either side of mouth that assist in feeding. (15-5A.)

Lateral teeth. Projections along hinge line that parallel hinge line and fit into sockets on the opposing valve. See "heterodont." (15-10F.)

Left. Direction to left of plane of symmetry. Determined by placing shell with hinge line up and anterior end pointing away from observer—left on pelecypod will correspond to observer's left.

Ligament. Elastic tissue attaching valves along hinge line and serving to open the valves, either by tension (external ligament) or by expansion (internal ligament, resilium). (15-5B.)

Ligament area. Scarred area between beak and hinge line that served for attachment of the ligament. May include most of the cardinal area. (15-5D.)

Ligament groove. Groove (typically one of several) in the area between beak and hinge line that served for attachment of the ligament. (15-5D.)

Lunule. Flat or curved area between the beak and the hinge line and in front of the beak. Corresponds to anterior portion of cardinal area and is distinguished from remainder of shell surface by sharp change in angle.

TABLE 15-4 (continued)

Mantle. A pair of folds from the dorsal body wall that extend laterally and ventrally over the sides of the animal. External cell layer forms shell and is attached to its inner surface. (15-5A, B.)

Pachyodont. Hinge line structure characterized by one or more very large, thick teeth. (15-10G.)

Pallial line. Line a short distance inside shell margins marking inner border of the relatively thick edges of mantle. Typically marked by a groove or ridge and by a change in texture of shell material. (15-5F.)

Pallial sinus. Inflection of posterior-ventral portion of pallial line that extends anteriorly or anterio-dorsally away from proximal shell margin. Marks position of siphons. (15-5F; 15-6B, C; *et al.*)

Plica. Fold, involving full thickness of shell, that extends radially from beak to shell margin. Shows on inner surface of shell as well as outer. (15-7C; 15-10E.)

Posterior. Direction in plane of symmetry opposite position of head. Determined in shell by posterior position of pallial sinus and by anterior projection of beaks. (15-5G.)

Posterior adductor muscle. Adductor muscle (see which) in posterior portion of pelecypod. (15-5A.)

Posterior adductor muscle scar. Area of attachment of posterior adductor. Generally a semicircular area depressed below surface of valve interior and variously scarred. (15-5F; 15-10).

Protobranch. Gill type. Simple leaflike gill with short simple filaments. (15-8A.)

Right. Direction to right of plane of symmetry. Determined by placing shell with hinge line up and anterior end pointing away from observer—right on pelecypod will correspond to observer's right.

Resilifer. Depression inside margin of hinge line which holds (or held) internal ligament (resilium). (15-5E; 15-10D, E, H.)

Resilium. Portion of ligament within hinge line. Compressed by hinge plate when valves are closed.

Schizodont. Type of hinge line structure characterized by teeth diverging sharply from beneath beak. (15-10C.)

Septibranch. Gill type. Consists of a perforate diaphragm extending horizontally across mantle cavity and dividing it into upper and lower moieties. (15-8D.)

Socket. Depression along hinge line that receives projecting tooth from opposite valve. (15-5F; *et al.*)

Taxodont. Type of hinge line structure. Characterized by numerous small, subequal teeth along hinge line. (15-10B.)

Umbo. Elevated and relatively convex portion of valve adjacent to the beak—essentially the "humped" part of the shell. (15-5G.)

Valve. One of the two convexly curved (rarely flat or concave) calcareous plates that lie on either side of the visceral hump and foot and that are articulated along a dorsal hinge line.

Ventral. In general the direction in the plane of symmetry toward the substrate. Defined in the pelecypods by the dorsal position of the hinge line and the ventral position of the opening between the valves. (15-5G.)

---

leaflike gills much like those of the chitons and snails, but in most of them the gills are modified into sheets of filaments, variously folded and interconnected. Cilia on the gills draw water into the shell along the ventral posterior border behind the foot, over the gill filaments, and out again posteriorly. Food particles are filtered from the water and carried to the mouth on ciliated tracts. Modifications of mantle and gills separate incurrent and excurrent flow.

The head is rudimentary, lacking eyes, tentacles, and radula. An arterial system carries blood from the heart to the gills, mantle, foot, and internal organs, and a venous system returns it to the heart. In the gills and mantle the blood is confined to capillaries; elsewhere it flows through large cavities about the organs. Respiration is carried on through the surface of the mantle and probably to some extent in the gills. An Ordovician bivalve, *Babinka,* not only has multiple pairs of pedal muscle scars but what appear to be gill muscle scars (McAlester, 1966). In this character, *Babinka* shows affinities to the monoplacophorans, and McAlester suggests that the lucinoid bivalves among which *Babinka* apparently belongs represent an independent lineage from the protobivalve stock. Other than *Babinka,* no intermediates between the bivalves and the Cambrian molluscs are yet known.

Shell characteristics and adaptations

Those few bivalves that retain the primitive creeping locomotion have a relatively primitive flattened foot on which the animal glides. The foot is large, and when the valves are open it protrudes from the anterior ventral part of the shell. The gills occupy much of the posterior part of the shells. The general shell form (Figure 15-6) is *equivalve* (the valves symmetrical) and *equilateral* (the anterior and posterior portion of each valve nearly equal). The valves are about as high as they are long and may be roughly triangular, with the apex dorsal and the base ventral. Because the anterior and posterior portions of the valves are nearly equal, the anterior and posterior adductors (and their scars on the valve interior) are of equal size.

In burrowing bivalves, the foot is narrow and elongate. Rather than protruding ventrally, it extends anteriorly, and the anterior portion of the shell is long and shallow. A similar change occurs in the posterior part of the shell with the development of *incurrent* and *excurrent siphons* to direct water flow. These form as tubes or folds of the mantle and may protrude as a sort of snorkel above the surface of the sand or mud. Their presence may be marked on the internal surface of the valves by an inflection (*pallial sinus*) in the line of attachment (*pallial line*) between the mantle lobe and the shell. The shells are equivalve, but many are *inequilateral* because of unequal size of the foot and the gill-siphon space. In

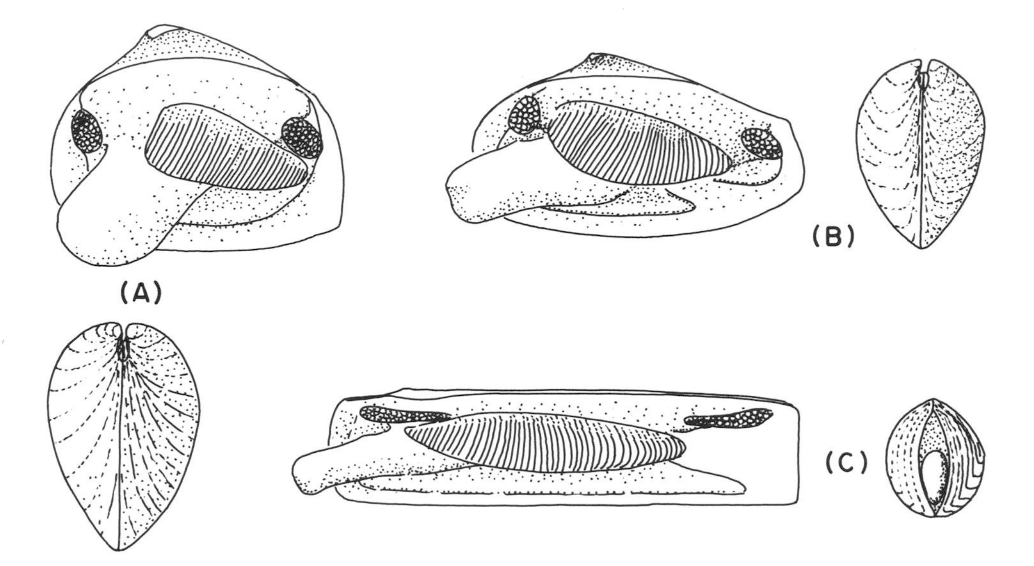

Fig. 15-6. BIVALVE ADAPTATION: CRAWLING AND BURROWING. **(A)** Lateral and anterior views of crawling bivalve, the former with the left valve removed. Equivalvular, equilateral, triangular form. **(B)** Same views of active burrower. Equivalvular, equilateral, quadrilateral. **(C)** Lateral and anterior views of a sluggish burrower. Equivalvular, inequilateral, quadrilateral. Note the gap between closed valves.

active burrowers the foot is large and the valves equilateral. In these the shells may be relatively deep but quadrilateral rather than triangular. The adductor muscles are, of course, nearly equal as are the adductor muscle scars. Other burrowing types live rather sedentary lives at the bottom of tubelike burrows. Their siphons are long, and the foot small. The valves are inequilateral and in many quite shallow. They are characteristically quadrilateral. Since they are protected by their burrows, they lose the ability to open and close their shells. The tooth and socket articulations of the hinge line are reduced in size. The valves gape on the anterior and posterior borders for protrusion of foot and siphon and remain closed on the ventral border to exclude sediment. Commonly, the shell is circular or nearly circular in cross section.

Because clams are protected by shells rather than speed, and because they feed by filtration of organic debris in bottom muds or of suspended particles and organisms from the water, they are predisposed to a sessile life (Figure 15-7). In some, part of the foot forms a fibrous threadlike or rootlike *byssus* that fastens the beast to the bottom or to other sessile organisms. The foot, deprived of its normal function, is greatly reduced. The anterior part of the shell fails to develop, and the byssus remains near the hinge line. With the degeneration of the anterior part of the shell, the anterior adductor is reduced or lost. An earlike projection of the hinge line with a notch below typically marks the presence of a byssus.

Some recent genera become free-living. Those with a reduced muscular foot can no longer crawl over the bottom. Instead, they swim by closing and opening the valves rapidly. This clapping motion forces water out between the mantle margins; the margins are closed along most of the periphery so that the jet is confined to a narrow, unoccluded gap. The organisms control the direction of movement by shifting the location of this gap. The single adductor muscle, the posterior, is very large and composed principally of striated muscles capable of very rapid contraction. Presumably, some similar fossil bivalves had a similar swimming habit.

Many bivalves with a byssal attachment rest on one valve. In adaptation to this change in position, one valve may be more convex than the other, and the shell *inequivalve*. Some cement the "lower" valve to rocks or other shells. Some of these cemented types evolved from those with byssal attachments; others from unattached sessile types that rest on one valve. Members of this latter group evolve heavy, deep "lower" valves to stabilize the animal and maintain it above the depositional surface. An extreme development of this tendency is shown by the coraliform rudistids—paralleling a similar trend in late Paleozoic brachiopods.

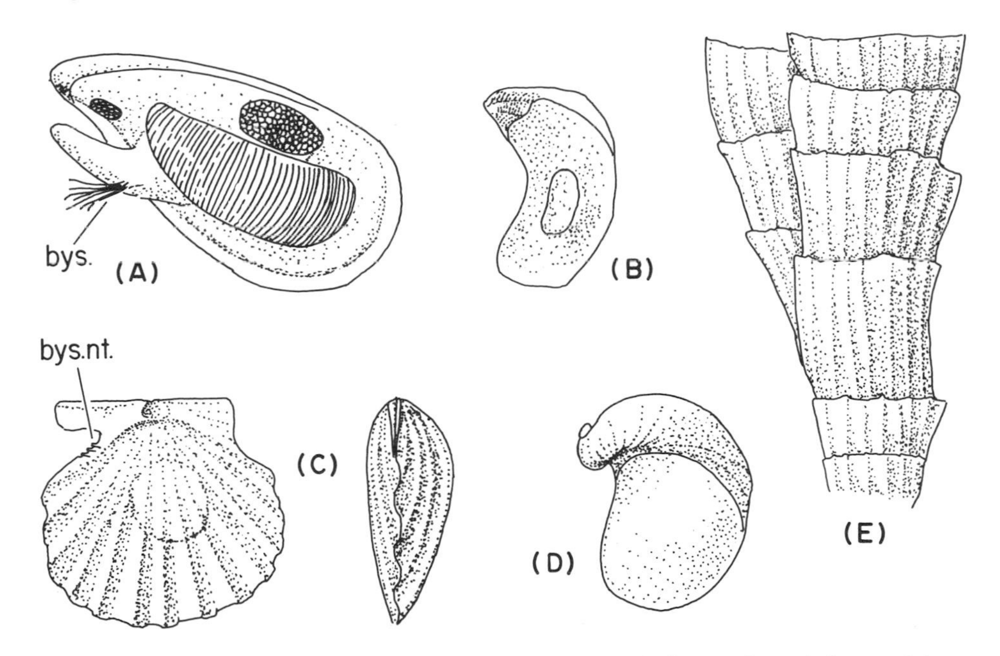

Fig. 15-7. BIVALVE ADAPTATION: SESSILE. **(A)** Partly sessile form with weak foot and byssus for attachment. Lateral view with one valve gone. **(B)** Sessile form. Internal surface of right valve. Irregular shape of valve reflects cementation to substrate. **(C)** Sessile form—attached by byssus. Lateral view of interior of right valve and anterior view. Lateral view demonstrates inequilateral form; anterior view shows inequivalvular form. Some members of this group swim by clapping valves together. **(D)** Sessile. The animal rests on the strongly coiled right valve. The left valve is small and nearly flat. **(E)** Sessile. One valve elongated into coral-like form. Some members of group form aggregates of loosely attached individuals.

Shell characteristics and classification

Since these modifications of shell form are relatively simple changes in an already existing plan and are clearly adaptations to specialized modes of life, they are of questionable value in interpretation of broader phylogenetic relationships. Zoologists, in their classification, give the greatest weight to variations in gill structure. The complexity of the gill makes it less likely that similar types evolved independently in two or more lines. Once a division is made on this basis, other characteristics can be found that accord with the division, and some that disagree. As in most single criterion classifications, there is a risk of mistaking characteristics produced by parallelism and convergence for those inherited from a common ancestor.

Four types of gills are recognized (Figure 15-8). The *protobranch* gill is a simple, leaflike structure similar to that of nonbivalve molluscs. The *filibranch* gill consists of a double row of closely spaced filaments, each row forming a sheet. These are attached to the gill bases and hang downward in the gill cavity, but the free end of each sheet is folded back upward, parallel to the descending sheet. In the *eulamellibranch* gill, adjacent filaments join at intervals to divide the space between the sheets into

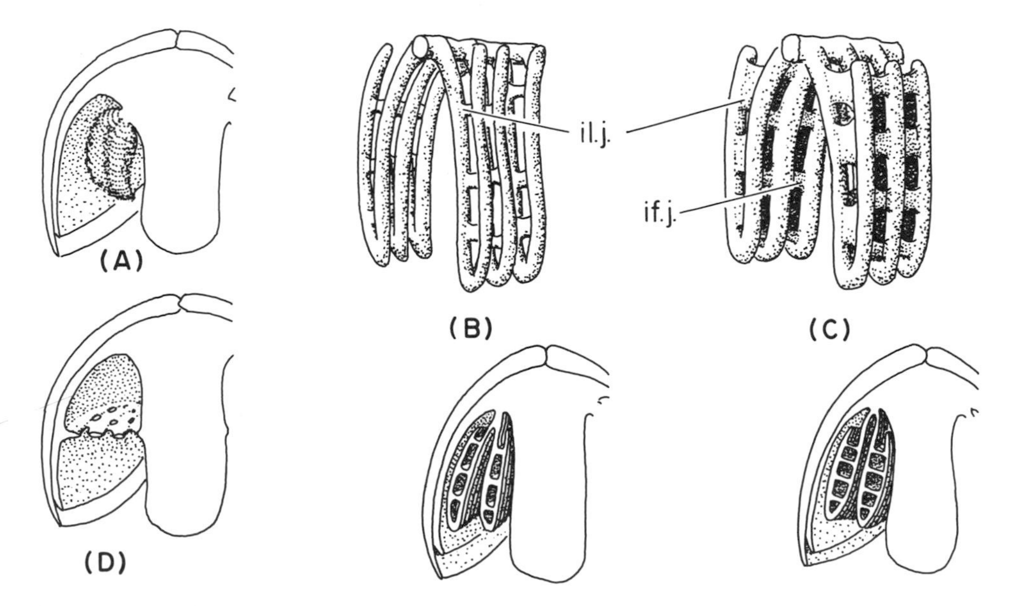

Fig. 15-8. BIVALVE GILL STRUCTURE. **(A)** Simple leaf-shaped gills, protobranch. View in right mantle cavity. **(B)** Filibranch gills, detail and view in right mantle cavity. Each filament is folded, and the descending and ascending portions may be connected (il. j.). **(C)** Eulamellibranch, detail and view in right mantle cavity. Adjacent filaments connected (if. j.) as well as branches of same filament. **(D)** Septibranch gill, view in right mantle cavity. Gill forms a horizontal, perforate partition.

a series of canals. In the *septibranch* type, the gill forms a perforate horizontal partition between the upper and lower parts of the mantle cavity.

Paleontologists naturally would prefer a classification based on shell characteristics. The only one that seems to involve a fundamental change of character is that of shell microstructure (Figure 15-9). Some pelecy-

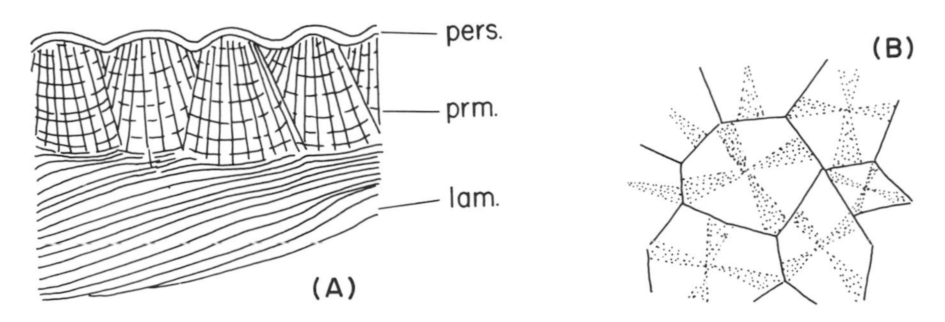

Fig. 15-9. BIVALVE SHELL MICROSTRUCTURE. **(A)** Cross section, thickness 0.5 mm., *Unio* shell. **(B)** Transverse section through prismatic layer. Width of prisms about 40 microns. Abbreviations: lam. laminated layer; pers. periostracum; prm. prismatic layer. (*After* Piveteau, ed., *Traité de Paléontologie*. Copyright, Masson et Cie. Used with permission.)

pods have a thick layer of calcite prisms at right angles to the shell surface and a thinner layer of sheets of calcite or aragonite beneath it. Others lack the prismatic layer. This division forms the two bivalve subclasses Prionodesmacea and Teleodesmacea (Table 15-1). Within these subclasses, many paleontologists utilize variations of the articulation between the valves (Figure 15-10) as taxonomic criteria. It is highly probable that this feature is not of major evolutionary significance but yields rather a key including parallel or convergent lineages. Therefore the orders employed here (Table 15-1) *are not phyletic units.* The Palaeoconcha lack distinct teeth and sockets on the hinge line, and the valves are held in articulation by the ligament above the hinge and by the adductor muscles. Recent genera have simple leaflike gills, the protobranch type.

A second hinge type comprises many small *teeth* (and corresponding *sockets*) along the articular surface, and defines the order Taxodonta. Two suborders are recognized, one with protobranch gills, the other with more complicated filibranch gills. If the filibranch gills evolved but once, all the remaining bivalve orders developed from this suborder of taxodonts.

The Schizodonta have a small number of large hinge teeth which diverge radially from the beak. All have filibranch gills. The Dysodonta typically lack teeth but possess highly specialized ligaments. A portion of the ligament, the resilium, lies inside the hinge line and is compressed when the valves are drawn shut. When the adductor muscles relax, ex-

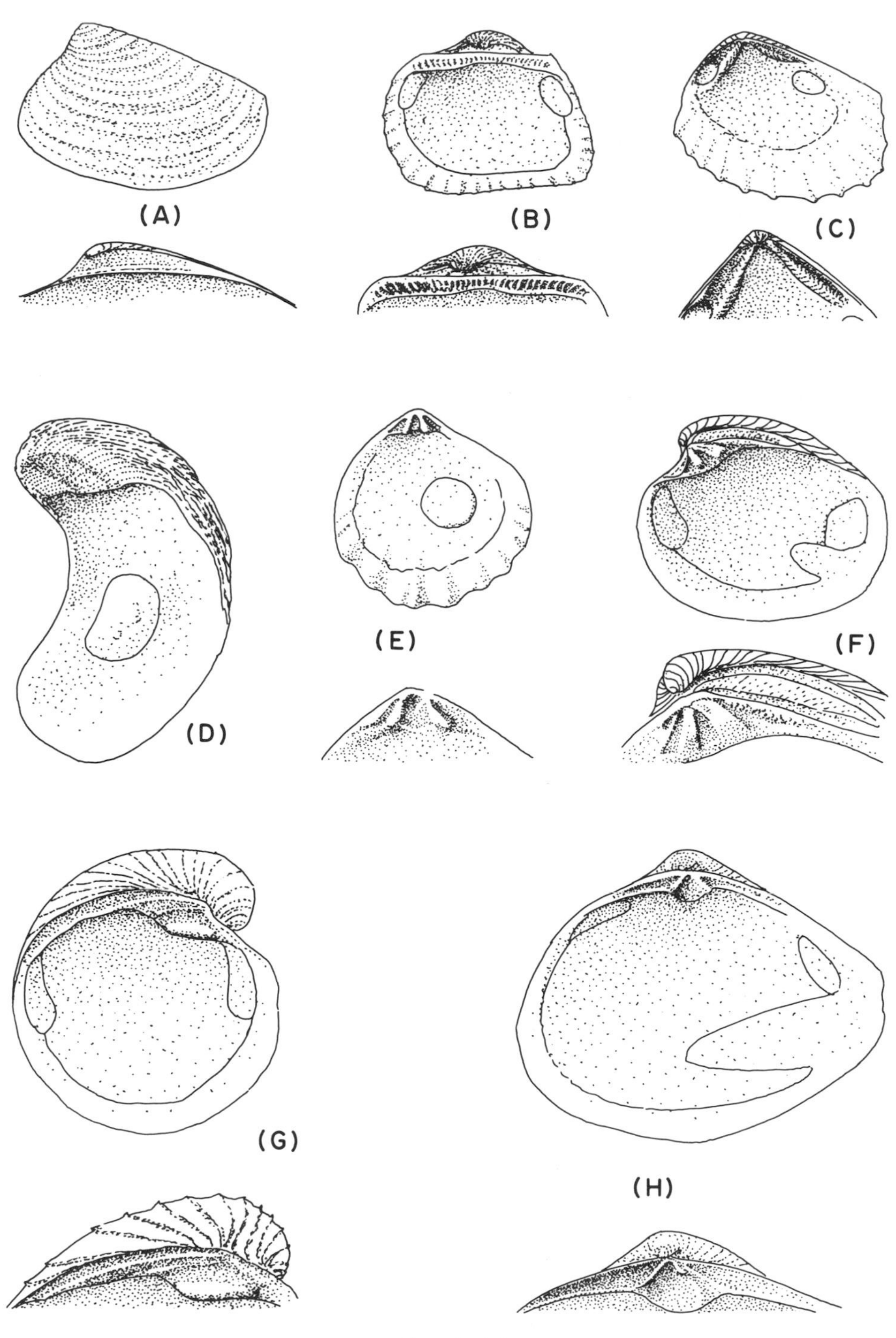

Fig. 15-10.

pansion of the resilium forces the valves open. Most dysodonts are sessile; many have a byssus; all have the anterior part of the shell reduced; and most have lost the anterior adductor. Some dysodonts have filibranch gills, but others, among them the oysters, have the eulamellibranch type. This creates a problem in phylogenetic interpretation, for the nondysodonts with eulamellibranch gills almost certainly did not have dysodont ancestry—their hinge structures can hardly have been derived from the already specialized dysodont pattern. If the dysodonts with more specialized gills had filibranch dysodont ancestors, then the eulamellibranch evolved independently in at least two lines. If this gill structure evolved but once, then dysodont hinge structures evolved independently in two or more lines.

The isodonts, with two subequal hinge teeth, filibranch gills, and a resilium, show considerable similarity to the more primitive dysodonts.

The Heterodonta are characterized by a small number of large hinge teeth, differentiated into two groups. One group, the *cardinals,* lie directly beneath the beak, perpendicular to the hinge line. The *lateral teeth* on either side are nearly parallel to the hinge line. The gill structure is eulamellibranch.

The Pachyodonta have a hinge and gills much like that of the Heterodonta, but their teeth are very massive. Pachyodonts are sessile, lying on one valve. This valve is larger and more convex than the other. In a group of late Mesozoic pachyodonts, the rudistids, the lower valve is pyramidal or conical and the individual grew from the bottom like a coral. In these, the upper valve is reduced to a flat plate covering the open end of the cone.

The Desmodonta are united by their possession of eulamellibranch gills (in one suborder, septibranch) and of an internal ligament borne on the hinge plate or on a partly separated spoonshaped structure. They are primarily burrowers and borers with similar shell adaptations. The hinge teeth are small or absent, but some have a heterodont dentition.

Fig. 15-10. BIVALVE HINGE STRUCTURE AND CLASSIFICATION. **(A)** *Palaeoconcha,* no hinge teeth. *Edmonia;* Devonian to Pennsylvanian; lateral view of left valve exterior and detail of right valve hinge line; length of upper figure 3.7 cm. **(B)** *Taxodonta,* taxodont dentition. *Arca;* Jurassic to Recent; lateral view of right valve interior and detail of hinge line; length of upper figure 3.6 cm. **(C)** *Schizodonta,* schizodont dentition. *Neotrigonia;* Miocene to Recent; interior of right valve and enlarged view of hinge line; length of shell 1.9 cm. **(D)** *Dysodonta,* dysodont hinge structure; single adductor muscle. *Ostrea;* middle Cretaceous to Recent; right valve interior; length 8.3 cm. **(E)** *Isodonta,* isodont dentition; single adductor muscle. *Plicatula;* Triassic to Recent; view of the interior of right valve and detail of dentition; length 2.3 cm. **(F)** *Heterodonta,* heterodont dentition. *Venus;* Jurassic to Recent; right valve interior and dental detail; length 10.9 cm. **(G)** *Pachyodonta,* pachyodont dentition. *Chama;* late Cretaceous to Recent; interior of left valve and detail of hinge area; length 4.2 cm. **(H)** *Desmodonta,* desmodont dentition. *Lutraria;* Miocene to Recent; interior of right valve and detail of dental structure; length of shell 6.8 cm.

In geologic perspective

As I have already indicated, the bivalves are almost entirely benthonic. They are primarily marine, though some inhabit fresh waters. Their adaptations of shell form for different niches are easily recognized in most cases, though some with intermediate characteristics are not as easily diagnosed. The scanty evidence now available (see bibliographies in the *Treatise on Marine Ecology and Paleoecology,* 1957) indicates the association of particular species and genera with particular depositional environments. Stenzel was able to associate local variation within a species population of oysters with environmental differences (see p. 89). The occurrence of recent genera in Mesozoic and Cenozoic rocks permits interpretation of their paleoecology in terms of the limiting factors recognized in recent ecologic studies (e.g., Addicott, 1966).

At present, pelecypods are useful in defining the large units of geologic time, but are with some exceptions of little help in the correlation of finer subdivisions. This deficiency is in part inherent, because their evolution, as reflected in shell characteristics, was quite slow, and most common genera and species have long stratigraphic ranges. Very detailed analyses of large samples collected at carefully determined stratigraphic levels show, however, a gradual transformation of a population succession. Thus, the work of Stenzel on *Ostrea* and of Newell (1942) on *Myalina* not only illustrates evolutionary principles, but provides a practical guide to dating important rock sequences.

REFERENCES

Addicott, W. O. 1966. "Late Pleistocene Marine Paleoecology and Zoogeography in Central California," *United States Geol. Sur., Prof. Paper,* 523-C.

Burnaby, T. P. 1965. "Reversed Coiling Trend in *Gryphaea arcuata,*" *Geological Jour.* vol. 4, pp. 257-278. Most recent in series of papers by various authors on evolution of a Cretaceous pelecypod.

Dodd, J. R. 1964. "Environmentally Controlled Variation in the Shell Structure of a Pelecypod Species," *Jour. of Paleontology,* vol. 38, pp. 1065-1071.

Fisher, D. W. 1962. "Small Conchoidal Shells of Uncertain Affinities," in Moore, R. C. (Ed.) *Treatise on Invertebrate Paleontology, Pt. W.,* pp. 91-141. New York: Geological Society of America.

Gould, S. J. 1966. "Allometry in Pleistocene Land Snails from Bermuda: The Influence of Size upon Shape," *Jour. of Paleontology,* vol. 40, pp. 1131-1141.

Ladd, H. S. (Ed.) 1957. "Treatise on Marine Ecology and Paleoecology," vol. 2, Geol. *Soc. of America, Memoir 67.* See pp. 817-893 on the molluscans.

Lemche, H. 1957. "A new living deep-sea mollusc of the Cambro-devonian Class, Monoplacophora," *Nature,* vol. 179, pp. 413-416.

March, L. and E. L. Yochelson. 1964. "Paleozoic Mollusk: *Hyolithes*," *Science*, vol. 146, pp. 1674-1675.

McAlester, A. L. 1965. "Systematics, Affinities, and Life Habits of *Babinka*, a Transitional Ordovician Lucinoid Bivalve," *Palaeontology*, vol. 8, pp. 231-246.

Moore, R. C. (Ed.) 1960. *Treatise on Invertebrate Paleontology, Pt. I, Mollusca 1*. New York: Geological Society of America. A general review of the Mollusca plus sections on scaphopods, chitons, monoplacophorans, and some gastropods.

Morton, J. E. and C. M. Yonge. 1964. "Classification and Structure of the Mollusca," in Wilbur, K. M. and C. M. Yonge, *Physiology of Mollusca, Vol. 1*, pp. 1-58. Book also includes sections on growth, shell development, locomotion, etc.

Newell, N. D. 1942. "Late Paleozoic Pelecypods, Mytilacea," *Kansas Geol. Surv.*, vol. 10, pp. 1-123.

Nicol, D. 1953. "Period of Existence of Some Late Cenozoic Pelecypods," *Jour. of Paleontology*, vol. 27, pp. 706-707.

Rodda, P. U. and W. L. Fisher. 1964. "Evolutionary Features of *Athleta* (Eocene, Gastropoda) from the Gulf Coastal Plain." *Evolution*, vol. 18, pp. 235-244.

Yochelson, E. L. 1966. "Mattheva, a Proposed New Class of Mollusks," *United States Geol. Sur., Prof. Paper*, 523-B.

16 To correlate: The Cephalopoda

If obscure generals of long-forgotten wars are honored by outsize bronze statues, how much more deserving are the cephalopods—not only for their intrinsic merit but also for their influence on stratigraphic paleontology.

William Smith first developed the principle of faunal correlation from the Jurassic sequence of Britain, faunas dominated in number and variety by ammonite cephalopods. D'Orbigny introduced the concept of stages in subdividing the Jurassic of western Europe, stages characterized by ammonites. Oppel developed the idea of faunal zones in these same rocks, zones indexed by ammonite species.

To succeed as a guide fossil

The cephalopods, particularly the ammonites, possess unusual qualifications to justify their fame. First, they evolved rapidly within a lineage and split into a large number of distinct lines. Second, many of them were pelagic swimmers with wide ecological tolerance. As a consequence, individual species have a wide distribution and occur in many different sedimentary environments. Third, they are relatively easy to identify, many of them from external characteristics. Fourth, they are relatively common fossils. Finally, some phylogenetic sequences can be established with a degree of certainty.

The ideal guide fossils would come from populations that were ubiquitous on the earth's surface, were abundant, were easily fossilized, and evolved at such a speed that successive generations could be distinguished. The cephalopods come as close as any group of organisms to that ideal.

TABLE 16-1

The Cephalopoda (*after Treatise on Invertebrate Paleontology*)

Late Cambrian to Recent. Tentacles. Body bent so that anus lies below head; viscera enclosed by conical, muscular mantle. Locomotion by expulsion of water through ventral hyponome. Multi-chambered shell—rudimentary in some. Marine, vagrant benthos and nekton; predaceous.

Subclass Nautiloidea
Late Cambrian to Recent. Calcareous shells—straight, curved or coiled. Siphuncle small to moderately large, variable in position. Septa plane or gently curved. Siphuncular deposits do not consist of calcareous conical sheaths nor do they contain specialized canals, etc.

Subclass Endoceratoidea
Early Ordovician to Silurian. Orthoconic to, rarely, cyrtoconic. Siphuncles large, generally marginal. Endosiphuncular deposits consist, typically, of conical sheaths.

Subclass Actinoceratoidea
Middle Ordovician to Pennsylvanian. Typically orthoconic; cameral deposits; siphuncular deposits contained radial canals, etc.

Subclass Bactritoidea
Ordovician to Permian. Orthoconic or weakly cyrtoconic. Septa concave anteriorly; no siphuncular or cameral deposits; marginal siphuncle. Sutures simple with at least one small, shallow V-shaped ventral lobe.

Subclass Ammonoidea
Devonian to Cretaceous. Typically coiled; marginal siphuncle. No siphuncular or cameral deposits. Sutures with several lobes and saddles—typically with secondary lobes and saddles.

Subclass Coleoidea
Mississippian to Recent. Shell reduced and enclosed in mantle or absent. Only the belemnoids with heavy calcareous rostrum are common fossils.

The anatomy of success

Recent cephalopods, the squids, octopus, and nautilus, are all active predators, either of the marine nekton or of the benthos. What is known of fossil cephalopods suggests that most, if not all, had a similar way of life. Unfortunately for the paleontologist who would interpret cephalopod functions, one subclass of cephalopods is represented by only one recent genus, four have no recent representatives, and the sixth has few fossil representatives.

The structure (Figure 16-1) of the nautilus is closest to that of the majority of fossil cephalopods, and so can introduce us to cephalopod morphology. The posterior end of the cephalopods has been rotated to lie below the anterior parts. The head bears tentacles arranged about the mouth and large eyes. The major nerve ganglia of the head are of a size

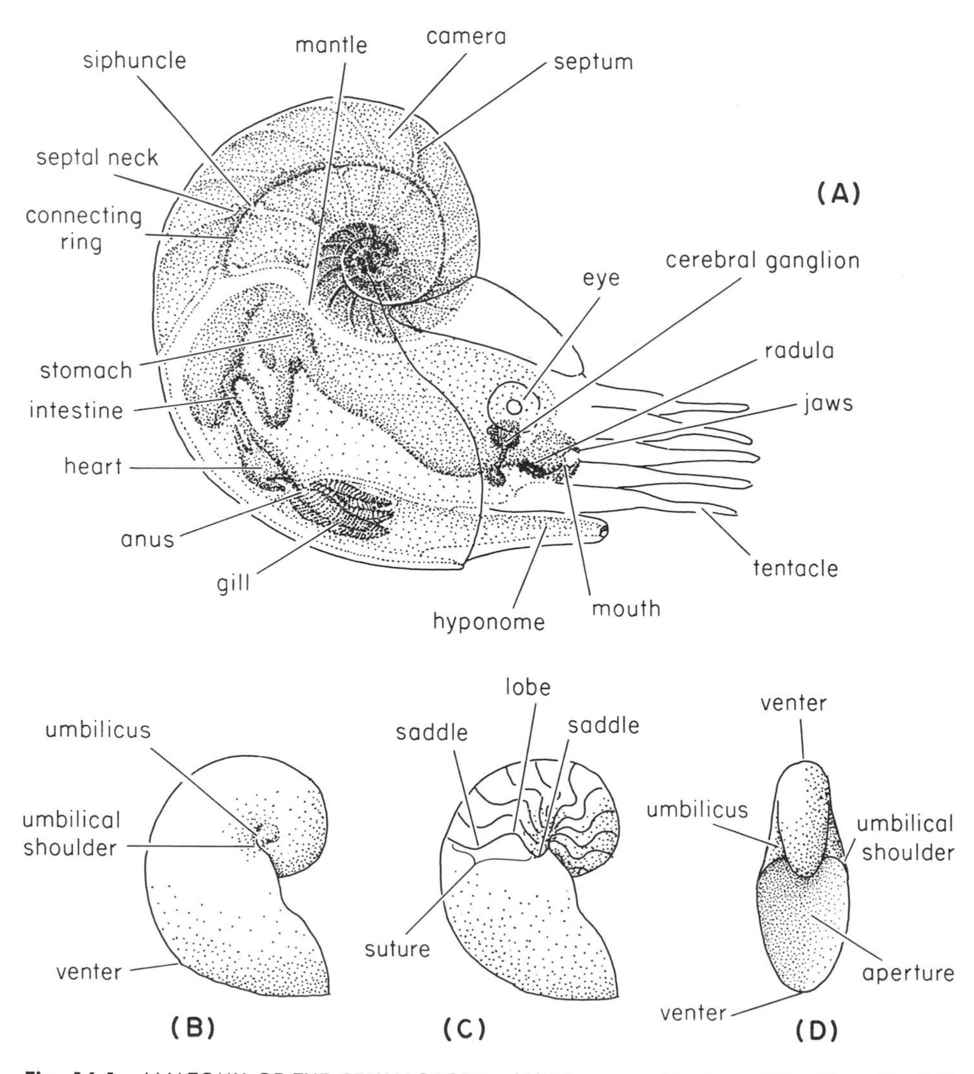

Fig. 16-1. ANATOMY OF THE CEPHALOPODA. **(A)** Diagrammatic view of *Nautilus* with shell cleared to show internal structure. **(B)** Lateral view of *Nautilus* shell. **(C)** Lateral view of fossil *Nautilus* shell. The outer layers of shell material have been removed to show the sutures, i.e., the line of intersection of shell and septum. **(D)** Apertural view of *Nautilus* shell.

TABLE 16-2

Glossary for the Cephalopoda (*figures in parentheses refer to pertinent illustration*)

Advolute. Type of coiled shell in which the outer whorl touches but does not cover any part of the adjacent inner whorls. (16-2I; 16-4D, E.)

Ammonitic suture. Type of suture characterized by complex fluting. Smaller secondary and tertiary lobes and saddles developed on larger primary set. (16-6C.)

Anus. In all known cephalopods the viscera are doubled back so that anus lies below head. (16-1A.)

Aperture. External opening of living chamber from which the head and tentacles are extended. (16-1D; *et al.*)

Ascocone. Type of shell. Earlier portion is slender and curved; later short and wide with chambers above the living chamber. Initial part of shell may be detached. (16-2F; 16-4B.)

Beak. Pair of horny jaws on either side of mouth. (16-1A.)

Brevicone. Shell type with short, blunt form and rapid taper from wide living chamber to initial chamber. Typically curved. (16-2E; 16-4A.)

Camera. Chamber in shell representing a portion of an earlier living chamber now closed off by a septum. (16-1A; 16-2C to G; *et al.*)

Cameral deposits. Calcareous deposits on septa and/or walls of camera. (16-3E. diagonal lined area)

Ceratite suture. Type of suture characterized by presence of small lobes and saddles on major lobes. (16-6D.)

Cerebral ganglion. See Table 15-1. (16-1A.)

Chamber. See camera.

Conispiral. Shell form characterized by a spiral coiled on a cone—whorls not in a single plane. (16-2P; 16-7G, H.)

Connecting ring. Calcareous ring forming wall of siphuncle between septa. (16-1A.)

Convolute. Type of coiled shell in which part of outer whorl extends in toward center of coil and covers inner whorls. (16-1B; 16-2K; 16-5D; 16-7D.)

Crescentic cross-section. Refers to crescent shape of whorl cross-section in some coiled cephalopods. (16-2L, M.)

Cyrtocone. Type of shell. Slender, curved cone. (16-2C.)

Dibranchiate. Refers to cephalopods bearing a single pair of gills. Includes all recent cephalopods except *Nautilus.* (16-8B.)

Endocones. Conical calcareous deposits inside siphuncle. Apices of cones point toward shell apex. (16-3D.)

Goniatitic suture. Type of suture characterized by simple fluting consisting of single series of lobes and saddles. (See ceratitic and ammonitic.) (16-6E.)

Gyrocone. Type of coiled shell in which adjacent whorls do not touch each other. (16-2H.)

Heteromorph. Ammonoid shell type. Includes those forms which are not planospiral and/or those in which the walls of the coil are not in contact.

TABLE 16-2 (continued)

Hood. Tough fleshy structure that lies above head of *Nautilus* and covers aperture when head is withdrawn into living chamber. In some fossil cephalopods this structure or one similar bore a pair of calcareous plates called the aptychus. (16-1A.)

Hyponome. Muscular tube just below head which extends externally from mantle cavity. Serves to confine and direct jet of water forced from that cavity. (16-1A; 16-8B.)

Involute. Type of coiled shell in which part of outer whorl extends in toward center of coil and covers part of adjacent inner whorl. (16-2J; 16-5A, B; *et al.*)

Lituiticone. Type of shell which has a coiled initial portion and a straight mature portion. (16-2G.)

Living chamber. Chamber in which soft parts of animal are housed. The last-formed chamber, it is bounded at the back by a septum and opens at the front through the aperture. (16-1A; *et al.*)

Lobe. A flexure of the suture line away from the aperture (toward the apex). (16-1C; 16-6.)

Mantle. See Table 15-1. Fold of the body wall enclosing the viscera and forming a sac or cone about them. (16-1A.)

Orthocone. Shell type consisting of a slender straight cone. (16-2D; 16-3C, D.)

Phragmocone (phrag). Portion of shell consisting of the camerae. (16-1A.)

Planispiral. Type of coiled shell in which whorls lie in a single plane—that one of bilateral symmetry. (16-1D; 16-2L through O; *et al.*)

Proostracum. Calcareous blade projecting anteriorly from dorsal border of belemnoid phragmocone. (16-8A, B.)

Protoconch. Initial chamber of shell—located at apex of cone or at center of coil. (16-1A.)

Quadrate cross-section. Shell form characterized by quadrate whorl cross-section. (16-2N.)

Rostrum. Structure occurring in belemnites. A thick conical or subconical calcareous deposit enclosing the phragmocone. (16-8A, B.)

Saddle. Flexure of suture line toward aperture. (16-1C; 16-6.)

Septal neck. Flexure of septum along siphuncle forming a short tube or funnel. (16-1A; 16-3C, E; *et al.*)

Septum. Calcareous partition transverse to walls of shell and separating camerae.

Siphonal deposits. Calcareous deposits along siphuncle. In some nautiloids of considerable thickness. (16-3E; 16-4A.)

Siphuncle. Tube extending from back of living chamber through septa to protoconch. (16-1A; 16-3B, C, D, E; *et al.*)

Subcircular cross-section. Shell form characterized by subcircular whorl cross-section. (16-2P; 16-7G, H; *et al.*)

Suture. Line of intersection between septum and inner surface of shell wall. (16-1C; 16-6; *et al.*)

Tentacle. One of several slender arm-like appendages that surround the mouth and extend anteriorly in front of head. (16-1A; 16-8B.)

TABLE 16-2 (continued)

Tetrabranchiate. Cephalopod bearing four gills—two pairs—in mantle cavity. Known certainly only in *Nautilus.* (16-1A.)

Umbilical perforation. Opening between inner sides of innermost whorl in position of axis of coiling. (16-1A.)

Umbilical plug. Calcareous deposit filling umbilicus. (16-1B.)

Umbilical shoulder. Portion of shell bordering umbilicus and forming its outer margin. (16-1C, D; 16-5A, B; 16-7A to F; *et al.*)

Umbilicus. Depression (or in some an opening) in axis of coiling formed by diminishing width of whorls toward axis.

Venter. Portion of whorl farthest from axis of coiling. (16-1B, D; *et al.*)

V-shaped cross-section. Type of shell form characterized by v-shaped whorl cross-section. (16-2O.)

to deserve the name of brain. The heart, too, is large, and the circulatory system efficient. The mantle forms a cone about the viscera. Below the head, the mantle is rolled into a tube, the *hyponome,* extending into the mantle cavity and opening externally. Within the mantle cavity lie two pairs of gills.

The mantle, or part of it, secretes the shell. Because calcium carbonate is deposited more rapidly on the ventral and lateral sides of the mantle than on the dorsal, the shell develops in a spiral—one coiled in a single plane, that of bilateral symmetry. The axis of coiling is above the visceral mass. The mantle deposits successive partitions (*septa*) across the shell, and the lines of juncture of the septa and inner layers of the shell form distinct *sutures.* The series of gas-filled chambers (*camerae*) forms the *phragmocone.* The shell thus comprises a phragmocone, i.e., all the camerae, and a *body chamber* housing the visceral mass. A slender strand of tissue passes back from the visceral hump through the septa. The surface of each septum turns back sharply to enclose this strand in a short calcareous tube, the *septal neck.* These are linked by delicate *connecting rings,* which form with the septal necks a continuous tube, the *siphuncle.*

The nautilus feeds on the bottom. The tentacles are used to seize and hold the food. It propels itself through the water with a water jet (Figure 16-2)—generated when water, which enters the mantle cavity on either side of the head, is forced out of the hyponome by rhythmic muscular contractions.

Evolutionary trends

The knowledge of fossil cephalopods is confined almost exclusively to the structure and form of their shells (Figure 16-2). The nautilus shell,

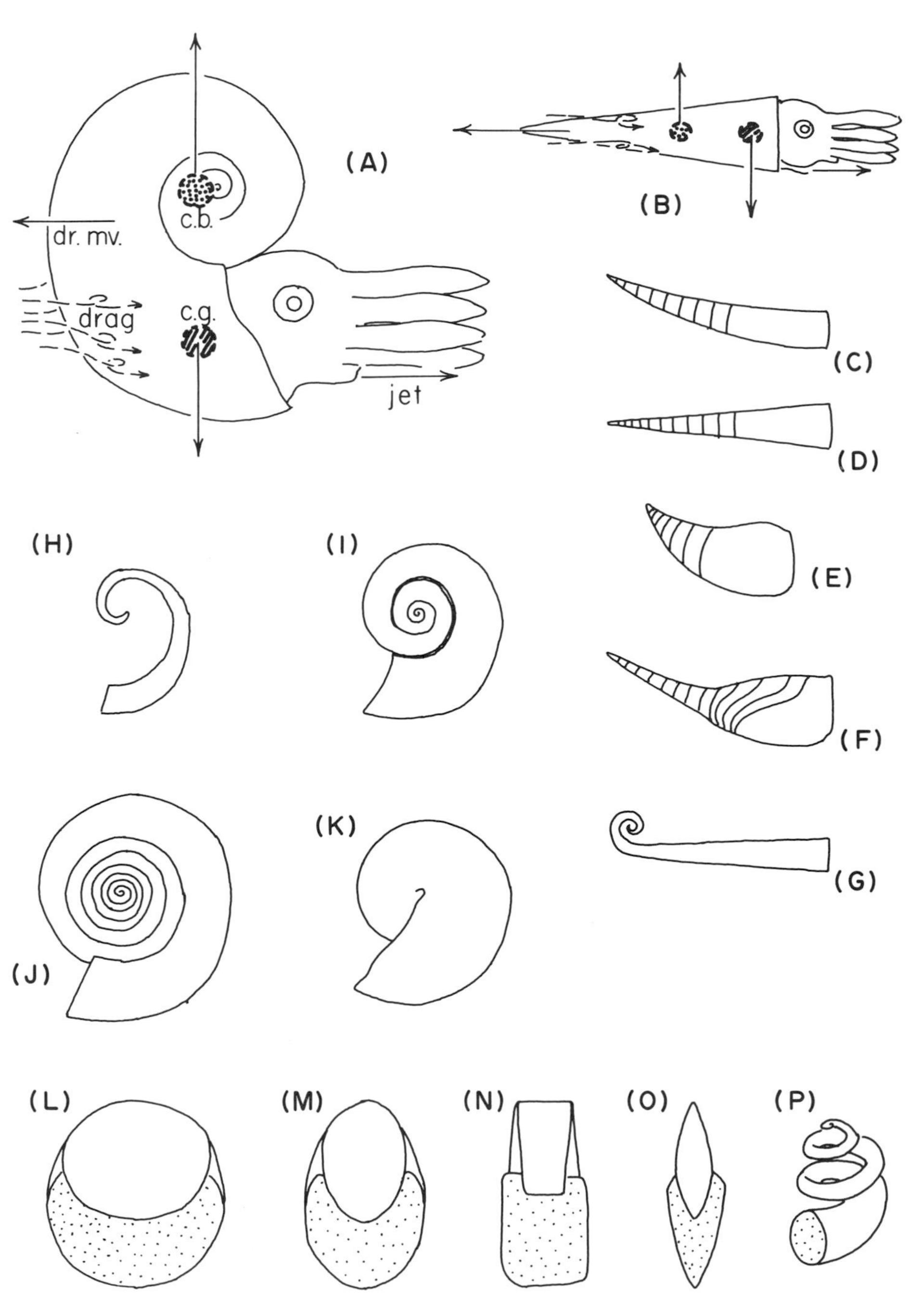

Fig. 16-2.

of course, serves as protection, but, more important, it buoys up the animal because of its gas-filled chambers.* The shape of the shell, its streamlining, determines the amount of drag. An ornate shell would interfere with locomotion, particularly swimming. Too light a shell would rupture under rapid pressure changes. If the center of buoyancy were far behind the animal's center of gravity, it would turn the animal so that the head, tentacles, and funnel would be down, and the shell up. Locomotion, either crawling or swimming, would hardly be assisted by this awkward position. Knowing these problems in adaptation, one could predict many of the evolutionary trends which show in the differentiation of the subclasses and orders and within the particular evolutionary lineages.

Subclasses Nautiloidea, Actinoceratoidea, Endoceratoidea

The oldest cephalopods are tiny (25 mm. or less) straight or slightly curved shells from upper Cambrian rocks. The shells are compressed from side to side and divided internally by closely-spaced septa. The siphuncle is large and in a ventral position; the connecting rings are thickly calcified. The abundance of septa and size of the siphuncle suggest that the shell had negative buoyancy and that these animals were benthonic. The late Cambrian forms are the earliest representatives of the subclass Nautiloidea and its basal order, the Ellesmerocerida; they form satisfactory ancestors for the Ordovician lineages. In many later ellesmerocerids the ballast of connecting rings and septa is somewhat reduced.

By middle Ordovician, the first radiation of the cephalopods (Figures 16-3, 16-4) was complete, with seven orders of the subclass Nautiloidea, two of the subclass Endoceratoidea, and one of the subclass Actinoceratoidea. The differentiation was based primarily on different methods of

* The lack of information on the living nautilus hinders functional interpretation of fossil shell form. Apparently some fluid is present in the camerae, and, probably, this fluid is added or removed through the siphuncle. In the cuttlefish, with a similar structural pattern, the camerae contain both gas and fluid; the fluid is pumped—apparently by osmotic gradients—into and out of these spaces to raise or lower the total density. It seems probable that the camerae of some fossil cephalopods contained liquid.

Fig. 16-2. DYNAMICS OF THE CEPHALOPOD SHELL AND VARIATION IN FORM. **(A)** Coiled type. The jet from the hyponome drives the animal through the water (direction of movement, dr. mv.). The buoyancy of the empty chambers (center of buoyancy, c. b.) lifts the animal; gravity pulls it downward (center of gravity, c.g.). Drag of the water around the shell opposes its movement. **(B)** Straight shell type. The difference in location of the center of gravity (forward in the living chamber) and of buoyancy (back in the empty shell chambers) produces a rotational couple that must be opposed by the animal. **(C)** Cyrtocone shell. **(D)** Orthocone. **(E)** Brevicone. **(F)** Ascocone. **(G)** Lituiticone. **(H)** Gyrocone. **(I)** Advolute. **(J)** Involute. **(K)** Convolute. **(L)** through **(O)** apertural views showing cross-sectional shape of shell. **(L)** Subcircular. **(M)** Flattened ovoid. **(N)** Rectangular. **(O)** Discoidal. **(P)** Conispiral shell.

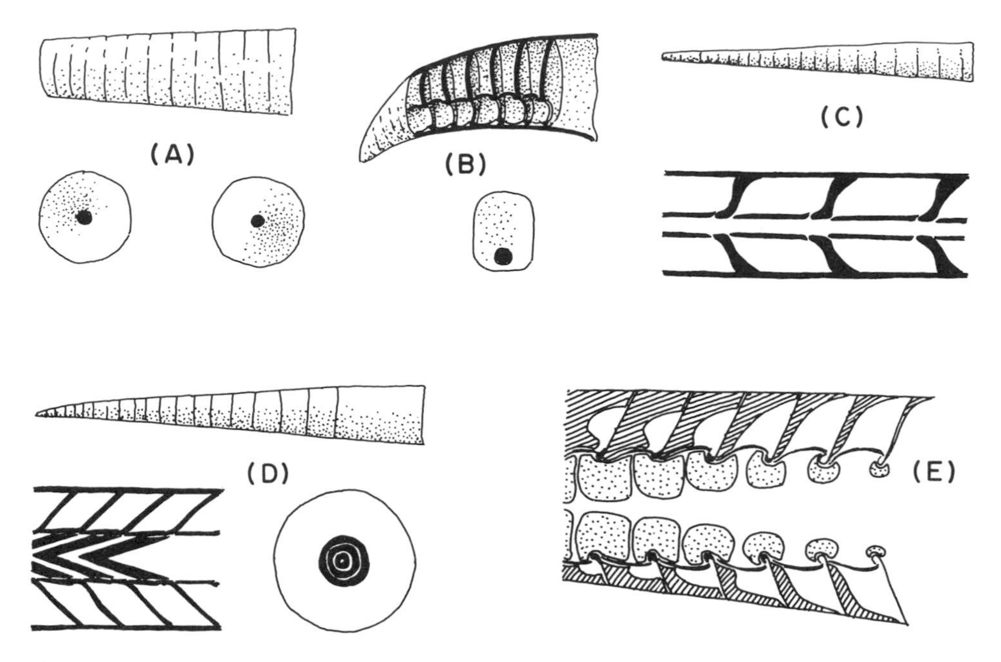

Fig. 16-3. EARLY CEPHALOPODS. **(A)** *Volborthella* assigned by some to the cephalopods; early and middle Cambrian; lateral view at top, apertural view of septal face at lower left, apical view at lower right; length 2.8 mm. **(B)** Ellesmeroceroid *Plectronoceras;* late Cambrian; longitudinal section above, apertural view below; length of specimen 6 mm. **(C)** Michelinoceroid *Michelinoceras;* middle Silurian to Triassic(?); lateral view above and diagrammatic cross-section below; length 6.2 cm. **(D)** Endocerid *Endoceras;* Ordovician; lateral view above, cross-section at right, longitudinal section at left shows endocones in longitudinal section. **(E)** Actinocerid *Actinoceras;* Ordovician; diagrammatic longitudinal section. Siphonal deposits of calcite stippled, cameral deposits lined. [(**A**) and (**B**) *after* Schindewolf; (**C**) *after* Barrande; (**D**) and (**E**) *after* Flower.]

compensating for torque induced by phragmocone buoyancy. The endoceratoids, the actinoceratoids, and, among the Nautiloidea, the discosorids and some oncocerids counter-balanced the shell with calcareous deposits of various types within the siphuncle. The nautiloid orthocerids developed calcareous deposits on the cameral walls as well as in the siphuncle. All of these adaptations brought the center of gravity back toward the center of buoyancy at the expense of reduced or negative buoyancy and increased average density. In all, it seems probable that these were largely animals of the benthos, swimming a short distance above the bottom and resting on it between bursts of activity. The situation resembles that in early fishes and in .primitive birds with well-established and permanently fixed hydrodynamic—or aerodynamic—stability and, consequently, little maneuverability.

The majority of oncocerids and many discosorids adopted a somewhat different compensator. The immature part of the shell is relatively short and the living chamber occupies most of the shell. Commonly the shell is

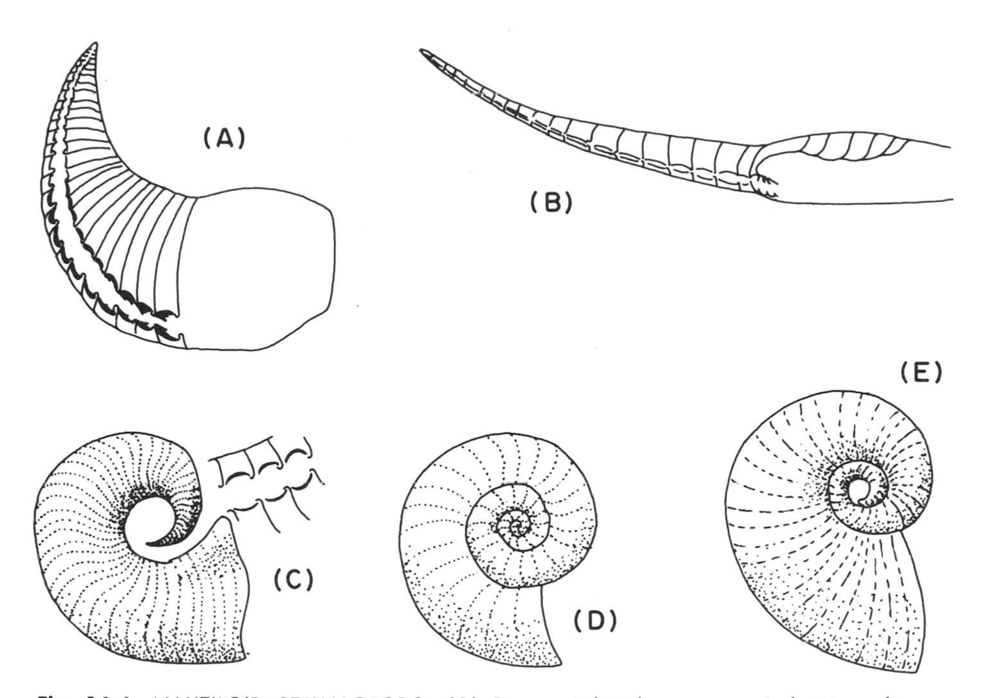

Fig. 16-4. NAUTILOID CEPHALOPODS. **(A)** Oncoceroid *Valcuoroceras;* Ordovician; diagrammatic longitudinal section. Siphonal deposits shown in black. **(B)** Ascoceroid *Ascoceras;* Silurian; diagrammatic longitudinal section. The slender conical portion of the shell broke off the mature portion at maturity. **(C)** Discosorid *Phragmoceras;* Silurian; lateral view and section of siphuncle; greatest diameter about 15 cm. **(D)** Tarphyceroid *Tarphyceras;* Ordovician; lateral view. **(E)** Barrandeoceratid *Barrandeoceras;* Ordovician to Devonian; lateral view. [**(A)** and **(B)** *after* Flower; **(C)** *after* Zittel.]

also curved so that the center of buoyancy is brought forward toward the head. Again stability is obtained but again at a price of reduced buoyancy, speed, and maneuverability.

Another nautiloid order, the Ascocerida, pursued still another adaptive path. The phragmocone, which has very thin walls, consists of a slender, straight or slightly curved immature portion—very like the ellesmerocerid shell—and a stubby mature portion in which the camerae extend forward above the living chamber. Typically, the early straight part of the shell broke free and was lost. The result is a short, streamlined shell of rather low buoyancy but with the center of buoyancy directly above the center of gravity. The arrangement is similar in its essentials to that of the recent cuttlefish and suggests similar habits.

The fourth adaptive pattern occurs in the remaining nautiloids, the Barrandeocerida, the Tarphycerida, and, in Silurian time, the Nautilida proper (Figure 16-5). In these, the shell coiled strongly so that the axis of coiling, and thus the center of buoyancy, came to lie above the living chamber. There was no longer any need to ballast the shell, so this trans-

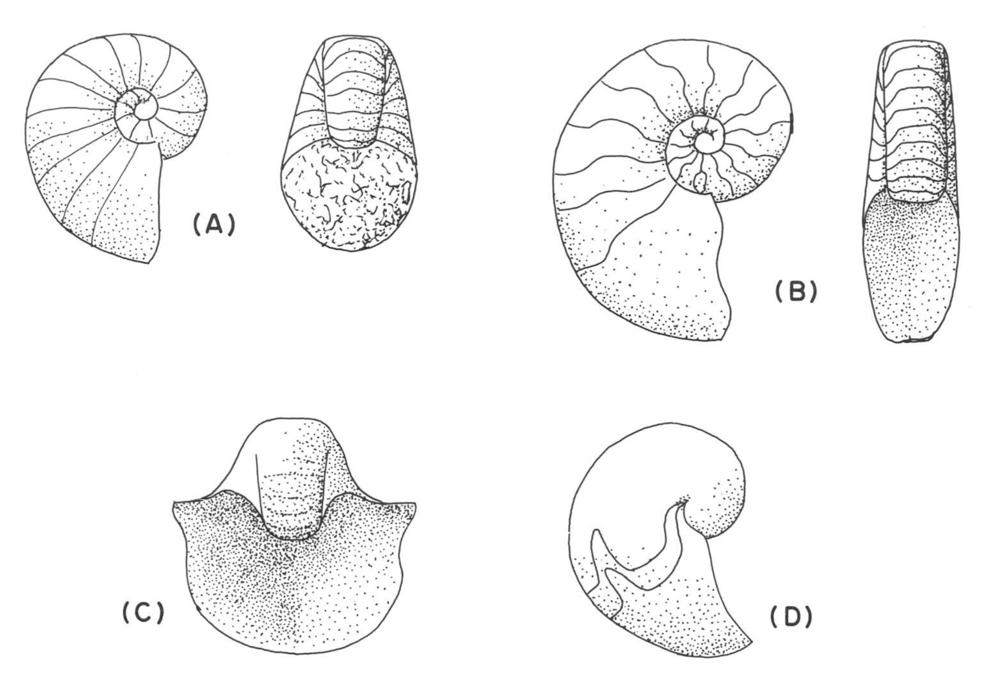

Fig. 16-5. NAUTILOID CEPHALOPODS. **(A)** Nautilid *Endolobus;* Mississippian to Permian; lateral and apertural views of internal mold. **(B)** Nautilid *Domatoceras;* Pennsylvanian to Permian; lateral and apertural views. **(C)** Nautilid *Solenocheilus;* Mississippian to Permian; apertural view. **(D)** Nautilid *Aturia;* Paleocene to Miocene; lateral view; maximum diameter 19 cm. [**(B)** and **(C)** *after* Miller; **(D)** *after* Miller and Downs.]

formation brought a maximum of stability with a minimum of weight. Individual genera vary in shape—some "fat" and others laterally compressed—and in tightness of coiling—some loosely coiled and others so tightly wound that the outer whorl hides the inner ones. Presumably these changes were adaptive to different ecologic niches, and some for better streamlining and greater stability.

Subclass Ammonoidea

Among the coiled Devonian cephalopods, several genera have a thin siphuncle placed near the ventral border of the shell and also have curved septa (Figure 16-6). The central part of their septa is flat or gently concave, but the margins are bent into several low folds. The suture, following the folds of the septum, curves also and traces a wavy line. These few genera are the earliest representatives of a new subclass, the Ammonoidea, although to a Devonian observer they would have been merely another twig in the nautiloid tree.

The ammonoids may derive from the orthocerid nautiloids through

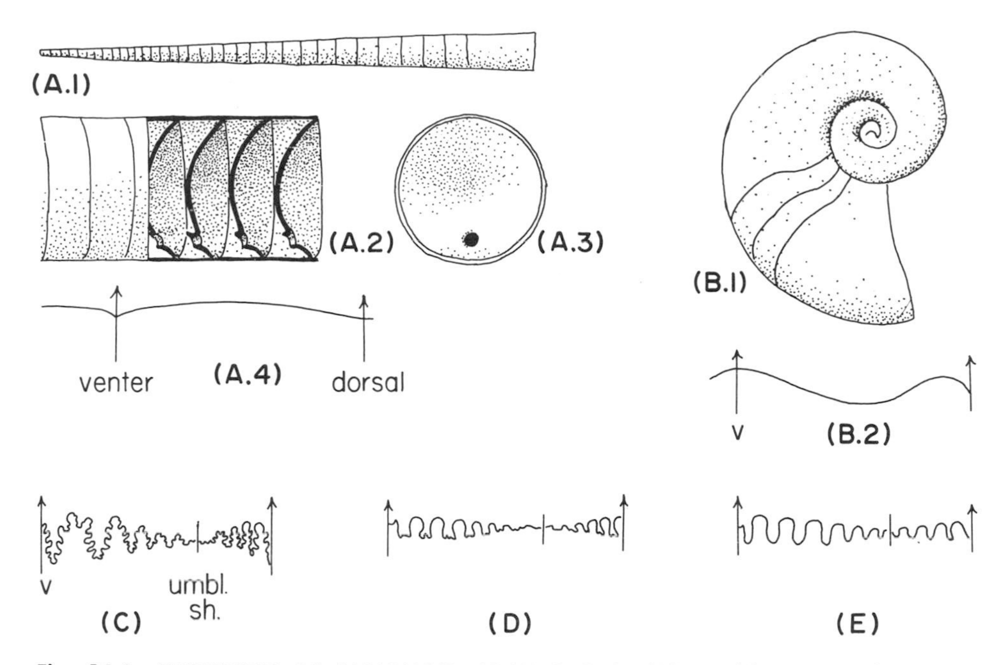

Fig. 16-6. EVOLUTION OF AMMONOID CEPHALOPODS. **(A)** Possible ammonoid ancestor, *Bactrites.* **(A.1)** Lateral view of shell. **(A.2)** Partial longitudinal section. **(A.3)** Septum—apertural face. **(A.4)** Trace of suture. Arrow points in direction of aperture. **(B)** Early and presumably primitive ammonoid, *Gyroceratites.* **(B.1)** Lateral view. **(B.2)** Trace of suture. **(C)** Complex suture—ammonite type—with primary saddles and lobes both subdivided. **(D)** Moderately complex suture of ceratite type. Lobes are serrate, but saddles are not. **(E)** Simple type, goniatite, in which neither saddles nor lobes are subdivided. Abbreviation: umbl. sh., umbilical shoulder —marked on suture by straight line without arrow. [**(B)** *after* Miller and Furnish.]

the Bactrididae although many specialists disagree. These are straight or curved shells with a thin ventral siphuncle and broadly curved suture line. They are essentially intermediate between the nautiloids and the earliest ammonoids in these characteristics, and, as you might expect, their classification is disputed. Are they an order of the subclass Ammonoidea? Or a suborder of the Orthocerida? Or representatives of a separate subclass, the Bactritoidea? All of these schemes are logically correct, but I have followed the *Treatise* here and placed them in a separate subclass.

In the ammonoids *senu strictu,* the warping of the septa and the accompanying inflection of suture line results in the formation of a series of curves toward (the *saddles*) and away from (the *lobes*) the open end of the shell. The uncomplicated version is called goniatitic, but as early as the Mississippian some ammonoids had evolved a more elaborate *ceratitic* suture distinguished by secondary folds (Figure 16-7). A third and still more complicated suture, the *ammonitic,* appeared in Permian ammo-

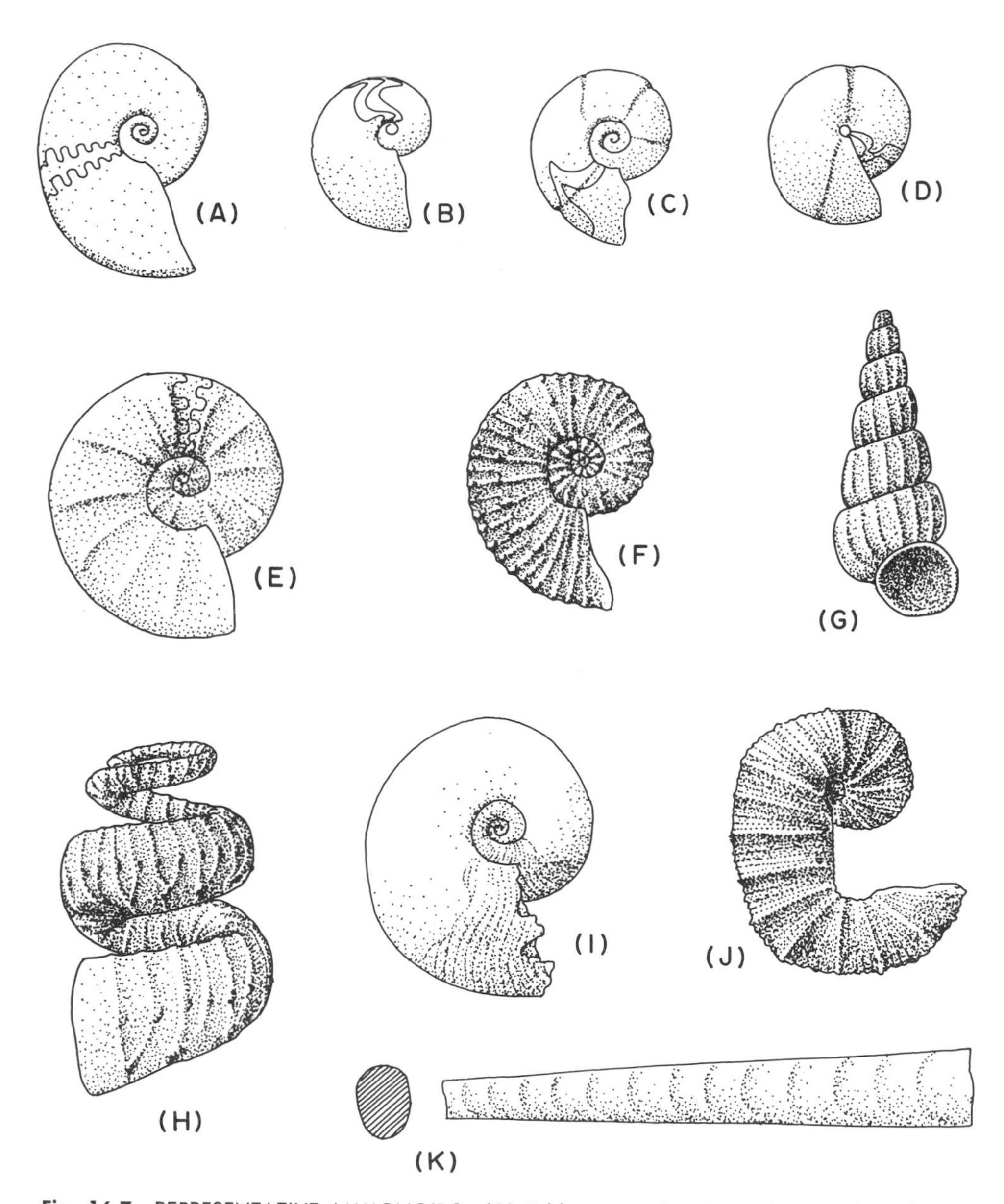

Fig. 16-7. REPRESENTATIVE AMMONOIDS. **(A)** *Uddenoceras;* late Pennsylvanian; lateral view with two sutures shown; greatest diameter 1.5 cm. **(B)** *Manticoceras;* late Devonian; lateral view with three sutures; greatest diameter 1.1 cm. **(C)** *Muensteroceras;* Mississippian; lateral view with two sutures; greatest diameter 3.6 cm. **(D)** *Sporadoceras biferum;* late Devonian; lateral view with two sutures shown; greatest diameter 3.3 cm. **(E)** *Ceratites;* middle Triassic; lateral view; greatest diameter 9.5 cm. **(F)** *Nevadites;* middle Triassic; lateral view; greatest diameter 6 cm. **(G)** *Cochloceras;* late Triassic; apertural view; height 2.7 cm. **(H)** *Emperoceras;* late Cretaceous; height 15 cm. **(I)** *Oxynoticeras;* early Jurassic; lateral view; greatest diameter 3.4 cm. **(J)** *Scaphites;* Cretaceous; lateral view; greatest diameter 4.4 cm. **(K)** *Baculites;* late Cretaceous; cross-section and a lateral view of incomplete specimen; length of specimen 17.5 cm. [(**A**) *after* Miller and Furnish; (**F**) *after* Smith; (**G**) *after* Zittel.]

noids. Secondary folds appeared on the saddles as well as on the lobes, and in some the margin of the septa crinkles so strongly that a third set of folds appears.

The adaptive value of these crenulations is not known. Pfaff suggested many years ago that the major folds in a septum represented a folded structural shell—analogous to those now employed in reinforced concrete structures—designed to resist hydrostatic stress on the shell. A similar argument has been advanced more recently by Ruzhencev (1946) and Westermann (1958). Presumably, an animal with such a braced shell could ascend and dive without compensation for pressure differentials between the water and the camerae. Raup, however (1966), has reported that *Nautilus* shells resist rather high hydrostatic pressures without fluting, and most of the ammonoids apparently lived in shallow seas where hydrostatic variations would be slight.

The ammonoids underwent several adaptive radiations, one Devonian, one late Paleozoic, one Triassic, and at least one in the Jurassic and Cretaceous. In each, a variety of shell forms (Figure 16-7) appeared, some loosely coiled, others tightly; some globose in cross-section, others flattened and discoidal. In the Mesozoic radiations, even odder types appeared. A few coiled like snails in a conical spire. Several genera have the immature part of the shell coiled and the adult portion straight. Others formed a coil early in development, then grew straight for a while, and finally formed a loop. Those variations that produced a streamlined shape like that of some discoidal genera may have belonged to strong swimmers. Many of the variations, though, seem as inexplicable in function as do the ribs, spines, and nodes that occur in some ammonoids.

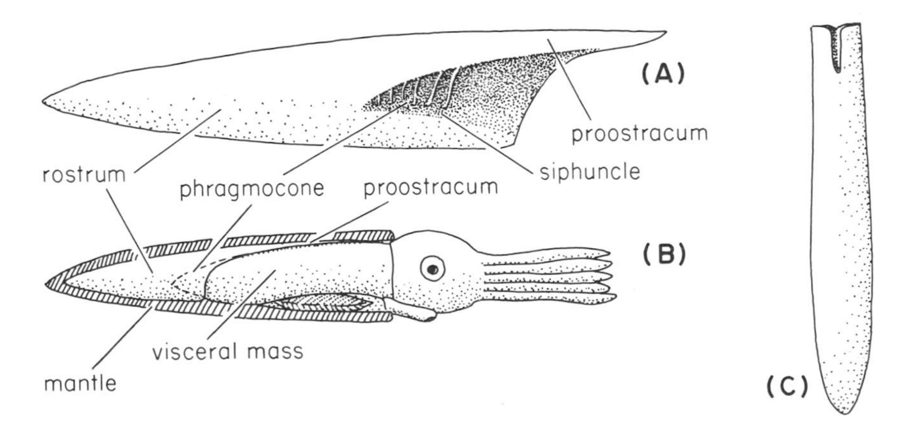

Fig. 16-8. BELEMNOIDS. **(A)** Lateral view of shell, cleared to show phragmocone. **(B)** Lateral view of reconstruction of belemnoid. The mantle has been sectioned to show the rostrum, gills, and visceral hump. **(C)** *Belemnitella;* late Cretaceous; ventral view; length 9.5 cm.

Subclass Coleoidea

The coleoid cephalopods either have no shell or have an internal one. In those that have a shell, the phragmocone is relatively small, and the living chamber is incomplete, projected forward as a slender blade, the *proostracum.* The mantle enfolded the shell and in some cases deposited concentric laminae of calcite on the tip of the phragmocone to form the *rostrum* (Figure 16-8). Superficially, this coleoid shell differs greatly from that of any pre-Mississippian cephalopods, but if the rostrum were removed and the proostracum expanded laterally to form a living chamber, it would be very like that of the bactritids—even to the presence of a ventral siphuncle. The rostrum was large in the extinct belemnoids but was reduced or absent in the lines that led to the recent cuttlefish, squid, and octopus.

Cephalopod stratigraphy

I suggested in several of the preceding chapters that interpretations of phylogeny and paleoecology depended, in large measure, on interpretations of function and adaptation. Without such knowledge, the paleontologist must hesitate in separating characteristics acquired by evolution in response to similar environments from those inherited from a common ancestor. Without such knowledge, he falters in the analysis of the limiting ecological factors in the fossil environment. When a stratigrapher asks if *A* or *B* is the older species, or if they lived in different environments at the same time, he can give only a very approximate answer.

Fortunately, these defects are not so serious for cephalopod stratigraphy as for cephalopod paleontology. The reason for this happy circumstance lies in the peculiarities of cephalopod geographic distribution. Some species were pelagic and, like recent marine vertebrates, the fishes, the sharks, and the whales, must have migrated widely across the oceans and into the confines of the epicontinental seas. Not all did so—many, such as the benthonic Paleozoic nautiloids, had narrow environmental limits—but a few, at least, left "home." These wide-ranging species are among the most important of zonal fossils. Admittedly, they, too, had ecologic limits and almost certainly appeared in different places at different times, but they have been found to succeed one another in local stratigraphic sections in invariable order.

Paleoecology

In spite of continued interest in cephalopod habits and environments, relatively little is known of their paleoecology. As with many fossil groups, some of the men most interested confined their studies to the museum and their speculations to an armchair. This has resulted in extensive literature on adaptation but very little on specific habitats. Some cephalopods show only very general associations with sedimentary environments and were surely nektonic; ammonoid, nautilid, and belemnoid species occur in a variety of lithologic types. On the other hand, many cephalopods, particularly the early Paleozoic types, were clearly restricted in distribution to a few habitats.

Cephalopod evolution: A postscript

Arkell, Kummel, and Wright (in Moore, 1957) point out that, in the past century, there have been changing fashions in interpretation of cephalopod phylogeny. These fashions have followed particular trends in evolutionary thought and have in turn served as examples or evidence for the favored theory. Thus, there are phylogenies based on the concept of "iterative evolution," others derived from a strict application of the "biogenetic law" that individual development recapitulates phyletic changes, and still others employing Dollo's law of the irreversibility of evolution. I have no space for discussion of all these ideas, but I refer you to the paper just cited.

A striking feature of cephalopod evolution is rapid morphologic change. This contrasts markedly with the slow evolution of the pelecypods. Why should a whole group of lineages undergo stronger selection than another group? Possibly these rapidly evolving lines enter into a feedback network. Thus, change in a predator modifies selection on the prey species; the changes induced in the prey modify, in turn, selection on the predator. Competitive predators and alternate prey are also affected. Such feedback relations seem most likely among terrestrial and nektonic animals which prey selectively on elusive food and/or avoid predators by speed or behavior. On the other hand, a stronger brachiopod pedicle does not change the strength of the waves; a better pelecypod gill filter system does not appreciably change the number of microorganisms in the incoming water current. (I have, of course, oversimplified, for a heavier shell, for example, may require a predaceous snail with a stronger radula to bore into it.) But an ammonite—or fish—population can hardly change without affecting other animal and plant populations and the selective pressures on them.

REFERENCES

Brinkmann, R. 1929. "Statistisch-biostratigraphische Untersuchungen an Mitteljurassischen Ammonitea, Ueber Artbegrift und Stammensententwicklung," *Abh. Ges. Wiss. Gottingen, Math Physik, Kl.*, N.F., Bd. 13, t. 3.

Erben, H. K. 1964. "Die Evolution der Ältesten Ammonoidea," *Neues Jahrb. für Geol. und Paläon.*, Bd. 120, s. 108-212.

Flower, R. H. 1955. "Trails and Tentacular Impressions of Orthoconic Cephalopods," *Jour. of Paleontology,* vol. 29, pp. 857-867.

Kummel, B. and R. M. Lloyd. 1955. "Experiments on Relative Streamlining of Coiled Cephalopod Shells," *Jour. of Paleontology,* vol. 29, pp. 159-170.

Moore, R. C. (Ed.), 1957. *Treatise on Invertebrate Paleontology, Pt. L. Mollusca 4.* New York: Geological Society of America.

———. 1964. *Ibid., Pt. K. Mollusca 3.*

Mutvei, H. 1964. "On the Shells of Nautilus and Spirula with Notes on the Shell Secretion in Non-Cephalopod Molluscs," *Arkiv für Zool.* Bd. 16, nr. 14.

Raup, D. 1966. "Experiments on Strength of Cephalopod Shells," *Geol. Soc. of America, Program 1966 Annual Mtg.*, pp. 172-173.

———. 1967. "Geometric Analysis of Shell Coiling: Coiling in Ammonoids," *Jour. of Paleontology,* vol. 41, pp. 43-65.

Reyment, R. A. 1955. "Some Examples of Homeomorphy in Nigerian Cretaceous Ammonites," *Geologiska Fören. I Stockholm Förhdl.*, Bd. 77, pp. 567-594.

———. 1958. "Some Factors in the Distribution of Fossil Cephalopods," *Stockholm Contrib. Geol.*, vol. 1, pp. 97-184.

Ruzhencev, V. E. 1946. "Evolution and Functional Significance of the Septa in Ammonites," *Akad. Sci. URSS, Inst. Paleont., Bull., Cl. sci. biol.*, vol. 6, pp. 675-706. In Russian with English summary.

Westermann, G. E. G. W. 1958. "The Significance of Septa and Sutures in Jurassic Ammonite Systematics," *Geological Mag.*, vol. 95, pp. 441-455.

———. 1964. "Sexual-Dimorphisms bei Ammonoideen und Seine Bedeutung fur die Taxionomie der Otoitidae," *Palaeontographica,* Bd. 124, Abt. A, s. 33-73.

17

Arms, stems, and spines: The Echinodermata

In the Precambrian, near the time when the ancestors of the annelid-mollusc stock evolved a coelom, another metazoan managed the same trick but in a different fashion—at least the difference in coelom development in recent animals suggests a different origin. In the annelid-mollusc stock, the coelom forms as a split within the mesodermal mass which budded off the gut wall. In the other group, the coelom begins as pouches from the sides of the embryonic gut (Figure 11-1, p. 267). These separate from the gut; the cavities of the pouches form the coelomic spaces, and their walls form the mesoderm. This mode of coelom formation is obscured in some members of the group by developmental adaptations, but is nearly universal among the simpler (less evolved?) types. Further, they differ from the annelid-mollusc group in the way in which the egg divides in initial development, in the position of the mouth and anus in the larva, and in the form of the larva. Their skeletal elements form internally in mesodermal tissue rather than externally in the *ectodermal* layers.

Just how far back the common ancestry of this and the annelid-mollusc line lies is uncertain. The two are distinct at their first appearance, and the profound differences in embryology may indicate a long period of divergence. Like the annelid-mollusc line, this one also divided early in its history into two major stocks—so early that some authorities won't concede that the two are closely related. One of the major branches, the Chordata, has contained principally active animals and paralleled the annelids in the evolution of segmentation. The other branch, the Echinodermata, adopted a sessile or sluggish benthonic habit. They evolved radial symmetry and specialized canals (the *water vascular system*) from parts of the coelom, but have retained primitive characteristics in the circulatory and nervous system. Unlike the chordates, which penetrated or evolved in fresh and brackish waters, the echinoderms have been entirely marine in habit.

The mysterious pentagram

The most distinctive feature of all recent and most fossil Echinodermata is fivefold radial symmetry (Figure 17-1) in the skeleton of the body, in appendages, and in the nerve cords, a symmetry that reflects the fivefold division of the water vascular system. This system develops from the anterior coelomic spaces and consists of the *ring canal,* which encircles the mouth; the *stone canal* which connects the ring canal to an external opening, the *hydropore;* and the *radial canals* (typically five), which extend laterally from the ring canal. The radial canals underlie ciliated grooves, the *ambulacra* (sometimes covered), which extend across the body surface. Among all recent echinoderms, minute closed branches, the *podia,* arise from the radial canals. Many suspension feeders, e.g., brachiopods and bryozoans, bear hollow ciliated tentacles. The spaces within the tentacles serve as a hydrostatic skeleton and as a system for absorption of oxygen and removal of carbon dioxide. It seems likely that the water vascular system evolved from the coelomic canals supporting a similar set of tentacles in early echinoderms. The hydrostatic skeleton is, in turn, preadaptive for locomotor functions. Evolutionary transformations in the structure of the water vascular system would then correspond to changes in feeding mode, respiratory requirements, and locomotion.

This explanation is at least logical, but if one inquires into echinoderm ontogeny, several difficulties appear. The larva itself has bilateral symmetry. During metamorphosis, part of the coelom on one side subdivides and differentiates to form the water vascular system. At the same time, the mouth shifts to this side, twisting the viscera so that bilateral symmetry is lost. The body reorganizes about the radial branches of the water vascular system to produce a fivefold radial symmetry. Possibly the water vascular system represents only one side of a circumoral ring of tentacles, but that fails to explain why the other side degenerates. Since some of these

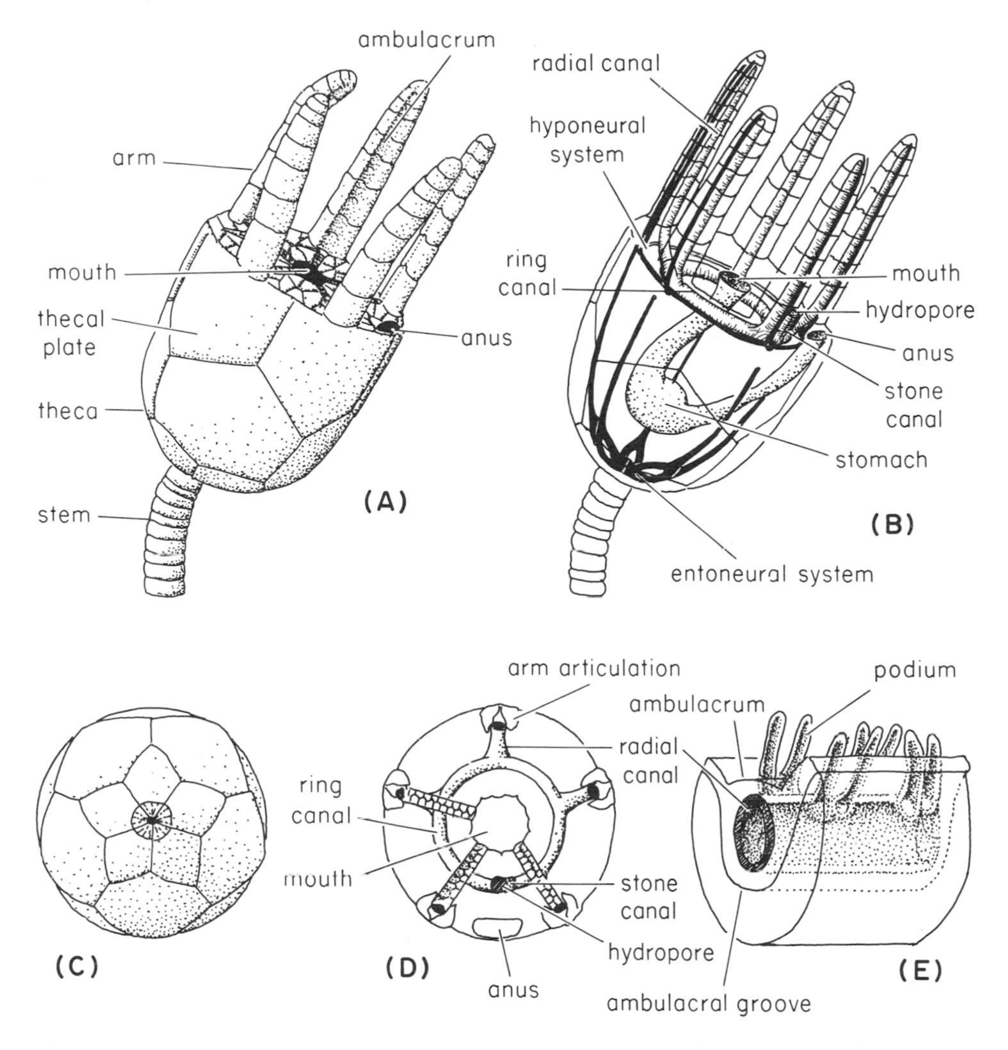

Fig. 17-1. GENERAL PLAN OF THE ECHINODERMATA. **(A)** Lateral view showing theca, arms, and portion of stem. **(B)** Lateral view with plates cleared to show internal anatomy. **(C)** Aboral view of theca. **(D)** Oral view with plates removed. **(E)** View of arm showing details of ambulacrum, radial canal, and podia.

peculiarities are probably adaptations in development, and others are remnants of an earlier phylogenetic stage, their significance may never be determined.

The adaptive radiation of the echinoderms depends largely on variation in locomotion and feeding. If the protoechinoderms employed ciliated arms for suspension or selective detritus feeding, they were probably sessile or sluggish benthonic animals; the oldest known echinoderms had very limited abilities for movement. The four basic adaptive patterns (Fell and

TABLE 17-1

Glossary for Echinodermata *(figures in parentheses refer to pertinent illustrations)*

Aboral. Direction opposite position of mouth.

Ambulacral groove. Groove in thecal or arm plates along course of ambulacrum. (17-1A, E; *et al.*)

Ambulacral plate. Calcareous plates forming floor of ambulacral tract. (17-1A; *et al.*)

Ambulacral pore. Pore in or between ambulacral plates for passage of podium or for connection of podium to ampulla. (17-2; 17-14A; 17-15A.)

Ambulacrum. Narrow tract or groove extending radially from mouth. Typically the tissue overlying the groove is thickly ciliated and is underlain by radial canal of water vascular system. May subdivide and extend on to appendages. (17-1A; *et al.*)

Arm. In general, appendage, typically one of several mounted on oral surface, which bears extension of ambulacrum. (17-1A, B, E.) Very strictly, limited to crinoids which have pinnules on appendages.

Columnals. Circular or polygonal, discoid plates that form the stem. (17-1A; 17-4A; 17-5A; 17-7A.)

Covering plate. Small plate roofing ambulacral groove or mouth. (17-2; 17-5E; 17-7B.)

Entoneural system. Nerve ring and radial branches developed in aboral body wall. (17-1B.)

Hydropore. External opening of water vascular system—located in one of the interambulacral areas. May be covered by a porous plate, the madreporite. (17-1B, D; *et al.*)

Hyponeural system. Nerve ring and radial branches lateral to ring canal and underlying oral body wall. (17-1B.)

Interambulacral. Referring to area between rays of ambulacra—particularly the plates in these areas.

Madreporite. Sieve-like plate covering external opening of water vascular system. (17-12A; 17-14B, D.)

Oral. Direction—toward mouth or on same surface as mouth.

Podia. Small tubes closed at tips and extending singly or in groups from sides of radial canal. (17-1E; *et al.*)

Radial canal. Tube extending radially from ring canal beneath ambulacrum. It is closed at its outer end and bears rows of closed branchlets, the podia. (17-1B, D, E.)

Rays. Radial directions established by position of ambulacra. (17-1A, D; 17-8E.)

Ring canal. Hollow tube forming a closed ring about mouth. Gives rise to radial branches —the radial canals—and is typically connected to exterior by a short tube, the stone canal. (17-1B, D.)

Stem. Series of disk-like plates mounted one on top of the other and attached to aboral end of theca. Typically the terminal end is fastened to the substrate. (17-1A.)

Stone canal. Tube (typically short) leading from ring canal to external opening, the hydropore. (17-1B, D.)

Theca. Sac-like skeleton of calcareous plates formed in lower layer of skin and enclosing viscera. (17-1A; *et al.*)

Thecal plate. Calcareous plate forming an element in the theca. Usually distinguished from ambulacral or arm plates. (17-1A.)

Moore, in Moore, 1966) were all established by the Ordovician; they include:

a). Homalozoan—early Paleozoic forms with an asymmetric or a bilaterally symmetric arrangement of skeletal plates. Typically unattached, flattened echinoderms with few or no arms and a plated "tail" rather than a stem. They were probably sluggish benthonic types and included both suspension and detritus feeders.

b). Crinozoan—forms with at least partial radial symmetry, several arms or rather similar *brachioles,* and a stem. They were (and are) predominantly attached forms and suspension feeders; a few recent and some fossil species were sluggish swimmers.

c). Asterozoan—forms with the body divided into radially directed rays and without a stem or "tail." They include sluggish (starfish) to active (brittle star) benthic forms and predators as well as suspension and detritus feeders.

d). Echinozoan—radially symmetrical, globose forms without arms and without stem or tail. The echinozoans include some sessile forms but are predominantly sluggish vagrant benthos and burrowers; some are predatory; others are browsers and detritus feeders.

Each of these patterns may map a phyletic unit; the *Treatise on Invertebrate Paleontology* divides the Echinodermata into four subphyla along these divisions, but some of them may actually be adaptive grades rather than phyletic assemblages.

Forerunners and failures

The earliest known echinoderms (at the base of the *Olenellus* Zone, lower Cambrian) include representatives of two subphyla, the Crinozoa and Echinozoa; a third subphylum, the Homalozoa, appears very shortly thereafter. The early Paleozoic was apparently a time of evolutionary exploration for the Echinodermata: of the sixteen classes recognized in the *Treatise* (Table 17-2) ten are found only in lower Paleozoic rocks and no class is abundant or diversified before the middle Ordovician.* This radia-

* The number of classes, eighteen, used in the *Treatise* seems excessive *relative to other phyla.* The vertebrates, classified by the same philosophy, i.e., on unbridged differences in basic plan and monophyletic origin, would comprise fifteen to twenty classes rather than eight. Such a classification, of course, emphasizes vertical relationships.

TABLE 17-2

The Echinodermata *(after* Treatise, Part U.)

Early Cambrian to Recent. Coelomate but generally of simple organization except for water vascular system derived from coelomic cavities. No heart or circulatory system. Nervous system rather diffuse. Skeleton of porous calcite plates developed in mesodermis. Typically fivefold radial symmetry; some asymmetric or bilaterally symmetrical. All marine, predominantly sessile and vagrant benthos.

SUBPHYLUM HOMALOZOA

Cambrian to Devonian. Asymmetric or bilaterally symmetrical skeleton. Unattached, flattened forms; few or no arms; plated "tail" or stele rather than a stem.

SUBPHYLUM CRINOZOA

Cambrian to Recent. Asymmetrical or radially symmetrical. Most with stems and attached. Arms or brachioles borne on theca. Suspension feeders.

CLASS EOCRINOIDEA

Cambrian to Ordovician. Radially symmetrical theca bearing biserial brachioles and supported by stem. Thecal plates nonperforate. Stem consisting of superposed disks.

CLASS PARACRINOIDEA

Ordovician. Asymmetrical theca bearing uniserial brachioles. Thecal plates with pores.

CLASS CYSTOIDEA

Ordovician to Devonian. Asymmetric or imperfect radial symmetry. Biserial brachioles. Thecal plates with pore rhomb or double pore system.

CLASS BLASTOIDEA

Ordovician to Permian. Theca with three circlets of radially distributed plates. Numerous brachioles. Pore system modified to form hydrospire.

CLASS CRINOIDEA

Ordovician to Recent. Theca with three or more circlets of radially arranged plates. Pinnulate, uniserial arms. Theca plates without pores.

SUBPHYLUM ASTEROZOA

CLASS STELLEROIDEA

Ordovician to Recent. Stellate body with fivefold symmetry. Numerous small plates, loosely articulated. Ambulacra and anus typically on substrate; anus and madreporite on upper surface. Radial canals in open or partly covered grooves. Vagrant benthos with ambulatory tube feet or motile arms.

SUBPHYLUM ECHINOZOA

Cambrian to Recent. Globose, cylindrical or discoid body without arms or brachioles. Radial symmetry—typically fivefold. Attached or vagrant benthos. Radial canals covered.

CLASS HELICOPLACOIDEA

Early Cambrian. Fusiform, free-living. Spirally pleated, flexible test. Two spirally arranged ambulacra—apparently without tentacles or brachioles.

CLASS EDRIOASTEROIDEA

Cambrian to Mississippian. Discoid theca with five regularly arranged ambulacra and small irregular plates in interambulacra. Typically attached on lower surface with central mouth on upper.

TABLE 17-2 (continued)

CLASS OPHIOCISTIOIDEA

Ordovician to Silurian. Pentaradiate, free-living forms with dome-shaped body. Ventral mouth surrounded by five very large armored podia.

CLASS CYCLOCYSTOIDEA

Ordovician to Devonian. Small, discoidal theca, probably attached to flattened aboral side. Numerous plates arranged in rings. Branching ambulacra.

CLASS ECHINOIDEA

Ordovician to Recent. Subspherical to flattened test with mouth on substrate. Five regularly arranged ambulacra. Radial canals internal to test. Radial or bilateral symmetry. Typically vagrant benthos.

CLASS HOLOTHUROIDEA

Devonian to Recent. Sac-like theca with small, loosely attached plates in leathery skin. Mouth at one end surrounded by ring of tentacles formed from podia. Vagrant benthos, rarely sluggish swimmers.

tion produced a variety of types, each developing some aspect of the basic echinoderm adaptation. They competed with one another and with the other benthic animals. They were eaten and parasitized by still other animals. The majority of the lines exploited only a limited series of niches and environments; a minority evolved a structure (and/or physiology) with wider potential, and initiated their own adaptive radiations. Evolution proceeds, as illustrated here, by a selection of species in an evolving natural society as well as by selection of individuals in a population.

Rather than follow out the evolution of each subphlyum in turn, the following discussion deals with the initial late Precambrian-Ordovician radiation and then examines the history of the six classes that dominated the subsequent history of the Echinodermata.

Edrioasteroids

Among the oldest fossil echinoderms (Figure 17-2) are several early Cambrian species that had a discoid sac-like skeleton composed of small calcareous plates and that attached broadly on one surface or lay free on the bottom. On the upper surface of this skeleton, the *theca,* are five arm-like tracts that radiate from the center. Typically, each arm, or, better, *ray,* consists of a double row of paired plates. These plates are depressed to form a groove where they join along the midline of the "arm." Typically, a small pore opens between adjacent plates in a row. Another series of small plates covers the groove externally. At the center of the rays, these grooves join and open, by a single large aperture, into the theca. In

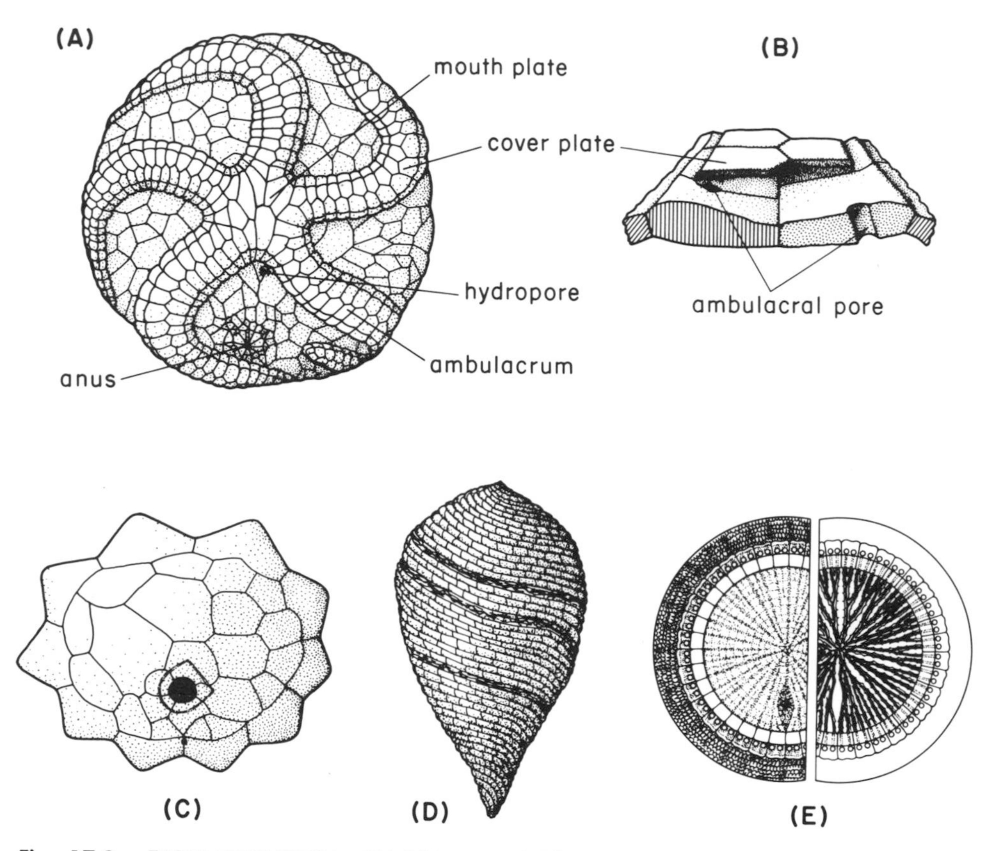

Fig. 17-2. EARLY ECHINOZOA. **(A)** Edrioasteroid *Edrioaster;* Ordovician; oral surface; diameter about 3.0 cm. **(B)** Detail of edrioasteroid ambulacrum; oblique view. **(C)** Ophiocistioid *Volchovia;* Ordovician; aboral surface; diameter about 5.0 cm. **(D)** Helicoplacoid *Helicoplacus;* Cambrian; lateral view; length about 2.5 cm. **(E)** Cyclocystoid *Cyclocystoides;* Ordovician to Devonian; oral surface; diameter about 2.0 cm. View at left with ambulacral cover plates; at right without. [**(C)** *after* Regnell; **(D)** *after* Durham; **(E)** *after* Kesling.]

one interray area is another smaller opening and a small plate perforated by a slit.

What can one make of the animal's anatomy from these structures? The theca is comparable to the calcareous skeletons that surround the viscera of recent echinoderms, and, presumably, was partly covered by a layer of living tissue. The number and arrangement of the "arms" suggests some sort of relation to the radial canals of the water vascular system. As I have already noted (Figure 17-1), one finds in recent echinoderms similar arm-like tracts, the ambulacra, in some actually borne on distinct arms. Beneath the midline of each of the ambulacra is a radial canal. In some, the canal lies directly beneath the skin and rests on the surface of the calcareous plates which may be slightly or very deeply grooved for its reception. If, as in some groups, the podia are simple, small tubes, they leave no impress on the skeleton. Some echinoderms, however, have a large sac, an *ampulla,* connected to the base of each podium. These sacs lie either in a deep pit in the ambulacral plates or inside the skeleton at the end of a short canal. In the latter case, rows of pores within or between the ambulacral plates mark the position of the podia. In other echinoderms, the radial canals extend in a channel through the ambulacral plates or rest against their internal surfaces. In these, the podia, whether they bear ampullae or not, reach the exterior through pores.

The ambulacra, typically, have a central *ambulacral groove,* roofed over in some, but open in others, and covered by ciliated skin. These grooves lead into the mouth. The water vascular system opens externally through a pore or perforated plate, the madreporite, in one of the *interambulacral areas.* The position of the anal opening is variable, but it also lies in one of the interambulacral areas.

Clearly, in the edrioasteroids the basic echinoderm structure had already developed, the theca of calcareous plates, the central mouth, the anus in an interray position, the radiating ambulacra, the water vascular system with podia and madreporite, and the internal torsion that brings the mouth to the center of the ambulacra. Whether the radial canals lay on or beneath the ambulacral plates is not clear. If the former, then the podia must have had internal ampullae to account for the pores.

In spite of their generalized characteristics and early appearance, many authorities regard the edrioasteroids as a sterile side-branch from the basal echinoderm trunk. Others believe, however, that they belong on the echinozoan line, particularly because they, like the modern representatives of the subphylum, have pores between the ambulacral plates.

Helioplacoidea, ophiocistioidea, cyclocystoidea

Included in the early Cambrian echinoderm finds is the strange, twisted *Helicoplacus* (Figure 17-2). The fusiform skeleton consists of small, loosely bound plates—presumably the body was rather flexible. Two ambulacra wind spirally away from an apical mouth. No attachment area nor stem is present; neither madreporite or anal opening have been observed. General similarities to the extinct edrioasteroids and to the sea cucumbers (Holothuroidea) and sea urchins (Echinoidea) suggest their placement in the Echinozoa. Their habit was probably that of a sluggish benthic form—most probably a detritus feeder.

Equally strange are the ophiocistioids (early Ordovician to late Silurian) and cyclocystoids (early Ordovician to Devonian) (Figure 17-2). The former have a discoidal, pentaradiate skeleton with tentacle-like appendages. These, however, are not arms but are enlarged tube feet covered by small calcareous plates. The positioning of the anus on the aboral surface and the arm-like podia suggest an active benthonic existence like the recent brittle stars; the development of "jaws" implies a predatory and/or browsing habit like the sea urchins. The cyclocystoids on the other hand were probably sessile suspension feeders. Their form is discoidal; the presumed upper surface bears a central mouth, a lateral anal opening, radially directed ambulacra roofed by covering plates, and a circlet of numerous faceted plates which probably bore short arms.

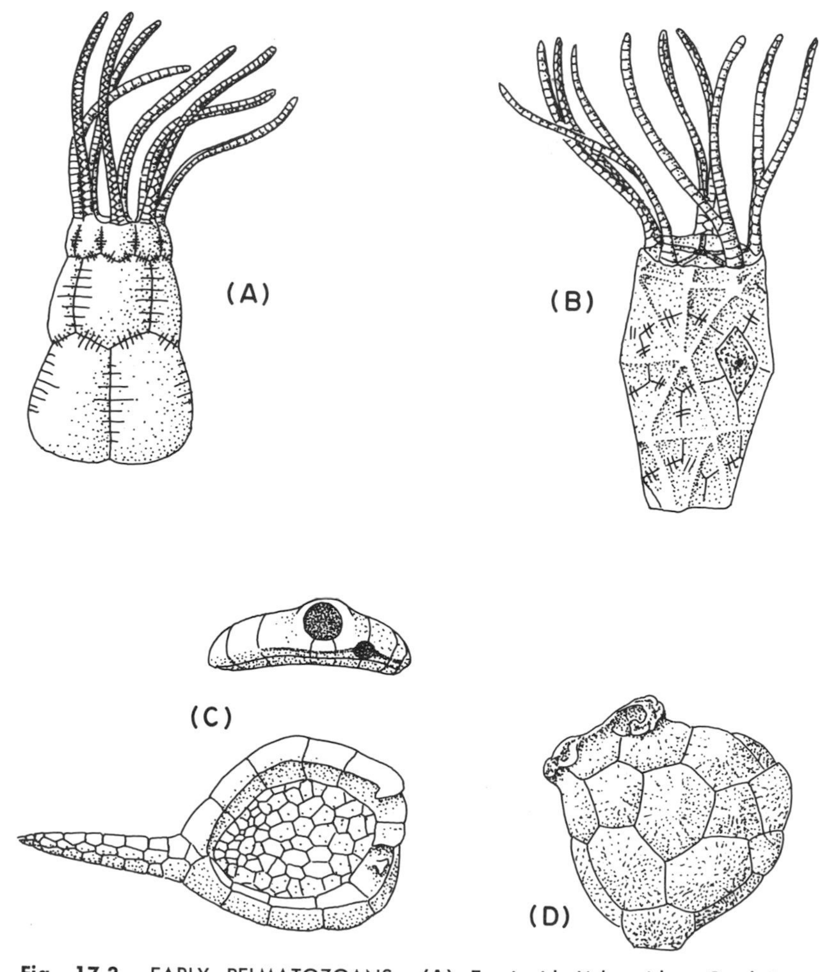

Fig. 17-3. EARLY PELMATOZOANS. **(A)** Eocrinoid *Lichenoides;* Cambrian; lateral view; height of theca 3.9 cm. **(B)** Eocrinoid *Macrocystella;* Cambrian; lateral view; height of theca 2.4 cm. **(C)** Carpoid *Trochocystis;* Cambrian; views of oral face and of one side. **(D)** Paracrinoid *Canadocystis;* middle Ordovician; lateral view; height of theca 4.0 cm. [**(A)** *after* Jaekel; **(B)** *after* Bather; **(C)** *after* Gislen; **(D)** *after* Hudson.]

Eocrinoidea

The chief trends in evolution of sessile echinoderms were the formation of an elongate stalk, a reduction in the number of thecal plates, the development of free arms, and the symmetrical arrangement of the thecal plates. All of these appear in a small, precocious group of Cambrian echinoderms, the Eocrinoidea (Figure 17-3).

The eocrinoids were attached by a distinct *stem** opposite the mouth and bore free arms on a globular *theca.* The thecal plates are arranged in a series of circular rows and in some have fairly regular radial symmetry. An "arm," a *brachiole,*† arises from each of the five plates in the uppermost circlet, and, in some genera, bifurcates a short distance above its base. Each brachiole consists of a double series of plates (*biserial*). The ambulacral grooves extend along the inner surface of the brachioles and are roofed over by small plates. In all eocrinoids the thecal plates themselves lack pores but are incised into lobate form by intersutural pores, i.e., pores between the plates. The arrangement of thecal pores and the symmetry are characteristics in common with crinoids. Links with either cystoids or crinoids have not, however, been found, or, if found, have not been recognized as such. They are uncommon fossils and apparently became extinct after the middle Ordovician.

Carpoids

One other group of echinoderms occurs in middle Cambrian deposits (Figure 17-3). They have a polyplacate *stele,* i.e., "tail," a theca of calcareous plates, and a few have brachioles. On these grounds they are obviously echinoderms—but they show hardly any of the other expected characteristics. Since the tip of the stele tapers sharply and terminates in a point or an irregular break, the adult must have lived unattached. The theca is flattened, and the animal very probably lay on one of these flattened sides. The thecal plates in some are large and completely asymmetrical. In others they are bilaterally symmetrical, the plane of symmetry perpendicular to the flattened surface. Typically, one of these surfaces is more convex than the other, and the plate arrangement differs on the "up" and "down" sides. The mouth and anus may be on the margin opposite the stem, or one or both may be shifted back along the sides. In several genera, neither mouth nor anus have been found; probably they were covered with movable plates. Several seem to lack a mouth

* Caster and Pope (1960; also see Robison, 1965) describe a polyplacate stele or "tail" rather than a stem (of superposed disks) in the early Cambrian *Eocystites* and suggest relationship with the holothuroid echinozoans.

† In echinoderm terminology, *arms* have rays, *pinnules,* arising laterally from each plate; brachioles are "arms" without pinnules.

altogether and have in its stead a row of inhalant pores like the chordate gill slits (see p. 423). The superficial appearance is very much like that of primitive armored fish, with a broad flat head shield and a slender tail.

Without knowledge of early ontogeny and of soft parts, interpretation of carpoid adaptations and relationships is nearly impossible. Possibly they had tentacles and/or some sort of filter for suspension and/or detritus feeding. If the stele was movable, they may have used it in locomotion or forced it into the substrate as an anchor. Possibly they were sessile; at best they moved slowly along the bottom. The absence of radial symmetry may be a primitive character or a highly specialized one. The bilateral symmetry could also have originated in either manner. The similarities to the early vertebrates, though intriguing, may be only adaptations to a similar mode of life. Never common, they survived into the lower Devonian.

Paracrinoidea

The paracrinoids (Figure 17-3) are an assemblage of middle Ordovician forms with pores in the thecal plates, with a theca composed of small irregular plates, and with a stem. Some have free appendages consisting of brachioles with a single row of plates (*uniserial*) bearing uniserial branches. In others, the arms are recumbent on the surface of the theca. The pore system is a cystoid characteristic; the uniserial arms are a crinoid one. They may well be a branch of the cystoid stock that evolved crinoid-like arms. In that case, they would probably rank as a cystoid subclass or order.

A quick success: The Cystoidea

A diversity of attached echinoderms with biserial brachioles and thecal pores appeared in the middle Ordovician. They are divided into two groups: the cystoids with irregular plate arrangements and imperfect radial symmetry, and the blastoids with perfect radial symmetry. This division is, in a sense, unequal, for the Cystoidea encompass a greater variety of form than the Blastoidea. The blastoid line, however, is easier to follow than the lines within the cystoids, and underwent its major radiation after the extinction of the cystoids. The distinction is a convenient one.

The general cystoid form (Figure 17-4) is a spheroidal sac. The number and arrangement of thecal plates varies greatly from genus to genus, but, typically, they are more numerous and irregular than in crinoids or

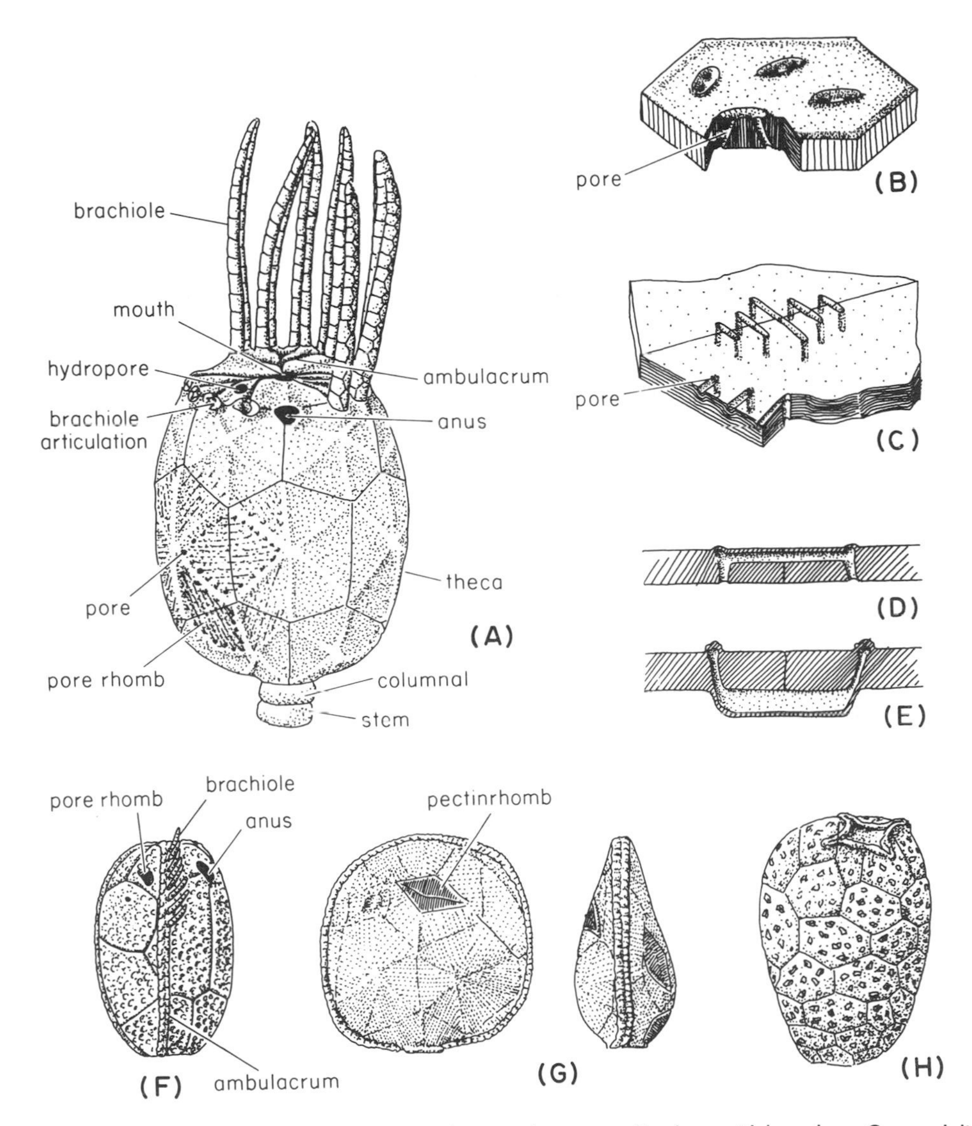

Fig. 17-4. THE CYSTOIDS. **(A)** Lateral view of a generalized cystoid based on *Caryocrinites,* a mid-Ordovician to mid-Silurian genus. Note three-fold symmetry of ambulacra. The brachioles are missing from the ray to the left. **(B)** View of thecal plate sectioned to show diplopore system. **(C)** View of thecal plates showing pore rhomb system. External grooves extend across adjacent plates and connect to interior by pores. **(D)** and **(E)** Cross-sections of two types of pore rhombs. External surface up. **(F)** *Jaekelocystis;* Silurian; lateral view; height of theca 1.5 cm. Several brachioles shown. **(G)** *Pseudocrinites;* Silurian and Devonian; lateral views, ambulacral and interambulacral, showing compressed theca; height of theca 2.5 cm. **(H)** *Megacystis;* Silurian; lateral view. A diploporid. [**(H)** *after* Jaekel.]

TABLE 17-3

Glossary for Cystoidea *(figures in parentheses refer to pertinent illustrations)*

Ambulacrum. See Table 17-1. Many genera have three distinct ambulacral rays—though some have more or less. (17-4A, F, G.)

Biserial. Type of arm or brachiole consisting of two rows of plates from thecal articulation upward. (17-4A.)

Brachiole. Arm-like appendage arising from thecal plates at ends or along sides of ambulacra. Bears extension of the ambulacral grooves but no lateral rods (pinnules). (17-4A, F.)

Brachiole articulation. Portion of thecal plate modified for attachment of brachiole. (17-4A.)

Diplopore. Pore system consisting of pairs of pores set in small depressions on the thecal plates. Each pore opens into the interior of the theca. (17-4B.)

Pectinirhomb. Specialized type of pore rhomb system consisting of a compact, rhomboidal structure of closely spaced grooves (comb-like). Typically set in a distinct depressed area on thecal plates. (17-4F, G.)

Pore. Small opening from exterior through thecal plates. May be closed off by deposition of calcareous material at one end or other. (17-4A, B, C, D, E.)

Pore rhomb. Pore system consisting of a rhomb-shaped structure on surface of theca plates. It comprises a number of grooves in plate surface at right angles to the sutures between adjacent plates and also the pores that connect ends of the grooves to thecal interior. Grooves may be roofed by thin calcareous sheet. (17-4A, C, D, E, F.)

blastoids. The mouth is at the summit of the theca. Ambulacra, typically three or five, radiate from the mouth. These may branch, and in most, they extend up short *brachioles.* If podia were present, they and the radial canals must have lain on the external surface of the plates, since there are no pores for them. The brachioles are *biserial,* i.e., they comprise a double row of plates and may bear biserial branchlets. The ambulacra and brachioles characteristically show a poor grade of radial symmetry, but some genera have almost perfect fivefold symmetry; others have bilateral symmetry, and still others are asymmetric. A high degree of symmetry commonly correlates with a reduced number and more regular order of thecal plates. The anus and the madreporite lie in an interambulacral area on the upper part of the theca. Many cystoids have an additional opening, presumably for the genital system. The stem varies greatly, long in some, absent in others. Some with stems were not attached to the bottom.

The distinctive feature of the cystoid is the thecal pores (Figure 17-4). Two general types are recognized. The *diploporid system* consists of tubes oblique or perpendicular to the surface of the plate. These are arranged in pairs, opening externally in a common pit or short groove. The *pore rhomb system* is more complicated and comprises a series of short parallel grooves arranged in a rhombic pattern on the external surface and a number of tubes that connect the grooves with the interior of the theca. Each rhomb appears on two thecal plates, half on each plate, and the grooves are perpendicular to the suture between the plates. In some genera, the center part of the rhomb is roofed over by a thin calcareous layer so that only the ends of the grooves are left open; in others, the roofing is complete.

Except for the pore system, the adaptations of the cystoids are comparable to the crinoids. Presumably, the ambulacral grooves were ciliated and small food particles falling on or near the brachioles were swept down the grooves to the mouth. An increase in the number of brachioles and of their branchlets increased the area of this net of filters. Stemless forms may have rested directly on the bottom and controlled their position by movements of the brachioles. Those that did not attach their stems to the substrate may have used the stem as a drag or as a prehensile tail to hold to shells or seaweed. Did the cystoids compete directly with crinoids? Direct paleoecologic studies have not produced any certain data on this question, but the increase of the crinoids in the later Ordovician and Silurian corresponds with a rapid decline in the cystoids.

In the absence of recent cystoids, paleontologists can only speculate about the significance of the pore system. Possibly, they served for direct exchange of gases and metabolic wastes between the coelomic fluids and the surrounding sea water. However, in those that had the pore rhombs roofed over by a calcareous sheet this function must have been impaired.

The irregularity of plate arrangement and the imperfections of the radial symmetry have induced some paleontologists to regard the cystoids as near the ancestry of the echinoderm stock. The absence of pore systems in some Cambrian echinoderms, the high degree of symmetry in some Cambrian species, and the orderly plate arrangements in some Cambrian eocrinoids* detract considerably from this theory. Until Cambrian cystoids are found, it seems best to regard the class as one of several lines arising from an ancestor who probably possessed radial symmetry and may have had a regular plate arrangement.

The relation of the diploporid and pore rhomb cystoids and the relation of the cystoids to other echinoderms is more a question of assumptions than of data. Some Cambrian echinoderm species have external grooves on their thecal plates. Did the pore rhomb evolve with the development of pores connecting these grooves to the interior? If so, the eocrinoids could be ancestral to the cystoids. But what does that do to the diploporid system that seems simpler and therefore more primitive? Are the grooved plates that occur in some crinoids and edrioasteroids as

* The eocrinoids are thought by some to be cystoids. If so, the irregularities in Ordovician cystoids are almost certainly specialized rather than primitive characters.

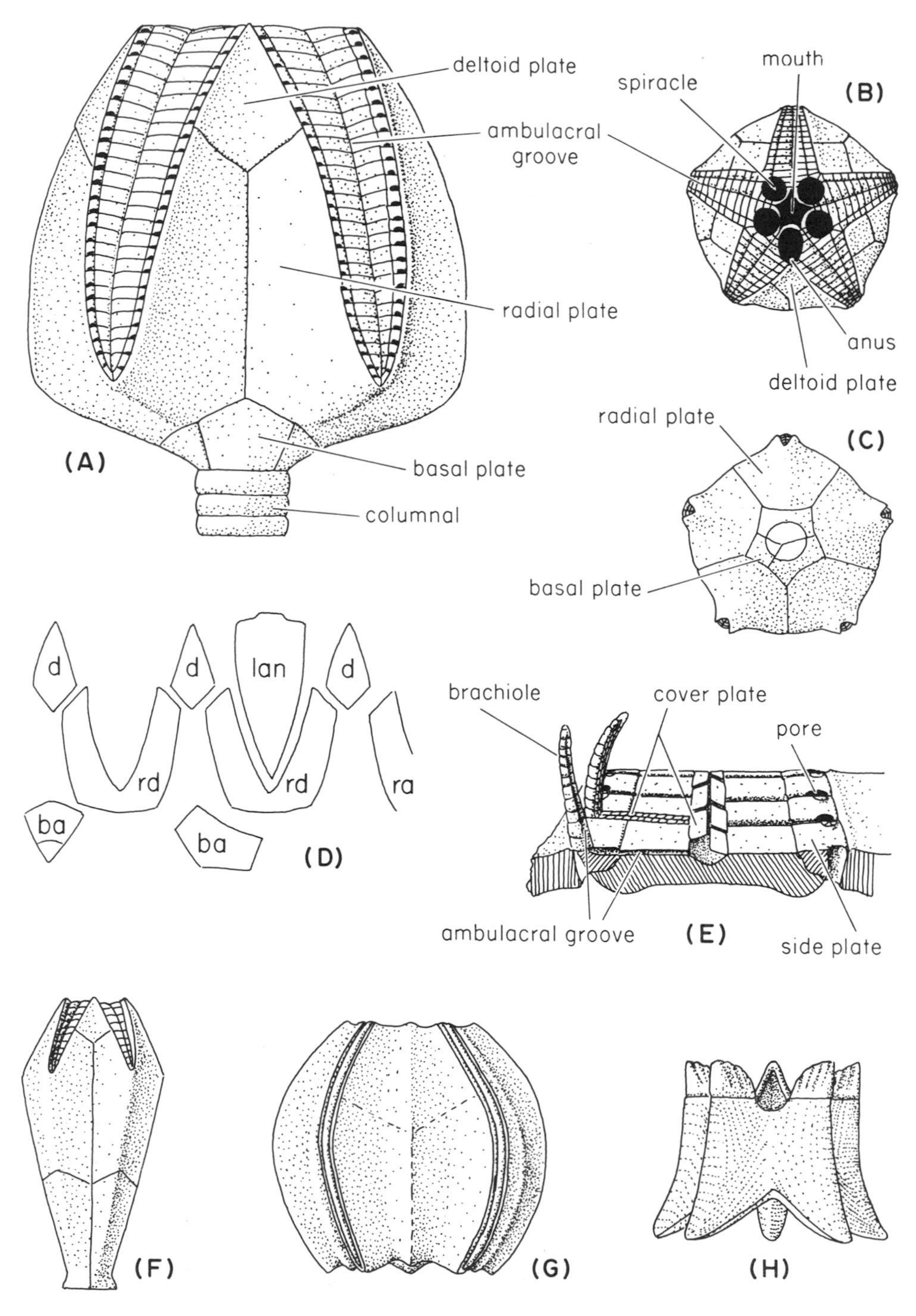
deltoid plate
ambulacral groove
radial plate
basal plate
columnal
(A)
mouth
spiracle
(B)
anus
deltoid plate
radial plate
(C)
basal plate
d
d
lan
d
rd
rd
ra
ba
ba
(D)
brachiole
cover plate
pore
ambulacral groove
(E)
side plate
(F)
(G)
(H)

Fig. 17-5.

well as eocrinoids the vestiges of a well-developed pore rhomb system?. Or, for that matter, is the difference between brachioles (cystoids and eocrinoids) and arms (crinoids) significant in phylogeny? No acceptable answer can be expected until additional Cambrian echinoderms close the gaps.

First flowering: Blastoidea

Some cystoid lines reduced the number of thecal plates and arranged the remainder in several circlets with radial symmetry. These changes correspond to an increase in the size and symmetry of the ambulacra. These tendencies culminated in the class Blastoidea (Figure 17-5). Here the ambulacra are typically large and petaloid, and each has a large central plate, the *lancet,* bordered by a number of small *side plates.* The theca consists of only 13 or 14 plates in just three circlets, *basal, radial,* and *deltoid.* A large number of small brachioles arise in a row on either side of each ambulacrum. Each brachiole bears a small side branch from the ambulacral groove. The absence of pores in the ambulacral plates for the tube feet indicates that the radial canals were external. The mouth is at the apex of the theca. The anus lies nearly adjacent in one of the interambulacral areas. The mouth, ambulacra, and grooves on the brachioles were roofed by a series of delicate covering plates. Because of their fragility, the brachioles and covering plates are only rarely found in place. Nearly all blastoids have a stem composed of discoidal columnals, stacked one on top of another.

Along either side of each ambulacrum, between the lancet plate and adjoining radial and deltoid, is a thin sheet of calcite. This sheet is folded into accordion-like pleats that parallel the ambulacral borders and extend deep into the interior of the theca (Figure 17-6). One end of the sheet is attached to a thin plate below the lancet; the other is attached to the edges of the deltoid and radial plates. These structures, called *hydrospires,* are formed by extensions of the deltoid and radial plates and, not so incidentally, the folds are approximately perpendicular to the sutures between these two plates.

Fig. 17-5. MORPHOLOGY OF THE BLASTOIDEA. **(A)** Generalized blastoid, based on *Pentremites.* Lateral view. **(B)** The same in oral view. **(C)** The same in aboral view. **(D)** Relationship of thecal plates. Plates in two rays rotated into the same plane. **(E)** Detail of ambulacrum with brachioles and covering plates. Oblique view. **(F)** *Troostocrinus;* middle Silurian to late Mississippian; lateral view; height of the theca is 1.8 cm. **(G)** *Cryptoblastus;* Mississippian; lateral view; height of theca is 1.8 cm. **(H)** *Timoroblastus;* Permian; lateral view; height of theca 1.5 cm. [(H) *after* Wanner.]

TABLE 17-4

Glossary for Blastoidea *(figures in parentheses refer to pertinent illustrations)*

Ambulacral groove. See Table 17-1. Passes down center of ambulacral area sending off lateral branch to each brachiole. (17-5A, E.)

Ambulacrum. See Table 17-1. Typically one of five petal-shaped areas. (17-5A, B, C, E, F, G.)

Basal plate. One of a circlet of plates composing the aboral end of the theca and articulated to the stem aborally and to radial plates on oral borders. (17-5A, C, D, F.)

Brachiole. See Table 17-3. Though not often found in fossils, the brachioles formed a row on either side of the ambulacral area. (17-5E.)

Brachiole articulation. Small area along border of ambulacrum that is modified for attachment of brachiole. (17-5E.)

Covering plate. See Table 17-1. (17-5E.)

Deltoid plate. One of a circlet of plates at oral end of theca lying between ambulacra. Each deltoid terminated orally and laterally by ambulacra and aborally by radial plates. (17-5A, B, D.)

Hydrospire. Folded sheet of calcite in interior of theca beneath and parallel to ambulacral border. Axes of hydrospire folds also parallel ambulacral borders. (17-6.)

Lancet plate. An elongate triangular plate that floors the ambulacral area. (17-5D.)

Mouth. Opens at free end of theca, at center of ambulacral "petals." (17-5B.)

Pore. One of numerous small openings between side plates along ambulacral border. Connect space enclosed by hydrospire with exterior. (17-5A, E; 17-6D.)

Radial plate. One of circlet of plates on sides of theca. Each deltoid embraces an ambulacral area and in consequence consists of two prongs connected below the aboral end of the ambulacrum. Bounded orally by deltoids and aborally by basals. (17-5A, B, C, D, F, G.)

Side plate. One of numerous small plates, arranged along ambulacral borders, that cover space between lancet plate and adjacent deltoid and radial plates. (17-5E; 17-6D.)

Spiracle. Large opening at oral end of ambulacral border, adjacent to mouth. Opens into space enclosed by hydrospire. Spiracles on adjacent borders of adjacent ambulacra coalesced in some blastoids to form single opening at oral end of each interambulacral area. (17-5B; 17-6D.)

The functional pattern

The habits of the blastoids must have been much like those of cystoids. Currents along the ambulacral grooves carried food particles, microscopic plants and animals, from the brachioles to the mouth. Most genera were attached—those living on soft bottoms by a spreading rootlike structure. Some, however, may have been free-living, anchored by a prehensile stem or resting on the bottom. The hydrospire, perhaps, was a respiratory structure, or, rather, it supported a thin layer of respiratory tissue that functioned in respiration. Since there is no direct evidence of the water vascular system—podia, radial canals, madreporite—no one can be certain of its character or whether it was fully developed. The hydrospire may have assumed part of the function of the water vascular system.

The evolution of a pattern

Whatever the function of the hydrospire, it apparently evolved from a pore rhomb system (Figure 17-6). The position of the folds of the hydrospire, perpendicular to the suture between the plates, and the contribution of each plate to its formation are precisely the character of the grooves in the pore rhombs. In the middle Ordovician *Blastoidocrinus,* pore rhomb grooves were so deep that the inner surface of the deltoid and radial plates was thrown into narrow folds, each corresponding to a

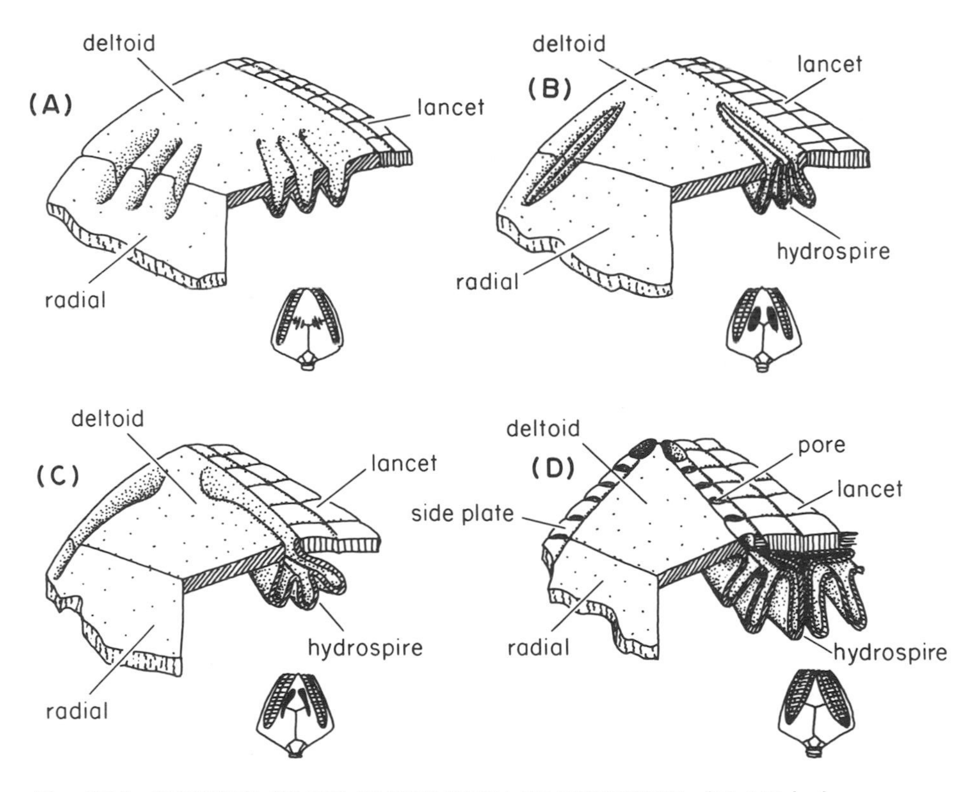

Fig. 17-6. PRESUMED STAGES IN EVOLUTION OF HYDROSPIRE. **(A)** Initial, deep grooves in deltoid and radial plates like cystoid pore rhomb. **(B)** Grooves have begun to "sink" into interior of theca. **(C)** A definite hydrospire formed by internal, folded sheet of calcite. Each hydrospire underlies the edge of adjacent lancet plate and connects to the exterior by a long, narrow slit. **(D)** Each hydrospire lies largely below the lancet close to the hydrospire from the other side. Side plates cover external slit except for pores and spiracle.

groove. In later blastoids, the folds are concentrated along the ambulacral borders to form the hydrospire. In primitive genera, each fold opens by a separate slit like the pore rhomb grooves. In more specialized forms, the borders of deltoid and radial plates extend in toward the ambulacral border and roof over the hydrospire except for a narrow slit. In some, still more specialized, the side plates close over this slit and leave only a few small pores along the edge of the ambulacrum and a large opening, the *spiracle,* at the oral end. In some blastoids, the spiracles of the two hydrospires in each interray coalesced, reducing their number in half. In early blastoids, the anal interray lacks hydrospires, but in later genera, a fifth pair of hydrospires develops in the anal interambulacral area with a shift of anus toward the mouth. In the most specialized genera, with an elaborately folded hydrospire, with pore openings, and with spiracles, the primitive rhomb pore structure is hardly distinguishable.

Most of these changes occurred early in blastoid history, and conservative and advanced forms lived side by side. Specialization of theca went still more rapidly, for the blastoid features were fixed by Silurian time. *Blastoidocrinus* has four circlets of plates: five basals, five radials, a large number of small irregular plates in the next circlet, and five large deltoids between the ambulacra. Silurian blastoids, like all the later ones, have but three basals and have lost the plates between the radials and deltoids. Other than modifications in the hydrospire apparatus, blastoid evolution after the Silurian was limited to changes in thecal shape (Figure 17.5) and size of the ambulacra.

The geologic record

Since the blastoids were undergoing a considerable radiation by Devonian time, some feature of their anatomy must have held an important selective advantage. They reached their maximum diversity in the Carboniferous and Permian but largely on the basis of small modifications of a conservative pattern. They did not survive the end of the Permian.

Cline and Beaver (in Ladd, 1957) conclude that blastoids were exclusively marine, thrived best in clear or only moderately turbid, quiet waters, and required a reasonably stable substrate for attachment. They commonly occur in concentrations and in association with crinoids, solitary corals, fenestellid Bryozoa, and brachiopods. Future work should delimit environments and associations for individual species.

The golden age: Crinoidea

Although the cystoid-blastoid group achieved some success, another of the echinoderm "investments," the Crinoidea, paid a larger return. About

140 genera of cystoids and blastoids have been described, in comparison to 750 genera of crinoids. Among the attached forms, only the crinoids survived beyond the Paleozoic, and the recent marine fauna includes over 600 species. They are primarily animals of the benthos, some attached by a slender stem, and others that crawl or swim. Most of the latter, however, spend most of their life fastened to objects on the sea bottom or to seaweed. Like the other pelmatozoans, they were—and are filter feeders; gentle currents along the ambulacra move small animals and plants to their mouths. Extensive branching of the arms and numerous pinnules can create a broad filter surface.

Structural plan

Like the cocrinoids, the crinoids have a cup-shaped body* (the *calyx*) supported on an aboral *stem* (Figure 17-7). The calcareous plates that form the calyx are typically imperforate, but they are marked in some genera by superficial grooves resembling the pore rhombs of the cystoids. The aboral portion of the calyx bears the attachment for the stem, two or more symmetrical circlets of plates, and a ring of arms about its upper border. The top of this *aboral cup* is closed over by the plates of the *tegmen.* The mouth lies in or near the center of the tegmen. The ambulacral grooves diverge radially from the mouth, and continue up the arms. The grooves are ciliated and bear small closed tubes (podia) which arise from the radial canals of the water vascular system. The podia arc also ciliated, and the beat of the cilia maintains currents of water along the grooves toward the mouth. In some fossil crinoids the ring canal had a conventional external pore but in all recent forms it opens into the coelom which, in turn, opens externally by tiny pores in the tegmenal plates. The mouth and ambulacral grooves are covered in many crinoids by small plates. The anus also opens on the tegmen in an interambulacral area.

The basal portion of an arm consists of a single row of plates (is uniserial). The more distal portions are either uniserial or biserial. In many crinoids, the arms branch one or more times, and each branch may bear fine lateral branchlets, the *pinnules,* one to each arm (*brachial*) plate. The ambulacra branch with the arms and send off side branches to each pinnule. The stem attaches to the aboral end of the theca and consists of a column of disks (or *columnals*) with a central canal. Some of the columnals may bear side branches, the *cirri.* The end of the stem may be attached directly to the substrate or may expand to a rootlike or disklike anchor. Some recent and fossil crinoids have their stems reduced or

* The terminology for crinoid parts varies somewhat from that in other pelmatozoans, partly because of morphologic differences but partly because of conventional usage. These changes are confusing, but are so thoroughly embedded in crinoid paleontology as to be unavoidable.

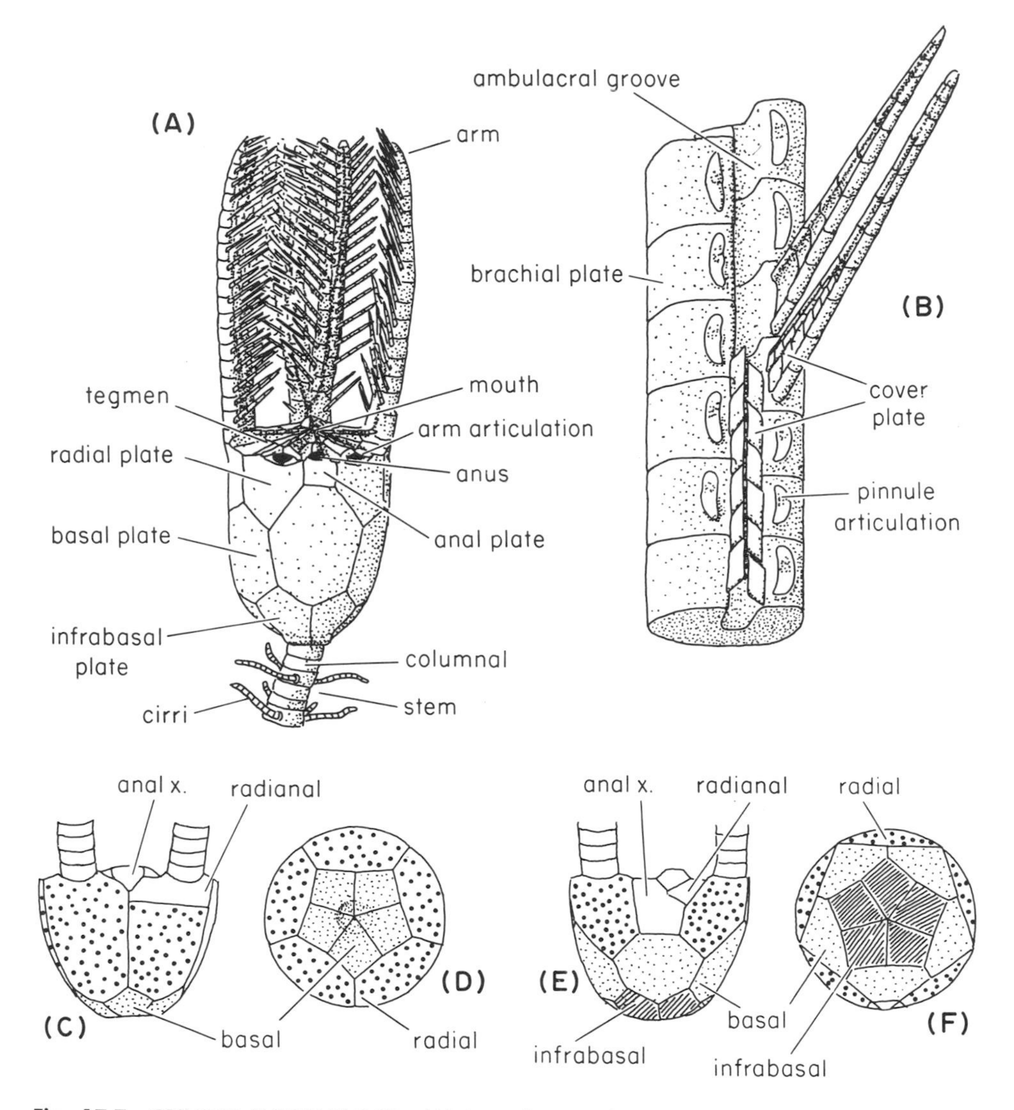

Fig. 17-7. CRINOID MORPHOLOGY. **(A)** Lateral view of crinoid with two arms and most of stem removed. **(B)** Detail of arm with most of pinnules removed. An oblique view of ambulacral face. **(C)** Monocyclic crinoid—with single cycle of plates below radials (heavy stipple). A lateral view of calyx. **(D)** The same in an aboral view. **(E)** Dicyclic crinoid—with two cycles of plates below radials. A lateral view. **(F)** An aboral view of the same calyx.

lost altogether. These swim or crawl by motions of their arms and may attach temporarily by prehensile cirri.

Recent crinoids have calyx, tegmen, arms, and stem partly covered by a leathery and/or muscular skin. The plates may be rigidly joined, articulated flexibly, or held loosely by the skin and muscular connections. The

TABLE 17-5

Glossary for Crinoidea *(figures in parentheses refer to pertinent illustrations)*

Aboral cup. Cup-shaped portion of crinoid skeleton that forms the aboral and lateral walls about the viscera. Does not include free arms, tegmen, or stem. (17-7A, C, D, E, F; *et al.*)

Ambulacral groove. See Table 17-1. Grooves cross tegmen and continue up arms giving off branches to pinnules. (17-7A, B.)

Ambulacral ray. A radius and area defined by direction of an ambulacrum radiating from mouth. (17-8).

Anal tube. Tubular elevation on tegmen bearing anal opening at end. (17-10C2.)

Anal x plate. Plate in posterior interambulacral area—adjacent or next adjacent to tegmen (oral surface). Above and to left of radianal plate. (17-7A, C, E; 17-9A to C.)

Arm. Free extension of body above aboral cup bearing lateral rods (pinnules). (17-7A; 17-10D, F.)

Arm articulation. Portion of radial plate modified for attachment of arm. (17-7A.)

Basal plate. One of a circlet of plates just below (aboral to) the arm-bearing radials. In interambulacral, "interarm," position. (17-7A, C, D, E, F; 17-9.)

Brachial plate. One of the plates that form the arms. Some of lower brachials may be incorporated in aboral cup. (17-7A, B, C, E; 17-9.)

Calyx. Portion of skeleton surrounding viscera. Comprises aboral cup and the tegmen that roofs cup. Does not include the free arms or stem. (17-7A; *et al.*)

Cirri. Appendage attached to side of stem and formed of small articulated plates. (17-7A.)

Crown. Portion of skeleton above stem—the aboral cup, tegmen, and arms. (17-7A.)

Dorsal. The aboral direction. (17-7D, F; 17-8B.)

Dorsal cup. See aboral cup.

Dicyclic. Cup type characterized by two circlets of plates aboral to the radials. (17-7E, F; 17-9, B, C, E.)

Infrabasal plate. Circlet of plates in crinoid cup below (aboral to) basal plates—the second circlet below the arm-bearing radials. Each plate in an ambulacral ray. (17-7E, F; 17-9.)

Interambulacral ray. Radius and area between two ambulacral rays. (17-8B.)

Interbrachial. Plates occurring in aboral cup between brachial plates of same ambulacral ray.

Interradial. Plates occurring in aboral cup between brachial plates of adjacent ambulacral rays. (17-9D, E.)

Monocyclic. Cup type characterized by single circlet of plates aboral to the radials. (17-7C; 17-9A, D.)

Oral. Direction and surface defined by position of mouth. (17-8B.)

Pinnule. One of several rods arranged on either side of arm or arm branch. (17-7A, B; 17-10D.)

Pinnule articulation. Portion of brachial plate modified for attachment of pinnule. (17-7B.)

TABLE 17-5 (continued)

Radial. One of circlet of cup plates to which arms are articulated. Located in ambulacral rays and in line with arm axes. The lowermost plate of the arm ray. (17-7A, C to F; 17-9.)

Radix. Rootlike structure at lower end of stem—attached to substrate.

Tegmen. Oral surface of body. May include calcareous ambulacral and interambulacral plates or be composed entirely of soft tissue. (17-7A.)

Tergal. Plate in posterior interambulacral ray of aboral cup just above basal plate and between radials. (17-9E.)

Uniserial arm. Arm at base consisting of a single row of plates. (17-7A, B.)

latter type usually go to pieces with the death of the animal and are rarely fossilized except as loose plates. Typically, the plates of the arms are articulated so that they can move by muscular contractions. The columnals and cirri are held together by ligamentous tissue.

The determination of the morphologic directions, anterior and posterior, dorsal and ventral, in echinoderms is primarily a matter of convention. For that reason, I have avoided the problem, as do most paleontologists, by use of "oral" and "aboral" and by location of the anal interray. The problem becomes somewhat clearer in the crinoids because the morphologic orientation of the embryo can be determined and applied to the adult condition.

The larva is initially free-swimming, but after a short period, it settles to the bottom and attaches by its anterior end. The viscera rotate 90° so that the mouth, initially ventral, opens at the free, posterior end. The anus develops later than the mouth and finally breaks to the surface to the left of the mouth, in the same interray as the embryonic opening of the water vascular system. The morphological anterior has become the functional ventral, and the morphological posterior the functional dorsal. This rotation is further complicated by a shift of the mouth in relation to the coelomic cavities. The water vascular system that encircles the mouth forms largely from the left anterior cavity. Thus, the mouth not only moves from anterior to posterior but brings with it the morphologic left side. Presumably, the adult could be described in terms of either the embryonic morphologic directions or the adult functional. To avoid these difficulties, a system of conventional symbols is now used to designate the various rays as indicated in Figure 17-8.

Problems in phylogeny and evolution

Probably no other class of fossils creates more problems in phylogenetic interpretation and in classification. Moore (1954) has argued for the

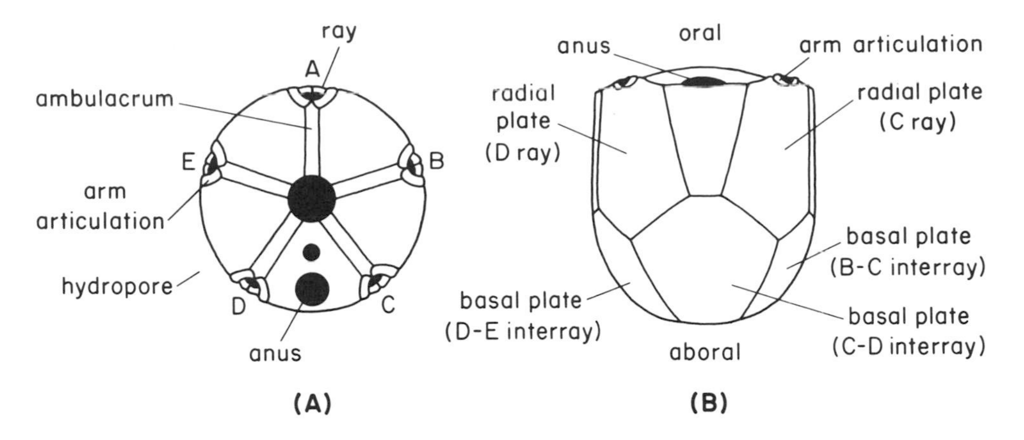

Fig. 17-8. CRINOID ORIENTATION. **(A)** Oral view of calyx. Letters indicate notation employed for different rays and interrays. **(B)** Lateral view of calyx from C-D (anal) interray.

derivation of both cystoids and crinoids from an eocrinoid base—or at least that they could have such ancestry. He has also suggested that the Crinoidea, as now defined, may include two or three separate evolutionary lines that paralleled each other in the development of crinoid characteristics.

The situation within the class is not much more satisfactory. Bather in the early 1900's erected two subclasses, one characterized by two circlets of plates in the aboral cup, the other by three. This division placed crinoid species with similar types of plate articulations, similar arrangements of plates in the calyx, and similar arm structure in different subclasses. More recent classifications (Figure 17-9) are based primarily on the mode of plate articulation and on incorporation of the interbrachial and brachial plates into the theca. These subclasses are further divided by reference to the arrangement of plates in the anal interray and to the presence of two or three circlets of thecal plates. In this interpretation, the two-circlet theca is presumed to have been derived by reduction of the lowest (*infrabasal*) circlet or by the fusion of the infrabasal plates with those of the next (*basal*) circlet.

Again, the fossil record is blank at this critical point. The oldest known crinoid, from the very early Ordovician of Europe, has a two-circlet cup, is relatively advanced in many characteristics, and could hardly have been ancestral to many—if any—of the later crinoids. Representatives of two subclasses (the inadunates and the camerates) occur in lower Ordovician and are quite distinct at first appearance. A third subclass, the Flexibilia, appear in mid-Ordovician and apparently derive from the inadunates. The fourth subclass was apparently derived from the surviving Paleozoic subclass in early Triassic time, but the phyletic links have yet to be identified.

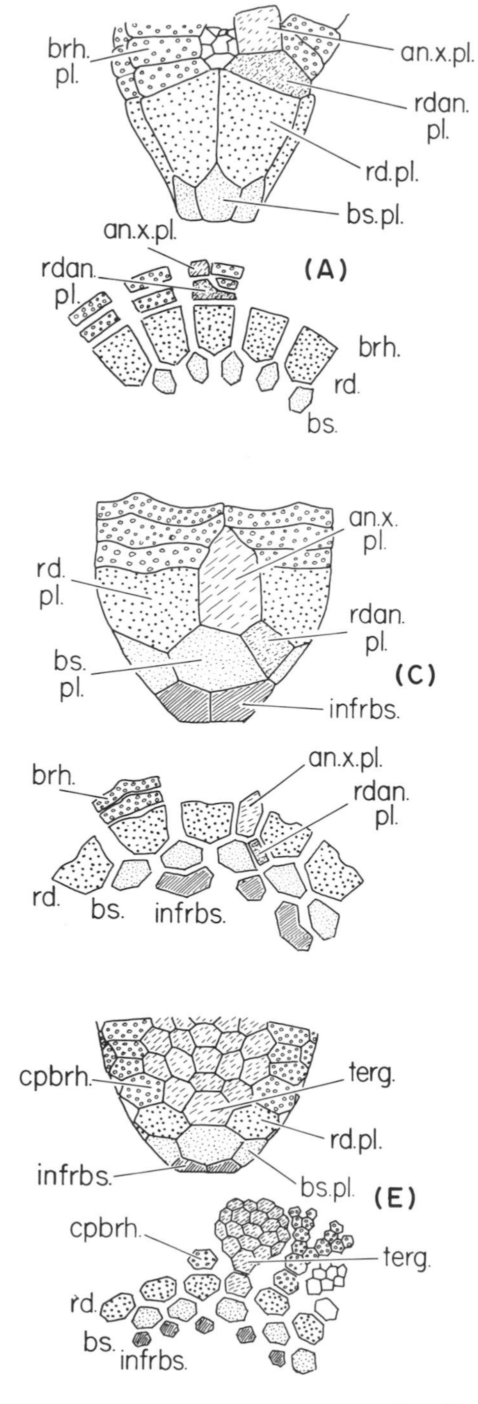
brh.
pl.
an.x.pl.
rdan.
pl.
rd.pl.
bs.pl.
an.x.pl.
rdan.
pl.
(A)
brh.
rd.
bs.
rd.
pl.
an.x.
pl.
rdan.
pl.
bs.
pl.
(C)
infrbs.
brh.
an.x.pl.
rdan.
pl.
rd.
bs.
infrbs.
cpbrh.
terg.
rd.pl.
infrbs.
bs.pl.
(E)
cpbrh.
terg.
rd.
bs.
infrbs.

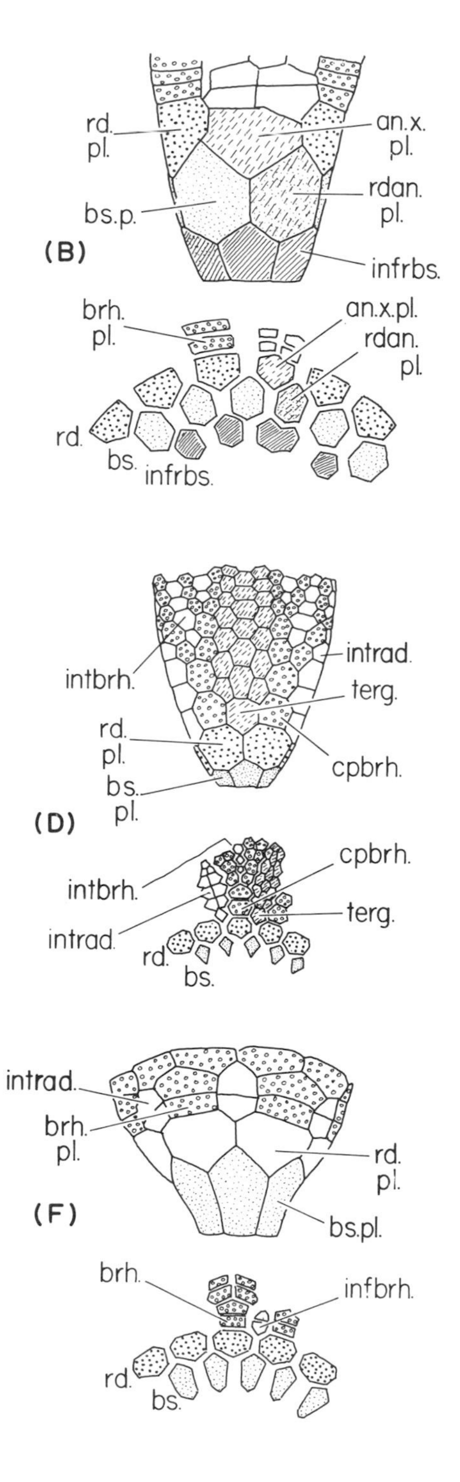
rd.
pl.
an.x.
pl.
rdan.
pl.
bs.p.
(B)
infrbs.
brh.
pl.
an.x.pl.
rdan.
pl.
rd.
bs.
infrbs.
intrad.
intbrh.
terg.
rd.
pl.
cpbrh.
bs.
pl.
(D)
cpbrh.
intbrh.
terg.
intrad.
rd.
bs.
intrad.
brh.
pl.
rd.
pl.
(F)
bs.pl.
brh.
infbrh.
rd.
bs.

Fig. 17-9.

The evolutionary trends within the orders (some of which are shown in Figure 17-10) include modification of nearly every part of the body. Many evolutionary sequences can be worked out in some detail, but, in spite of some knowledge of the adaptations in recent crinoids, paleontologists have acquired little information about the factors controlling these sequences. As Laudon (in Ladd, 1957) observed, correlated paleoecologic and evolutionary studies of crinoid assemblages might be very useful.

Presumably, the crinoids, like other animals of the sessile benthos, adapted to specific bottom habits. Many lineages tend to increase the branching of the arms and number of pinnules and, thus, increase the size of the filtering surface. Others with reduced arms may have been adapted to turbulent environments. Presumably, the character and length of the stem and its basal attachment evolved to fit different conditions of substrate, turbidity, and turbulence. The *anal tubes* or *chimneys* of some possibly carried indigestible particles away from the filters in quiet water environments. The various modifications of cup shape and plate arrangement may be related to changes in the viscera with different diet, to variation in arm structure, or to adjustment of coelomic cavities and water vascular system to the needs of circulation, respiration, and the reproductive organs.

Identification of crinoids

Although some crinoid thecae depart considerably from the simple cup shape, most retain it only slightly modified. This similarity of form through 700 or so genera does not help in identification—nor does the tendency to reduce the infrabasal plate circlet so that it is concealed below the first columnal or fused so that it resembles a columnal. These deceptive similarities make it more difficult to determine the order or subclass than to determine the genus for some crinoids. Identification depends on careful examination of the relations of thecal plates, particularly the accessory plates incorporated in the areas between the arm bases; on the determination of the plate arrangement, of the branching of the arms, and upon study of the tegmen and its variation.

Fig. 17-9. TYPES OF CRINOID PLATE ARRANGEMENTS. Lateral view of calyx and diagrammatic view of plates rotated into single plane. **(A)** Monocyclic cup; disparid inadunate crinoid; *Iocrinus.* **(B)** Dicyclic cup; cladid inadunate crinoid. **(C)** Dicyclic cup; flexible crinoid; *Lecanocrinus.* **(D)** Monocyclic cup; monobathrid camerate crinoid; *Glyptocrinus.* **(E)** Dicyclic cup; diplobathrid camerate crinoid; *Ptychocrinus.* **(F)** Secondarily monocyclic cup; articulate crinoid; *Dadocrinus.* Abbreviations: an. x pl, anal x plate; brh. pl, brachial plate; bs. pl, basal plate; cp. brh, cup brachial; intbrh, interbrachial; rd. pl, radial plate; rdan. pl, radianal plate; terg, tergal. [**(A)** to **(E)** *after* Moore; **(F)** *after* von Buch.]

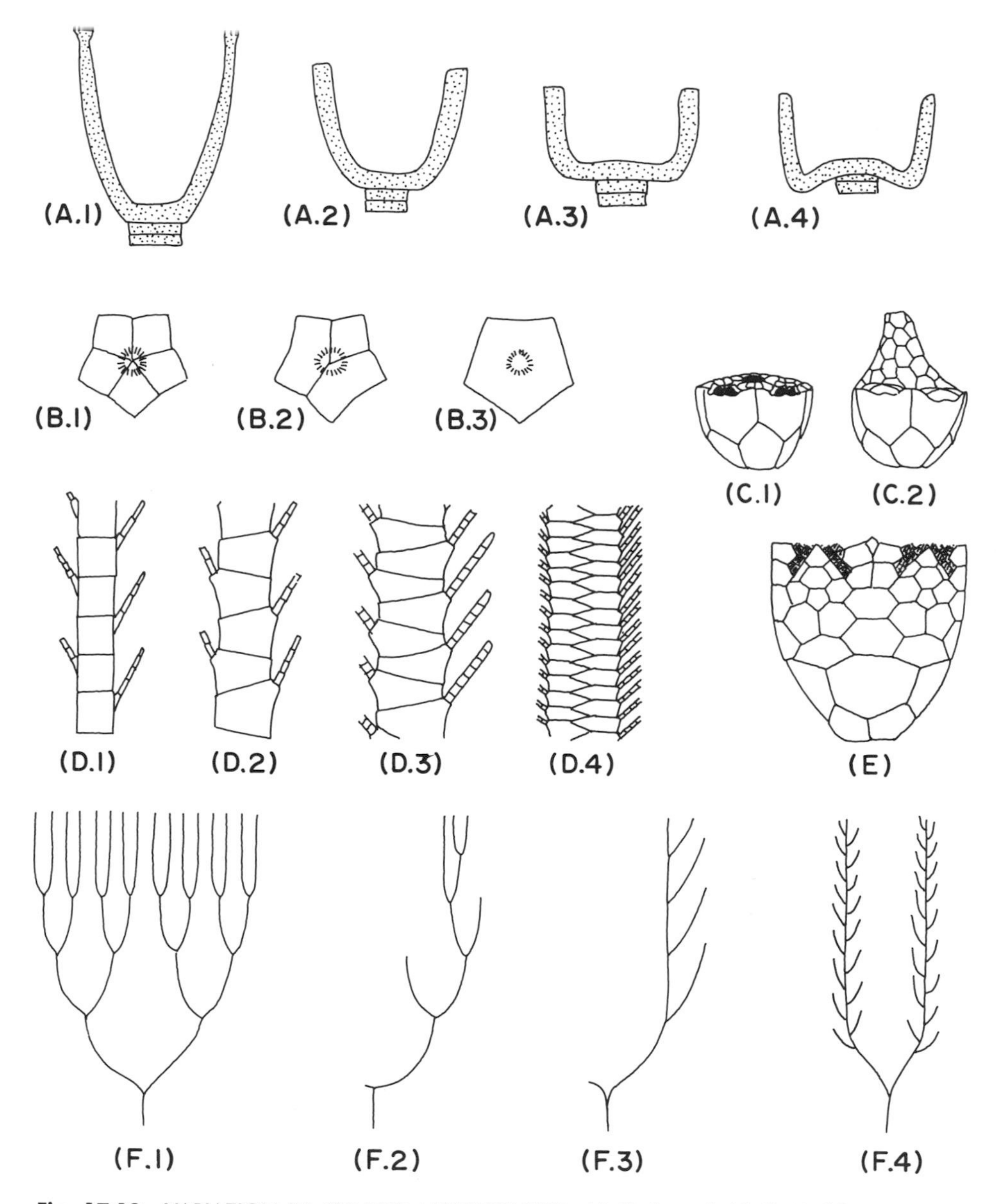

Fig. 17-10. VARIATION IN CRINOID MORPHOLOGY. **(A.1)** through **(A.4)** Modification in cup shape. The deep cup **(A.1)** the most primitive type; the depressed cup **(A.4)** the most specialized. **(B.1)** to **(B.3)** Steps in the fusion of plates in the infrabasal circlet. **(C.1)** and **(C.2)** Development of anal chimney. **(D.1)** through **(D.4)** Steps in the modification of arm structure. The primitive type **(D.1)** consists of a single row of plates with pinnules given off alternately. The most specialized has a double row of plates, each plate on each side giving rise to a pinnule. **(E)** Incorporation of brachial and interradial plates and of pinnules in the cup. **(F.1)** through **(F.4)** Various types of arm arrangement. **(F.1)** Equal branching. **(F.2)** Unequal branching, one branch of pair larger. **(F.3)** Unequal branching, series of secondary branches given off to one side of main branch. **(F.4)** Unequal branching, short side branches given off to alternate sides of main branch.

Problems in paleoecology

Of the recent crinoid species, only about an eighth are stalked; the remainder lose their attachments after the initial, fixed stage. The stalked crinoids are relatively deep-water types, and most occur in the 200 to 5000 meter range. The free-living crinoids, on the other hand, are common in shallow waters—in the littoral zone as well as deeper. Both types are found in polar seas, but the free-living types, at least, are more characteristic of warm, shallow seas.

This information is not of much value in interpreting fossil crinoid environments in rocks older than Cenozoic. The abundant stalked crinoids of the Mississippian are surely not deep water or cold water forms; the other elements of the fauna and the lithology are completely against it. Most species and genera occur in a limited range of sedimentary environments—only that seems certain. Analysis of their paleoecology depends on analysis of the complete faunal assemblages and the lithology, and, to quote Laudon (in Ladd, 1957): "With a few exceptions, work on the paleoecology of crinoids is of little consequence."

Crinoid correlation

In spite of their limited geographic distribution and the minuscule knowledge of the factors controlling distribution, many genera and species are valuable guide fossils. This value derives chiefly from their complicated arrangement of plates and arms. Since two species may be distinguished by differences in a single plate, these characteristics show very subtle changes in species populations. Even in slowly evolving lines, the calyx or arms show detectable changes in short time intervals. The complexity makes identification laborious but pays off in the end.

Unfortunately, crinoids are of less use as guides to the major time units than to the smaller ones. Not only is it difficult to determine subclass and order by casual study, but also the major categories so overlap in time that their occurrence is not diagnostic. Although the orders show marked evolutionary changes, conservative groups and some with secondarily primitive characters survived alongside the more progressive types.

Reversing the field: Free-living echinoderms

In the basal echinoderm radiation, a considerable number of the lineages had vagrant benthonic habits: the homalozoans, helicoplacoids, and a few edrioasteroids. In Ordovician rocks, additional types adapted to this mode of life appear. They include the ophiocistioids, described above, several starfish-like forms, and, before the end of the period, the first sea cucumbers and a number of echinoid lineages. The phyletic relationships of these groups to each other and to the attached echinoderms are fuzzy. Fell (1963) and Spencer and Wright (in Moore, 1966) suggest an origin of the various starfishes and brittle stars from the basal crinoid stock. Fell also argues for a common origin for the helicoplacoids, edrioasteroids, ophiocistioids, holothuroids (sea cucumbers) and echinoids (sea urchins). Finally he implies a rather closer relationship between the helicoplacoids and echinoids and between the edrioasteroids and holothuroids. On the other hand, alternate phylogenies can be supported from the evidence now available—for example the earliest known starfish are essentially contemporaneous with the earliest known crinoids.

TABLE 17-6

Glossary for Asteroidea, Somasteroidea, and Ophiuroidea (*figure numbers in parentheses refer to pertinent illustrations*)

Ambulacral plates. Plates arranged along midline of ambulacrum and flooring that structure. (17-12B, E.)

Ambulacral pore. See Table 17-1. (17-12B; 17-13B.)

Ampulla. Bulb or sac-like structure attached to internal end of podium. (17-12A, B; 17-13B.)

Ampullar cup. Cup-shaped depression in external surface of ambulacral plate which housed ampulla. (17-12E.)

Arm. Radial arm-like extension of body; bears ambulacrum.

Branchia. Slender, hollow finger-like extension of body wall. Interior connected to coelomic cavity. (17-12A, B.)

Central plate. Plate in center of aboral surface. (17-11A.)

Digestive glands. Paired digestive organs extending length of each arm.

Genital plate. One of circlet of five plates on aboral surface immediately around central plate. (17-11A, B.)

Madreporite. See Table 17-1. Located on aboral surface. (17-11; 17-12A.)

Terminal plate. Plate at end of arm. (17-11A.)

Virgalia. An articulated series of rods that extend outward from an ambulacral plate. (17-12D.)

The principal common feature of these free-living echinoderms, beside the lack of a stem, is the development of *ampullae* on the podia (Figure 17-12). These are bulblike pouches opening from the podia. Contraction of the ampullae forces water into the podia; relaxation allows the podia to relax. In many, the ends of the podia are disk shaped and act as tiny suction cups. By suitable rhythmic expansions and contractions of the podia, assisted by body musculature, the animals glide along the bottom. Only the brittle stars (ophiuroids) lack ampullae, and their podia are limited to respiratory functions. Fossil evidence indicates they lost the ampullae with a change in method of locomotion, reversed evolution, and so regained a primitive character.

Kinds of stars

The star-like free-living echinoderms, the Asterozoa, form a compact group with clear phyletic relationships and comprise the somasteroids (almost entirely extinct), the asteroids (the starfishes) and the ophiuroids (the brittle stars). The asterozoans show some rather striking resemblances to crinoids with free arms arranged radially about a central mouth. Further, in early forms, these arms bear short lateral branches resembling crinoid pinnules. Unlike the crinoids, however, the mouth is functionally ventral-opposed to the substrate—and the anus is functionally dorsal—on the aboral surface opposite the mouth.

During ontogeny, the mouth shifts, in the typical echinoderm torsion, to the left, posterior side of the body,* and the anterior part largely degenerates. The anus forms on the right posterior surface, somewhat to the ventral side. The young animal settles to the bottom with the mouth on the substrate and the anus near the center of the aboral surface. A central plate and two circlets of five plates each begin to form in the middle of the aboral surface (Figure 17-11). Additional plates are added to this initial group of *apical plates* on both oral and aboral surfaces as growth continues to shape the characteristic adult form.

Whether this developmental pattern recapitulates, in any way, the evolutionary history is uncertain. What is certain is that the essential characteristics had evolved before the start of the Ordovician and various lines had taken off in diverse directions to exploit the possibilities offered by the changed body orientation and ambulatory podia.

Somasteroidea

The earliest and most primitive genera are starfish-like types (grouped as somasteroids) from lower Ordovician rocks (Figure 17-12). In them, each ambulacral plate bears a *virgalium,* which consists of a series of rods extending laterally from the arms. These were connected by skin to form broad petaloid rays. The radial canals lie in deep grooves along the

* This shift does not occur in ophiuroids even though the right side of the body is reduced in normal echinoderm fashion. It would seem probable that this difference is an ontogenetic novelty rather than a profound phylogenetic divergence.

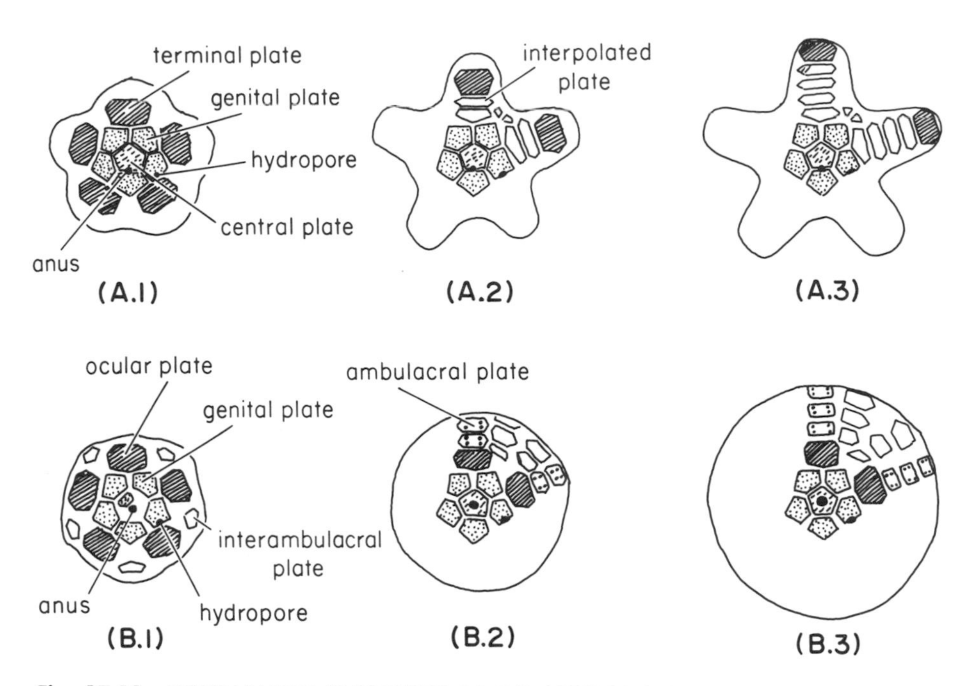

Fig. 17-11. COMPARATIVE ONTOGENY OF THE ASTEROIDS AND ECHINOIDS. Diagrammatic views of aboral surface. **(A.1)** to **(A.3)** Asteroid; plates added between inner and outer cycles of initial series. Ambulacral plates keep step on oral surface. **(B.1)** to **(B.3)** Echinoid; no plates added between inner and outer cycles. In consequence ambulacral plates appear on both oral and aboral surfaces.

midline of the ambulacral plates; the ampullae of the podia were protected in cup-shaped depressions along the borders of the ambulacral plates.

Asteroidea

True starfish, the asteroids, appeared by the middle Ordovician. They lack virgalia, and the arms are, typically, more distinct, though still enclosing large extensions of the body cavity (Figure 17-12). In primitive asteroids, the ampullae were exposed along the oral surface of the ambulacra, but in advanced types, they are recessed into the cavity of the arm and communicate with the tube feet through canals between the ambulacral plates. The anus, of course, opens on the aboral surface, as does the hydropore of the water vascular system, here through a perforated plate, the *madreporite*. The skeletal plates are (or were) held in a leathery skin. Typically, when the animal dies the skin decays and the plates separate. Only rapid burial in quiet water environments will preserve complete starfish skeletons; good specimens are rare.

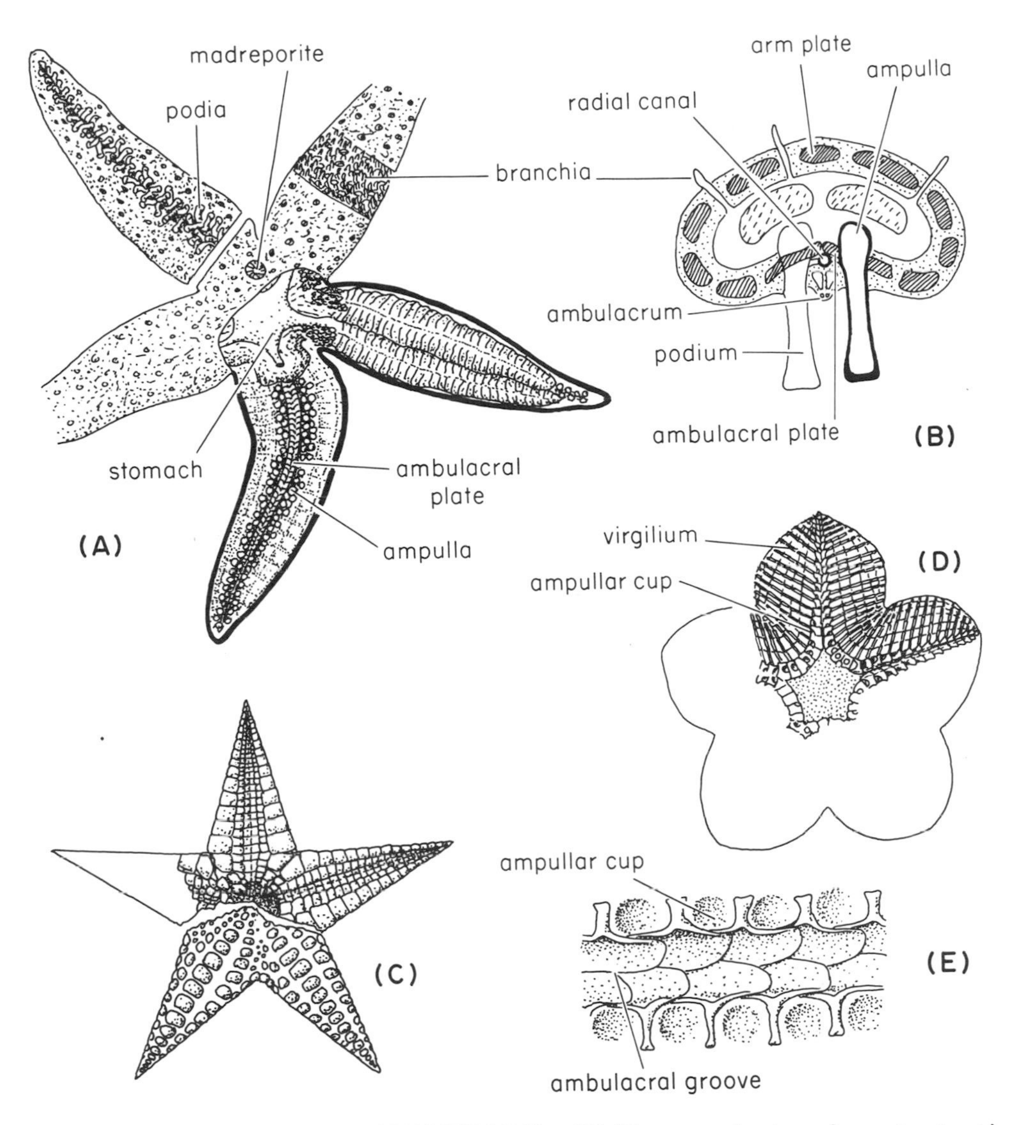

Fig. 17-12. ASTEROIDEA AND SOMASTEROIDEA. **(A)** Diagrammatic view of recent asteroid. Aboral view except that arm at upper left is rotated to show oral surface. Portion of upper right arm shows branchia. Arm at right has external surface removed to show digestive glands; digestive glands removed as well on lower arm to reveal inner surface of ambulacrum. **(B)** Diagrammatic cross section of arm. Plates cross-lined; body wall dotted; digestive gland dashed; walls of radial canal and podia in black. **(C)** Asteroid *Xenaster;* early Devonian; upper portion shows oral surface and the lower the aboral; length of arm from center of disk 5.0 cm. **(D)** Somasteroid *Villebrunaster;* early Ordovician; oral view; natural size. **(E)** Detail of somasteroid ambulacrum. Note that ampullae were set in cups and not closed off in the interior of the arm as in modern asteroids—see **(B)**. [**(C)** *after* Schondorf; **(D)** and **(E)** *after* Spencer.]

Ophiuroids

A third group of star-shaped echinoderms also appears in the early Ordovician. Their arms are more slender than those of the asteroids and are set off sharply from the body disk (Figure 17-13). In early genera,

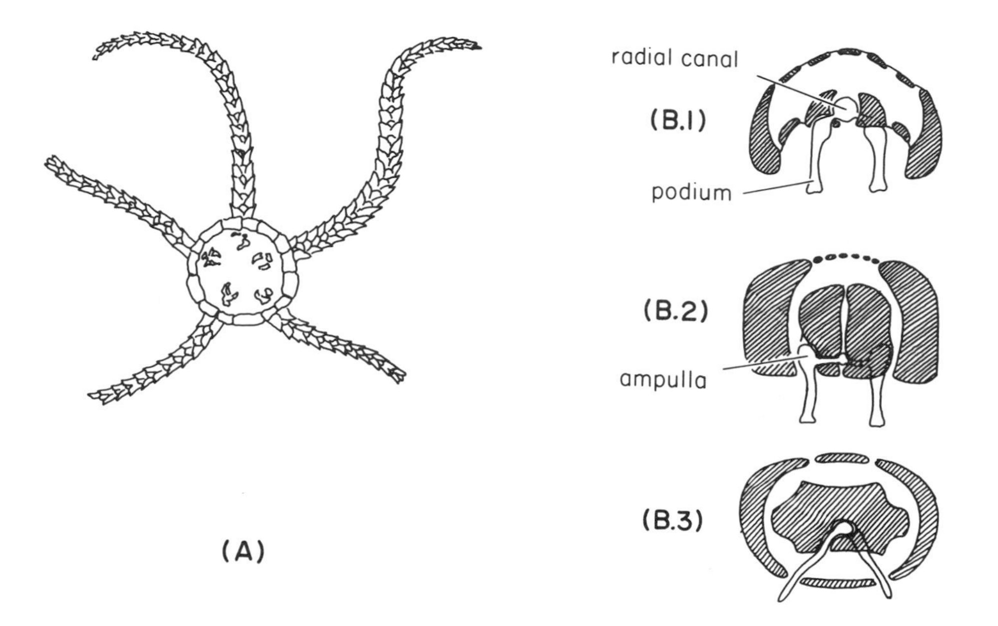

Fig. 17-13. OPHIUROIDEA. **(A)** *Ophiaulax;* late Devonian; aboral view; diameter of central disk 6 mm. **(B.1)** through **(B.3)** Evolution of orpiuroid arm structure. Cross-section of arm, plates patterned by diagonal lines. **(B.1)** Primitive condition. Radial canal not completely enclosed; podia with ampullae. **(B.2)** Intermediate condition. Radial canal enclosed between paired ambulacral plates; ampullae set in lateral cavities. **(B.3)** Advanced stage. Radial canal passes through center of solid arm ossicle; podia lack ampullae. [**(A)** *after* Ubaghs.]

the radial canal is partly enclosed by ambulacral plates; in later ones, it is completely covered. The early genera also appear to have had ampullae, at least of a sort; these were lost in subsequent evolution, and the podia lost their adaptations for locomotion. The madreporite opens on the oral surface, and in most if not all genera there is no anus. The modern ophiuroids, the brittle stars, replaced the slow, creeping locomotion possible to the asteroids by rapid movements of their arms. Many species burrow, and most are sedentary except when disturbed. In spite of these locomotor specializations, they are predominantly suspension feeders.

The symmetry—habitat problem

The stelleroids create a difficult problem in interpretation of evolutionary adaptations. An important generalization deals with the relation of general symmetry and mode of life: sessile benthonic animals evolve radial symmetry, and vagrant benthonic evolve bilateral symmetry. Some obvious exceptions exist; the stelleroids are a prominent one. In pre-echinoderm days, their ancestors had bilateral symmetry. This was lost when the proto-echinoderm became a sessile organism, and radial symmetry was superposed. Why haven't the stelleroids redeveloped bilateral symmetry? Is the whole concept wrong? Is the interpretation of metazoan origins reported back on p. 264 a fairy tale? Are the interpretations of habit from symmetry in fossil groups erroneous? Or have I overlooked an explanation for stelleroid symmetry?

The holothurians

The second successful lineage of free-living echinoderms, the sea cucumbers or holothuroids, offer quite a different adaptive pattern from the asterozoans. The body is sac-like and shows bilateral symmetry; the skeleton is typically limited to scattered plates and spicules, although two recent families, the placothuriids and ypsilothuriids possess complete tests of closely packed plates. The animal rests on one side, the mouth and anus at opposite ends. Some of these characteristics may be primitive; others, such as the symmetry, may be secondary—readoptions of characteristics once lost. The structure of the water vascular system is extremely specialized, for it is sunken into the interior and the ambulacral grooves are covered. Some of the podia are ambulatory and bear ampullae; others retain or have resumed their primitive respiratory and sensory adaptations.

The holothuroid fossil record consists largely of isolated spicules and plates; the oldest reported are from the Ordovician. Only three species are known from entire specimen: one is from the early Devonian, and two are from the Jurassic Solenhofen limestone.

A one-track mind: The Echinoidea

If the starfish and brittle stars disappoint my expectations by their failure to develop bilateral symmetry, another group, the echinoids, did so within the discernible fossil record. Superficially, the globular or discoid

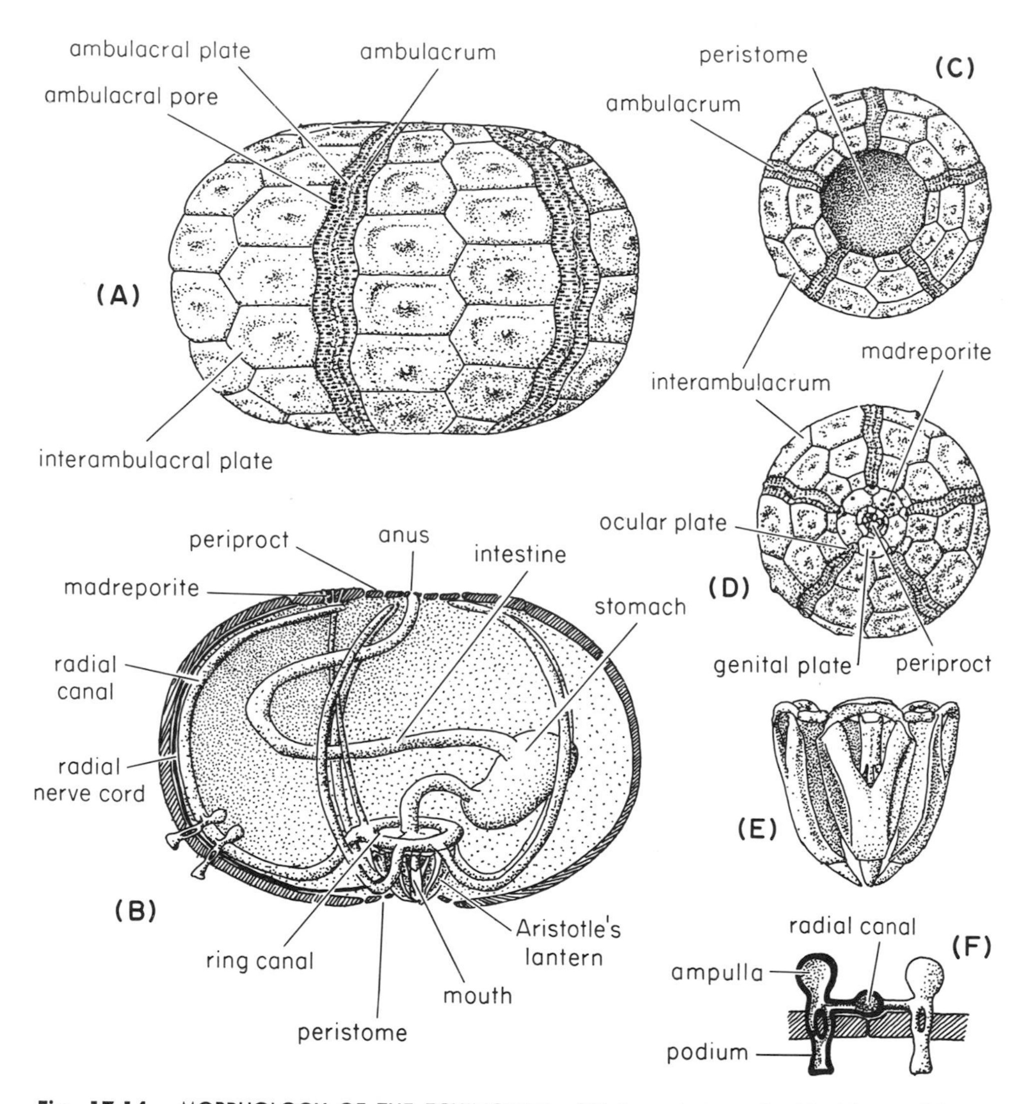

Fig. 17-14. MORPHOLOGY OF THE ECHINOIDEA. **(A)** Lateral view of echinoid test. **(B)** Diagram of soft morphology, lateral view, one side of test removed. In a living echinoid the space around the viscera shown here is partly filled by digestive glands and gonads. **(C)** Oral view. **(D)** Aboral view. **(E)** Echinoid "jaw" structure—Aristotle's lantern. **(F)** Diagrammatic cross-section of ambulacrum showing radial canal and two podia.

echinoid seems quite different from either asteroid or ophiuroid, but comparison of adult form (Figure 17-14) and ontogeny reveals some similarities. They, like the asteroids, have ambulatory podia with ampullae. This, of course, is an ancient echinoderm adaptation and means little in itself. Their radial canals are covered by overgrowths of the ambulacral plates. The podia then protrude through pores in the plates. This is reminiscent of the ophiuroid structure, though in that group the podia have lost their locomotor function. The mouth lies on the undersurface of the body; the anus, in early echinoids at least, lies near the center of the aboral surface. The ambulacra radiate from the mouth and terminate on the aboral surface very near the center.

The relationships of the echinoids to other echinoderms are not obvious. Durham (in Moore, 1966) suggests a derivation from a Precambrian source in the original echinoderm radiation. On the other hand, the structure of the tube feet and some features of skeletal development suggest a relationship to the asterozoans. The critical similarity is in the group of plates in the center of the aboral surface. They include a *central plate* (absent in many echinoids) and two cycles of five plates each (Figure 17-11). The plates of the inner cycle lie along the midline of the interambulacral areas; they each have a genital pore and so form a *genital cycle;* one is perforated by canals of the water vascular system, and thus is a madreporite. The plates of the outer, *ocular* cycle alternate with the genital plates and, consequently, lie in the midline of the ambulacra, in fact, terminate the ambulacra on the aboral surface. This is, essentially, the arrangement of plates in the apical plate system of very young starfish and brittle stars already described.

Since these plates are the first to form in the echinoderm, the immature of the two classes are very similar. After this stage in development, however, the starfish and the sea urchin begin to diverge. In the starfish and brittle stars, additional plates are inserted in a series between the ocular and the genital cycles. As the plates form, they displace the end of each ambulacrum laterally. Addition of ambulacral plates keeps pace, but few interambulacral plates are added. In consequence, free arms extend from a central disk, each arm bearing on its oral surface an ambulacral groove. In the echinoid, no plates are inserted inside the apical system, so ocular plates and the ends of the ambulacra remain near the center of the aboral surface. Plates are added in the ambulacra and interambulacral areas at equal rates so that a spheroidal skeleton is formed, growing out and down from the apical plate system.

If the asterozoans derive from a primitive crinoid, the echinoids would represent a second adaptive step. Such an explanation is consistent with the appearance of the echinoids distinctly later than the somasteroids in the Ordovician. It requires, however, the separation of the helicoplacoids and the edrioasteroids from any immediate common ancestry with echinoids.

Water vascular system

Some Ordovician echinoids show a transition between the external radial canal of the primitive echinoderm and the internal canal of the

TABLE 17-7

Glossary for Echinoidea *(figures in parentheses refer to pertinent illustration)*

Ambulacral pore. See Table 17-1. (17-15A.)

Ampulla. See Table 17-6. (17-14F; 17-15A.)

Anterior. In echinoids the direction of the ambulacral ray opposite the anal interray —where the anus has shifted from the apex of the skeleton. (17-17A, B, G; *et al.*)

Anus. May be located in periproct in center of aboral surface or be shifted down onto oral surface. (17-14B; 17-17B, E, G.)

Aristotle's lantern. Complex system of calcareous elements that surround mouth and function as jaws. (17-14B, E.)

Bourrelets. Elevated areas adjacent to peristome—located in interrays. (17-15C.)

Genital plate. Plate belonging to inner circlet in center of aboral surface. Arranged in interray positions. In some echinoids they surround periproct. (17-14D.)

Interray. Radius between two rays—corresponds to radius of interambulacral area.

Madreporite. See Table 17-1. Located in right anterior genital plate. (17-14B, D.)

Ocular plate. One of a circlet of plates around genital circlet in center of aboral surface. Each marks the aboral terminus of an ambulacrum. (17-14D.)

Oculogenital ring. Two circlets (ocular and genital) of plates in center of aboral surface. Represent initial plates of skeleton. (17-14D; 17-16B; 17-17A.)

Periproct. Area surrounding anus—covered by skin in which small plates are embedded. Plates lost in most fossils. (17-14B, D; 17-16B.)

Peristome. Area surrounding mouth. Covered with leathery skin supplemented in some by small calcareous plates. The latter are typically lost in fossils. (17-14B, C; 17-15C.)

Petal. Petal-shaped portion of ambulacrum on aboral surface. In some depressed below level of skeletal surface. (17-15B; 17-17F, G.)

Phyllode. Depressed petaloid portion of ambulacrum adjacent to mouth. Bears specialized podia. (17-15C.)

Radial nerve cord. Nerve cord just outside radial canal and parallel to it. (17-14B.)

Ray. Radius defined by location of ambulacrum.

Spine articulation. Tubercle to which spine is attached. (17-14A to D; 17-16B.)

later echinoids (Figure 17-15). In these primitive echinoids, the canal still lies in the ambulacral groove, but the groove is roofed by the growth of the ambulacral plate toward the midline. The ambulacral plates thus enclose the canal completely. In later echinoids, the internal flanges of the ambulacral plates are first reduced to a ridge, so that the canal lies in a groove on the internal surface of the ambulacrum, and finally are lost altogether.

In the primitive types, the bases of the podia and the canals connecting them in the radial canal lie between successive ambulacral plates and are partly covered by them. The ampullae were already completely inside the skeleton. In more advanced echinoids, the edges of the ambulacral plates grew around the basal part of the podia so that they projected through pores within the plates. Some Ordovician echinoids had a single

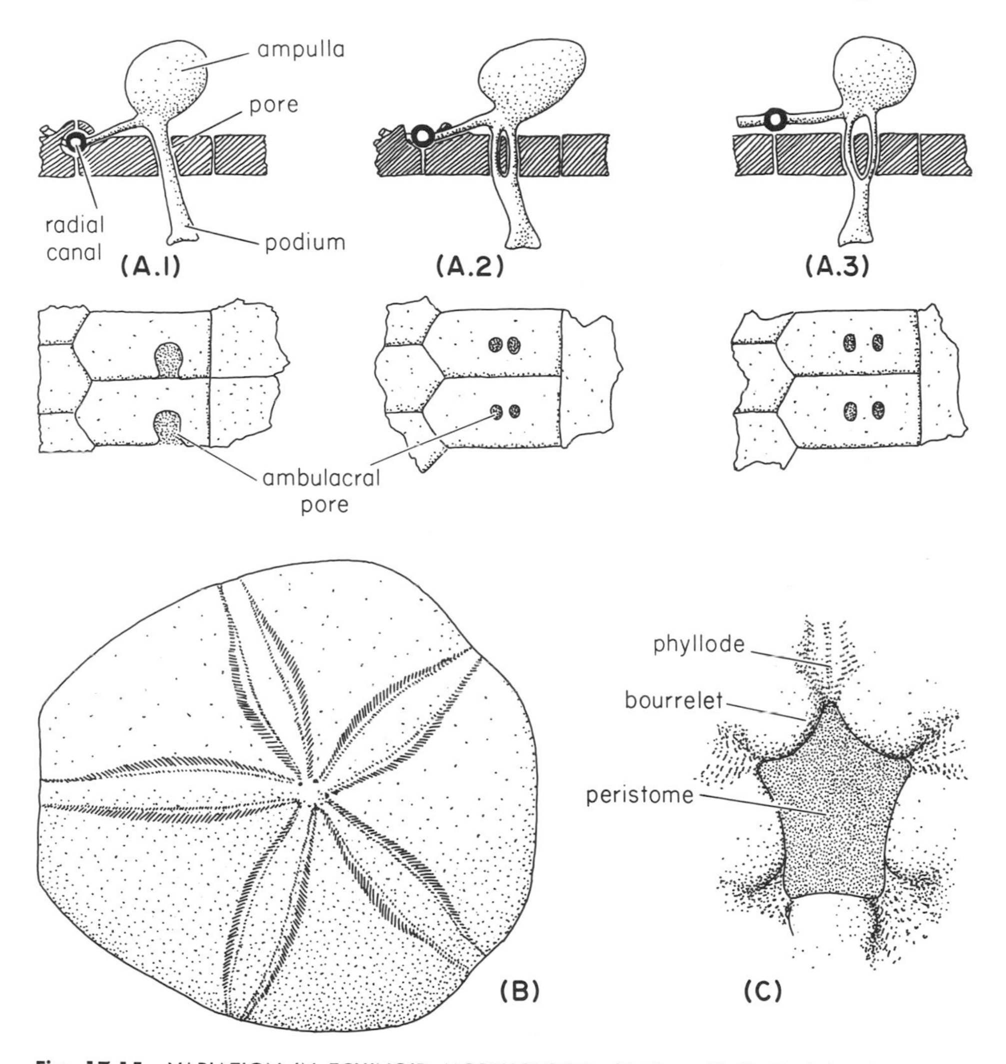

Fig. 17-15. VARIATION IN ECHINOID MORPHOLOGY. **(A.1)** to **(A.3)** Evolution in structure of podia and radial canal. Upper views, sections through ambulacrum; lower views of external surface. **(A.1)** Condition in early echinoids. Radial canal enclosed in ambulacral plates; podium opens through single pore; pore opens between ambulacral plates. **(A.2)** Intermediate stage. Radial canal in groove, and branch to podium passes through short canal in plate; podium opens through pair of pores; pores open within ambulacral plate. **(A.3)** Advanced stage. Radial canal lies entirely inside ambulacral plates, and connection to podium is not enclosed in canal. **(B)** Irregular echinoid, aboral view. Ambulacra petaloid. **(C)** The same, oral view of the peristome and adjacent surface of test. Grooves at ends of ambulacra, the phyllodes, and intervening elevations, bourrelets, form flower-like structure, the floscelle.

pore for each podium, but all later ones had a double pore. The principal result of these changes—presumably the cause for them—is protection of the water vascular system by the skeleton.

Paleozoic echinoids and their more conservative descendants had ambulatory podia that were extremely similar throughout the ambulacral areas. The more advanced Mesozoic and Cenozoic genera tended to modify this simple setup by specialization of the podia on the aboral surface for respiration and of those about the mouth for assistance in feeding. In the correlated modifications of the ambulacra, the aboral portions form elongate *petals* with slit-like, closely crowded pores, and the oral surface may have short petaloid grooves (*phyllodes*), with enlarged pores about the mouth. Many of these advanced forms tend to lose the locomotor adaptations of the podia altogether and to replace them functionally by movable spines. In these, podia other than those used in respiration and food getting are reduced in size and serve minor sensory functions.

The skeleton

That part of the echinoid skeleton beyond the apical plate system is distinguished as the *corona.* In Paleozoic and some later echinoids it is nearly spherical, with mouth and anus at opposite poles, and shows regular fivefold symmetry (Figures 17-14). The mouth lies in a circular area of leathery skin, the *peristome,* on the lower surface. The anus lies in a similar area, the *periproct,* on the upper surface. If the central plate is reduced, the anus may lie in the center of the apical system; if not, it is displaced toward the right posterior ray, but still within the apical system. Each ambulacral and interambulacral area comprises two parallel rows of plates, rather loosely articulated.

One group, the Regularia, retain the primitive shape and symmetry throughout their career (Figure 17-16). In general, they evolve a firmer articulation between plates so that the corona becomes a rigid structure. Some lines increase the number of rows of plates in the interambulacra areas and/or in the ambulacra. If the latter, the number of podia are increased correspondingly. Small plates typically appear within the peristome and periproct. Many genera appear with large spines articulated to the coronal plates. Internal calcareous supports for the viscera and jaws are developed. The jaws themselves are formed into an elaborate apparatus called *Aristotle's lantern*—"lantern" for obvious reasons (Figure 17-14). The strengthening of the test and development of spines is presumably protective. The spines also play a part in locomotion and in burrowing, for the animals can move them by muscles attached to their base. Changes in jaws and jaw supports are, presumably, related to adaptation to different types of food. The increase in coronal plates and in the

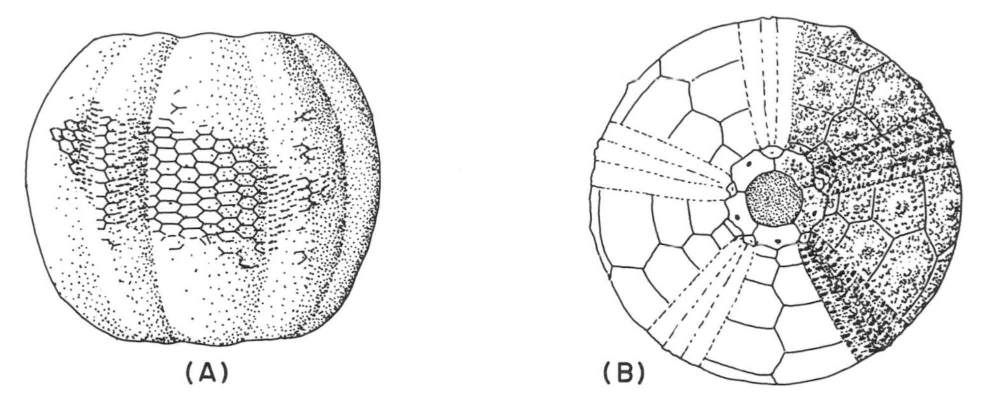

Fig. 17-16. REGULAR ECHINOIDS. **(A)** *Melonechinus;* Mississippian; lateral view; length 11.2 cm. **(B)** *Acrosalenia;* early Jurassic to early Cretaceous; aboral view; length 3.1 cm. Only two rays shown in detail.

number of podia is less obviously adaptive, and little is known of its significance.

Bilaterality

These modifications, however, are far less striking than those in body symmetry (Figure 17-17). Among some regular echinoids, the anus (and the periproct) has shifted to one border of the apical system. In the lower Jurassic, echinoids appeared with the periproct outside of this ring, shifted into one of the interareas. Similarities of these Jurassic forms to certain regular echinoids suggest that this change evolved in two or three separate lines of regular echinoids. The shift of the periproct produces, of course, a bilateral symmetry. The theca is characteristically somewhat flattened. In a few Jurassic forms—and many later genera—the periproct appears far down the upper surface of the theca, on the margin of the flattened lower surface or the lower (oral) surface itself. In some, the mouth, in turn, shifts forward so that it lies near the anterior margin. In many, these changes are associated with great flattening of the test and/or with anterior-posterior elongation; some lose their jaw structures.

These modifications are also associated with modification of the podia on different parts of the ambulacra for respiration and food getting and with the development of tracts of fine spines, the *fascioles,* which maintain a current over the body surface, even when the animal is in a burrow.

Why did the echinoids, apparently successful as radially symmetrical forms, develop bilateral symmetry? The answer may be change of habit. Most irregular echinoids burrow or else crawl along the bottom with only

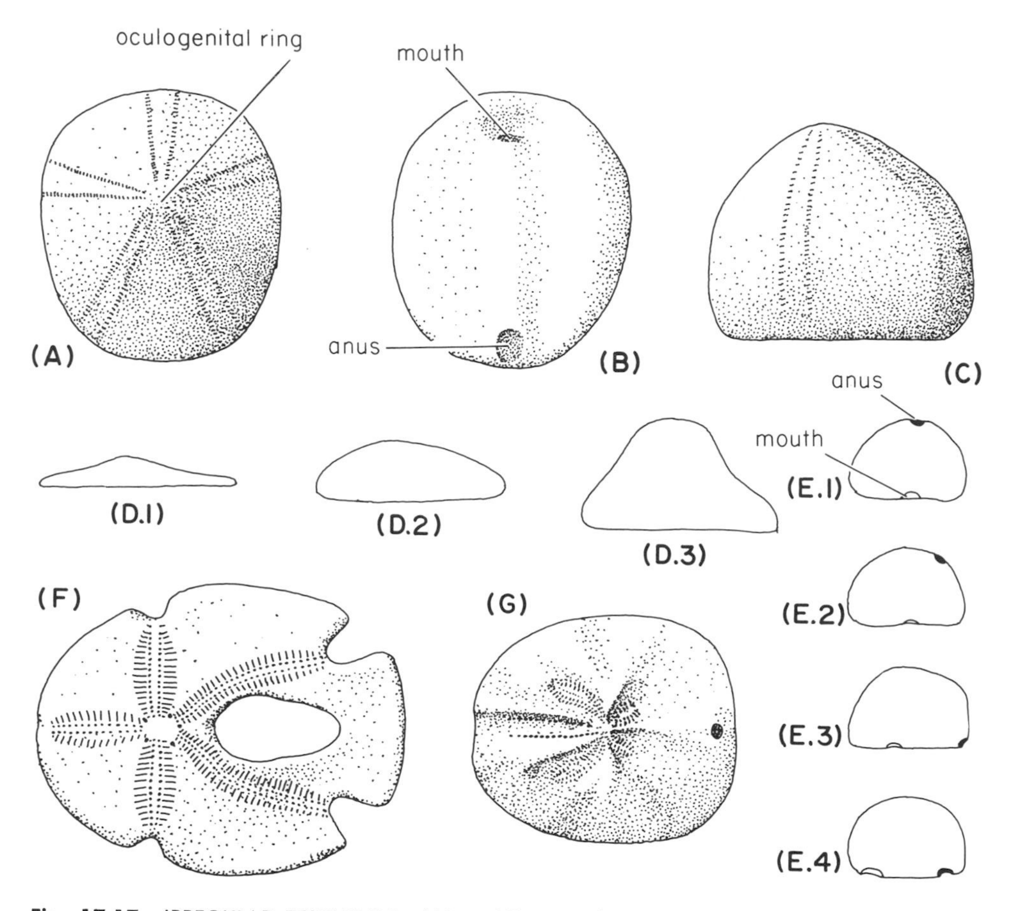

Fig. 17-17. IRREGULAR ECHINOIDS. **(A)** to **(C)** *Ananchytes;* aboral, oral, and lateral views. The anus opens on the oral surface rather than through the oculogenital ring. As a consequence the test assumes bilateral symmetry, the plane of symmetry along the mouth-anus axis. **(D.1)** to **(D.3)** Lateral views showing variation in body shape from flattened burrower to high-domed surface dweller. **(E.1)** through **(E.4)** Variation in position of anus and mouth. **(E.1)** Anus in oculogenital ring but not central; mouth in center of oral surface. **(E.2)** Anus on aboral surface but not in oculogenital ring. **(E.3)** Anus on border of oral and aboral surface; mouth shifted out of center. **(E.4)** Anus on oral surface. **(F)** *Encope;* Miocene to Recent; aboral view; length 7.5 cm. **(G)** *Hemiaster;* Cretaceous; aboral view; length 4.0 cm.

the upper part of the test exposed. Animals with these habits are characteristically flattened and many are elongate. An anus on the upper surface of the body would hardly be satisfactory for the discharge of wastes. If it were shifted down (and by its shift producing the direction "back"), it could function much more efficiently. Once shifted, the proper functioning would occur only if the animal crawled or burrowed in just one direction. Elongation and modification of the arrangement of the ambulacra would follow as an adjustment to this single direction of movement. Since some irregular echinoids are closely related to regular species, a study of fossil environments in relation to the change in form might provide a test of this hypothesis.

The geologic record

Though abundant in some rocks and with short stratigraphic ranges, most species and genera of echinoids were so limited by ecologic factors that they are of little use as guide fossils. This implies that they would be excellent subjects for paleoecologic work, but Cooke (in Ladd, 1957) cited only fourteen papers that deal with the ecology of post-Paleozoic echinoids; Cooper (in Ladd, 1957), eight papers on the ecology of the much rarer Paleozoic echinoids.

REFERENCES

Brower, J. C. 1966. "Functional Morphology of Calceocrinidae with Description of Some New Species," *Jour. of Paleontology,* vol. 40, pp. 613-634.

Caster, K. E. and J. K. Pope. 1960. "Morphology and Affinities of *Eocrinus,* An Archetype of the Echinodermata," *Geol. Soc. of America,* vol. 71, pp. 1840-1841.

Durham, J. W. 1966. "Evolution Among the Echinoidea," *Biological Reviews,* vol. 41, pp. 368-391.

———. 1967. "Notes on the Helicoplacoidea and Early Echinoderms," *Jour. of Paleontology,* vol. 41, pp. 97-102.

Fell, H. B. 1963. "The Evolution of the Echinoderms," *Smithsonian Inst., Ann. Rept. 1962,* pp. 457-490.

Hyman, L. H. 1955. *The Invertebrates: vol. 6, Echinodermata.* New York: McGraw-Hill.

Kier, P. M. 1965. "Evolutionary Trends in Paleozoic Echinoids," *Jour. of Paleontology,* vol. 39, pp. 436-465.

Ladd, H. S. 1957. "Treatise on Marine Ecology and Paleoecology," vol. 2, *Geol. Soc. of America, Memoir 67,* pp. 955-982.

Lane, N. G. 1963. "Meristic Variation in the Dorsal Cup of Monobathrid Camerate Crinoids," *Jour. of Paleontology,* vol. 37, pp. 917-930.

Macurda, D. B., Jr. 1965. "The Functional Morphology and Stratigraphic Distribution of the Mississippian Blastoid Genus *Orophocrinus,*" *Jour. of Paleontology,* vol. 39, pp. 1045-1096.

Moore, R. C. 1954. "Pelmatozoa," *Mus. Comp. Zoology, Harvard, Bull,* vol. 112, pp. 125-149.

———. 1966. *Treatise on Invertebrate Paleontology, Pt. U, Echinodermata 3.* New York: Geological Society of America. The asterozoans and echinozoans.

Philip, G. M. 1965. "Classification of Echinoids," *Jour. of Paleontology,* vol. 39, pp. 45-62. A dissent from the phylogenetic arrangement and classification in Moore, 1966.

Raup, D. M. 1958. "The Relation between Water Temperature and Morphology in *Dendraster*," *Jour. of Geology*, vol. 66, pp. 668-677.

———. 1962. "The Phylogeny of Calcite Crystallography in Echinoids," *Jour. of Paleontology*, vol. 36, pp. 793-810.

Robison, R. A. 1965. "Middle Cambrian Eocrinoids from Western North America," *Jour. of Paleontology*, vol. 39, pp. 355-364.

18

Brothers under the skin: Protochordata

Most of the great advances in animal organization—polarity, bilateral symmetry, the coelom, segmentation, cephalization—apparently occurred as adaptations to active benthonic or nektonic habits. The elaborate specializations that sessile organisms play upon a structure theme seem dead ends—so far as organizational innovation is concerned. The concept need not be belabored further; the evidence is particularly clear among the vertebrates and their relatives. Unfortunately, in this group (chordates and protochordates) as in so many others, paleontologists lack direct evidence. They have as fossils the vertebrates, first known from Ordovician rocks and thereafter reasonably common, and the graptolites, an extinct group whose relationships are questioned and whose specialized structure gives little information on their ancestry. The members of the group that could contribute most to a reconstruction of evolutionary pattern are known only from relatively specialized recent species. The next few pages, therefore, will be primarily zoological, will deal largely with details of ontogeny, and will be highly speculative.

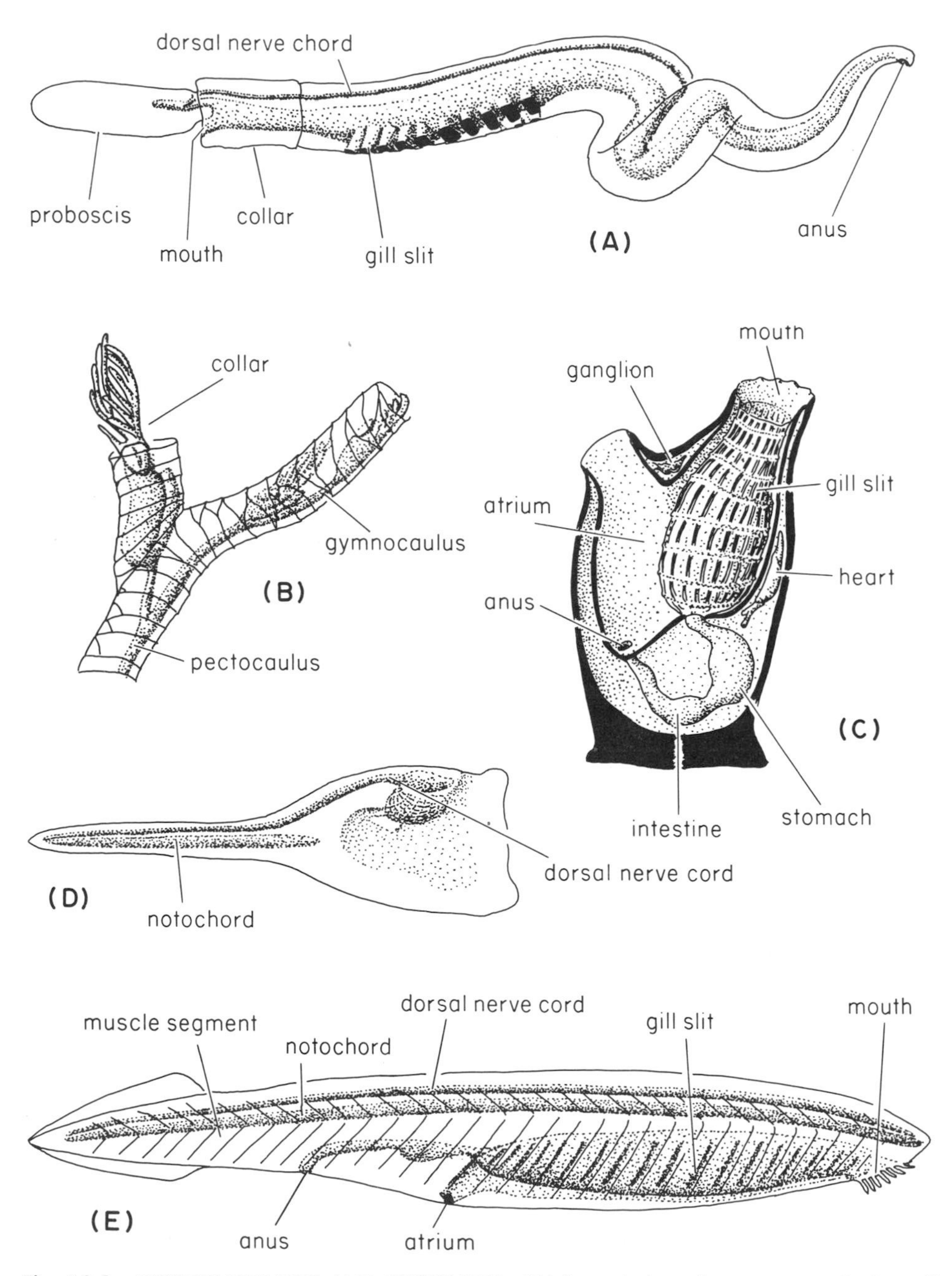

Fig. 18-1. PROTOCHORDATES AND CHORDATES. **(A)** Lateral view of acorn worm; body wall cleared to show part of viscera. **(B)** A portion of a pterobranch colony showing a mature individual to the left and an immature individual and the "parent bud" to the right. Skeleton cleared to show soft parts. **(C)** A diagrammatic lateral view of a sectioned sea squirt. **(D)** The larva of a sea squirt—body wall cleared to show viscera. **(E)** *Amphioxus*. A lateral view with body wall and muscle segments cleared.

Characteristics of the protochordates and chordates

Burrowers and tube dwellers

The acorn worms and the pterobranchs are in some ways the simplest and, presumably, the most primitive of the whole chordate-protochordate complex. The acorn worms have an elongate, bilaterally symmetrical body (Figure 18-1). The mouth is anterior, of course, and the anus posterior but not terminal. The body wall immediately behind the mouth is quite thick and distinct from the remainder; this structure is called the *collar*. A conical *proboscis*, stiffened by a short internal rod, lies in front of and above the mouth. A dorsal nerve cord extends backward from the base of the proboscis. The anterior end of the cord is swollen and hollow. The ventral cord begins farther back near the posterior margin of the collar. On either side of the body for a distance behind the collar, the *gill slits* pierce the body wall and open into the anterior part of the gut. A capacious coelom develops like that of the echinoderms as pouches from the embryonic gut. The ciliated planktonic larva also resembles that of the echinoderm. They burrow by use of the collar and proboscis and feed on organic particles in the bottom muds. They include only a few species, are entirely marine, and are entirely unknown as fossils.

The pterobranchs differ from the acorn worms principally in the presence of branched arms on the collar and in the small size of the proboscis (Figure 18-1). Unlike the acorn worms, the pterobranchs are sessile tube builders. They, too, are marine, but feed on particles and organisms filtered from the water by the ciliated arms. Probably as an adaptation to this sessile life, the gut turns upward and forward so that the anus lies above the head. Pterobranchs bud asexually to form colonies. They seem to connect the protochordate stock with the extinct colonial Graptolithinia. I'll cover this problem at greater length in a succeeding section.

Sea squirts

The acorn worm with its elongate shape and bilateral symmetry at least suggests a vertebrate form. The adult sea squirts could hardly be less like a vertebrate (Figure 18-1). They are sessile sac-like animals. The upper end of the sac is drawn into two tubes—one opening into the mouth, the other into a large chamber, the *atrium*. The gut for a distance below the mouth is much expanded and perforated by gill slits that empty into the atrium. The anus likewise empties into the atrium. The beat of cilia around the gill slits maintains a current of water in through the mouth, through the gill slits into the atrium, and back out through the atrium's external opening. The animal filters small food particles from the water as it passes through the gill slits. There is no trace of a coelom. Remarkably, the external covering of the animal is a cellulose-like material. No fossil sea squirts are known.

As in most sessile marine organisms, the sea squirts have a planktonic larva. These somewhat resemble a tadpole with a large "head" and a slender muscular tail. At metamorphosis, the larva attaches by the anterior end of the head; the tail degenerates; because of the relative growth, the mouth shifts toward the posterior free end; the atrium forms, and an external atrial opening develops. The most interesting feature of the larva, however, is the structure of the tail. In its dorsal part is a nerve cord. This extends into the "head," where it is enlarged and hollow. Below the nerve cord is a stiff rod, the *notochord,* which ends before entry to the head.

Almost an ancestor

There is one other recent group of animals, the cephalochordates, that show affinities to the vertebrates. Unlike the preceding two, they look a good deal like vertebrates, for they are small, slender, fish-like animals (Figure 18-1). Like the others, they have an extensive series of gill slits. These slits open out like those of the sea squirts into a chamber, the atrium, which has an excurrent pore in front of the anus. Again, like the sea squirts, they filter food particles from water passing through the gill slits. The anus itself is subterminal so that a distinct tail lies behind it. The animal has a hollow dorsal nerve cord running the length of the body. Below it and extending into the tip of the head is the notochord. This latter structure consists of a tough cover and a nearly gelatinous interior, and functions as a stiff but flexible beam on which the body muscles can pull. The muscles themselves are divided into a series of transverse segments, the *myomeres.*

Although capable of swimming, the animal usually buries itself in the bottom sediment. The beat of cilia in the anterior part of the gut maintains an inflowing current through the mouth; food particles are filtered out as the water passes through the gill slits into the atrium.

The reinforced notochord

This brief description of the cephalochordates need not be altered much to fit the vertebrates (Figure 18-2). The vertebrate notochord does not reach the tip of the head but ends some segments behind. Otherwise they (we) have a hollow dorsal nerve cord and at least rudiments of

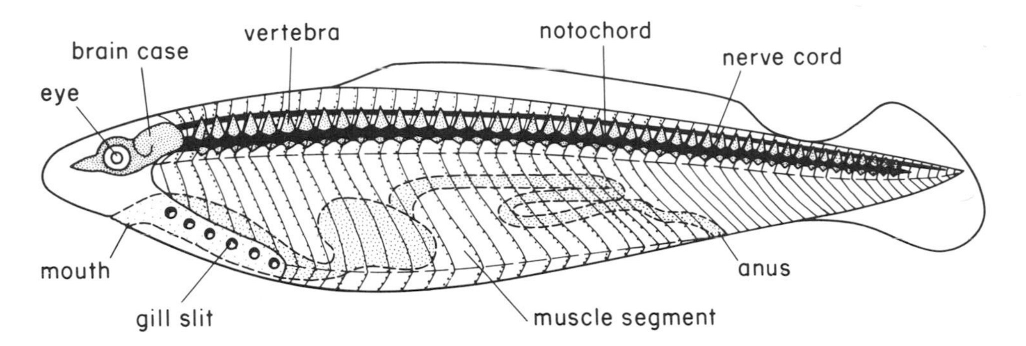

Fig. 18-2. GENERAL PLAN OF THE VERTEBRATA. Lateral view of a hypothetical vertebrate. Body wall cleared to show internal structures.

gill slits. In all living vertebrates and most, if not all, fossil ones, the notochord is supplemented or replaced by bony or cartilaginous blocks, the *vertebrae,* that form around it. Since the muscles in each segment attach to two adjoining vertebrae, they produce more efficient and controlled motion than the muscles working on an unsegmented notochord.

Structural plans and ancestors

All of these various friends and relatives of ours possess gill slits, a notochord or some pretense of one, a hollow dorsal nerve cord, and a distinct tail behind the anus. These form the major elements in the structural plan of the protochordate-chordate group. Presumably, their common ancestors had these characteristics; each, by itself, might be considered a parallel development, but, because they have a constant topographic relationship within the body and a similar mode of development, more than likely they were inherited.

What were the other characteristics of the common ancestor, and his mode of life? Everyone concerned with vertebrate origins wishes he knew. The acorn worm has specialized for burrowing with the proboscis-collar structure; the pterobranch for a sessile tube dwelling habit with "arms"; the sea squirt for a sessile filter-feeding existence; the cephalochordate and vertebrates for locomotion with segmentation. Perhaps the common ancestral species was like an acorn worm which lacked a proboscis and collar and which lived crawling or swimming along the bottom. But gills seem to have developed as a filter-feeding mechanism—a feature of sessile animals. If so, perhaps the ancestral protochordate was like the sea squirts but without their peculiarities. The other protochordates and chordates would have evolved by retardation of development so that the larval form was retained in maturity.

TABLE 18-1

Protochordata and Chordata

PHYLUM PROTOCHORDATA

Cambrian to Recent. Coelomate; level of organization approximates that of Annelida. Possess some trace of notochord and gill slits. Nonsegmented. Distinguished by muscular collar that may bear tentacles. Marine; burrowing, benthonic or planktonic.

CLASS ENTEROPNEUSTA

Recent. Solitary; worm-like; collar simple; muscular proboscis. Burrowing. Mud strainer.

CLASS PTEROBRANCHIA

Ordovician to Recent. Colonial; chitinoid skeleton; collar bears tentacles. Lack specialized initial cup, the sicula. Sessile benthonic. Filter feeders.

(?) CLASS GRAPTOLITHINA

Cambrian to Mississippian. Branching colonial; chitinoid skeleton; soft parts largely unknown Sicula. Planktonic and/or sessile benthonic. (?) Filter feeders.

PHYLUM CHORDATA

Ordovician to Recent. Coelomate; level of organization approximates that of Arthropoda. Notochord and gill slits present, at least in embryo.

SUBPHYLUM TUNICATA

Recent. Larval has notochord in tail region. Notochord lost but gill basket much enlarged in adult; body enclosed in sac-like, cellulose tunic. Marine, predominately sessile benthos. Filter feeders.

SUBPHYLUM CEPHALOCHORDATA

Recent. Notochord large, extends to tip of head. Gill basket large; body segmented and fish-like. Vagrant benthos. Marine and brackish. Filter feeders.

SUBPHYLUM VERTEBRATA

Ordovician to Recent. Notochord primitively large but not extending to tip of head. Skeletal elements develop in mesenchyme. Notochord supplemented or replaced by series of cartilaginous or bony vertebrae. Marine, freshwater, and terrestrial. A wide variety of adaptations.

CLASS AGNATHA

Ordovician to Recent. Notochord large. Gill chambers. Elongate, fishlike, but without paired appendages or jaws. In most, head is covered by bony armor. Most lineages extinct after Devonian. Freshwater and marine. Sluggish to active, benthonic and nektonic. Filter feeders and mud grubbers.

CLASS PLACODERMI

Silurian to Mississippian. Notochord large. Gill chambers generally reduced. Elongate, fish-like; with bony or cartilaginous jaws and some type of paired appendages. Freshwater and marine, nektonic. Predators and herbivores.

CLASS CHONDRICHTHYES

Devonian to Recent. Skeleton entirely of cartilage. Notochord reduced; cartilaginous vertebral elements. Jaws propped against skull by hyoid gill arch. Two pairs of paired appendages. Predominantly marine, nektonic, and predaceous.

TABLE 18-1 (continued)

CLASS OSTEICHTHYES

Devonian to Recent. Bony skeleton. Notochord reduced. Jaws variously attached to skull. Two pairs of paired appendages. Marine and freshwater; nektonic; predators, and herbivores.

Subclass Actinopterygii
Middle Devonian to Recent. Fins supported by slender rays. End of vertebral column turned into upper lobe of tail. Nostrils have no internal openings.

Subclass Choanichthyes
Early Devonian to Recent. Paired fins supported by stout bony rods. Nostrils have internal openings. Lungs variously developed. End of vertebral column is straight with lobe of tail above and below. Predominantly freshwater, predaceous.

CLASS AMPHIBIA

Late Devonian to Recent. Internal nostrils and typically well-developed lungs. Paired limbs; most have ossified vertebrae; except in primitive forms, no fish like tail. Eggs laid and develop in fresh water only. Freshwater and terrestrial; predominantly predaceous, some herbivorous.

CLASS REPTILIA

Pennsylvanian to Recent. Well developed limbs and limb girdles; vertebrae bony and well articulated. Gill slits absent in adult. Eggs laid on land with protective shell and membranes. Marine, freshwater, terrestrial, and aerial. Predators and herbivores.

CLASS AVES

Jurassic to Recent. Forelimbs modified into wings. Feathers. Endothermal. Terrestrial and aerial; secondarily aquatic. Predators and herbivores.

CLASS MAMMALIA

Latest Triassic to Recent. Hair; mammary glands. Endothermal. Most bear young partly developed; some have marsupium; others have internal embryonic development with placenta. Terrestrial, aerial, aquatic. Predators and herbivores.

Classification of this phyletic complex varies with the classifier. One extreme is to place each group in a separate phylum; the other is to lump them all under Chordata. Or the primitive protochordates may be shoved together, and the advanced protochordates and vertebrates thrown into the phylum Chordata. Many zoologists feel that the sea squirts are closer to cephalochordates than to acorn worms and place them, also, as a subphylum of the Chordata. Table 18-1 expresses this last viewpoint, which seems to me to be the most nearly correct.

A problem in multiple personalities: Graptolithina

The witches of *Macbeth* with their "Double, double, toil and trouble" prophesied the paleontological career of the graptolites. Probably no other common group of fossils has produced so much argument about its taxonomic affinities. The work of Kozlowski on the fine details of their morphology may have settled the problem. We can hope so anyway.

The morphologic problem

As commonly seen, the graptolite skeleton is relatively simple (Figure 18-3). It consists of cuplike proteinaceous *thecae* arranged in series along a branch, the *stipe*. The stipes may be solitary or form a branching net. The stipes may also be connected by cross bars, the *dissepiments*. In some, the colony, the *rhabdosome,* was supported at one point by a thread, the *nema*. The stipes branch from the end of the nema. The first cup, the one at the end of the nema, is termed the *sicula*. Presumably, each theca bore an individual animal, and the sicula represents the initial individual of the colony.

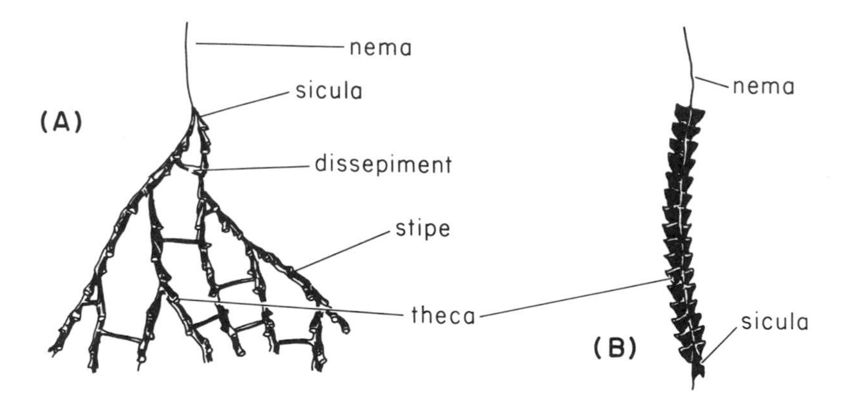

Fig. 18-3. GENERAL FORM OF THE GRAPTOLITES. **(A)** Dendroid graptolite. **(B)** Graptoloid graptolite.

So much can be observed in the ordinary graptolite preservation in black shales. Compaction leaves the graptolite only as a carbonized streak hardly distinguishable from a pencil mark. According to burial position, individual thecae are more or less distinct.

Occasionally graptolites are found in limestone or chert matrix. Because of the massive character of the rock, the fossils are only slightly distorted. Because of the difference in composition between matrix and fossil, the paleontologist can use acids that dissolve calcite or chert but do not attack the fossil. Freed from the rock, the graptolite can be studied directly in a free mount or can be imbedded in plastic or paraffin and sectioned.

The earliest graptolites, from late Cambrian deposits, have a many-branched (dendroid) structure. Free mounts and sectioned specimens show three distinct kinds of thecae (Figure 18-4). A single individual, the *budding zooid,* gave rise to three zooids at each division; each zooid built a distinct type of theca.* One of these is the stolotheca formed by the new budding zooid. Since this, in turn, gives rise to another set of zooids, stolothecae form the axis of the stipe. The other two zooids construct thecae that open laterally from the stipe. These zooids differed in some unknown way, for one formed a larger theca, the *autotheca,* than did the other, the *bitheca.* Probably each one was adapted for a

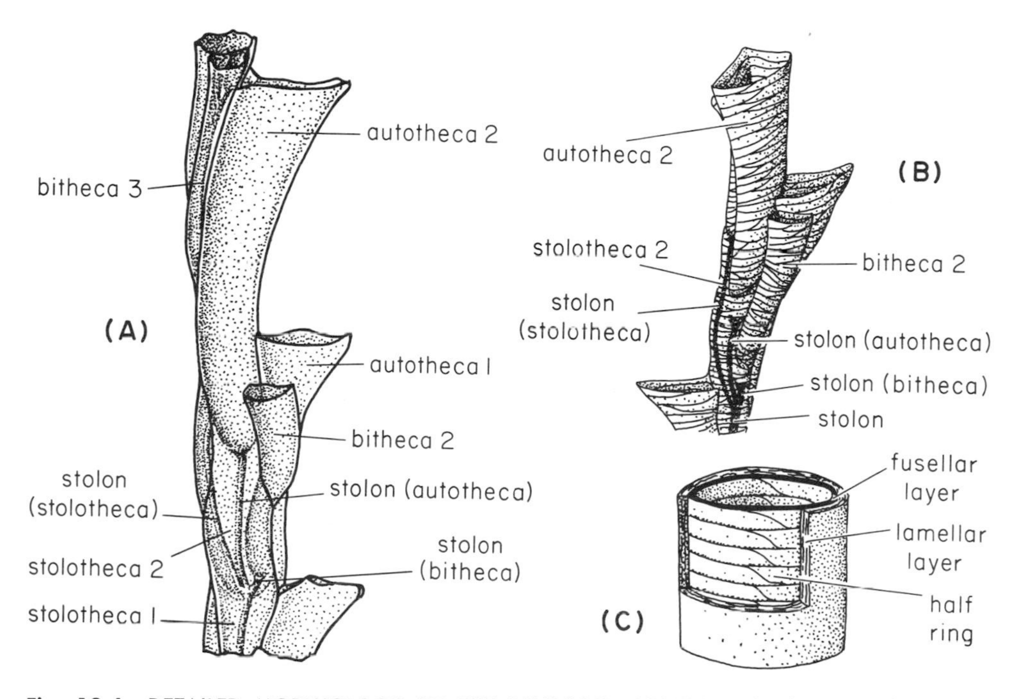

Fig. 18-4. DETAILED MORPHOLOGY OF THE DENDROID GRAPTOLITES. **(A)** Lateral view of portion of stipe. Three buds arise from stolotheca 1. One of these forms stolotheca 2, one forms bitheca 2, and one forms autotheca 2. A tube, the stolon extends along the stolotheca and gives off side branches to the bitheca and to the autotheca. **(B)** Lateral view of portion of stipe showing fusellar half rings that form part of skeleton. Labeling of thecae corresponds to that in **(A)**. **(C)** Detail of skeleton showing inner and outer layers. (*Based on* Kozlowski, 1947.)

different function in either food getting, protection, respiration or reproduction. The thecae are connected internally by a delicate tubule, the *stolon,* which extends through the stolotheca and divides into three branches, one to each theca of the next generation. The wall of the theca consists of two layers. The inner is composed of half rings that

* Decker and Gold, 1957, described two other thecal types, gonotheca, which they attribute to female individuals, and nematotheca, which they believe housed nematocysts of coelenterate type. The specimens, preserved as compressions in shale, are very poor, and these features have not been observed in well preserved, 3-dimensional specimens.

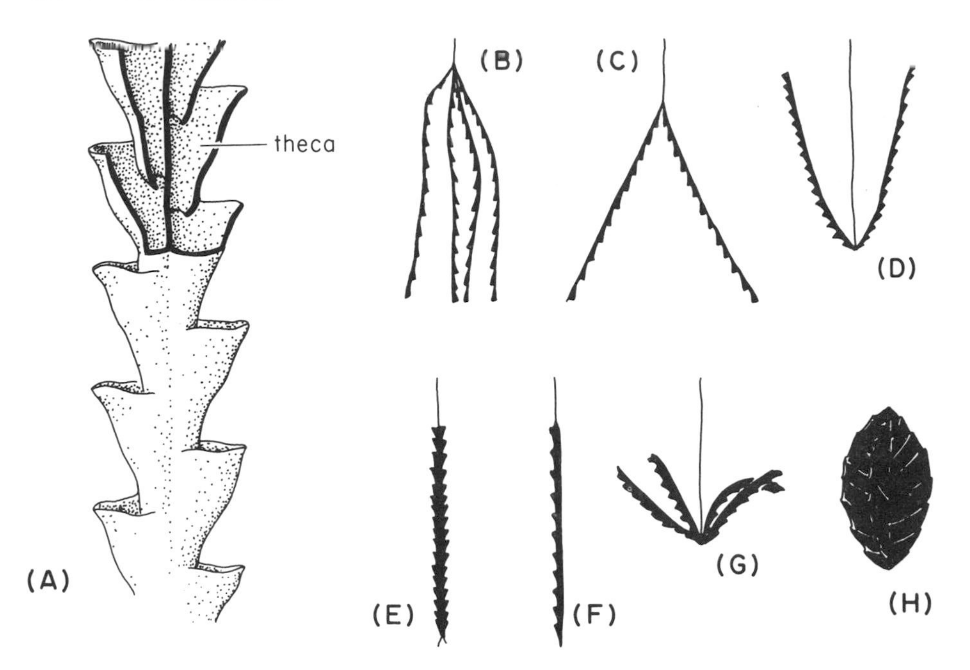

Fig. 18-5. FORM AND VARIATION OF GRAPTOLOIDS. **(A)** View of detail of stipe—upper portion sectioned. **(B)** *Tetragraptus;* early Ordovician. Four-branched type; stipes pendant. **(C)** *Didymograptus,* early to middle Ordovician. Two-branched type; stipes pendant from nema. **(D)** *Dicellograptus;* middle to late Ordovician. Two-branched type—stipes turned upward. **(E)** *Diplograptus.* Two stipes—turned upward along and attached to nema. **(F)** *Monograptus;* Silurian. Like *Diplograptus,* but reduced to single row of thecae. **(G)** *Tetragraptus* (see **B**); stipes turned upward. **(H)** *Phyllograptus.* Four stipes turned upward and attached to nema.

dovetail to form zigzag longitudinal sutures. The outer layer consists of thin concentric sheets deposited on top of one another.

In the early Ordovician, graptolites appeared without stolons and with only one kind of theca (Figure 18-5). Although their simpler structure would suggest that they were more primitive, they are connected to the earlier types by intermediate forms so that the simplicity is secondary. Characteristically, the rhabdosome in these specialized types had fewer stipes, and the stipes, rather than growing away from the nema, turned outward or even grew back along it. Internally, each theca opens into a common canal.

The perfect crime

The relationship of the graptolites to other animals was, and perhaps still is, a major mystery. They have been considered inorganic structures and remnants of plants. Even after their animal origin was agreed on, they

TABLE 18-2

Glossary for the Graptolithina

Autotheca. Largest tube of three produced at each budding in the development of a dendroid colony. (18-4A, B.)

Bitheca. Small tube formed at each budding in the development of a dendroid colony. (18-4A, B.)

Dissepiment. Crossbar uniting adjacent branches (stipes) of dendroid colony. (18-3A.)

Fusellar layer. Inner layer of the two that compose the graptolite tube. Consists of "half" rings. (18-4B, C.)

Half ring. One of series of chitinoid half rings that are joined to form the inner wall of the graptolite tube. (18-4B, C.)

Lamellar layer. Outer layer of graptolite tube. Consists of concentric laminae. (18-4C.)

Nema (nm). Delicate tube to which the base of graptolite colony is attached. (18-3A, B; 18-5.)

Rhabdosome. The entire graptolite colony. (18-3A, B; 18-5.)

Sicula. Tube formed by initial individual in colony. (18-3A, B.)

Stipe. Branch of graptolite colony comprising a series of tubes (thecae). (18-3A; 18-5B to H.)

Stolon. Dense chitinoid tubule extending through successive stolothecae and sending off branches to base of each autotheca and bitheca. (18-4A, B.)

Stolotheca. Tube (theca) which contains stolon and which gives rise to three new thecae (autotheca, bitheca, and stolotheca) by budding. (18-4A, B.)

Theca. Individual tube in graptolite colony. (18-3; 18-4; 18-5.)

---

were assigned to the sponges, the hydroid coelenterates, the alcyonnarian coelenterates, the bryozoans, the cephalopods, and the pterobranchs.

Certainly they are colonial and are not sponges or cephalopods. Ulrich and Ruedemann (1931) argued for their affinities to the Bryozoa, but most specialists now believe that the similarities are the consequence of convergent evolution. They seem closest to the hydroids among the coelenterates in that they have a tubular skeleton of organic material and several different types of specialized polyps.

Unlike the hydroids, however, the graptolite colony and the individual thecae display a degree of bilateral symmetry (Kozlowski, 1966). Further, the stolon that connects the thecae corresponds to that of the pterobranch rather than the hydroid structure. The half-ring skeletal elements, with their zigzag sutures, are nearly identical with those of the pterobranchs (Figure 18-1), and the chemical composition is also quite similar. On the other hand, the mode of formation of the outer, lamellar layer of the skeleton, and the process of budding differs greatly.

On the balance of the evidence, Kozlowski places the graptolites with the pterobranchs among the protochordates. Bohlin (1950), Decker and Gold (1957) and Hyman (1959) have questioned this assignment, but their arguments, except as they are based on the very questionable reports of gonothecae and nematothecae, demonstrate only that the graptolites are not pterobranchs (which Kozlowski never claimed). Their argument for coelenterate affinities seems far weaker than that of Kozlowski for pterobranch affinities.

The elephant graveyard

African legend holds that dying elephants travel to a secret common graveyard. The distribution of graptolites very nearly produced a similar legend among paleontologists. Graptolite fossils are common in fine-grained black shales and extraordinarily rare in other rock types. The characteristics of these shales suggest that the water above the bottom had little or no oxygen and was loaded with hydrogen sulphide. Very few animals can tolerate this anaerobic environment for even short periods. Fossils, except graptolites, are rare in these shales. Early paleontologists were struck by this peculiar distribution, and their interest increased after the environmental significance of the black shales was recognized.

Did the graptolites live in an anaerobic environment? Certainly their occurrence would suggest this. Even though their skeletons were light, it seems improbable that currents otherwise capable of transporting only fine clay could have carried them into this area of deposition. But this conclusion rests on an assumption that graptolites were benthonic animals. The early dendroid types may have been, for their structure and associations indicate they lived as upright branching colonies like the "erect" bryozoans. The evidence is not so clear, however, for later, more specialized species. In these, the nema is so delicate that it could hardly have supported the rhabdosome above the bottom, but the rhabdosome could have dangled downward from the nema, which, in turn, was attached to floating seaweed. Some apparently had a bladder-like float to which the nema attached.* As long ago as 1865, Hall suggested that the graptolites were planktonic, and the direct evidence for a planktonic life is supported by the wide geographic distribution of many graptolite species. Nektonic or planktonic species commonly inhabit wide areas; benthonic forms rarely do so.

On this evidence, many graptolites were planktonic. They floated and lived in the aerated surface waters, and after death sank into the poisonous black mud environment. "Fine," you say, "but if they floated over wide areas why are they so rare except in black shales?" Consider now the composition of their tests. A proteinaceous skeleton is particularly susceptible to attack by organisms and is generally unstable in an oxidizing

* Bulman (1964) suggests that many relied on internal flotation mechanisms—gas or oil bubbles—rather than external aids.

environment. Arthropods and other animals with similar skeletons are rare fossils except for those that deposit calcium carbonate or phosphate in the organic matrix. On the other hand, the chitinoid and proteinaceous compounds are relatively stable in reducing environments. On bottoms high in oxygen, organic skeletal material would be destroyed. On anaerobic bottoms, the kinds and activities of organisms are limited, and the organic compounds are preserved.

Note that special factors produced a rock-fossil association that led to an erroneous paleoecologic interpretation. These factors were identified, and the error recognized through a careful study of the morphology and its functional significance.

The importance of being planktonic

Within a limited time range, Ordovician and Silurian, and within a limited series of rock types, black shales, the graptolites are useful guide fossils, for they evolved rapidly and single species appeared in widely separated seas. Since they are the only common fossil of the black shale environment, they permit correlation of rocks otherwise dated only from questionable interpretations of physical stratigraphy.

REFERENCES

Berry, W. B. N. 1962. "Graptolite Occurrence and Ecology," *Jour. of Paleontology*, vol. 36, pp. 285-293.

Bohlin, B. 1950. "The Affinities of the Graptolites," *Bull. Geol. Instit., Univ. of Uppsala*, vol. 34, pp. 107-113.

Bulman, O. M. B. 1964. "Lower Paleozoic Plankton," *Quart. Jour. of the Geol. Soc. of London*, vol. 120, pp. 455-476.

Decker, C. E. and I. B. Gold. 1957. "Bithecae, Gonothecae, and Nematothecae on Graptoloidea," *Jour. of Paleontology*, vol. 31, pp. 1154-1158.

Dillon, L. S. 1965. "The Hydrocoel and the Ancestry of the Chordates," *Evolution*, vol. 19, pp. 436-446.

Hyman, L. H. 1959. *The Invertebrates, Vol. V, Smaller Coelomate Groups*. New York: McGraw-Hill.

Kozlowski, R. 1966. "On the Structure and Relationships of Graptolites," *Jour. of Paleontology*, vol. 40, pp. 489-501.

Moore, R. C. 1955. *Treatise on Invertebrate Paleontology, Pt. V, Graptolithina*. New York: Geological Society of America.

Ulrich, E. O. and R. Ruedemann. 1931. "Are the Graptolites Bryozoans?" *Geol. Soc. of America, Bull.*, vol. 42, pp. 589-603.

19

The coming of the vertebrates: Fish and amphibians

As the arthropods stand at the apex of one of the great lines in animal evolution, so the vertebrates stand at the apex of another. In individuation and intelligence, those things that we self-centered vertebrates consider the most important measures of evolutionary advance, they have risen far above the arthropods.

Despite their long separate evolutionary history, the vertebrates and arthropods resemble one another to a surprising degree. The vertebrates also show primary segmentation modified by specialization of segments and by their fusion in functional regions. They, like arthropods, have a jointed skeleton extending to the limbs, though the vertebrate skeleton is internal rather than external. Most vertebrates, like arthropods, have paired limbs. All have a definite head, with brain and sense organs. These similarities must result from similar selective pressures acting on quite different initial forms, because the proto-arthropod was a worm with paired appendages and a stiff, external cuticle, and the proto-vertebrate a larval sea squirt with gill slits, an internal stiffening rod, and a muscular tail for swimming. The similarities evolved in spite of these differences.

Vertebrates are rare fossils—anyone who has hunted them will testify to that—but despite this rarity, their phylogeny is better known than any of the far more abundant invertebrate groups. This knowledge is partly due to our intense interest in our forebears and all our uncles, cousins, and aunts. You have only to inspect a natural history museum to see the results of this interest. Since they are rare and of slight importance in correlation except of continental deposits, they have been collected primarily for their evolutionary significance. This is hardly true of any invertebrate group. Finally, the vertebrate skeleton is intimately related to many of the animal's functions. Therefore, the vertebrate paleontologist interprets adaptive trends and separates adaptive similarities and differences from phylogenetic ones with at least some certainty.

The origins of the vertebrates

The reconstruction of the ancestral vertebrate derives, in part, from modern protochordates, but it is not too different from the oldest known vertebrates (Figure 19-1). Some vertebrate bone fragments have been found in Ordovician rocks, but the first good specimens come from the lower Silurian. These "fishes" consist basically of a large set of gills protected by bony armor and of a tail hung onto the back end of the gills. The mouth was a simple slit, without jaws, at the anterior ventral margin—hence the name Agnatha, "no jaws," for the class. In some the gills were supported by skeletal bars more or less fused to the head armor. The gill slits opened either separately along either side of the head or into an internal pouch which had a single lateral opening.

The eyes were small. The nostrils, or better, *nares,* were simply a pair of olfactory pits at the anterior end of the head. One group is characterized by a single median nostril. Some also had a medial opening in the armor toward the back of the head, the pineal opening. This housed the pineal "eye," which may have detected light and dark but not much more. The brain, of course, was not fossilized, but portions of the head armor were molded around the brain and nerves. The support below and at the sides of the brain was, typically, cartilaginous, but was sufficiently ossified in some so that it too was preserved. As the result, a good deal is known of the brain structure from the form of the cavity it occupied.

The trunk and tail were unarmored or covered with small bony plates. No traces of the internal skeleton have been found—what there was, presumably, was cartilaginous. The notochord was probably large and functioned in support of the body and as a base for muscle action. Al-

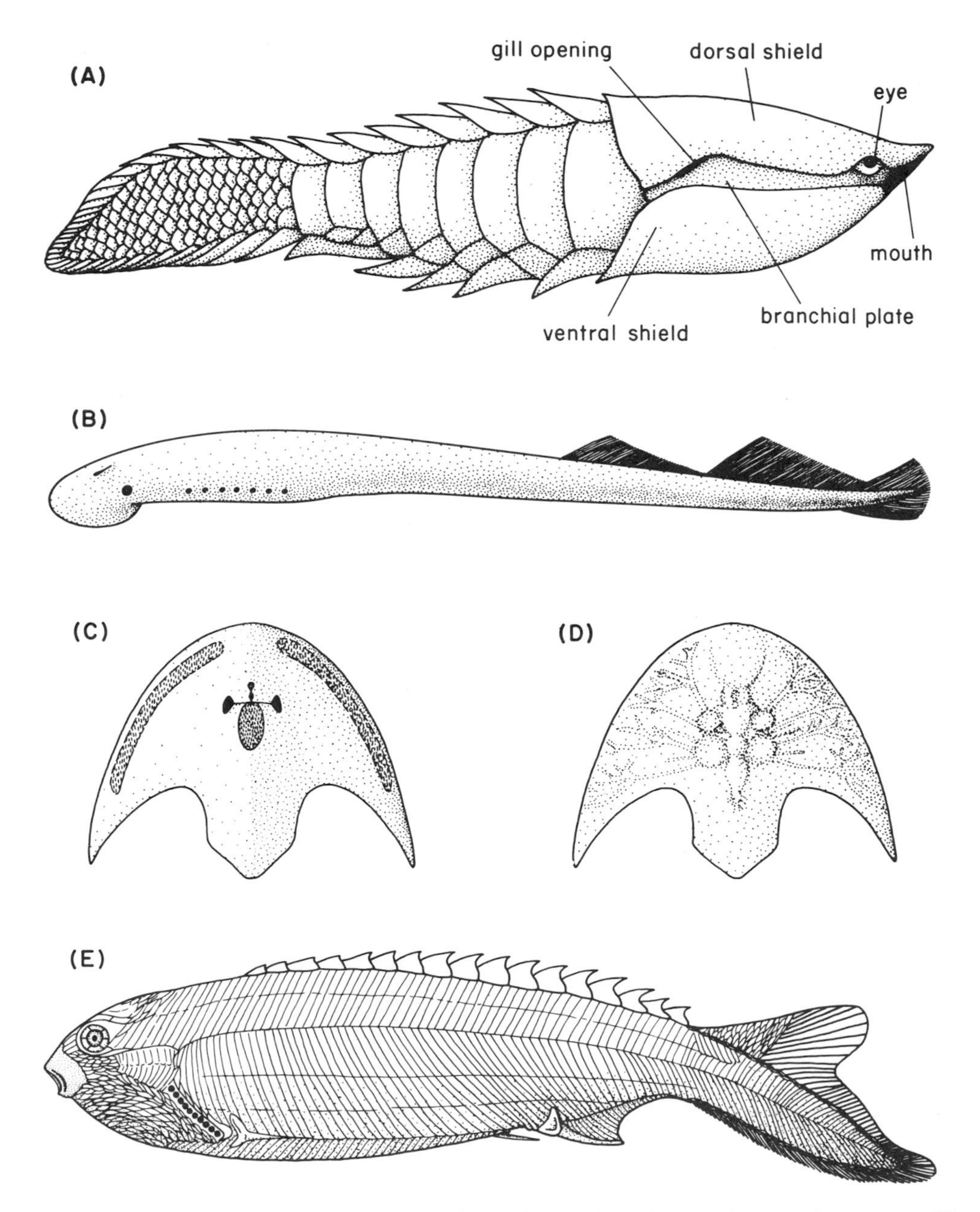

Fig. 19-1. THE AGNATHA. **(A)** *Poraspis;* late Silurian; lateral view of restored specimen; 15 cm long. **(B)** *Pteromyzon;* Recent. **(C)** *Cephalaspis;* late Silurian to early Devonian; dorsal view of head shield; 10 cm long. **(D)** Same with internal mold of brain case. **(E)** *Pterolepis;* 10 cm long. [**(A)** *after* Kiaer and Denison; **(C)** and **(D)** *after* Wänqsjö; **(E)** *after* Kiaer.]

though several genera have a pair of lateral flaps back of the head, these apparently lacked an internal skeleton and were not true paired limbs. The tail was relatively small.

The lack of jaws and the large size of the gill chambers implies that the Agnatha were particle feeders. Denison (1961) suggests that some with ventral mouths, large gill chambers, and flattened heads fed by mass ingestion and filtration of bottom muds. Others were probably suctorial feeders like the living hagfish; still others fed selectively on detritus, picking individual particles from the bottom. Most were probably poor swimmers and passed most of their life lying in bovine contemplation on the bottom or grubbing placidly in the mud.

Although the agnathans resemble the cephalochordates and probably had a similar way of life, they are clearly much advanced over the cephalochordate condition. Unfortunately, the specializations of the living cephalochordates and agnathans confuse this comparison. The former are probably a good deal more sedentary, and their sense organs, brain, and circulatory system may have regressed in complexity. The recent agnathans have unpleasant semipredatory-semiparasitic habits: The lampreys attach themselves to a fish by a specialized sucking mouth and burrow into its flesh with a rasp-like tongue; the hagfishes feed on dying fish or scavenge their carcasses.

The vertebrates did not leap to the center of the evolutionary stage as suddenly as did the trilobites and other invertebrates, for these fragments in Ordovician rocks give warning of their imminent appearance. The several lineages, however, are sharply separate at their first appearance and indicate an earlier, unrecorded radiation. Why is there no record of this radiation? The problem centers on the normal environment of early vertebrates. After careful consideration of the lithology and paleontology of Ordovician and Silurian fish localities, Romer and Grove (1935, also Romer, 1955) concluded that the early vertebrates lived only in freshwater habitats. Denison (1956), after an equally careful study, concluded that they were not freshwater but marine—or at least brackish. The faunal associates of these early fish are not typically marine—if by typical, neritic is meant. On the other hand, paleontologists know very little of early Paleozoic brackish faunas, and it is uncertain whether they can always distinguish them from freshwater faunas. Either interpretation would account for the absence of vertebrates from the well-known Ordovician deposits, which are predominantly of neritic origin. Almost surely someone, somebody, will find well-preserved fish in Ordovician rocks to end this argument and to establish more clearly the relation of the vertebrates to the more primitive chordates.

Whether estuarine or fresh water, many early fish species lived in waters with strong currents. Even though they were sluggish beasts who grubbed for food or filtered it from the passing water, there was a selective premium in swimming. Only by swimming could they move in those currents and adjust to the shifting environments of streams or tidal passes. Because of these environmental factors, they could not evolve, as do most filter feeders, into sessile animals. The bony armor about the head protected them against predators—possibly the eurypterids which are found in some of the same beds—but armor by its weight and rigidity interfered with swimming. Caught between fires the agnathans tended to reduce or even lose their armor and to protect themselves by swifter flight—at least the majority of the later members of the class have smaller and thinner head shields. The concurrent reduction of eurypterid numbers and variety may be either cause or effect.

A time for experiments

For most animals, filter feeding became an evolutionary blind alley. The vertebrates probably escaped by the circumstance of their physical environment. Their success, however, depended on an extraordinary evolutionary event—the development of jaws. The annelid-arthropod line accomplished it by transformation of some paired appendages, but the vertebrates had no such resource.

Zoologists demonstrated the development of the vertebrate jaw over seventy-five years ago in a series of remarkable embryological studies. The jaw consists initially of four elements—one above and one below on either side (Figure 19-2). In the embryo, these are cartilaginous and are replaced by other bony elements in the adult, except in sharks. The jaw elements are in line with and resemble the bars between the gill slits. Still more important, the nerves and major blood vessels of the jaw resemble those of the gill arches in arrangement. Continued in later development,

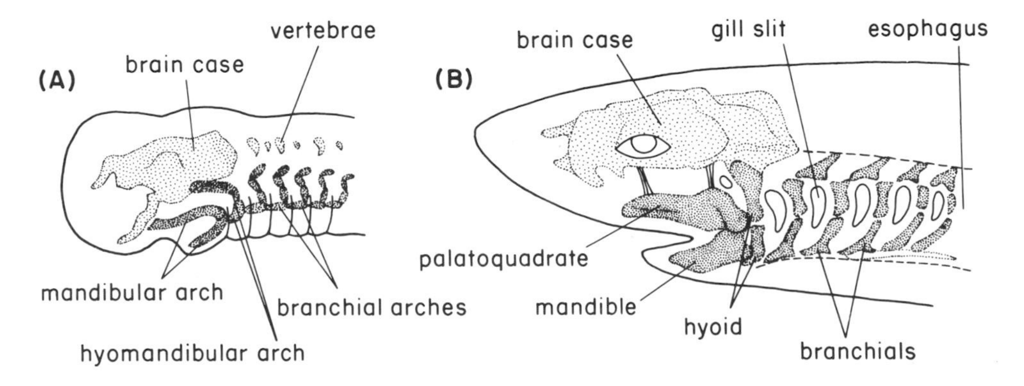

Fig. 19-2. DEVELOPMENT OF THE JAWS IN SHARK ONTOGENY. **(A)** Embryo. **(B)** Adult. The gill arches and jaws are shown in close stipple; the brain case in open stipple. [**(A)** *after* Goodrich; **(B)** *after* Watson, *Paleontology and Modern Biology*. Copyright Yale University Press, New Haven. Used with permission.]

the jaw is braced against the brain case by the succeeding gill bar. The gill slit between the jaw and this specialized bar is reduced to a tiny hole, the *spiracle*. Therefore, their ontogeny indicates that the jaws evolved from a set of gill bars.*

If this interpretation is correct, paleontologists should observe the intermediate steps between gill bar and jaw in fossils. Unfortunately the connecting forms are "missing links," though some early jawed fish appear to have a normal large gill slit just behind the jaw. Can we reconstruct the evolutionary sequence and predict the characteristics of these missing links?

Some agnathans had their gill bars free of the head shield rather than fused into it. Very probably, the muscles attached to the bars pumped the whole gill basket like a bellows to force the water—or mud—out. Some had small movable plates around the mouth that could have been used to nibble on plants or small invertebrates. The musculature of the gill bars assisted in moving these plates. In this fashion, the anterior bars became part of a biting mouth. Selection could then incorporate variants with larger and stronger anterior gill bars into the incipient jaw, and the more effective gill bar jaws replaced the "nibbling plates."

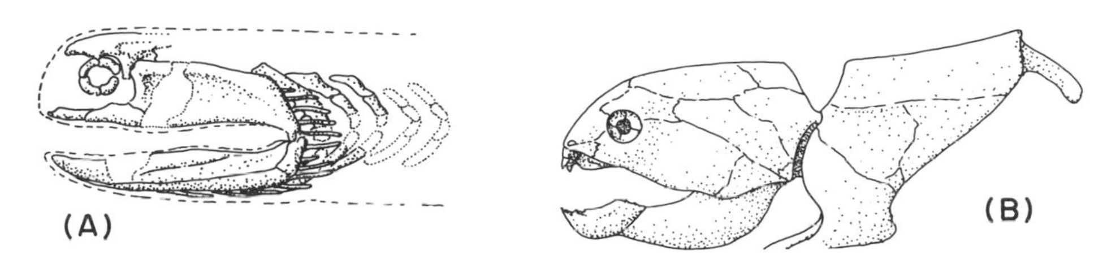

Fig. 19-3. JAW STRUCTURES IN PRIMITIVE FISH. **(A)** *Acanthodes,* an early Permian acanthodian. The gill arch (hyoid) immediately behind the jaw bears a full set of gill rakers. These indicate the hyoid slit was a normal functional slit rather than the reduced pore-like opening (the spiracle) known in sharks and bony fishes. **(B)** *Dinichthyes,* a late Devonian placoderm. The skull is articulated to the shoulder girdle with a ball-and-socket joint and may have moved up and down on a stationary lower jaw. [**(A)** *after* Watson; **(B)** *after* Heintz.]

Jawed fish need not burrow in the mud nor filter food from the water. They can browse on aquatic plants or devour large invertebrates and other fish. By late Silurian time the critical steps had been taken, and the radiation of jawed fishes was well under way. The striking feature of this radiation is its experimental, almost tentative, nature. A variety of jaw types (Figure 19-3) appeared—some of them dissimilar adaptations to quite similar functions.

* This is not an illustration of "ontogeny recapitulates phylogeny," for the jaws are never functional gill bars as they must have been in the ancestral population. Rather the ontogeny of the recent forms resembles the ontogeny of the ancestors. Many, if not most, of the examples given for the rule are of this sort.

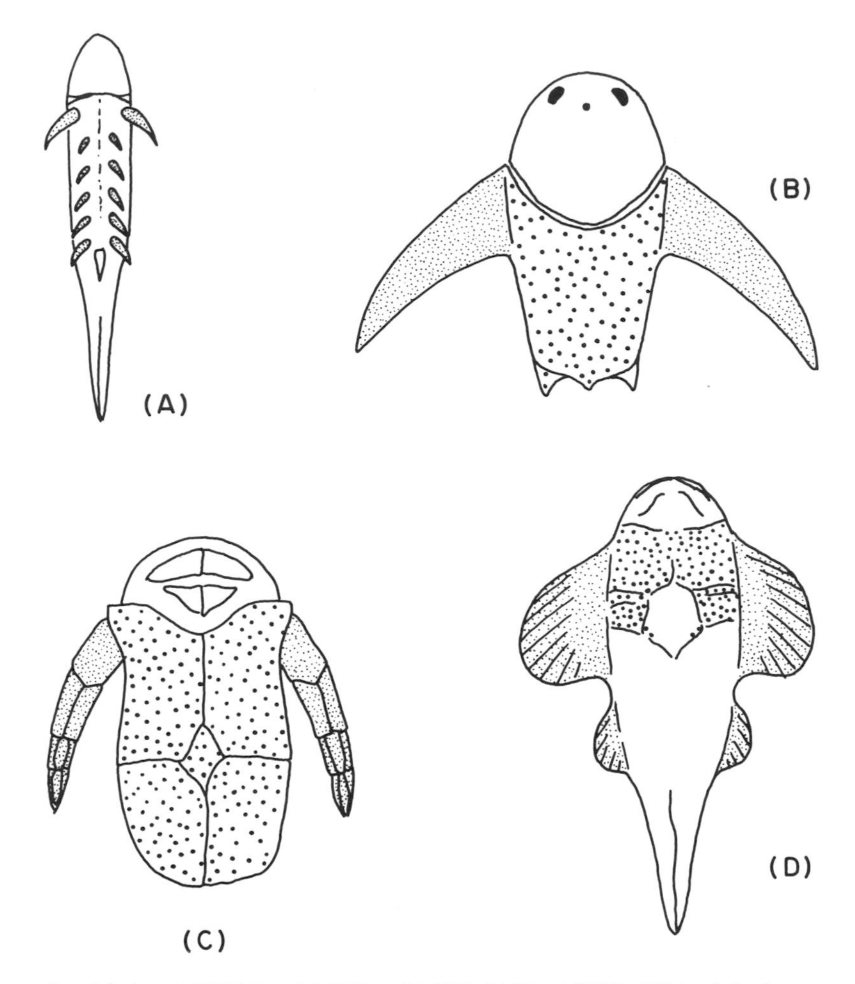

Fig. 19-4. VARIATION OF FORM IN THE EARLY JAWED FISH. Paired appendages shown by light stipple; limb girdles by heavy stipple. **(A)** *Climatius,* an early Devonian acanthodian with seven pairs of spines on either side of the ventral midline. **(B)** *Arctolepis;* early Devonian; dorsal view of head and shoulder armor. A pair of hollow spines extend laterally from the shoulder girdle and are fused to the girdle. **(C)** *Pterichthyodes;* middle Devonian; ventral view of head and shoulder armor. A pair of externally armored limbs are articulated to the sides of the heavy thoracic armor. **(D)** *Gemuendina;* early Devonian; ventral view. The body is rather broad and flat and bears both anterior and posterior paired appendages. [(**B**) *after* Heintz; (**C**) *after* Traguair; (**D**) *after* Broili.]

The jawed fishes first appear with agnathans in those controversial brackish or freshwater associations. By the Devonian, however, they occur with characteristic marine organisms and in undoubted freshwater habitats. Some apparently retained the ancient bottom-feeding habit, for they have a flattened body and heavy cranial armor like the agnathans (Figure 19-4). Others have a deep, torpedo-shaped body complemented by powerful jaws. These were carnivores of the open waters like the recent pelagic sharks and bony fishes. Some lost their cranial armor altogether; others had it transformed to small, relatively thin plates, and still others elaborated and enlarged it and supplemented the head shield by heavy bony scales on trunk and tail.

To catch smaller fish or to escape from larger ones, these first jawed fishes had to swim more swiftly and maneuver more agilely. Most agnathans were like airplanes without ailerons or elevators. Pitch, roll, and yaw were damped by the flattened head and by medial fin folds. The result was a highly stable but inflexible configuration—one maneuvered only by twisting the propeller, i.e., the posterior trunk and the tail. A more flexible alternative would be lateral fins that by their movements could direct movements by controlling pitch, roll, and yaw. This shift would require substantial improvement in directional preceptors of the nervous system and in neuro-muscular coordination.

The initial steps in the evolution of a flexible hydrodynamic configuration are not well known, but several different types appeared in early fishes (Figure 19-4). Some had a lateral fold on either side extending part of the length of the body; others had paired lateral spines—some with as many as eight or ten pairs; still others had a pair of appendages at the back end of the head. These latter, supported by bone developed just below the skin, were jointed and articulated in a manner somewhat like that of arthropod legs.

Alternate adaptations of jaws and paired fins form seven distinct groups which may represent several separate lineages from agnathan ancestors. Six of these are grouped as orders within the Class Placoderma—characterized positively by the presence of jaws and of varying amounts of bony armor over the anterior trunk region. One, the Petalichthyida, may contain the ancestry of the sharks (Romer, 1966). The seventh group, the Acanthodii, is placed as a subclass of the Class Osteichthyes, i.e., the "higher," bony fishes. The evolutionary tenure of the placoderms was short, for only a few survived beyond the Devonian, and even these went down to extinction before the beginning of the Mesozoic.

Vertebrates without bones

Out of this welter of experiments in the late Silurian and early Devonian evolved two lines that found the correct combination—though this correctness was not immediately apparent. These two, the sharks and the bony fishes, were distinct at their first appearance in the fossil record. If the relationships suggested above, i.e., that of the sharks to the petalich-

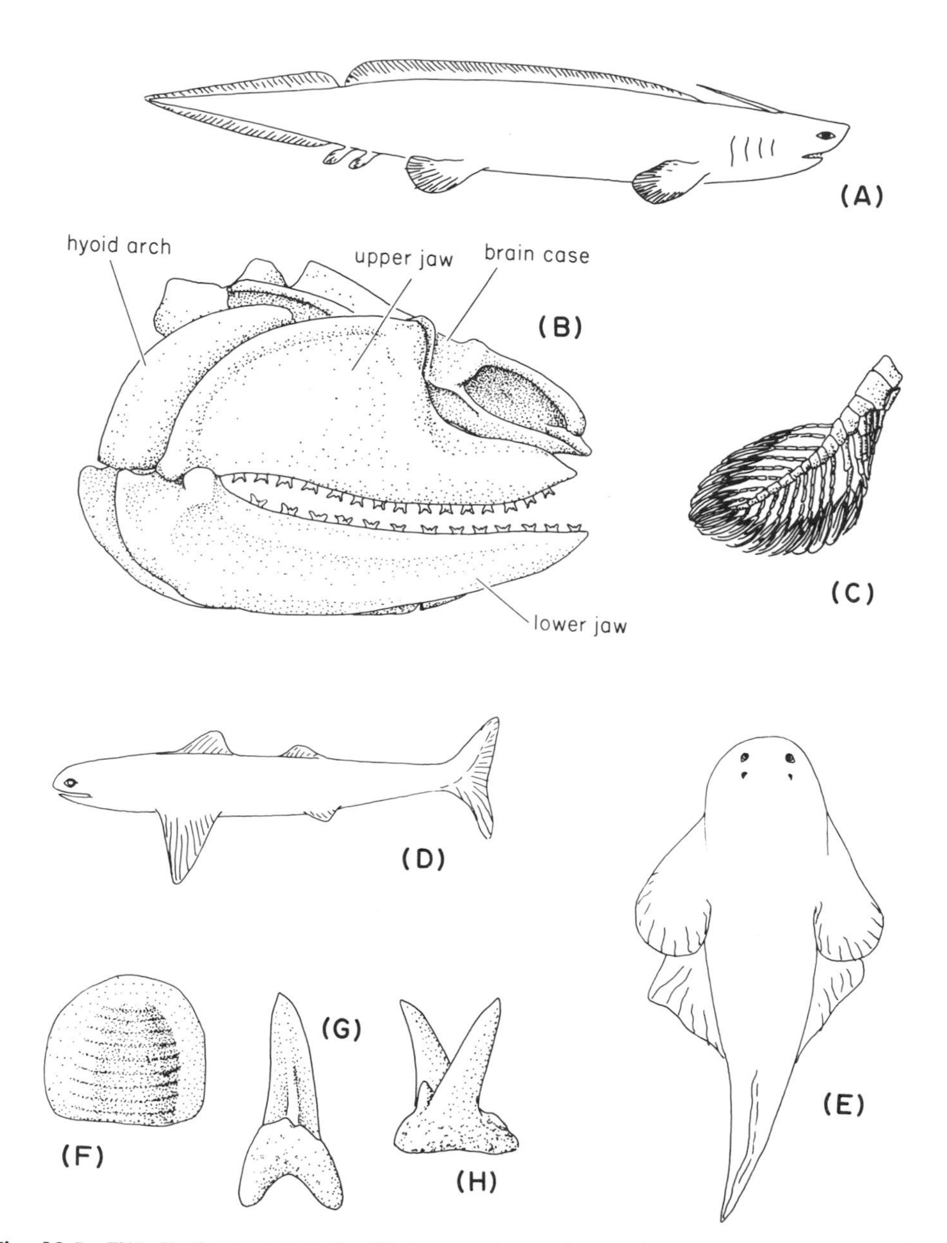

Fig. 19-5. THE CHONDRICHTHYES. **(A)** General body plan—a late Paleozoic freshwater shark, *Pleuracanthus*. **(B)** Brain case and jaws of *Pleuracanthus*; lateral view. **(C)** Skeleton of anterior (pectoral) paired fin of *Pleuracanthus*. **(D)** *Cladoselache*; late Devonian; lateral view. A slender, fusiform shark—probably an active predator. **(E)** *Rhina*; late Jurassic to Recent; dorsal view. A flattened, bottom feeder. **(F)** Tooth of *Ptychodus*; Cretaceous; crown view. A broad flat tooth probably used to crush molluscs and arthropods. **(G)** *Isurus*; early Cretaceous to Recent; lateral view of tooth. **(H)** A pleuracanth tooth; Carboniferous to Permian; lateral view of tooth. [(**B**) and (**F**) *after* Romer, *Vertebrate Paleontology*; reprinted by permission of the University of Chicago Press. Copyright 1966, University of Chicago Press, all rights reserved.]

thyids and that of the bony fishes to the acanthodians, are correct, the common ancestry of the two probably lies in an agnathan stock, and many of the similarities between sharks and bony fish are the result of parallelism or even convergence.

The ancestral sharks diverged from the placoderm condition in three essentials:

a). Two sets of flexible paired fins, the anterior *pectoral* and posterior *pelvic,* were established as a definitive and immutable plan;

b). Bone, both internal and external, was lost totally and replaced by cartilage;

c). The covering over the branchial area was lost so that the gill slits open separately to the exterior rather than into a common branchial or opercular cavity.

In early sharks, the second gill bar, the *hyomandibular*, braces the jaw against the back of the brain case as in modern sharks; however, the jaw was also connected to the front of the brain case. The paired fins had a broad skeletal base with only slight mobility, so that maneuverability was little advanced over the placoderm condition.

The radiation of the chondrichthyans is but poorly known. Cartilage decays rapidly; only when it is partially calcified or when it leaves its impression in fine-grained shales is it fossilized; otherwise fossil shark species are teeth and fin spines. The first sharks were probably pelagic marine carnivores. Only one group, the pleuracanths, penetrated to and evolved in fresh waters. Most sharks remained marine, though a few recent species tolerate brackish or even fresh waters. They largely retained the carnivorous pelagic habit, but some, like the modern skates and rays, foraged on bottom invertebrates. In adaptation to this habit, the latter have a broad, flattened body, and some evolved heavy flat teeth to crush thick-shelled animals. The sharks diversified less than the bony fish and had a subsidiary role in the drama of vertebrate evolution—the elderly character actor always on the stage but with only a few lines to speak. The chondrichthyes are a terminal group in themselves; they did not, as is sometimes implied in elementary zoology courses, evolve into bony fish —they evolved only into sharks.

The bony fishes

If the chondrichthyes have supporting roles, the bony fish, the Osteichthyes, and their weird side branches are the leads. It is commonplace to state that evolution is limited by previous adaptations—I have many examples in preceding chapters. I also have a few examples of adaptations that created new evolutionary possibilities. No group illustrates this phenomenon so clearly as the Osteichthyes. As good swimmers with paired fins they should have gone on, getting to be better and better "fish." And so they have, for over ninety percent of recent fish are bony fish, and, even more significant, they have eliminated many of their competitors to dominate the vagrant benthos and the nekton. But they had some poor relations who clambered onto land to share the terrestrial environments with the arthropods.

As defined by Romer (1966) the bony fish include three rather distinct subclasses, the acanthodians which appeared in the late Silurian, the sarcopterygians (fleshy-finned types), which first occurred in the early Devonian, and the actinopterygians (ray-finned types), which appeared in middle Devonian. The acanthodians are primitive in paired fin structure, possessing from two (pectoral and pelvic) to as many as seven pairs. The fins are unusual in that all except the caudal are stiffened by a large dermal spine; in some species only the spine is present, but in others the spine supports a web of skin. The jaw is primitive with a large hyoid arch and with some indication of a complete gill slit between the hyoid and the jaw. Each gill bar supported a small covering flap for the adjacent gill slit; in addition a large flap, the operculum, extended backward from the jaw. The scales are like those of the early actinopterygians. The internal skeleton was partially ossified, and the skull comprised a large number of plates more or less tightly sutured. The other subclasses never possess more than two pairs of lateral fins; the operculum arises from the hyoid rather than the jaw, and the first gill opening is reduced to a small circular opening, the *spiracle,* in front of the operculum. Their fins lack a dermal spine but have instead internal bony rays or rods for support. From their geologic occurrence and from the physiological adaptations of recent bony fish, an origin in upland lakes and streams seems probable.

To rule the seas

Among the best known of these Devonian fishes are two genera of similar form (Figure 19-6). A Devonian ichthyologist, impressed by these similarities and their common differences from placoderms, might have placed them in the same family. Modern vertebrate paleontologists classify them in different subclasses. The differences though small have been perpetuated through 300,000,000 years and through a rapid succession of evolutionary changes and radiations. Only the perspective of time permits this extreme separation.

What of the differences? One has blind pits for nostrils, and only slender rays supporting the fins. The other has nostrils that open into the

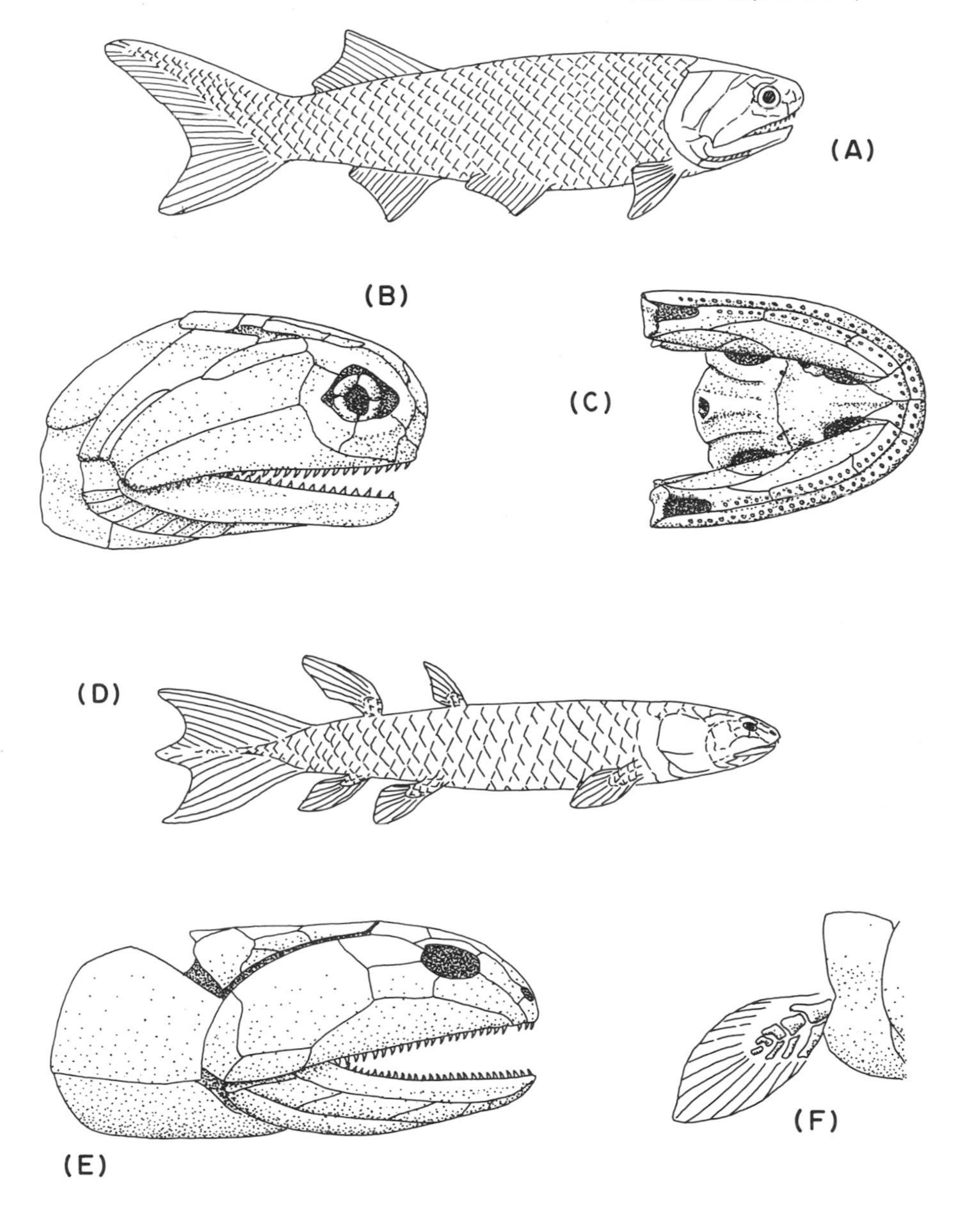

Fig. 19-6. EARLY OSTEICHTHYES. **(A)** *Cheirolepis;* middle Devonian; lateral view; length about 20 cm. **(B)** Lateral view of skull of *Cheirolepis.* **(C)** Ventral view of *Cheirolepis* skull —lower jaw removed. **(D)** *Eusthenopteron;* late Devonian; lateral view. **(E)** *Osteolepsis;* middle Devonian; lateral view of skull. **(F)** *Eusthenopteron;* view of pectoral fin and shoulder girdle. [**(A)** *after* Traguair; **(B)** *and* **(C)** *after* Watson; **(D)** *and* **(F)** *after various sources;* **(E)** *after* Save-Soderbergh.]

mouth cavity and heavy bony rods in the paired fins. They also differ minutely in the arrangement of bones in the skull. These structural differences imply an adaptive divergence, but a very slight one, for they both inhabited freshwater environments, both had slender torpedo-shaped bodies, and both possessed the sharp teeth of carnivores. If not directly competitive, they certainly excluded each other from niches both could otherwise have easily occupied.

The ray-fins (subclass Actinopterygia), although they started slowly in the Paleozoic, steadily shoved their vertebrate and invertebrate competitors into extinction. They diverged continuously into new niches and developed new forms for old niches. They include rapid swimmers of the open seas with elongate torpedo shapes, and flattened bottom-dwellers like the flounder. Many are still predaceous, but some have become herbivores or feed on the tiny animals and plants of the plankton. They include giants and dwarfs, eels and sea horses.

But throughout this almost continuous, explosive radiation runs a common evolutionary progression. Among the more important changes have been the reduction of the bony scales to thin horny plates; the shift of the pectoral fins up the side of the body and the pelvic fins forward below the pectorals; an increasing ossification of ribs and vertebrae; and a reduction of the thickness and rigidity of the dermal skull bones. Typically, they have large eyes, and the brain is modified by the correlative change of the optical centers.

These features have evolved in many lines, cross-cutting adaptive diversification. The hapless taxonomist can recognize three major radiations, late Paleozoic and early Mesozoic, Mesozoic, and late Mesozoic and Cenozoic, but each radiation is a phylogenetic thicket rather than a family tree. The species in each radiation are of a similar structural grade attained by parallel evolution. As more fossils are collected and more carefully studied, some of the central stems in the thicket can be discerned and separated from sterile shoots, but this effort has scarcely begun.

Black sheep

The third subclass of bony fish includes a moderate number of Paleozoic, a few Mesozoic and Cenozoic, and four recent genera. They would hardly be worth mentioning—except that they gave rise to the terrestrial vertebrates. But why did a group otherwise obscure produce such important offspring? The answer may lie in two of the characteristics that distinguish them from the ray-fins. The internal nares permitted them to breathe when their mouths were closed—breathe, that is, by intake of water or air. The rods in their paired fins were sufficiently sturdy to bear their weight out of water. This, combined with two characteristics, thick scales and paired sacs attached to the esophagus, held in common with the primitive ray-fins, permitted them to survive on land. The scales protected them from drying; the sacs formed lungs into which air could be drawn for respiration. Finally, the early sarcopterygians were freshwater forms and may have been driven to land because of the drying of the streams or lakes or in a search for food. The early stages may simply have involved moving from one channel or pond to a nearby one.

Presumably, only one or two sarcopterygian lineages evolved into tetrapods. The remainder plodded along into obscurity or extinction. Even in the early Devonian, two separate lines can be distinguished (see also pp. 149*ff.*). One group, marked by elongation of the body to an eel-like form and by reduction of the paired fins, leads to the modern lungfishes, the Dipnoi. The dipnoans adapted to life in the intermittent streams and ponds characteristic of a monsoonal river delta. Two of the recent genera, *Protopterus* and *Lepidosiren,* burrow into the mud during the dry season and aestivate. This habit appears to be of ancient origin, because Permian lungfish occur in vertical "pipes" that seem to represent sediment-filled burrows. The other recent genus does not aestivate but survives in pools of muddy water by breathing air. The dipnoans feed principally on small invertebrates and on plants and evolved crushing and cutting plates to handle this fare. The skull, in turn, changed greatly to support these plates and the powerful jaw muscles.

The other group of sarcopterygians (Figure 6-12) is distinguished primarily by broad-based, paired fins with a sturdy skeleton—hence the name Crossopterygia, lobe-fin. The crossopterygians also differ from the Dipnoi in arrangement and number of skull bones. The structure of the paired fins foreshadows that of the tetrapod limb, and the skull structure is close to that of the earliest amphibians. The Crossopterygia are important members of late Paleozoic freshwater faunas, at least in number of fossils. In the Devonian, they were probably the dominant predators. If the proto-amphibian was chased out of the water, it was probably by a crossopterygian cousin.

The freshwater crossopterygians are unknown after the Permian, but a distinct order, the coelocanths, managed with some success in the Mesozoic marine environments. Long thought to be extinct, a living coelocanth, very like the Cretaceous ones, was collected off the African coast in the late 1930's.

An excess of oxygen

A Danish expedition to eastern Greenland in 1931 made a dramatic discovery in a fossiliferous red sandstone. Most of the vertebrates were typical late Devonian forms, including placoderms, lungfish, and crossopterygians. Among them, however, were partial skulls and isolated limb bones of two species of amphibians. Though fragmentary Devonian amphibian material had been collected some years before, this was the first opportunity to learn the details of their anatomy. Comparison of Carboniferous amphibians with Devonian fish had indicated that amphibian ancestry lay in the crossopterygians. If this interpretation were correct, the Devonian amphibians would be intermediate in form.

The study of these Devonian amphibia, though not yet finished, has borne out this expectation (Figure 19-7). The limbs are of the typical tetrapod form, though short and undoubtedly clumsy. The pelvic and pectoral girdles are also more amphibian than fishlike, but these structures

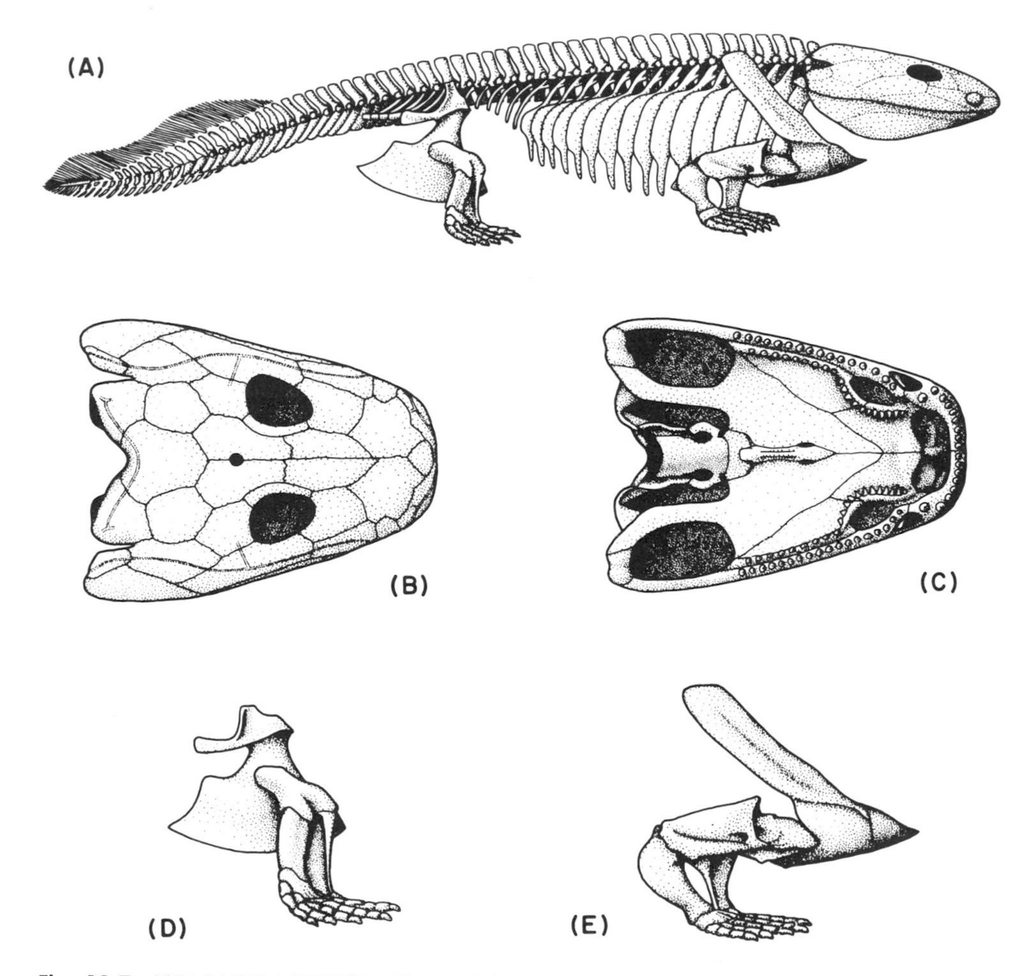

Fig. 19-7. THE EARLY AMPHIBIA. *ICHTHYOSTEGA*. **(A)** Lateral view of reconstructed skeleton. **(B)** Dorsal view of skull. **(C)** Palatal view of skull. **(D)** Hind limb and pelvis. **(E)** Forelimb and pectoral girdle (*after* Jarvik).

are easily derived from those of the crossopterygians. The vertebrae are essentially like those of the crossopterygians except that the dorsal (*neural*) arches that covered the spinal cord are more closely articulated. The body of each vertebra consists of three pieces, a pair of small ones (the *pleurocentra*) that lay on the upper surface of the notochord and a larger one (the *intercentrum*) that wrapped around its lower part and extended up its sides to articulate with the neural arch. The notochord was large and still bore a share of the load on the back.

The pattern of bones in the skull, the arrangement of bones in the palate, and the structure of the lower jaw also resemble those of the crossopterygians. On the other hand, the head looks more amphibian than fishlike. The snout is long; the part of the skull behind the eyes is short. The opercular bones are gone, except for two small elements. The teeth are simple sharp cones borne on the bones of the palate as well as on the margins of the jaw. The enamel around the base of the tooth is infolded to form a labyrinthine pattern in a cross section. These characters of the teeth are primitive, since the crossopterygians have similar dentition, but they are also typically amphibian and were retained in many later species.

The general form of body reminds one of a fish. The neck is nonexistent, for the back of the skull lies almost against the pectoral girdle. The tail is long and, most remarkably, the vertebrae bear short ventral and dorsal spines. These spines are clearly the rays supporting a fishlike tail—one very like that of the Crossopterygia.

The mixture of fishlike characteristics with those of the tetrapods demonstrates that these Devonian amphibians were near the origin of the tetrapods. It also demonstrates that the evolution of tetrapod characteristics went on at different rates in different parts of the body. What makes a land animal? Obviously, legs and lungs. These early amphibians have legs; the lungs cannot be seen, but, since the operculum is gone, the gills must be, too. If the gills are lost, the lungs must have replaced them. Besides, the legs are too good for animals that only ventured from the water momentarily. Legs and lungs probably evolved under intense selective pressure, but once these critical adaptations were made, the other changes, the refinements as it were, could go on more slowly, and indeed some were not completed until late in the evolution of the reptiles.

The great invasion

The early terrestrial vertebrates had a new world to exploit. The floodplains were covered by a variety of plants. The insects and other terrestrial arthropods had begun their radiation. With this opportunity, the Amphibia began to diverge in form almost before they became amphibians. Even the primitive Devonian genera represent two distinct families. Since specializations of skull structure debar any of the three well-known genera from a direct ancestry of Carboniferous amphibians, there must have been still other Devonian types. Finally, a major group of Paleozoic amphibians, characterized by spool-like vertebrae, have no apparent relationship to the other early amphibians. They probably represent a divergent line from the protoamphibian stock or, less likely, a parallel lineage from the crossopterygian level.

Fossilization in the terrestrial environments

The origin of the class is, as I have shown, fairly well known; the evolutionary paths within the class are known only imperfectly. The reasons for this latter situation lie both in the nature of fossilization in the terrestrial environment and in the peculiarities of the geologic record of the Carboniferous.

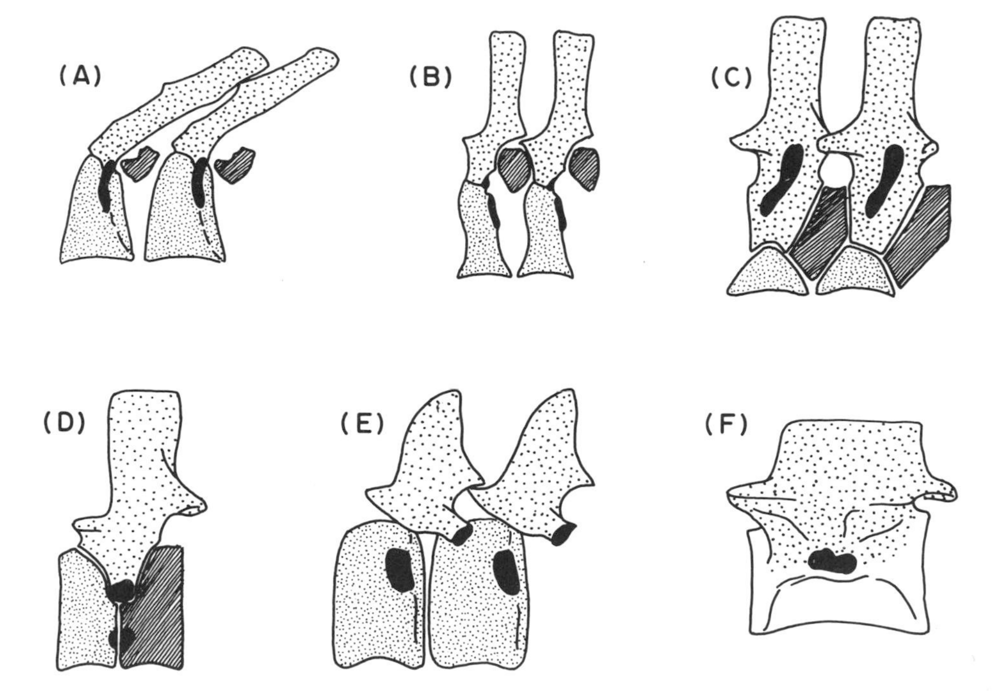

Fig. 19-8. AMPHIBIAN VERTEBRAE. Neural arch shown by heavy stipple; pleurocentrum by diagonal lines; intercentrum by fine stipple; and rib articulations by solid black. **(A)** Vertebrae of crossopterygian fish. **(B)** Vertebrae of primitive amphibian (ichthyostegalian). **(C)** Vertebrae of moderately advanced amphibian. Rhachitomous type of vertebra. **(D)** Vertebra of moderately advanced amphibian. Embolomerous type. **(E)** Vertebrae of advanced amphibian. Stereospondylus type. **(F)** Vertebra of lepospondylus amphibian. Body of vertebra formed by deposition of bone around notochord, not by ossification of intercentral and pleurocentral cartilages. [**(A)** *after* Gregory; **(B)** *altered from* Jarvik; **(C)** and **(D)** *after* Williston.]

First of all, fossils are rare in all sedimentary rocks of terrestrial origin. The reasons are several: terrestrial animals, even aquatic ones, are rarely as abundant in any one place as are marine animals; if threatened, they can all pack up and leave, but marine faunas have many sedentary species; terrestrial environments of deposition, because of their very local and shifting occurrence, rarely provide rapid permanent burial; and the alternate wetting and drying and high oxygen concentrations destroy much of what is buried. In major river channels vertebrate skeletons are disarticulated and the bones broken, worn, and scattered. On the floodplains a dying animal is attacked by scavengers, and the indigestible remains left to the bacteria and the corrosive atmosphere. Occasionally, swamp deposits contain accumulations of vertebrate bones, although the chemistry of most swamp environments is unfavorable for fossilization. Where can vertebrate skeletons accumulate and be buried in numbers? Around ephemeral lakes and ponds where animals bog down and are buried in mud holes or where their carcasses sink undisturbed to the bottom of a pond. Or in the backwaters of small channels, where isolated bones or whole skeletons accumulate.

An extensive exposure of marine rocks will almost always yield fossils to a patient collector. An extensive exposure of terrestrial rocks will rarely do so unless it cuts into one of the extremely localized ponds or minor channels. If vegetation and soil cover most of the bedrock, the

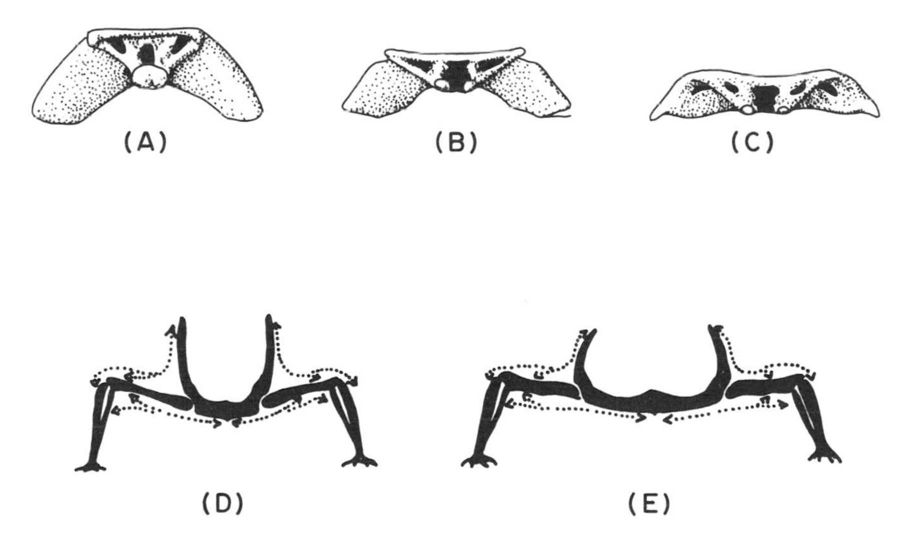

Fig. 19-9. EVOLUTION OF SKULL AND BODY FORM. **(A)** through **(C)** posterior view of amphibian skulls. **(A)** Primitive, deep skull; Carboniferous embolomere. **(B)** Moderately flattened; Permian rhachitome. **(C)** Strongly flattened; Triassic stereospondyl. **(D)** and **(E)** Diagrammatic cross-section of front limbs and pectoral girdle. Bones in solid black; muscles dotted. **(D)** Primitive form with deep narrow body. **(E)** Advanced form with wide, flat body. [**(A)**-**(C)** *after* Watson.]

chances of finding vertebrate fossils become almost infinitesimal. For example, lake and pond deposits are unusually abundant in the Dunkard (Permo-Carboniferous) beds of West Virginia, Ohio, and Pennsylvania. Yet in some seven field seasons, only about ten significant vertebrate fossil localities have been found in an area of over 7000 square miles. The search, of course, was limited to road cuts because of the extensive mantle cover.

Mississippian rocks in most parts of the world are of marine origin. Many of those that are continental, such as the Pocono and Mauch Chunk formations in the Appalachian basin, are poorly exposed. The amphibian species known from the Mississippian represent, then, only a small fraction of those that existed during the period. Paleontologists can recognize the fact of an extensive amphibian radiation but not the details nor the origins of the separate lines. The Pennsylvanian record is somewhat more satisfactory but limited principally to one type of environment—the coal swamp.

Long after most of the major events in amphibian evolution, paleontologists find numerous amphibian fossils in the extensive Permian and Triassic deltaic deposits of the western United States, Russia, and South Africa. Any description of amphibian evolution must be based in large part on a reconstruction of pre-Permian events from Permian fossils.

Amphibian evolution

The most widespread and consistent trend in the Amphibia was for modification of the vertebrae to reduce the stresses on the notochord and

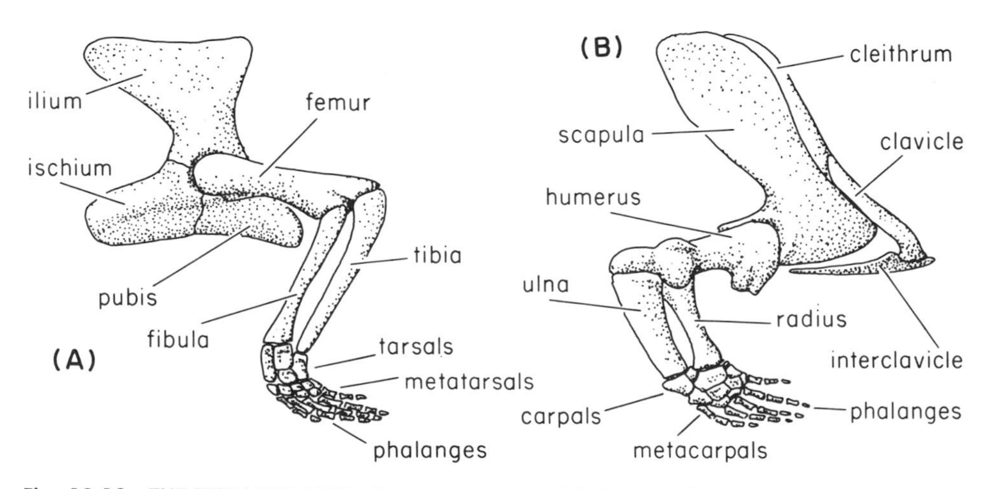

Fig. 19-10. THE TETRAPOD LIMB. General structure of limbs in early amphibians and reptiles. **(A)** Posterior (pelvic) limb and girdle. **(B)** Anterior (pectoral) limb and girdle.

on back muscles and to place the static and dynamic loads on the bony arches and centra. This was accomplished differently in different primitive stocks (Figure 19-8), and for this reason vertebral characteristics form the principal basis of amphibian classification. These modifications of the vertebrae were certainly functional—Panchen (1967) suggests that they express the relative importance of the body musculature in locomo-

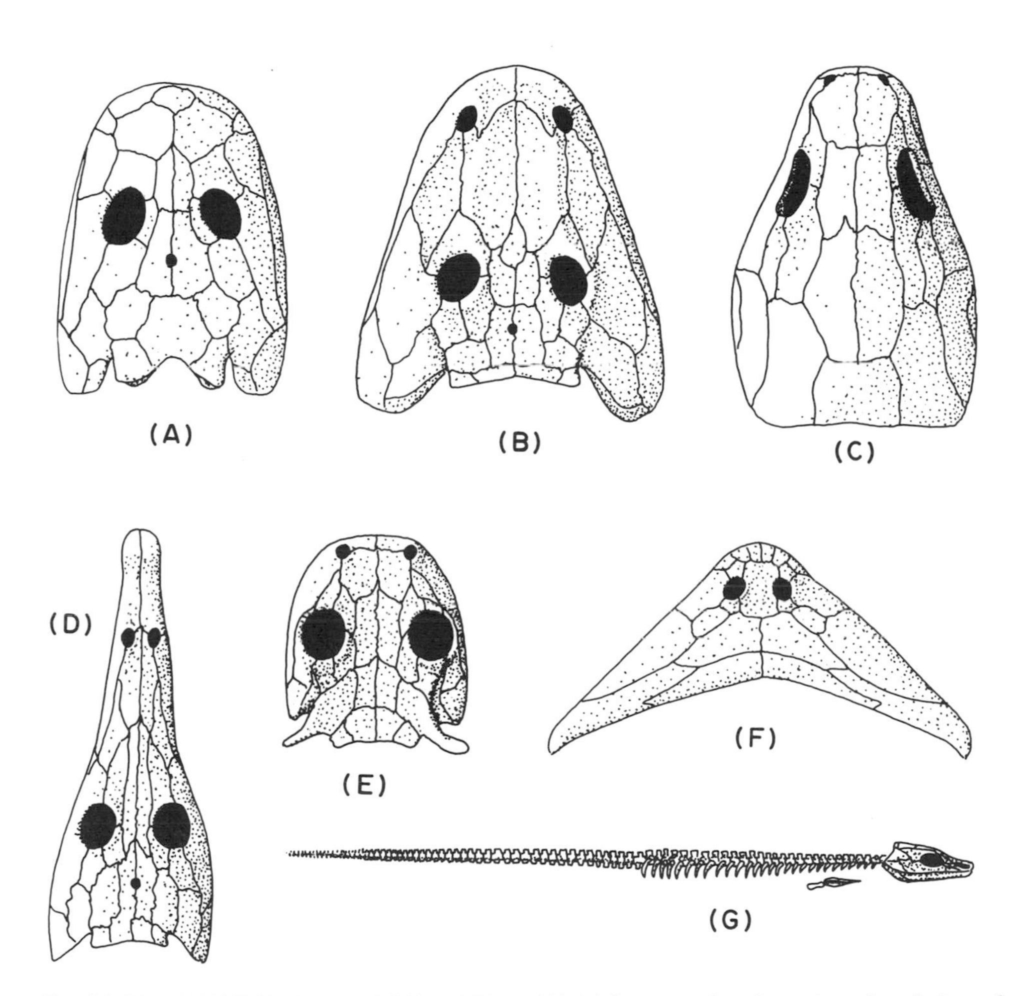

Fig. 19-11. VARIATION IN AMPHIBIAN FORM. **(A)** *Ichthyostega;* late Devonian; dorsal view of skull; length 20 cm. A relatively deep, fish-like skull. **(B)** *Eryops;* late Pennsylvanian to early Permian; dorsal view of skull; length about 50 cm. Flattened skull with eyes high on head and far back. **(C)** *Euryodus;* early Permian; dorsal view of skull; length about 2.5 cm. Short snouted form. **(D)** *Platyops;* middle(?) Permian; dorsal view of skull. Long snouted form. **(E)** *Stegops;* Pennsylvanian; dorsal view of skull. Note short "horn" at posterior corners of skull. **(F)** *Diplocaulus;* late Pennsylvanian to early Permian; dorsal view of skull; width about 27 cm. A "long-horned" amphibian. **(G)** *Sauropleura;* Pennsylvanian; lateral view; length 18 cm. Snake-like amphibian with rudimentary limbs. [**(A)** *after* Jarvik; **(B)** *after* various sources; **(C)** *after* Olson; **(D)** *and* **(E)** *after* Bystrow; **(F)** *after* Williston; **(G)** *after* Steen.]

tion. Most lines also show reduction of the bones of the palate—related probably to the change in skull form discussed below.

The skull of early amphibians, like that of their fish ancestors, is deep and compressed from side to side. Nearly all later amphibians have the skull flattened dorso-ventrally so that it is wide but shallow (Figure 19-9). The functional significance of the change is not clear but would seem related to the flattening of the body. The change in body shape from narrow to flat correlates with the position of the limbs and their use. The proximal limb segments, the humerus in the forelimb, the femur in the hind, stuck out laterally and nearly parallel to the ground. The next segment, comprising the ulna and radius bones in front and the tibia and fibula behind (Figure 19-10), were at right angles to the ground and actually carried the animal above it. The gait must have been awkward and sprawling. The load rested on the long muscles of the arm and leg working over the levers at the shoulder and hip and at the elbow and knee. These muscles would work best if long and nearly parallel to the proximal limb segments. They would be longest and most nearly parallel if the body were flat rather than narrow. With these short, wide-spraddled legs, a flat body and head would also give more clearance above the ground.

Adaptive radiation

Left to themselves until the evolution of the reptiles, the amphibians entered a variety of aquatic, semiaquatic, semiterrestrial, and terrestrial niches. The terrestrial types are poorly known but include short-bodied types, some of which had bony armor on the back. Those found so far are small forms that must have preyed on insects and smaller terrestrial vertebrates. The semiaquatic and semiterrestrial amphibians corresponded ecologically to alligators and predaceous turtles, lived along the banks of streams and lakes, and preyed on fish, aquatic amphibians, insects, and unwary terrestrial vertebrates that came to drink or to lay their eggs. Some were of large size and certainly capable of holding their own with the early carnivorous reptiles.

Some of the most important subsidiary trends are shown in Figure 19-11 as adaptations to particular environments or modes of life. Many of these appeared independently in otherwise distinct lines and are the result of parallelism and convergence.

REFERENCES

Denison, R. H. 1956. "A Review of the Habitat of the Earliest Vertebrates," *Fieldiana: Geology*, vol. 11, pp. 361-457.

———, *et al.* 1961. "Evolution and Dynamics of Vertebrate Feeding Mechanisms," *American Zool.*, vol. 1, pp. 177-234. A symposium dealing with a wide range of fossil and recent vertebrates.

———, *et al.* 1966. "The Vertebrate Ear," *American Zool.*, vol. 6, pp. 368-466. A symposium on ear evolution and function.

Jarvik, E. 1960. *Théories de l'Évolution des Vertébrés*. Paris. Presentation of an important minority viewpoint on early vertebrate evolution.

Miles, R. S. 1965. "Some Features in the Cranial Morphology of Acanthodians and the Relationships of the Acanthodii," *Acta Zool.*, vol. 46, pp. 233-255.

Nursall, J. R., *et al.* 1962. "Vertebrate Locomotion," *American Zool.*, vol. 2, pp. 127-208. A symposium ranging from fish to bipeds.

Olson, E. C., *et al.* 1965. "Evolution and Relationships of the Amphibia," *American Zool.*, vol. 5, pp. 263-334. Symposium on origin, radiation, and relation to reptile origins.

Panchen, A. L. 1967. "The Homologies of the Labyrinthodont Centrum," *Evolution*, vol. 21, pp. 24-33.

Romer, A. S. 1955. "Fish Origins—Fresh or Salt Water?" *Deep-Sea Res.*, vol. 3 (suppl.), pp. 261-280.

———. 1966. *Vertebrate Paleontology*. Chicago: University of Chicago Press.

Watson, D. M. S. 1951. *Paleontology and Modern Biology*. New Haven: Yale University Press. Includes discussion of evolution of jawed fishes.

Westoll, T. S. (Ed.) 1958. *Studies on Fossil Vertebrates*. London.

Williams, E. F. 1959. "Gadow's Arcualia and the Development of Tetrapod Vertebrae," *Quart. Review of Biology*, vol. 34, pp. 1-32.

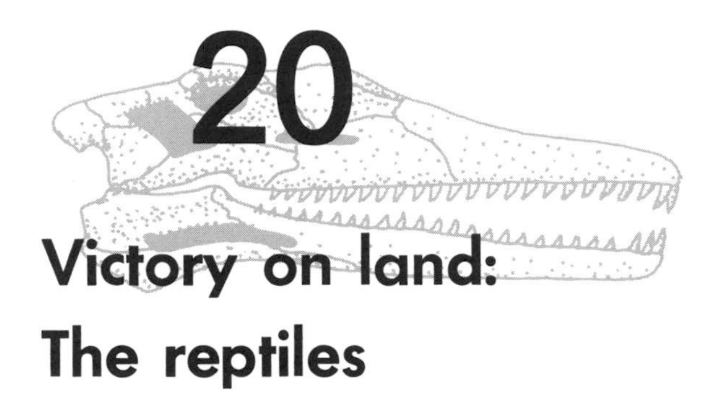

20

Victory on land: The reptiles

The large number of amphibian lineages that returned to entirely aquatic habits demonstrates the uncertainty of their terrestrial adaptations. Regardless of the efficiency of lungs or legs, they must return to the water to lay their eggs. Mechanisms evolved to circumvent this need are stopgaps successful only because the animals live in exceptionally humid environments. Only an amphibian lineage that evolved a nonaquatic egg could free itself from the uncertainties of an ephemeral, stagnant pond.

The problem is simple—just build a shell of low permeability. Oxygen enters freely, but the water inside, essential to development, cannot diffuse out rapidly. But what is to be done with excretory wastes spilled out into the fluid surrounding the embryo? How can sufficient oxygen reach the embryo in its watery envelope? To make matters more complex, the young must survive terrestrial conditions on hatching; that means a longer period of development before they leave the egg. The answer is the amniote egg with its respiratory membranes, its external bladder for storage of excretory products, its large yolk sac for food—plus accompanying physiological and morphological adaptations in the embryo.

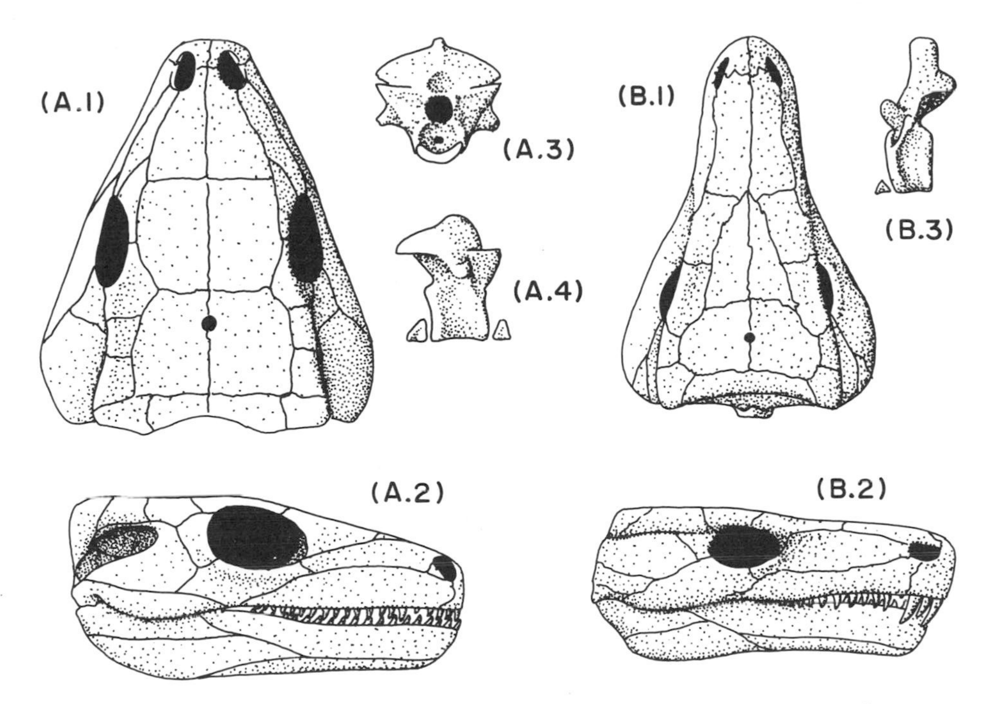

Fig. 20-1. PRIMITIVE REPTILES AND REPTILE-LIKE AMPHIBIANS. **(A)** *Seymouria;* early Permian; **(A.1)** dorsal view of skull; **(A.2)** lateral view of same; **(A.3)** and **(A.4)** anterior and lateral views of vertebra. A genus close to reptile-amphibian boundary and assigned by various paleontologists to one or other of these classes. **(B)** *Limnoscelis;* early Permian; **(B.1)** dorsal view of skull; **(B.2)** lateral view of skull; **(B.3)** lateral view of vertebra; length of skull 24 cm. A primitive reptile. [**(A)** *after* White; **(B)** *after* Williston.]

If paleontologists knew when and how this egg evolved, the separation of reptiles and amphibians would be easy. Fossil eggs are so rare that their absence is meaningless. It is customary to set the boundary between fossil reptiles and amphibians on details of the vertebrae and of the arrangement of bones in the skull roof. One feature seems to be completely diagnostic. Fish and many amphibians have—and had—grooves on the snout and sides of the skull that are the impressions of the *lateral line system.* This system consists of canals and associated nerves and blood vessels lying immediately below the skin. These lateral lines apparently perceive pressure changes and water movement. If lateral lines occur on a fossil skull, they are presumptive evidence of aquatic life or, in other words, the development of the young as larvae in a stream or pond. Unfortunately, the post-larval skull may lose any trace of the lateral-line grooves. Thus some skulls of reptile-like *Seymouria* preserve such traces (White, 1939) but others do not. In addition, early and middle Permian relatives of *Seymouria* have an array of characteristics—including gill-bearing larvae—that demonstrate that they were completely aquatic. Very well, *Seymouria* is an amphibian. But what of *Diadectes,* which is similar in vertebrae and in the notching of the back of the skull, but which shows other features which link it to the primitive reptiles?

The present consensus is that the reptiles derive from one of the less abundant amphibian stocks, the anthracosaurs. In these, unlike most amphibians, the paired, pleurocentral elements of the vertebrae (p. 449) are enlarged and, in some, fused to form a single major element. The anthracosaurs apparently gave rise to the embolomerids, in which both the intercentra and pleurocentra are large and subequal in size; the result is a pair of complete centra in every vertebra. On the other hand, they were probably ancestral also to the group of *Seymouria*-like amphibians which have very large pleurocentra. From somewhere near the base of this latter group the earliest reptiles evolved; the reptile-like characteristics of the *Seymouria*-like amphibians are explained as 1) retention of distinctive anthracosaur characteristics, and 2) parallel evolution of additional reptilian characteristics.

The oldest known reptile* is a small form, *Hylonomus,* from the early Pennsylvanian. By late Pennsylvanian a diversity of lines had evolved, including most of those from the early Permian—though few good specimens occur. The commonest and best known groups in this Permo-Pennsylvanian radiation are the surviving representatives of the primitive stock, the captorhinomorphs, and a more advanced assemblage, the pelycosaurs. The latter must have evolved very early in the radiation, for specialized genera appear in upper Pennsylvanian rocks. The skull details indicate derivation from the captorhinomorphs.

The pelycosaurian species share an extensive common structural plan. The most characteristic, though probably not most critical, feature of that plan is the presence in the skull of a single *lateral temporal opening.* The captorhinomorphs have a solid skull roof; the *dermal bones,* formed just below the skin, extend down and laterally from the midline to cover the cheek region; many of the jaw muscles attached to the inner surface of this roof; the brain is separated from these muscles by an inner wall of cartilage and bone (Figure 20-2). Turtles retain this structure, but all other recent reptiles have a different arrangement. One or two openings occur in the skull over the great muscles of the cheek region. When these muscles contract, they bulge laterally through the temporal openings. The middle Permian therapsids evolved from a pelycosaurian stock and, in turn, gave rise to the mammals; thus, pelycosaurs and therapsids are termed mammal-like reptiles. The details of this sequence are fairly well understood, and I'll discuss them further in the following chapter.

Reptiles, other than the captorhinomorphs and pelycosaurs, are represented in the lower Permian by only a few tantalizing specimens. For ex-

* The definition as a reptile is based positively on skull and vertebral characteristics and negatively on the absence of evidence of an aquatic larval stage in this or in any later related form.

ample, the genus *Araeoscelis* has been variously interpreted as a lizard ancestor, as a relative of the pelycosaurs, or as a member of a stock that gave rise to ichthyosaurs, plesiosaurs, and the diapsids (lizards, dinosaurs, birds, and so on). Like the pelycosaurs, it has a single temporal opening, but the opening is high up on the skull rather than on the lower cheek region (Figure 20-2). At the present time, these specimens help very little in interpreting the ancestry of more advanced reptiles.

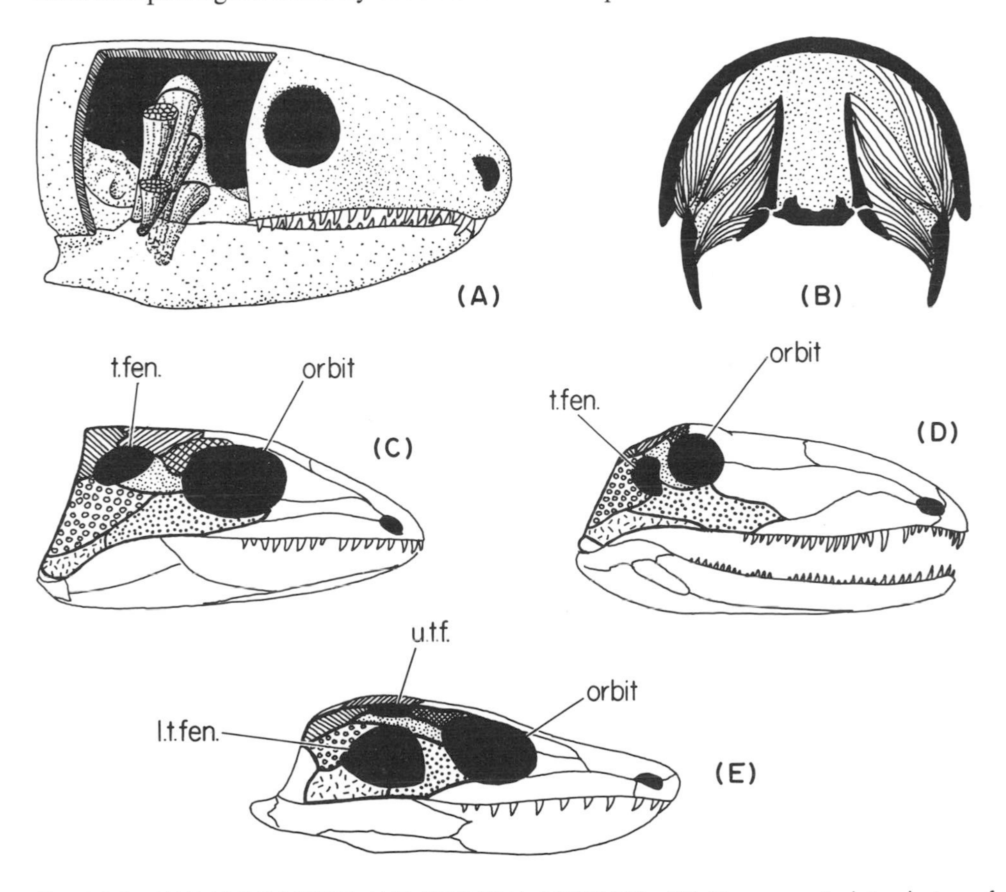

Fig. 20-2. JAW MUSCULATURE AND TEMPORAL OPENINGS. **(A)** Diagrammatic lateral view of primitive reptile skull. A window has been cut in the skull roof and cheek to show the muscles. In the animal the attachment of the muscles to the interior of the skull roof would be much wider. Note that part of them originate from the bones of the palate and at the sides of the brain case. **(B)** Diagrammatic cross-section of same. Bone shown in solid black. **(C)** through **(E)** are lateral views of reptile skulls that show various types of temporal openings. Different patterns indicate different bones of cheek and skull roof. **(C)** *Araeoscelis.* A single temporal opening surrounded by the parietal (lines diagonal to right), postorbital (light stipple), squamosal (circles), and supratemporal (lines diagonal to left) bones. The parapsid condition. **(D)** *Ophiacodon.* A single opening surrounded by the postorbital, jugal (heavy stipple), and squamosal bones. The synapsid condition. **(E)** *Youngina.* Two openings present; the diapsid condition. Abbreviations: l. t. fen., lower temporal fenestra; orbit, orbit; t. fen., temporal fenestra; u. t. fen., upper temporal fenestra. [**(C)** *and* **(D)** *altered from* Romer; **(E)** *altered from* Broom *and* Olson.]

Domination of the land

Act I; scene 1

The mists of Pennsylvanian swamps that largely conceal the origins of the reptiles began to dissolve near the end of that period. By mid-early Permian the scene was brightly illuminated. At that time, river deltas had filled the shallow seas that had covered Oklahoma and north-central Texas for most of the Pennsylvanian. A wide variety of reptiles and amphibians (see Fig. 5-5, p. 116) lived on the terrestrial portions of these deltas, and many were fossilized. The area subsequently was tilted gently, and, in the Cenozoic, a thin cover of young rocks was stripped away. Because of the low dip, these beds are exposed over a wide area. Because of the present semi-arid climate there are extensive outcrops. As a result, the badlands of the Wichita River Valley are a paradise for vertebrate fossil collectors.

The deltas were well vegetated, except possibly for the drier interfluves. The climate was warm with marked wet and dry seasons. The deltas were crossed by distributary channels of the rivers. On the flood plain were shallow lakes and, at least in the early part of the sequence, local swamps. The lakes probably were reduced to muddy ponds during the dry season; the distributaries, to trickles and pools. In the lakes and channels lived xenacanth sharks, lungfish, crossopterygians, and a variety of amphibians.* The xenacanths were sharklike in behavior as well as appearance, and their coprolites contain bones of fish and amphibians. The lungfish were well on their way to their recent form; they had elongate bodies, slender paired fins, and broad crushing plates in place of teeth. The crossopterygians and most of the amphibians were predators, though some, such as the "longhorned" *Diplocaulus* (Figure 19-10) and the tiny, limbless *Lysorophus* grubbed through the bottom muds for small invertebrates. Large predaceous amphibians and reptiles crawled along the banks and returned to the water like recent crocodiles to eat fish and aquatic amphibians. Several species of small captorhinomorph reptiles and a few amphibians foraged on the floodplains; a few probably ate plants; others, insects or smaller vertebrates. *Diadectes* and some of the pelycosaurs, including one (*Edaphosaurus*) with long dorsal spines, may have fed on plants or, possibly, on shelled invertebrates. Medium and large-sized pelycosaurs were the dominant terrestrial carnivores. The largest of these, *Dimetrodon,* also had long dorsal spines on their vertebrae and a "sail" of skin between the spines.

Olson (1952 and earlier papers) in a detailed study of some early Permian fauna in north-central Texas (see also pp. 116 and 183) devel-

* These interpretations of habitat and habit are based largely on the structure of limbs and teeth and on occurrence in various rock associations.

oped the paleoecologic interpretation summarized above, and he demonstrated gradual changes in the fossil assemblages as the climate became drier. Near the end of the early Permian, the Texas delta apparently became too dry for survival of a vertebrate fauna. A somewhat different assemblage of advanced pelycosaurs, of primitive therapsids, and of captorhinomorphs appears at the base of the upper Permian sequence. Though deposition continued until near the end of the Permian, no vertebrate fossils have been found in these upper beds.

Act I; scene 2

At about the same time that the vertebrate fossil record fails in North America, it begins in Russia and South Africa. Recent collecting at the

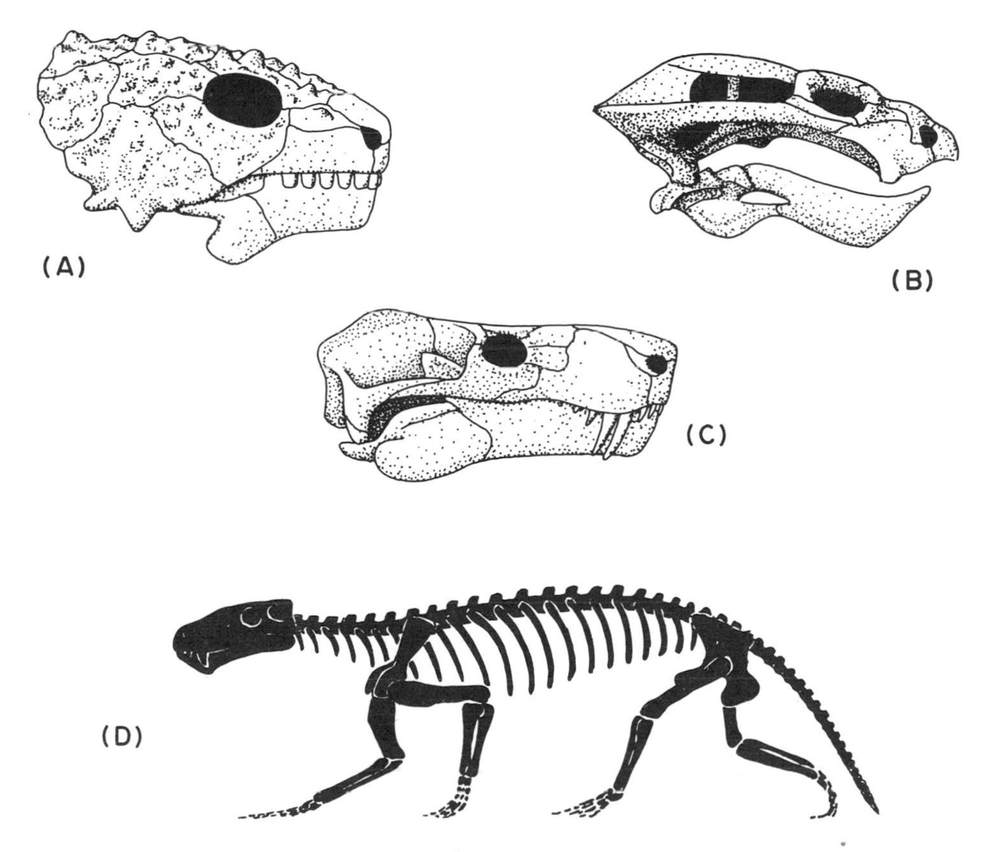

Fig. 20-3. REPRESENTATIVES OF THE PERMO-TRIASSIC RADIATION. **(A)** *Pareiasaurus;* late Permian; lateral view of skull. **(B)** *Dicynodon;* middle to late Permian; lateral view of skull. A herbivorous therapsid. **(C)** *Lycosuchus;* middle Permian; lateral view of skull. A carnivorous therapsid. **(D)** *Lycaenops;* late Permian; lateral view. A carnivorous therapsid. [**(A)**, **(B)**, *and* **(C)** *after* Broom; **(D)** *after* Colbert.]

top of the fossiliferous sequence in Texas and at the base of the Russian indicates the two may overlap in time. The oldest South African fauna is slightly younger. In both Russia and South Africa, deposition was nearly continuous from middle Permian through early Triassic.

So far as is known, the environment of the Russian and African deltas was much like that of the Texas one—warm with seasonal rainfall. The faunas, however, show a sharp change (Figure 20-3). The amphibians were reduced in variety. Those that remained were aquatic; some had flattened heads and reduced limbs, and others elongate snouts adapted to catching fish. Captorhinomorphs had nearly disappeared and did so before long. The procolophonids and pareisaurids, large, heavy-bodied reptiles, may have evolved as a herbivore specialization from the captorhinomorphs. The primitive herbivorous and carnivorous pelycosaurs had disappeared. They were replaced by herbivorous and carnivorous therapsids —both of which evolved from a carnivorous pelycosaur stock. The therapsids were the dominant members of the fauna and would seem to be the first thoroughly successful terrestrial vertebrates. They evolved with rapidity and radiated to different adaptations even more rapidly. Watson and Romer (1956) list over 180 genera of carnivorous therapsids, representing 40 families, and over 90 genera of herbivores in 7 families. They were of moderate to large size; the skull, jaw, and teeth were considerably modified for specialized diets; and the limbs were brought under the body rather than spraddled laterally as in early Permian types.

Because of the great thickness and extent of these beds, because of the abundance of fossils, and partly because of their remoteness from centers of paleontologic research, study of the Russian and South African faunas has hardly gone beyond the initial, descriptive stage. The phylogenetic interpretations are tentative and based on very tentative stratigraphy. Paleoecologic studies have hardly begun.

Act II

A dispassionate observer in the early Triassic could hardly have avoided concluding that the therapsids had taken the final step in tetrapod evolution and would live happily ever after. Of course, the diapsid reptiles were on the scene. But they were only subsidiary characters. But in the middle Triassic the diapsid villains (the archosaurs) got the upper hand, and the therapsids faded into obscurity. The reasons for this evolutionary upset are unknown.

Ancestors and conservatives. The early history of the diapsids is obscure. Certainly *Youngina* and a series of related genera, the eosuchians, from the late Permian of South Africa, are diapsids (Figure 20-2). They have the characteristic pair of temporal openings, upper and lower, on each side of the skull and retain a sufficient number of primitive reptilian adaptations to stand as satisfactory ancestors for the three advanced lines —the lizards, the rhynchocephalians, and the archosaurs. In general, these late Permian proletarians are small, probably lizard-like in habit as well as form.

The rhynchocephalians have been the most conservative diapsid lineage (Figure 20-4). Their teeth are fused to the edge of the jaws rather than

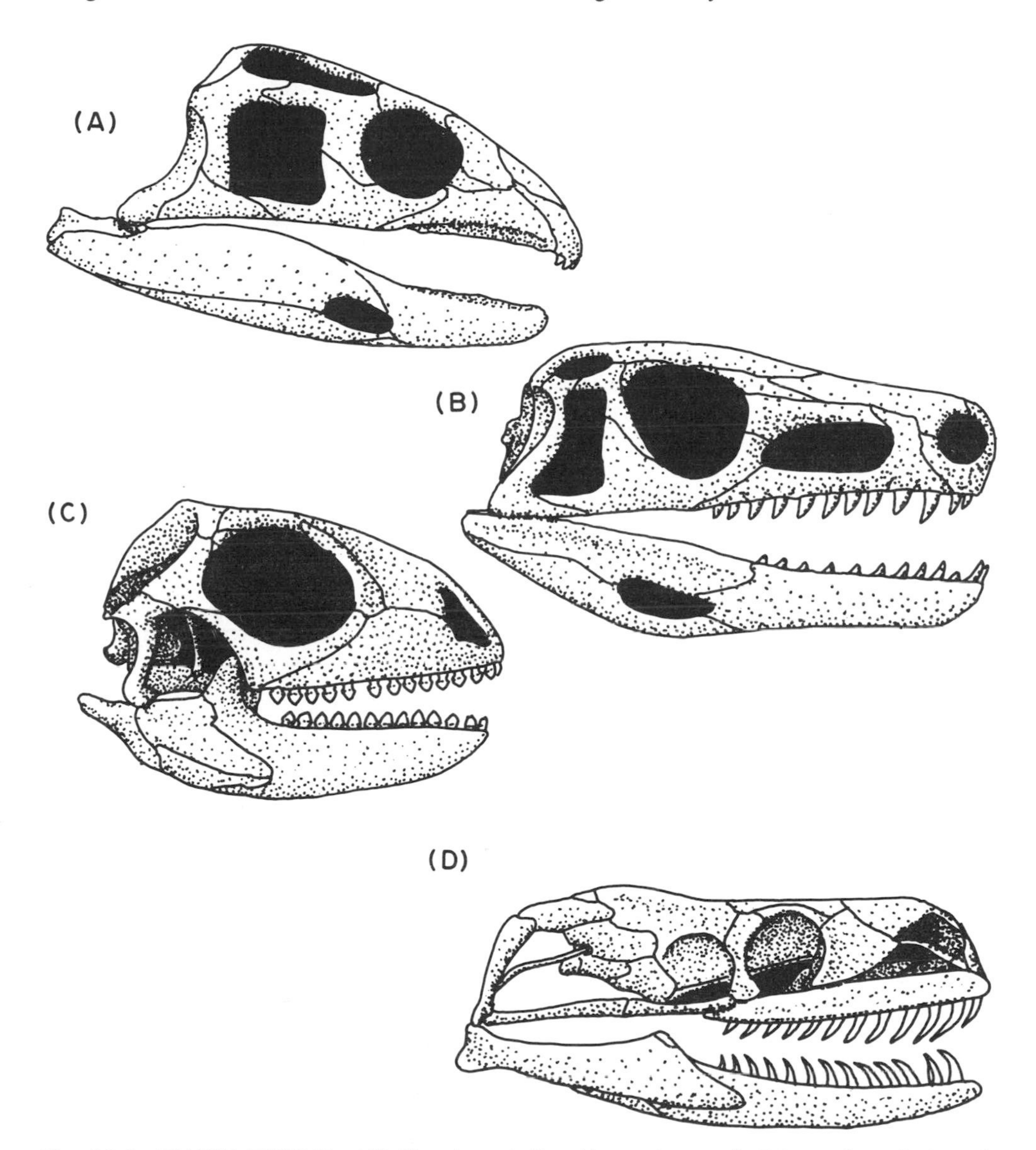

Fig. 20-4. DIAPSID REPTILES. **(A)** Rhynchocephalian *Mesosuchus;* early Triassic; lateral view of skull. **(B)** Thecodont *Euparkia;* early Triassic; lateral view of skull. **(C)** Squamate *Iguana;* Pleistocene to Recent; lateral view of skull. A herbivorous lizard. **(D)** Squamate *Python;* Pliocene to Recent; lateral view of skull. A snake. [**(A)** and **(B)** *after* Brown; **(D)** *after* Smith.]

set in sockets, and they have a small beak on the upper jaw. But they retain the pineal eye (see p. 435), and the palate has a movable articulation on the brain case like the ancestral crossopterygian fish. The history of the order has been modest; the oldest fossils date from the early Triassic; they underwent a small radiation in that period; and a single genus, *Sphenodon,* survives today in the protective isolation of New Zealand.

The lizards and snakes (order Squamata, suborders Lacertilia and Ophidia) have been the true heirs of the primitive diapsids. Most retain the pineal eye and the primitive palate (Figure 20-4) and resemble the late Permian form in size and adaptations. They are characterized by three major specializations—the development of ball and socket joints between the vertebrae, the loss of the bar below the lower temporal opening, and fusion of the teeth to the inner or outer margins of the jaw. The modification of the vertebrae provides a very strong but flexible backbone. The loss or reduction of the bony elements in the cheek frees the bone (the *quadrate*) that articulates with the lower jaw. The jaw is, in essence, double-jointed. The added flexibility assists in holding and swallowing large pieces of food.

The modification of vertebrae and jaw articulation has been carried furthest in the snakes. The ball and socket vertebrae provide an efficient framework for the body muscles, so efficient that a wriggling locomotion can substitute for limbs. The snakes also lost the bar between the upper and lower temporal opening so that the quadrate is very loosely attached to the skull. In addition, the palate and upper jaw move freely on the brain case; the lower jaw has a joint on either side midway of its length; and the two halves of the lower jaw are only loosely connected by a ligament.

The oldest fossil lizards are of Triassic age, and the oldest snakes late Cretaceous. Both are essentially modern groups and have survived and radiated widely in competition with the mammals.

Kings of the hill. The archosaurs have been the most spectacular successes—and failures—of the three diapsid lineages. Primitive archosaurs, the thecodonts, appear first in lower Triassic rocks. In upper Triassic rocks one can distinguish eight suborders and fifteen families. The primitive thecodonts retained the slender, lizard-like body of the eosuchians (Figure 20-5A) but otherwise departed from that plan (Figure 20-4B). The pineal eye was lost; the upper temporal openings were reduced in size; an opening was developed in the skull immediately in front of the orbits. Yet more significant of the archosaur future, some had relatively long hind legs. The legs, like those of the advanced mammal-like reptiles, were directly under the body rather than sprawling to the sides. The ventral part of the pelvis was extended fore-and-aft so that the muscles of the hip could swing the limbs fore-and-aft. The tail was long, large, and presumably heavily muscled, the muscles attached in part to the upper leg. By analogy with later archosaurs and modern bipedal lizards, they must have

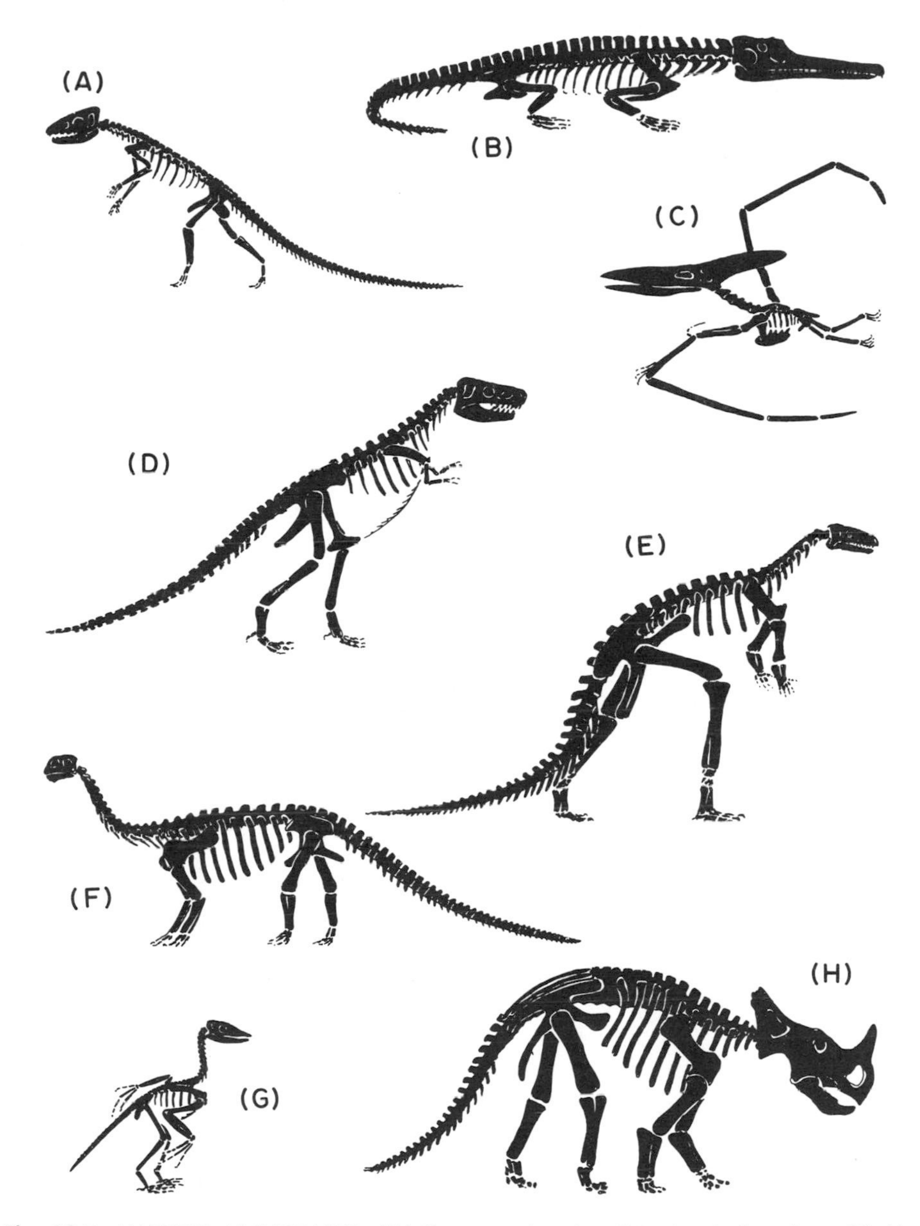

Fig. 20-5. VARIOUS ARCHOSAURS. **(A)** *Hesperosuchus;* late Triassic. A thecodont. **(B)** *Mystriosuchus;* late Triassic; length about 3.3 m. An aquatic thecodont—a phytosaur. **(C)** *Pteranodon;* late Cretaceous; wing span about 7.0 m. A pterosaur. **(D)** *Aublysodon;* late Cretaceous; length about 9.0 m. A carnivorous saurischian. **(E)** *Camptosaurus;* late Jurassic to early Cretaceous. An ornithischian. **(F)** *Camarasaurus;* late Jurassic to early Cretaceous; length about 5.5 m. A "small" herbivorous saurischian. **(G)** The primitive bird *Archaeornis;* Jurassic. **(H)** *Monoclonius;* late Cretaceous; length about 5.2 m. An ornithischian. [**(A)** *after* Colbert; **(B)** *after* McGregor; **(C)** *after* Eaton; **(D)** *after* Lamb; **(E)** *after* Gilmore; **(G)** *after* Heilmann, *Origin of the Birds.* By permission of Appleton Century, affiliate of Meredith Press. **(H)** *after* Brown.]

run on the hind legs, the body and tail balanced about the fulcrum of the hip joint. This characteristic adaptation, though suppressed in many later archosaur lines, shows even in the crocodiles and alligators—their hind limbs, though small relative to body size, are much larger than the forelimbs.

Some thecodont lineages abandoned the biped habit and adopted a sluggish aquatic life (Figures 20-5 and 20-6). There they competed with the giant predaceous amphibians and ultimately displaced them. The phytosaur lineage made the adaptive shift in the early Triassic, underwent a modest radiation in the middle and late Triassic, and disappeared by the end of the period. Crocodilian in general form, with a large, long-snouted skull and a heavy, short-limbed body, they differed considerably in detail from the crocodiles.

The other lineage of aquatic archosaurs included the crocodiles and alligators (Figure 20-6). Although anticipated by the phytosaurs, they outdid them in diversity and persistence and survived all their more spectacular archosaurian relatives. Some Jurassic crocodilians apparently expanded into marine environments.

Two groups of archosaurs also took up flying. The pterodactyls, who developed a flight membrane from a sheet of skin stretched between elongate fingers (Figure 20-5), were moderately successful but failed to survive the big kill at the end of the Cretaceous. The group appears at

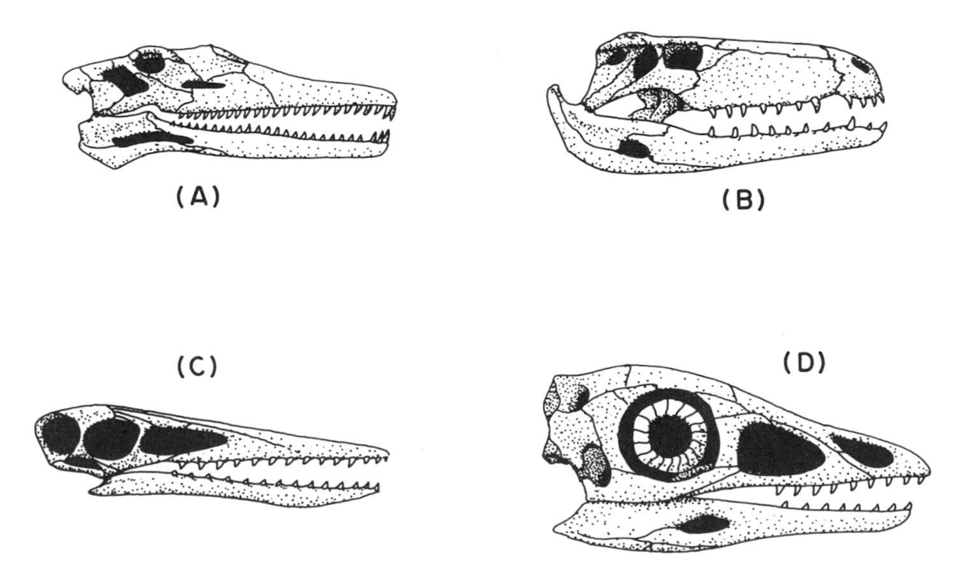

Fig. 20-6. ARCHOSAUR SKULLS. **(A)** Thecodont phytosaur *Machaeroprosopus;* late Triassic; length of skull about 1.0 m. **(B)** Crocodilian *Sebecus;* Paleocene to Miocene. **(C)** Pterosaur *Rhamphorhynchus;* late Jurassic; length of skull about 12 cm. **(D)** The primitive bird, *Archaeornis;* Jurassic; length of skull about 5.0 cm. [**(A)** *after* Camp; **(B)** *after* Colbert *and* Mook; **(C)** *after* Jaekel; **(D)** *after* Heilmann, *Origin of the Birds.* By permission of Appleton Century, affiliate of Meredith Press.]

the beginning of the Jurassic with well-developed wings—transitional stages should occur in the late Triassic.

The other flying archosaurs did sufficiently well to found a new company of their own, the class Aves. The birds are distinguished from the primitive archosaurs primarily by the presence of wings and feathers, the extreme shortening of the tail, and the loss of teeth. The primitive bird *Archaeopteryx,* known from Solenhofen lithographic limestone (late Jurassic), had wings and feathers, the latter preserved as impressions in the fine-grained limestone, but also had a long tail and teeth (Figures 20-5 and 20-6). In structure, *Archaeopteryx* is almost exactly intermediate between later birds and the primitive archosaurs—a "missing link" discovered before anyone thought of it being missing. I will write more about the evolution of birds in the concluding chapter.

The remaining archosaur lineages, the dinosaurs, formed the dominant elements of the later Mesozoic faunas and evolved into some of the largest and most bizarre of vertebrates, fossil or recent. To men, another group of bizarre vertebrates, they have been the most fascinating of fossils. They occupy the centers of museum halls and form the *pièce de résistance* of popular books on fossils. In spite of this—or perhaps because of it—their paleontology and evolution are poorly known. Their very size diminishes their scientific worth, for few museums or universities can afford to collect large numbers of dinosaurs. In recent years some paleontologists have undertaken the necessary careful studies of the dinosaurs, but less is known of them than of less common and perhaps less interesting animals.

Paleontologists recognize two distinct lines of dinosaurs, the saurischians and the ornithischians. The former appeared in the late Triassic as slightly aberrant thecodonts (Figures 20-5 and 20-8) and would surely

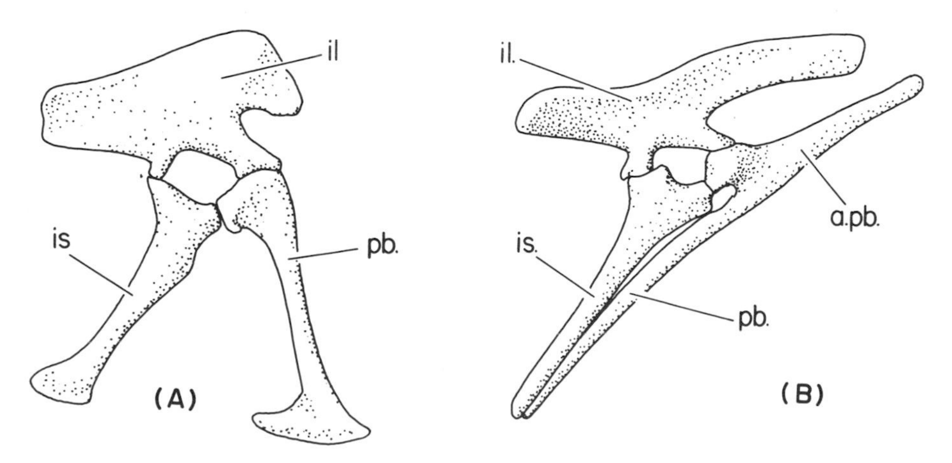

Fig. 20-7. DINOSAUR PELVES. **(A)** Saurischian pelvis. **(B)** Ornithischian pelvis. Abbreviations: a. pb., anterior process of pubis; il., ilium; is., ischium; pb., pubis.

be classified with them if later, more specialized saurischians were unknown. This line is characterized by the form of the pelvis: a broad plate, the *ilium,* above the hip joint, a slender bone, the *pubis,* extending forward and down and a similar element, the *ischium,* extending backward and down (Figure 20-7). The primitive saurischians were bipeds and probably swift runners. The smaller species have hollow bones somewhat like those of birds. Their teeth were sharp, serrate, compressed, and thereby fit for a diet of flesh.

Before the end of the Triassic, this primitive stock had begun a radiation (Figures 20-5 and 20-8). Some species retained the primitive characters and adaptations. Others evolved long, slender hind limbs, long delicate forelimbs fitted for grasping, and a long neck. Some of these have very small teeth or none at all and except for the long tail resemble an ostrich. Probably they ate small vertebrates and invertebrates and succulent plants. Still another series evolved large size. Their heads were large; their jaws still larger in proportion and set with sharp, recurved, cutting teeth. These types must have been the master carnivores of the Mesozoic as the great cats are at present. Representatives of all three lines thrived throughout the Jurassic and Cretaceous.

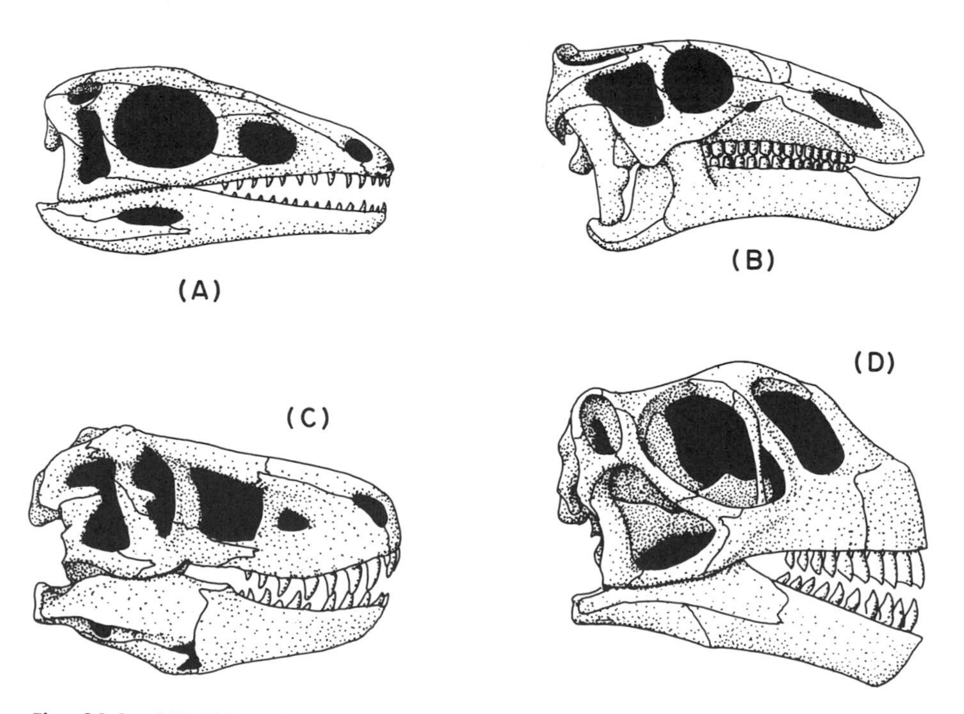

Fig. 20-8. DINOSAUR SKULLS. **(A)** *Compsognathus;* late Jurassic. A small predaceous saurischian. **(B)** *Iguanodon;* late Jurassic to early Cretaceous. An ornithischian. **(C)** *Tyrannosaurus;* late Cretaceous. A large carnivorous saurischian. **(D)** *Apatosaurus;* late Jurassic. A large herbivorous saurischian. [**(B)** *after* Hooley; **(C)** *after* Osborn; **(D)** *after* Gilmore.]

The saurischian stock also produced quadruped herbivorous dinosaurs. Most of these were giants, some up to eighty feet long. Tails and necks were long; the body short though massive; and the head diminutive. Because of their small heads, weak jaws, and reduced, peg-like teeth they must have fed on soft vegetation. They are best known from Jurassic deposits but range from late Triassic to late Cretaceous.

The other major dinosaur line, the ornithischian, also appeared in the late Triassic but unlike the saurischian had no close connection with the known thecodonts. In the ornithischian pelvis the pubis extends down and back parallel to the ischium rather than forward. The position and, in part, the function of the saurischian pubis are taken by a new element that arises as a process from the base of the pubis. As part of the adaptation to a plant diet, the teeth in the anterior end of the jaws are lost; a beak replaces the lost teeth, probably for cropping leaves; and a new bone, the *predentary* serves as the lower elements of this beak.

Although many of the ornithischians are bipedal, there appears a persistent tendency to return to the quadrupedal habit (Figure 20-5). They remained herbivores and exhibited the characteristic dinosaur trend toward great size. Some evolved large batteries of grinding teeth—probably to chew harsh plants. The quadrupedal types evolved either heavy body armor or large horns and a protective neck shield. The ornithischians, though common and varied in Jurassic deposits, are more characteristic of the Cretaceous and thrived to the very end of the period.

Return to the womb

The success of the reptiles is expressed in their invasion of a wide variety of habitats; the dinosaurs on the ground; the phytosaurs and crocodilians in lakes and streams; the pterodactyls and birds in the air. But equally striking are the variety of marine reptiles. Five distinct groups have been identified—the ichthyosaurs, the sauropterygians, the mosasaurs, the geosaurs, and the turtles. The first two appeared in the Triassic and underwent a moderate radiation through the Jurassic and into the Cretaceous. The third evolved from a lizard ancestry in the Cretaceous; the fourth from a crocodilian stock in the Jurassic.

The order Ichthyosauria comprises a group of large, very fishlike reptiles (Figure 20-9) characterized by a single large temporal opening high up on either side of the skull. At their first appearance they had already assumed most of the adaptations necessary for aquatic survival, a fishlike tail fin, a dorsal fin, paddle-like limbs, and an elongate, fusiform body. They were apparently active predators, or so their elongate jaws with sharp teeth, their large eyes, and their streamlined shape imply. Some apparently bore their young alive, for adult specimens have been col-

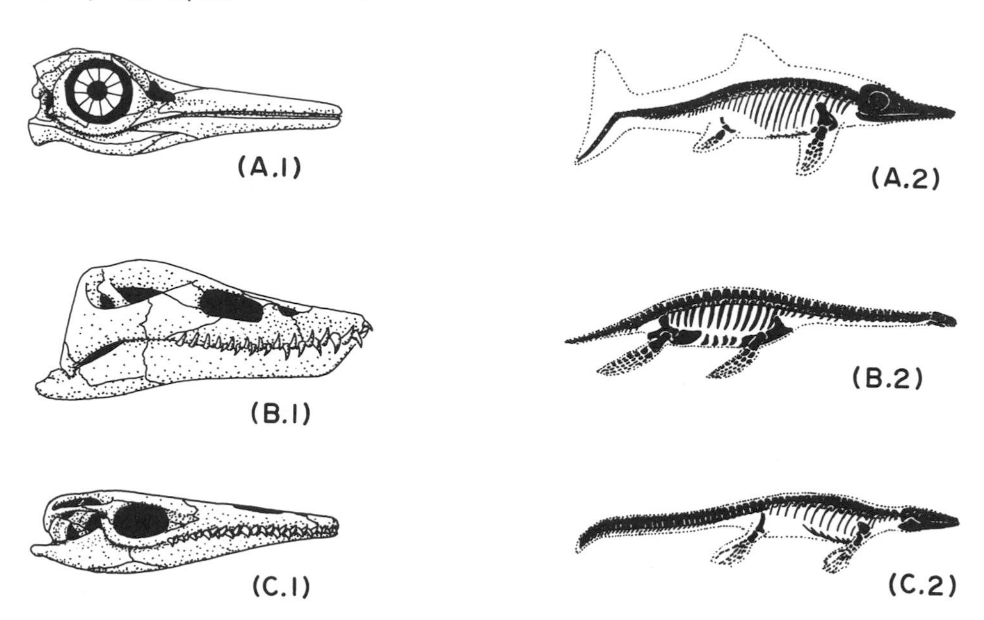

Fig. 20-9. MARINE REPTILES. **(A.1)** Ichthyosaur *Ophthalmosaurus;* middle to late Jurassic; length of skull 1.0 m. **(A.2)** Skeleton and body outline of advanced ichthyosaur. **(B.1)** Sauropterygian (plesiosaur) *Hydrotherosaurus;* late Cretaceous. **(B.2)** Skeleton and body outline of a plesiosaur. **(C.1)** Squamate (mosasaur) *Tylosaurus;* late Cretaceous. **(C.2)** Skeleton and body outline of a mosasaur. [**(A.1)** *after* Gilmore; **(B.1)** *after* Welles; **(C.1)** *after* Williston.]

lected with young ichthyosaurs lying in and emerging from the pelvic region.

Ancestors for the ichthyosaurs have yet to be collected. Their characteristics, other than the adaptations for marine life, are those of primitive reptiles. The temporal opening bears a different relationship to the bones of the skull than in other reptiles, so this tells nothing of their origin. They reached their acme of diversification in the Jurassic and slowly declined in the Cretaceous.

The sauropterygians form a somewhat more diversified group (Figure 20-9). The Triassic nothosaurs had begun adaptation to the aquatic environment and had long slender bodies, short limbs, and a bony secondary palate to separate the air passages from the mouth. The placodonts, also Triassic, were more heavy-bodied, had a bony armor, and bear wide, crushing teeth on jaws and palate. The plesiosaurs apparently evolved from an early nothosaur species and appear in Jurassic marine deposits. They have a broad, heavy body with a relatively short tail. The limbs are large, massive paddles. Two groups of plesiosaurs are recognized, those with small heads and long necks and those with large heads and short necks.

The marine crocodiles are less important and are limited to the Jurassic. Their most noteworthy specializations are paddle-like limbs and a fishlike tail fin. The mosasaurs occur only in late Cretaceous deposits and had diverged only slightly from their lizard ancestors by evolution of paddles and of long, laterally compressed tails. Both groups were probably predaceous, living on fish—and probably on cephalopods.

The marine turtles are an obvious extension of the typical turtle habit. The most marked changes are the loss of armor and the broadening of feet into paddles. Some are of large size—*Archelon* is 12 feet long. The turtles, alone of the marine reptiles, survived the end of the Cretaceous and persist to the present.

REFERENCES

Carroll, R. L. 1964. "The Earliest Reptiles," *Jour. of Linnean Society of London* (Zool.), vol. 45, pp. 61-83.

Charig, A. J., *et al.* 1965. "On the Origin of the Sauropods and the Classification of the Saurischia," *Proc. Linnean Soc. of London*, vol. 176, pp. 197-221.

Hughes, B. 1963. "The Earliest Archosaurian Reptiles," *South African Jour. of Sci.*, vol. 59, pp. 221-241.

Olson, E. C. 1952. See refs. p. 131.

Ostrom, J. H. 1966. "Functional Morphology and Evolution of the Ceratopsian Dinosaurs," *Evolution*, vol. 20, pp. 290-308.

Watson, D. M. S. and A. S. Romer. 1956. "A Classification of Therapsid Reptiles," *Bull. Mus. of Comp. Zoology*, vol. 114, pp. 37-89.

(*See also the references for Chapter 19.*)

21

The ascendency of the birds and mammals

To the end of the Cretaceous, the archosaurs dominated the terrestrial vertebrate associations; the ichthyosaurs, plesiosaurs, and aquatic lizards (the mosasaurs) shared the seas with the ray-finned fishes, the sharks, and the cephalopods.

The mammals formed an insignificant element in the terrestrial faunas throughout the Cretaceous. The ornithischian dinosaurs were represented by a series of horned quadrupedal genera, a group of large biped forms, and several armored types; the saurischians by a group of large carnivores and by a number of light-bodied ostrich-like forms. The latter have been variously considered herbivores, egg eaters, or, most probably, omnivores that ate mammals and lizards as well as eggs and succulent vegetation. The ranks of the insectivores and small herbivores numbered a variety of lizards as well as primitive mammals. Turtles and crocodilians occupied the lakes and rivers along with a number of small amphibians (frogs and urodeles). The birds, though very rare as fossils, were probably abundant and diversified. The fauna as a whole would seem to be well balanced, and most, if not all, major adaptive types were present.* The flora had a relatively modern aspect, including many flowering plants.

The climate of the late Cretaceous was warmer than at present; this is evidenced by the occurrence of cold-sensitive vertebrates in what are now cold temperate areas, and is further substantiated by temperature determinations based on the oxygen isotope method.

The fall of the mighty

In parts of western North America, continental Cenozoic overlies continental Cretaceous without any physical break. The systemic boundary is placed by the disappearance of a late Cretaceous assemblage characterized by the horned dinosaur *Triceratops* and the appearance of a quite different assemblage of early Paleocene mammals. In general, this boundary is distinct, but at Bug Creek, Montana, a partial, gradational sequence has been discovered (Sloan and Van Valen, 1965). The uppermost part of the Cretaceous reveals progressive reduction in dinosaur abundance, a progressive disappearance of typical Cretaceous mammals, an overlap of Cretaceous and Paleocene plant species, and most important, the occurrence of primitive representatives of Paleocene mammal groups. Sloan and Van Valen suggest that three distinct assemblages are represented in the Bug Creek deposits:

a). The aquatic and semiaquatic community that includes various fishes, amphibians, crocodilians, turtles, and birds and extended from the late Cretaceous into the Paleocene.

b). The "proximal" community that includes several mammals (multituberculates, insectivores, and condylarths) of Paleocene aspect.

c). The "distal" community that includes dinosaurs, lizards, and mammals (opossums and multituberculates) of Cretaceous aspect.

With this discovery, the discontinuity between Cretaceous and Cenozoic terrestrial vertebrate assemblages largely disappears, and the coexistence of distinct mammal-dominated and dinosaur-dominated communities is demonstrated.

This discovery also brings entirely new evidence on the nature—if not the ultimate causes—of dinosaur extinction and their replacement by

* No thorough study of paleoecology of Jurassic or Cretaceous terrestrial communities has been completed. Speculation about dinosaur extinction lacks the restraint of ecologic evidence.

mammals. What happened to the dinosaurs? Change of temperature? Change in the plants? Blasts of heat from a meteor, mammals eating dinosaur eggs, hyperpituiterism, change in oxygen concentration, over-specialization and senility of the dinosaur stalk, and so on . . . and on . . . and on? Some of these explanations seem absurd, but this illustrates the desperate straits into which paleontologists have been pushed by the extinction of the dinosaurs.

Some explanations disregard ecologic common sense—surely Cretaceous mammals ate dinosaur eggs, but so did Jurassic ones, and so did the therapsids back in the middle Triassic. A mammal may even have eaten the last dinosaur egg, but something must have already upset the balance between eggs laid and eggs eaten. If temperature grew hotter, they should have survived in the polar regions; if cooler, then in the equatorial. The critical evidence at hand amounts to this:

a). During the late Cretaceous, the number of different dinosaur assemblages decreases progressively and only one, the *Triceratops,* is known in very late Cretaceous deposits. This is admittedly negative evidence, but it suggests a slow, selective disappearance of dinosaur communities.

b). The late Cretaceous plants undergo a similar progressive change involving a rather complete replacement of plant communities.

c). The overlap of vertebrate assemblages (dinosaur and mammalian) at Bug Creek coincides with the overlap of plant assemblages.

d). The Bug Creek mammals include the two groups, insectivores and condylarths, which were probably ancestral to most later mammalian orders.

From this, I would suggest a one-by-one replacement of older plant communities and their associated and dependent dinosaur assemblages by new plant communities and their associated and dependent mammalian assemblages. But why were there no dinosaurs at all in these new assemblages? Is this a matter of chance? Or did the new plant communities themselves evolve in a feedback system with mammals and, probably, with the modern insect orders? This feedback arrangement might progressively exclude the dinosaurs if they were not originally a major element. Finally, in the competition of coexisting ecological systems, the modern plant-mammal-modern insect community(ies) proved superior. Neither the evidence nor evolutionary theory are sufficiently strong to demonstrate such a conclusion, but it is susceptible to further test, unlike little green men from Mars.

Birds

The fall of the dinosaurs left two vertebrate groups, the birds and the mammals, to exploit the provinces of the terrestrial environment. As already indicated they originated in the mid-Mesozoic and filled secondary roles in the Cretaceous terrestrial communities. They began the Cenozoic —marking its boundary—with an explosive adaptive radiation.

The fossil history of birds is fragmentary. The finding of the Jurassic fossil bird *Archaeopteryx* (Figures 20-5 and 20-6) must be considered an extraordinary stroke of luck. Only two Cretaceous birds are well known, a tern-like form and an aquatic bird with vestigial wings. The latter retains teeth; the jaws of the former are not certainly known. Otherwise they are "good" birds with a reduced tail, essentially modern wing structure, and an enlarged breastbone (*sternum*) for attachment of the flight muscles. Between these advanced forms and *Archaeopteryx* there are no intermediates; if *Archaeopteryx* had not been found, one might well imagine the bird lineage arising as some sudden "hopeful monster" within a Cretaceous stock.

A few puzzling and intriguing fossils suggest the evolution of toothless birds by late Cretaceous time. Over half the modern orders are known from fossils no younger than Eocene, and very probably many of the other orders are as old.

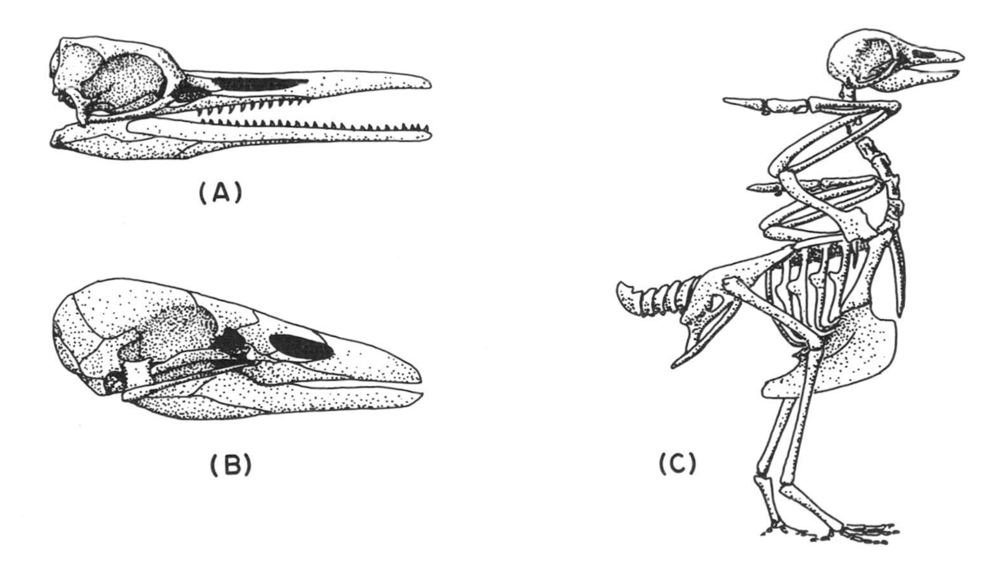

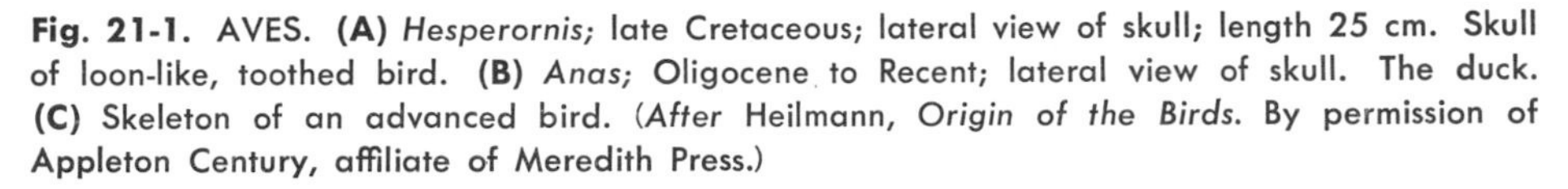

Fig. 21-1. AVES. **(A)** *Hesperornis;* late Cretaceous; lateral view of skull; length 25 cm. Skull of loon-like, toothed bird. **(B)** *Anas;* Oligocene to Recent; lateral view of skull. The duck. **(C)** Skeleton of an advanced bird. (*After* Heilmann, *Origin of the Birds.* By permission of Appleton Century, affiliate of Meredith Press.)

The birds achieved their basic adaptations to flight before the end of the Mesozoic and accomplished the major steps in their radiation before the end of the Eocene. Fixed in adaptations to flight and unable to compete with mammals on the ground,* birds have been without significant evolutionary issue in the later Cenozoic. Their evolution in structure (Figures 20-5 and 21-1) and behavior has been remarkable but apparently self-limiting, for each successful tour de force has closed the boundaries of specialization tighter about them.

Mammalian origins

The preceding chapter reported that the mammal-like reptiles faded into obscurity with the Triassic radiation of the diapsids. One must flash back in this story to find out how they were saved.

As noted, the therapsids composed the major part of the early Triassic terrestrial fossil assemblages. Some approached the mammals in characters of skull and skeleton, and only the most mammal-like of the mammal-like reptiles survived the archosaurian breakthrough in the middle Triassic. Two groups of therapsids, the tritylodonts and the ictidiosaurs, are known from the late Triassic (Figure 21-2). These might be classified as mammals except that the lower jaw still articulates with the skull through a pair of small bones, the *articular* (lower jaw) and the *quadrate* (upper jaw) rather than through the dentary-squamosal contact of the "true" mammals. In the ictidiosaurs, however, the dentary and squamosal share the articulation with the articular and angular, and this group may include the ancestry of the mammals—or at least some of them if the class had a multiple phyletic origin.

The earliest recognizable mammals occur in the very late Triassic in association with ictidiosaurs and represent at least three distinct phyletic lines. Late Jurassic rocks have yielded representatives of two of these lines and of three new lineages. Early Cretaceous rocks have produced little mammalian material, but teeth and jaw fragments of mammals of more modern aspect occur in the Trinity formation of Texas along with survivors of three of the Jurassic groups. In nearly all cases the specimens consist of isolated teeth and fragmentary jaws. At best they say mammals were here, were small, and were primarily carnivorous or insectivorous (excepting the herbivorous multituberculates). Little is known of the evolution of the cranial portions of the skull or of the post-cranial skeleton. Late Cretaceous mammals include multituberculates, primitive opossums, and at least two placental orders.

* The cursorial birds are an exception; but, by their small diversity, restricted distribution, and limited ecologic importance, they prove the rule.

The origin of a class

The fossil record does not impress the observer—but it does provide the documentation for the evolution of a major animal group, a class. And

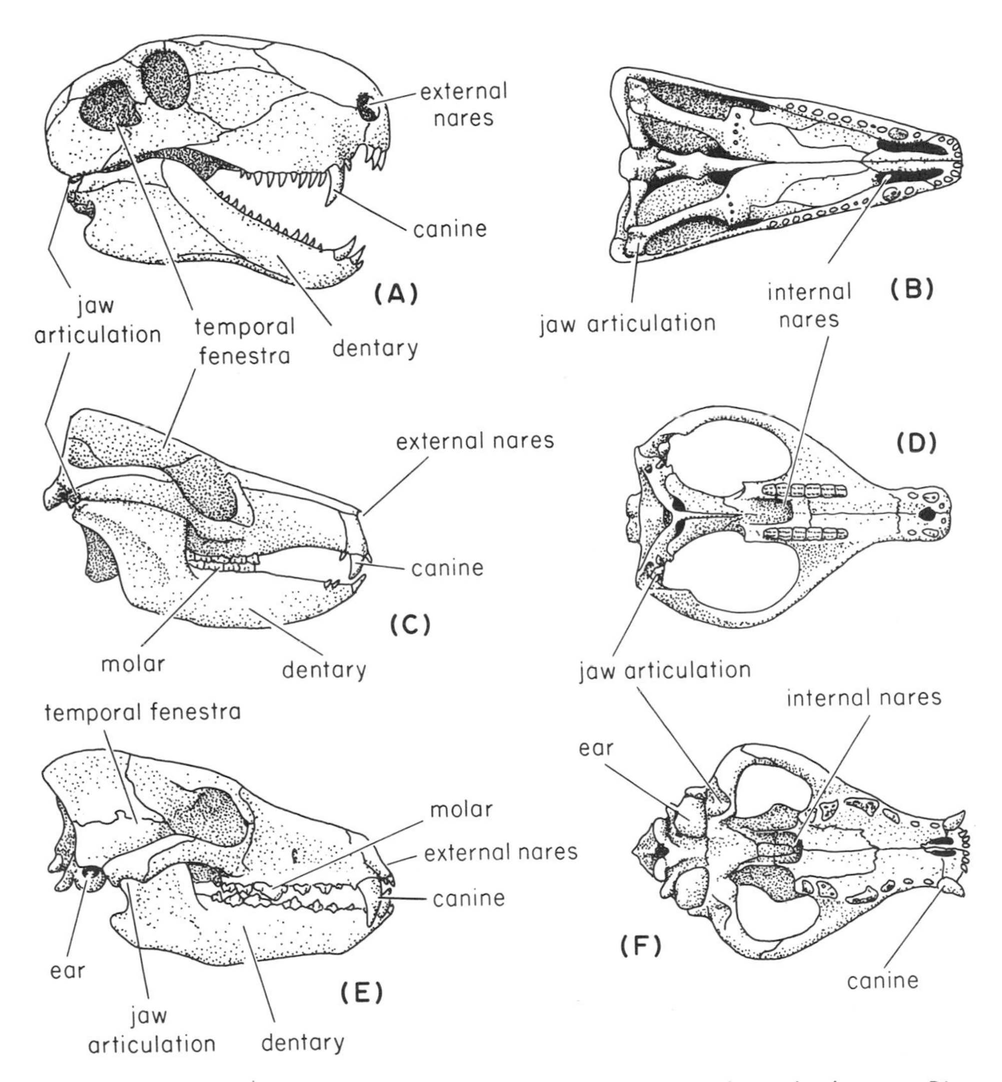

Fig. 21-2. THE ORIGIN OF MAMMALS—*THE SKULL.* **(A)** and **(B)** Advanced pelycosaur, *Dimetrodon,* an early Permian genus near the ancestry of the therapsids. Note differentiation of canine tooth and the anterior position of the internal nostrils. **(C)** and **(D)** Advanced therapsid, *Oligokyphus,* a late Triassic genus. Note the differentiation of the canine and cheek teeth, the presence of a secondary palate, and the relative size of dentary bone in the lower jaw. **(E)** and **(F)** A mammal, *Canis.* The jaw articulation has shifted forward to the squamosal and dentary; the bones that originally formed this articulation (the quadrate and the articular) have become part of the middle ear. [**(A)** *after* Romer; **(B)** *after* Kühne.]

the origin of the class Mammalia was a gradual process, not a sudden leap. The mammal-like reptiles show a gradual approach to the mammalian structure—particularly in the carnivorous types. The definition of the boundary between fossil reptiles and mammals depends on osteological characteristics observed in all modern mammals and correlated in them with the presence of mammary glands, hair, a four-chambered heart with the main artery passing to the left, and a constant body temperature. The osteological features include (Figures 21-2 and 21-3) the development of a bony secondary palate; the reduction of the bones in the lower jaw to a single element, the dentary on either side; the shift of the lower jaw articulation forward; the incorporation of the two bones of the old jaw articulation into the middle ear; adaptation of teeth in the front part of the jaws for biting and in the rear for chewing; rotation of the pelvic girdle so that the dorsal element shifted forward and the two ventral elements backward; and reduction of the size and, commonly, the number of the ventral elements of the pectoral girdle. These changes are all adaptive: the change in limb girdles to the fore-and-aft movement of the limbs *under* rather than at the sides of the body; the modification of the middle ear to better hearing (particularly if, as Watson [1951] believes, the

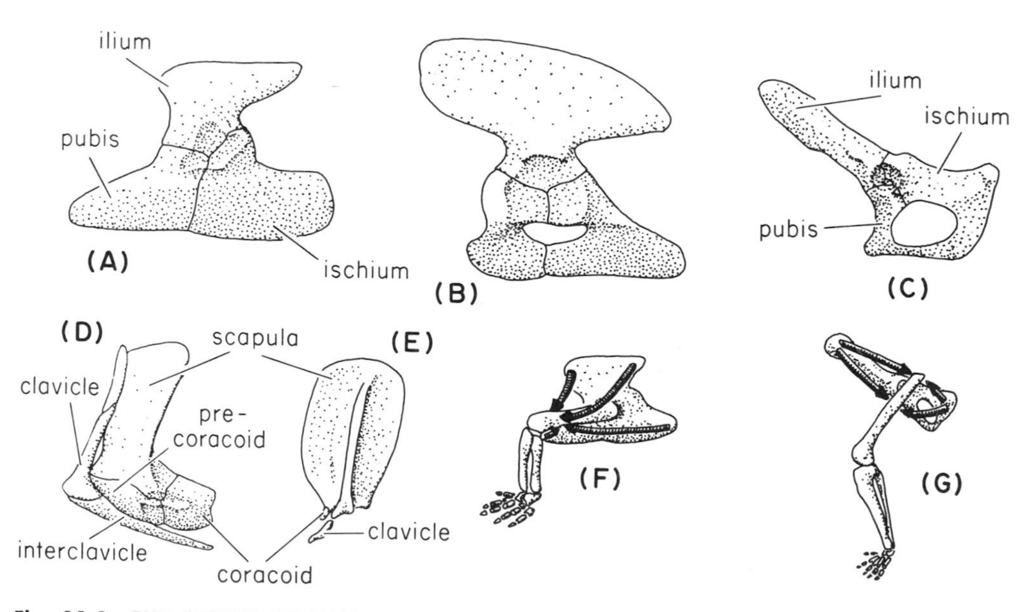

Fig. 21-3. THE ORIGIN OF MAMMALS—*THE LIMBS*. **(A)** Lateral view of pelvis of pelycosaur. Note relative size of ventral elements. **(B)** Lateral view of left pelvis of therapsid. Note reduction of ventral elements and expansion of dorsal. **(C)** Lateral view of pelvis of mammal. In this, the pubis and ischium have been rotated posterially. **(D)** Lateral view of pectoral girdle of pelycosaur. As in the pelvis **(A)** the ventral elements are relatively large. **(E)** Lateral view of pectoral girdle of mammal—ventral elements very much reduced. **(F)** Limb position in primitive reptile. The musculature (arrows) lies principally above and below the limb. **(G)** Limb position in a mammal. The musculature lies in front and behind the limb.

tympanic membrane was borne on the articular bone); the change in jaw articulation to changes in jaw musculature and the biting mechanism; and the formation of the secondary palate to permit uninterrupted passage of air above the palate to the back of the throat.

All of these features, except the shift of ear ossicles, began to appear in the mammal-like reptiles during the early Triassic and occurred in several lines, though at somewhat different rates and in somewhat different combinations. After vertebrate paleontologists studied the structures of these Triassic forms, they began to wonder whether the therapsids also approached the mammals in the character of the soft anatomy.

The evolution of the secondary palate is suggestive. In the amphibians and most reptiles the nostrils open into the mouth near the front end of the snout. When the animal has food in its mouth, breathing is difficult if not impossible. But this difficulty really doesn't matter since its metabolism is so low that a small amount of oxygen is sufficient—you need only observe a snake eating another snake. Aquatic reptiles, however, have special difficulties. Even when they raise their nostrils above the water, they cannot breathe if their mouths are open. For this reason, several aquatic reptile lineages evolved a bony secondary palate. This palate closes over the roof of the mouth and forms a separate air passage from the internal openings of the nostrils back to the throat.

Among recent terrestrial vertebrates, only the mammals have a secondary palate. In them the structure serves as an adaptation to continuous high metabolism. They deplete the oxygen in the cells, blood stream, and lungs so rapidly that they must breathe at short intervals. Mammals also chew their food into small pieces that can be quickly swallowed—for this they have an elaborate set of cutting and crushing teeth. Many reptiles bolt their food whole so that the throat is blocked for long periods. Breaking food into small pieces not only assists the digestive system, but it also reduces interruptions in respiration.

It seems reasonable to suggest that the carnivorous therapsids, with secondary palate and well developed chewing mechanisms (teeth and muscles), had an internally controlled body temperature—were *endotherms*. This is further confirmed by the development of bony scrolls in the nasal air passages,* since, in recent mammals, these bear mucous membranes that warm and filter the incoming air. If they had constant body temperature, they probably also had a four-chambered heart to separate oxygenated blood from the oxygen-depleted venous blood. Very possibly, they had an insulating layer (hair) to reduce the effect of external temperature variations.

To pile hypothesis on hypothesis, one can suggest that their eggs would have required incubation—a feature perhaps attained by carrying the eggs in an abdominal pouch or by retaining them within the body

* See Brink, 1957, for additional characters indicating endothermy.

during development. The latter adaptation would require development of a placenta to supply the embryo with oxygen and food and to remove metabolic wastes. The newly hatched young would require an abundant and constant supply of food that would be provided by mammary glands.

Very probably, some of these features appeared fairly early—temperature regulatory mechanisms perhaps as early as middle Permian. The frequency of independent developments of secondary adaptations to this primary change (the palate for example) implies that the unknown advanced pelycosaur that was ancestral to the therapsids may have had some control of internal temperatures. The other features surely appeared at different times in different lines and in different combinations. But regardless of where and when these events occurred, we cannot question the origin of mammals from one or more evolving populations of therapsid reptiles. So far as the fossil record ever "proves" anything, that evolution was gradual—extending over 45,000,000 years—and consistent with the concept of selection of slight variants within a species population. No "hopeful monsters" need apply.

Development of the mammals

The proto-mammals obeyed the adage and placed their eggs in several baskets; at least three distinct lineages, the symmetrodonts, the morganucodonts, and the triconodonts, appear at the very beginning of the

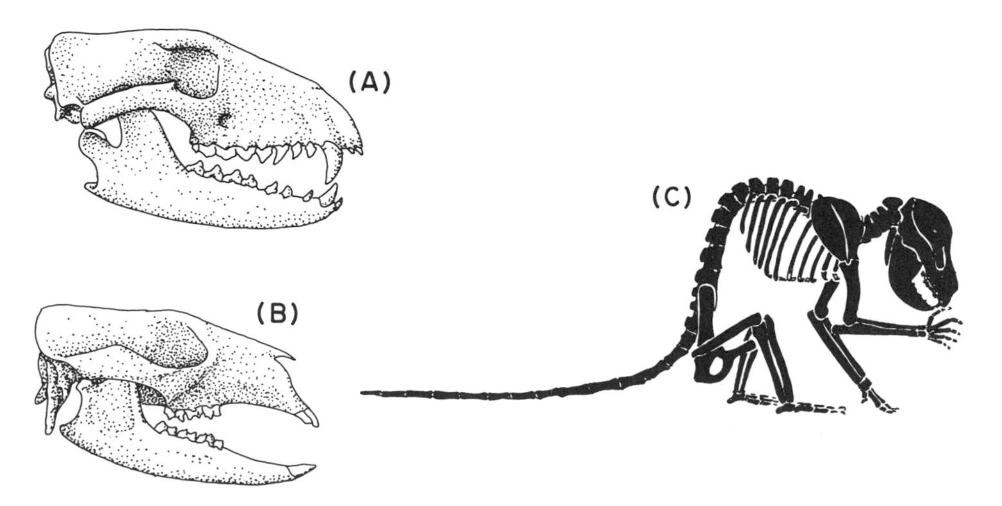

Fig. 21-4. MARSUPIALIA. **(A)** *Thylacinus;* Pleistocene to Recent; Australia; lateral view of skull. A marsupial carnivore. **(B)** *Macropus;* Pleistocene to Recent, Australia; lateral view of skull. A marsupial herbivore. **(C)** *Didelphis;* Pliocene to Recent, New World; skeleton. An unspecialized omnivore of primitive structure. (*After* Gregory, *Evolution Emerging.* Copyright, The Museum of Natural History, New York. Used with permission.)

mammal record. The additional Jurassic orders, the multituberculates, the docodonts, and the pantotheres, may represent further independent lineages, and the recent monotremes may form a sixth stock, all derived independently from different therapsid genera.

Of these experimenters in mammalness, only the pantotheres achieved a large success. They did so by evolving into marsupial and placental mammals. The marsupials (Figure 21-4) are characterized by an abdominal pouch (*marsupium*) in which the young are housed during the later stages of development and, in the skeleton, by marsupial bones to support the pouch, by the inward inflection of the back corner of the lower jaw, and by a distinctive pattern of cusps on the molar teeth. The placentals, on the other hand, carry their embryos to a full term of development, lack marsupium, marsupial bones, and the inflection of the jaw angle, and possess a different molar pattern. Generalized marsupial-placental style teeth in lower Cretaceous rocks and the marsupial and placental skulls and jaws in upper Cretaceous beds suggest that the two lines separated early in the Cretaceous.

The cause of our party

Before them lay an entire world and no competitors but the birds, which were already bound in adaptations to flight. The mammals, marvelously equipped for terrestrial life, could not fail. Adaptive radiation began in very late Cretaceous time. By the close of the Paleocene, eighteen orders had appeared. Nine of these, mostly large, precociously specialized herbivores, later became extinct, and most of the earliest suborders and superfamilies of the other nine were supplanted. The Eocene was a transitional stage, with evolution of the modern orders and suborders well under way. In general, the archaic orders are heavy-bodied and short-limbed. Their brains were small. The teeth of herbivores were short and had a simple pattern of cones or ridges not well adapted to tough, abrasive plants. The modern mammals typically are light-bodied (at least primitively), long-limbed, and large brained. The herbivores evolved high-crowned teeth with complex crown patterns for grinding.

The complexity of the mammalian radiations and the extensive knowledge of fossil mammals preclude all but the most general description here (Figures 21-5 and 21-6). The origins of the different mammalian orders, whether "archaic" or modern are not well known. The order Carnivora is represented in the late Paleocene by a primitive family, the Miacidae, probably derived from still more primitive Cretaceous or Paleocene condylarths. The archaic carnivores, the creodonts, apparently evolved independently from the basic insectivore stock in late Cretaceous. The condylarths, an archaic herbivore order, diverged from the placental stock

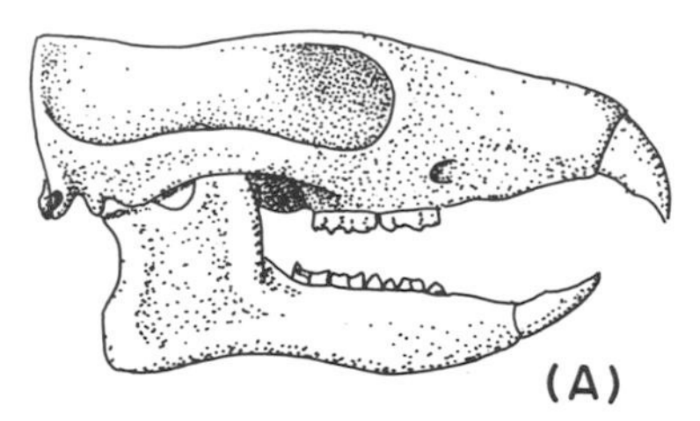

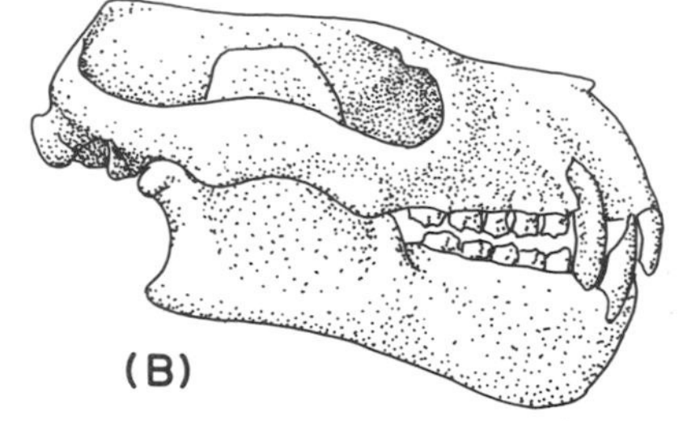

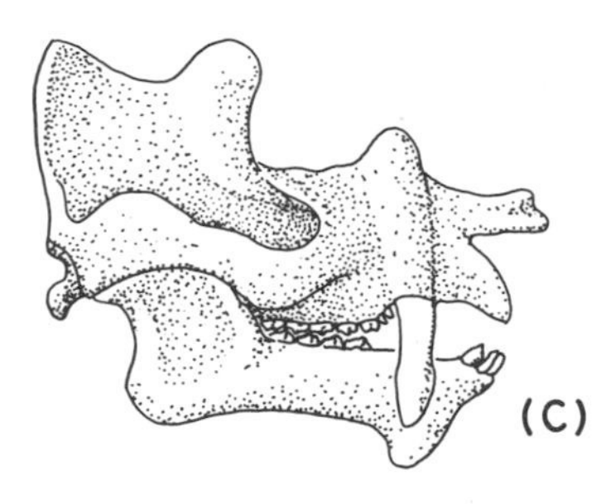

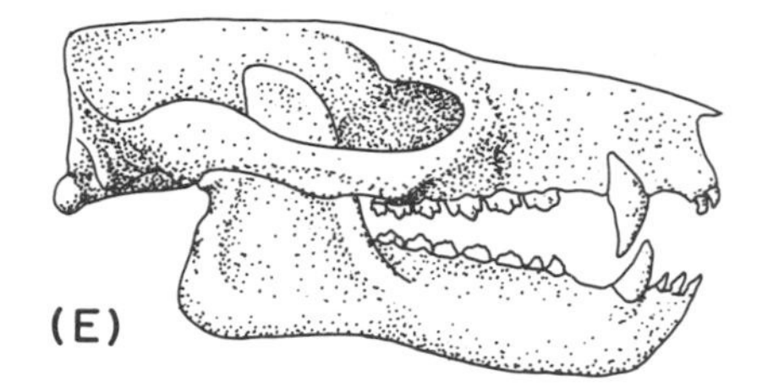

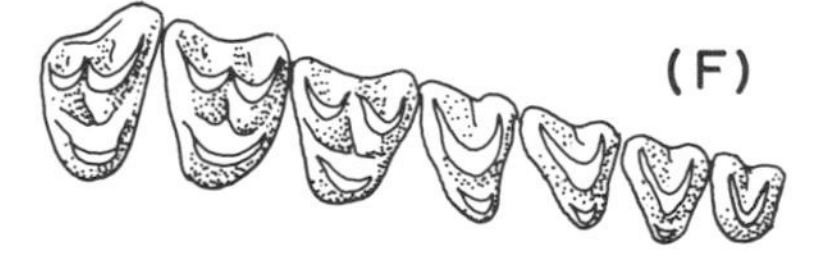

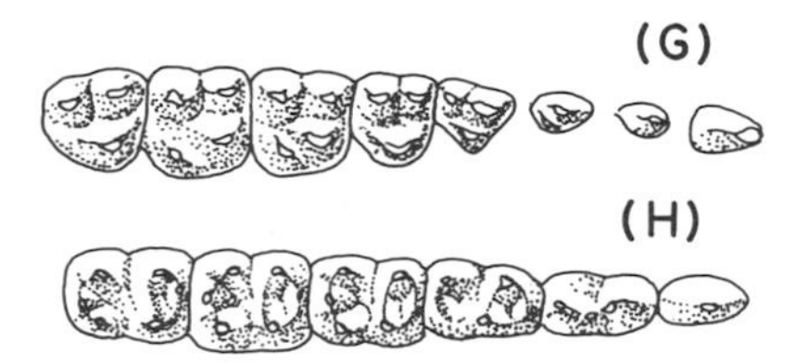

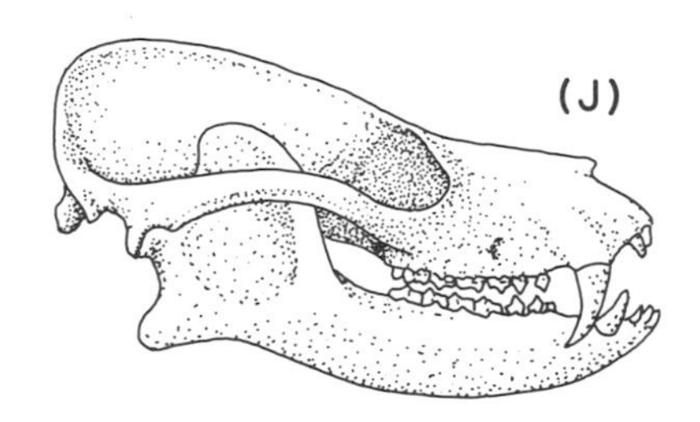

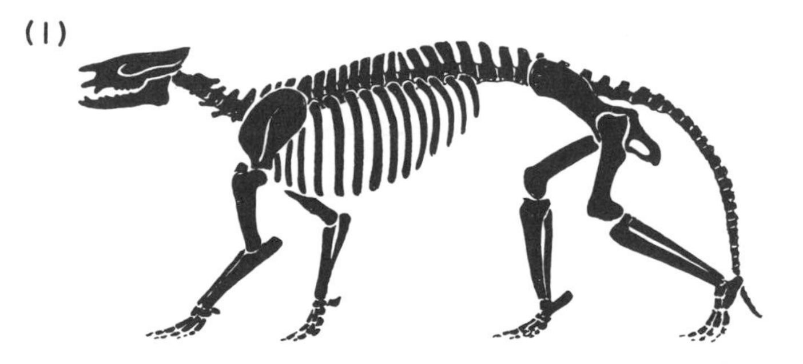

Fig. 21-5.

before the very late Cretaceous—perhaps from the leptictid insectivores. The odd-toed hoofed animals, the perissodactyls, may have evolved from the condylarth stock; the even-toed ungulates, Order Artiodactyla, almost certainly did. Probably some extinct orders of South American herbivores also evolved from the Condylarthra. The elephants (and some obscure relatives of theirs), the sea cows, and some extinct South American orders may form another group with a primitive condylarthran ancestry. Two archaic orders of herbivores known only from the early Cenozoic may have deived from the condylarthran stock, but are equally likely to be separate lines from the late Cretaceous radiation.

The primates, bats, edentates (anteaters, armadillos, and sloths both tree and ground), and perhaps the whales may have evolved as independent lines from the same Cretaceous stock, the leptictids, that produced the condylarths. The "modern" insectivores, hedgehogs, shrews, and moles, show somewhat greater affinities to the ancestral creodonts, the deltatherids, than to the leptictids. The rabbits and rodents appear suddenly at the end of the Paleocene without apparent affinities to each other beyond a possible source in the leptictid insectivores.

Paleoecology, paleogeography, and evolution

Cenozoic terrestrial deposits cover wide areas of western North America, central Asia, and southern South America. Less extensive but very significant deposits occur on the other continental areas. A very large number of fossils has been collected. Because the complex mammalian molar tooth is distinctive even for genera and species, fragmentary jaws and isolated teeth useless in any other vertebrate class are of great value in mammalian paleontology. As a consequence, some evolutionary lines, e.g., the horses, have been traced out in detail; enough specimens exist for similar studies in many other groups. This abundance of material

Fig. 21-5. THE ARCHAIC MAMMALIAN RADIATION. **(A)** A tillodont, *Tillotherium;* middle Eocene; lateral view of skull; length 32 cm. A herbivore with rodent-like incisors. **(B)** A taeniodont, *Psittacotherium;* middle Paleocene; length of skull 22 cm. A herbivore. **(C)** A dinocerate, *Uintatherium; middle* Eocene; length of skull approximately 75 cm. A large herbivore. **(D)** Right upper cheek teeth of *Uintatherium.* **(E)** and **(F)** A pantodont, *Pantolambda;* a moderate-sized herbivore. **(E)** Skull, length of skull approximately 1.5 cm. **(F)** Upper right cheek teeth. **(G)** through **(I)** A condylarth, *Phenacodus;* late Paleocene to early Eocene. **(G)** Upper right cheek teeth. **(H)** Lower left cheek teeth. **(I)** Skeleton of *Phenacodus*—length about 1.6 m. An extremely primitive herbivore near the ancestry of the carnivores, artiodactyls, perissodactyls, and several other herbivore orders. **(J)** A primitive carnivore, *Deltatherium;* middle Paleocene. [**(A)** *after* Marsh; **(B)** *after* Matthew *and* Wortman; **(C)** *after various sources;* **(D)-(H)** *after* Matthew; **(I)** *after* Gregory; *Evolution Emerging.* Copyright, The American Museum of Natural History, New York. Used with permission; **(J)** *after* Matthew.]

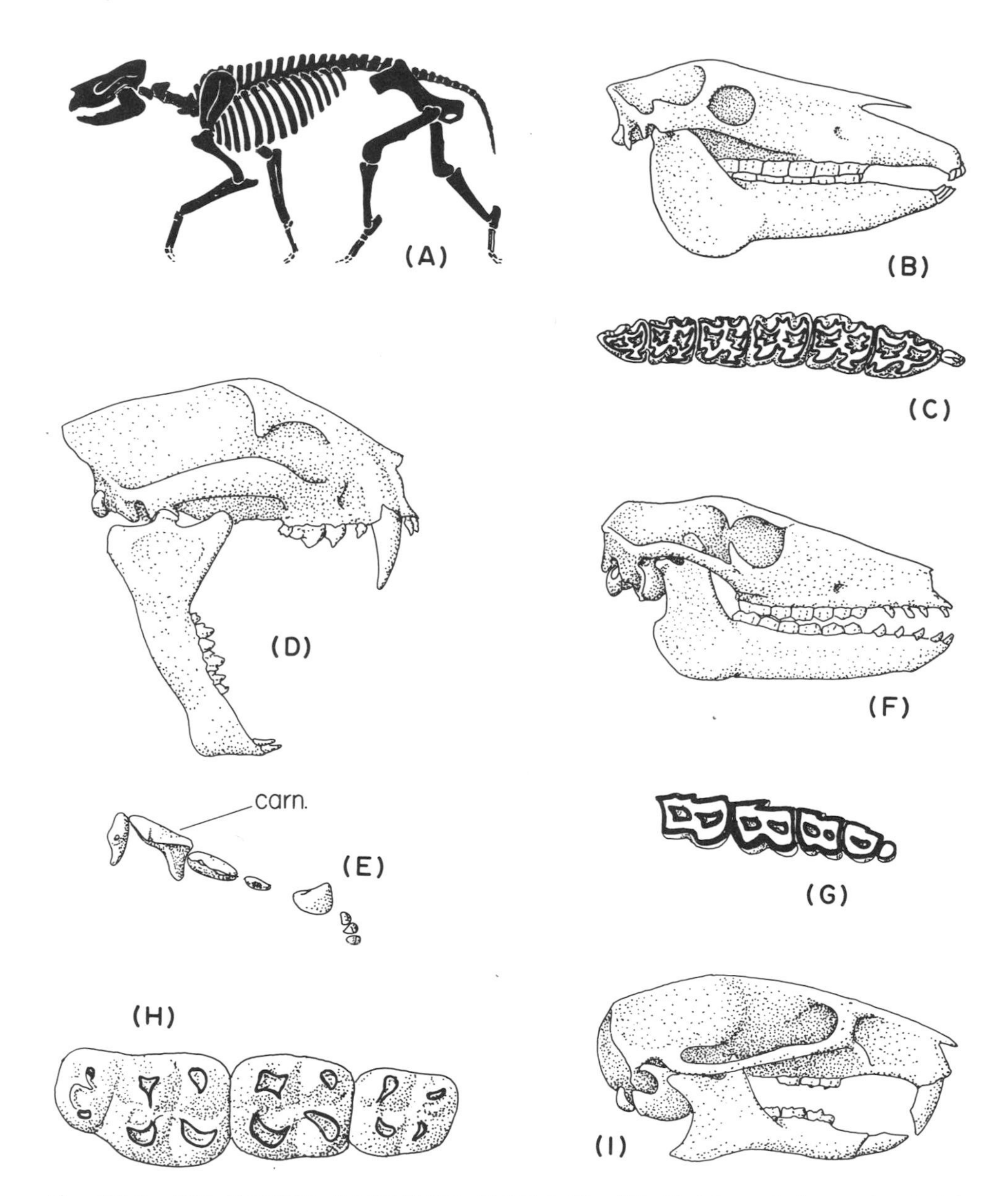

Fig. 21-6. THE MODERN MAMMALIAN RADIATION. **(A)** Skeleton of the perissodactyl, *Hyracodon,* an early and middle Oligocene herbivore. **(B)** and **(C)** *Equus,* another perissodactyl; late Pliocene to Recent; **(B)** is a lateral view of the skull, **(C)** is a crown view of the upper right cheek teeth; length of skull 55 cm. The enamel in the teeth is shown in black, and the dentine in white. **(D)** and **(E)** An Oligocene carnivore, *Dinictis.* **(D)** Lateral view of skull—length 13 cm. **(E)** Upper right dentition. **(F)** An artiodactyl, *Poebrotherium;* middle Oligocene; length of skull about 12.5 cm. A small primitive camel. **(G)** *Camelops;* Pleistocene; upper right cheek teeth. An advanced camel. **(H)** Upper left cheek teeth of a rodent, *Cricetops,* a late Oligocene genus. **(I)** Skull of *Cricetops,* length approximately 2.4 cm. [**(A)** *after* Scott; **(C)** *after* Simpson; **(D)** *and* **(E)** *after* Matthew; **(F)** *after* Wortman; **(G)** *after* Merriam; **(H)** *and* **(I)** *after* Schaub.]

taken with the great knowledge of functions, adaptations, and ecology of recent mammals has made the mammals the most useful group of animals for evolutionary studies.

Of particular importance, both for current knowledge of evolutionary processes and for future study, are the relations between these phylogenetic lines and changes in the physical environment, the natural societies, and continental relationships.

For example, North America and Eurasia were connected intermittently during the Cenozoic. Each gained immigrants during the connections. Some of these immigrants replaced native species; others filled unoccupied adaptive zones. This intensified selection by bringing together competitors that had evolved in different communities. As a consequence, when northern hemisphere mammals reached South America in the Pliocene and Pleistocene, they rapidly disposed of the South American forms, which had had less competition and thus were subjected to lower selective pressures.

Within the North American-Eurasian faunal system, we can recognize two or three major shifts related to climatic changes—warm and moist in the early Cenozoic, warm and arid during the middle Cenozoic, and cool and arid in the last epochs. The hunter returning from a safari on the modern African plain would find the Pliocene and Pleistocene mammals of North America quite commonplace, though they would be small and less specialized; but he would discover the Paleocene and Eocene faunas to be very strange.

By and large, the Paleocene and Eocene mammals appear to be forest

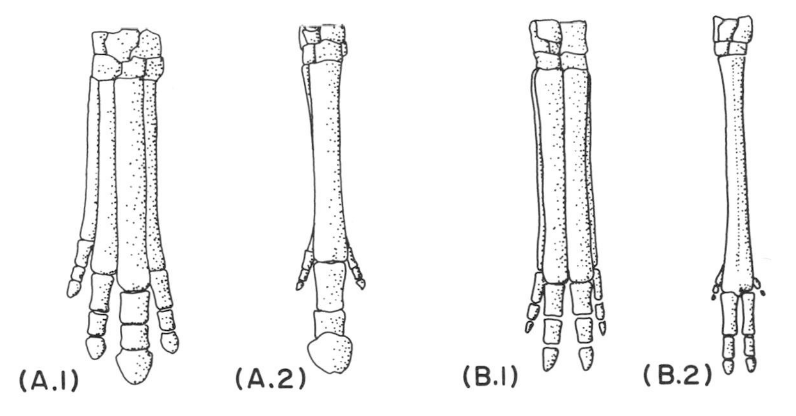

Fig. 21-7. EVOLUTION OF THE FEET IN THE HOOFED MAMMALS. **(A.1)** Forefoot of a primitive perissodactyl (and ancestral horse) *Hyracotherium.* **(A.2)** Forefoot of an advanced horse, *Merychippus.* **(B.1)** Forefoot of a primitive artiodactyl, *Leptomeryx.* **(B.2)** Forefoot of an advanced artiodactyl, *Merycodus.* [**(A.1)** *after* Cope; **(A.2)** *after* Osborn; **(B.1)** *after* Romer, *Vertebrate Paleontology,* reprinted by permission of the University of Chicago Press; copyright, 1966, University of Chicago Press. All rights reserved. **(B.2)** *after* Matthew.]

animals. Arboreal types like the primates are moderately abundant. The fossil reptiles, particularly the crocodilians, and the plants from the same beds indicate a moist, subtropical climate in western North America, as far north as Alberta. The expansion of the modern orders and suborders appears to connect with the spread of grassland in Oligocene and later times. The short-legged, splay-footed jungle browsers (leaf eaters) gave place to long-legged, hoofed grazers (grass eaters) as shown in Figure 21-7. The clumsy, small-brained carnivores fell before the agile, clever dogs and cats.

Less is known of mammalian evolution in the southern continents. Africa must have been isolated during parts of the Cenozoic, for several unique groups, most strikingly the elephants, apparently evolved there. On the other hand, the persistence of warm-humid and warm-arid environments in Africa provided a refuge for groups that disappeared from the northern continents.

Australia and South America were still more isolated from the main stream of mammalian evolution. A marsupial population reached Aus-

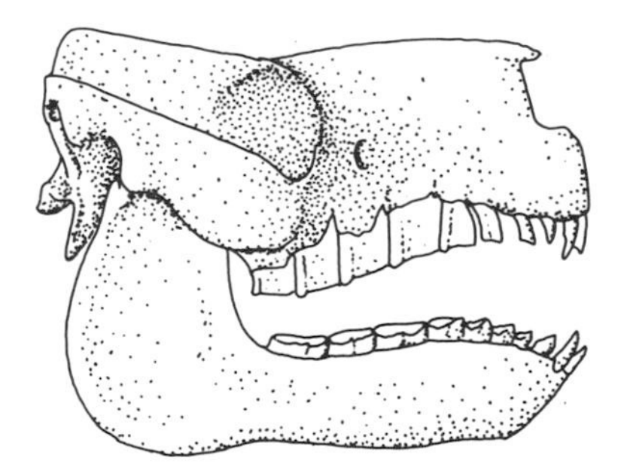

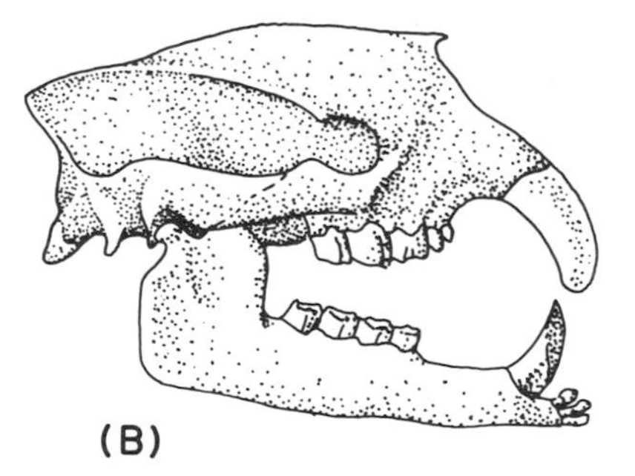

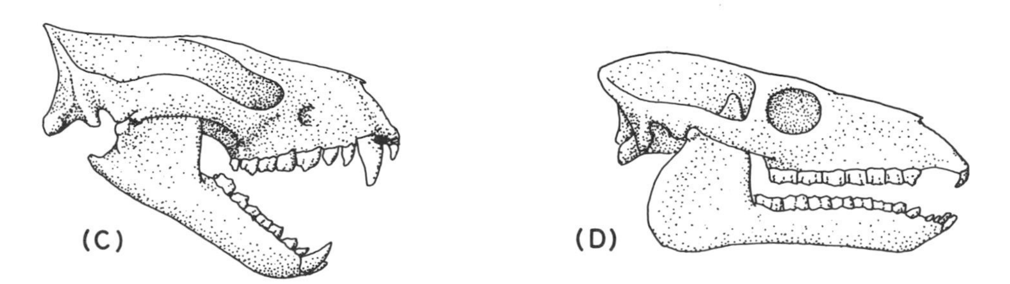

Fig. 21-8. THE SOUTH AMERICAN MAMMALIAN RADIATION. **(A)** A large herbivore (notoungulate) *Nesdon;* early to middle Miocene; length 38 cm. **(B)** A large herbivore (astrapothere) *Astrapotherium;* late Oligocene to late Miocene; length 67 cm. **(C)** A marsupial carnivore, *Borhyaena;* late Oligocene to early Miocene; length 22 cm. **(D)** A horse-like herbivore (a litoptern) *Thoatherium;* early Miocene; length 17 cm. [**(A)**, **(B)**, *and* **(D)** *after* Scott; **(C)** *after* Sinclair.]

tralia at an early date—a date by inference, for the earliest fossil localities are early Miocene—and evolved in isolation into a wide variety of adaptive types. South America received a partial sample of North American placental mammals in the Paleocene. In the subsequent isolation, these species, primitive condylarths and insectivores, evolved a wide variety of herbivores and some omnivores (Figure 21-8). Some of these resemble northern types; others are quite different. No placental carnivores appear in South America before the Pliocene, but their place was taken by carnivorous marsupials.

The sociable ape

The jungles of Africa, South America, and South Asia harbored, among other refugees, a variety of primates. The order is a very ancient one—part of the Paleocene deployment in a forest environment. Fossil primates are relatively rare, but the lineage is fairly obvious—from insectivore to lemur to monkey to ape (Figure 21-9). In response to selection in the arboreal environment, the line evolved prehensile hands and feet (and some, prehensile tails), large eyes with stereoscopic vision, and a notably large brain. Within the tropical forests they thrived, but an event more important to us occurred along the grassland-forest boundary. This environment placed in action a different selective factor. Primates could have become, like dogs, runners on the plains; but the other mammals, both herbivores and carnivores, had a long head start. Primates survived, therefore, by clinging to the forest border, to its plainward extension along rivers or to the "forest" of rocks on isolated inselbergs or cliffs. The monkey species that shifted to this marginal environment were initially quadrupeds and remained so. Their descendants are the baboons.

The apes, on the other hand, attained a partial bipedal habit, and their "ground ape" descendants emphasized this adaptively. The history of this adaptation is fragmentary. The oldest fossil apes are from the early Oligocene of North Africa. Probable ancestors of modern arboreal apes have been collected from the Miocene—these may include ground ape ancestors as well. *Australopithecus,* man-like apes from the early Pleistocene, probably includes human progenitors. True men, though of a different species and possibly a different genus, appeared well back in the Pleistocene; *Homo sapiens* probably antedates the last glacial age.

The human evolutionary lineage attained three striking structural changes: the modification of the prehensile simian foot to a more rigid walking foot with small stiff toes, an arch, and a heel; the rotation and expansion of the pelvis to support the body in an upright position; and the enlargement of the brain case. *Australopithecus* approaches *Homo* in foot and pelvic structure, though the brain is only about half the size of

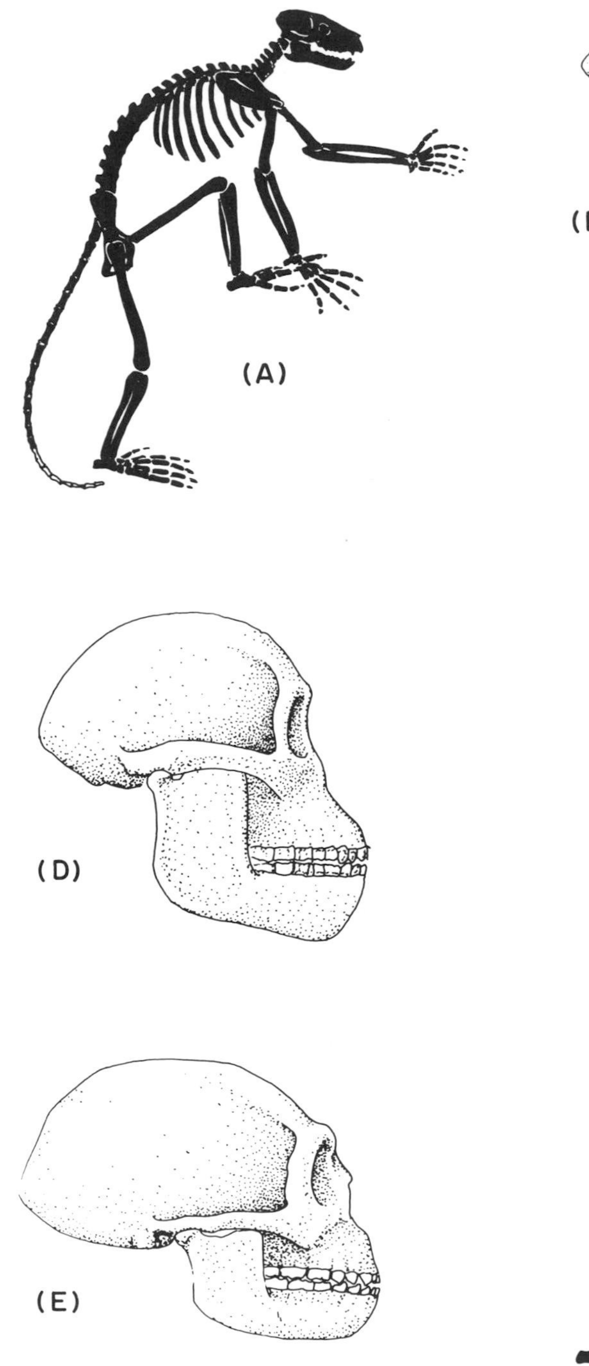

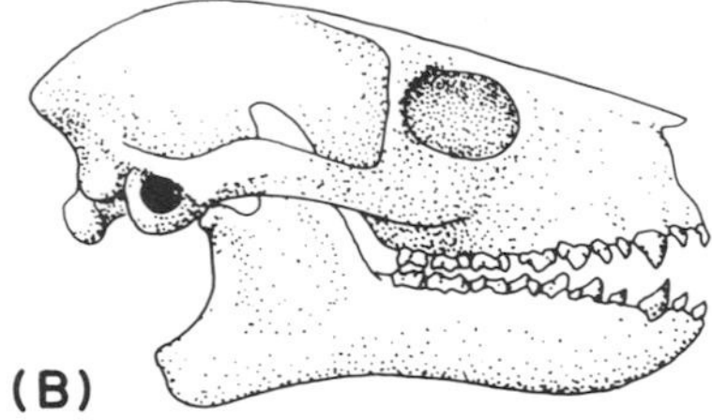

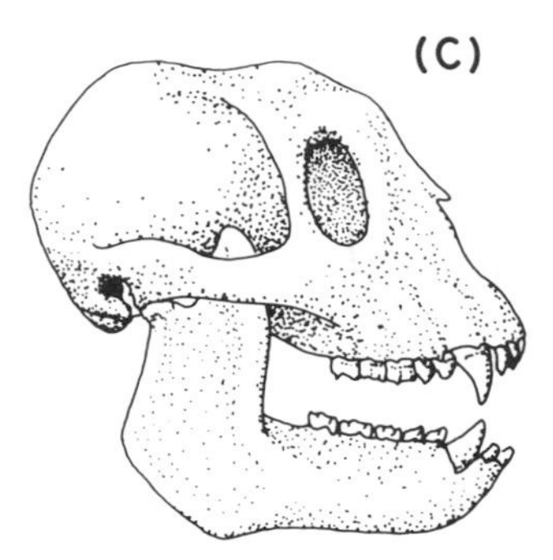

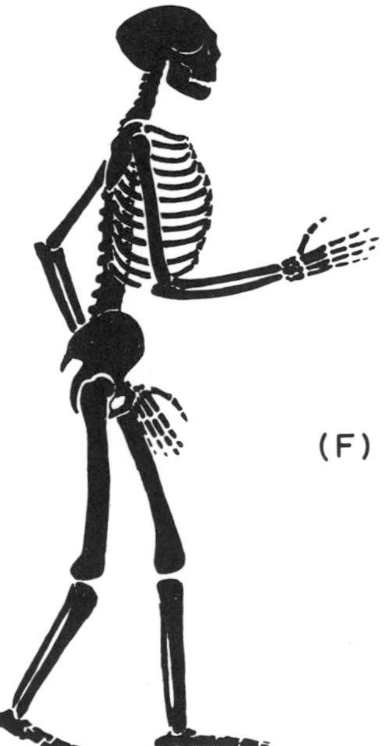

Fig. 21-9.

that in *Homo sapiens*. The middle Pleistocene fossil man *Homo* (*Pithecanthropus*) *erectus* has essentially a "modern" skeleton, though the brain is slightly smaller.

Australopithecus offers special opportunities and difficulties in the interpretation of human evolution. Nearly all of the material is from Africa—cave deposits in South Africa and lake border sediments in East Africa. A standard faunal sequence is not yet available to establish correlation between the two areas or even between the confusing stratigraphy of individual caves. Many "species" were based on archaic typological concepts rather than realistic population models. Phylogenies were derived from unweighted morphologic similarities and differences without reference to evolutionary models. In South Africa, however, *Australopithecus* comprised two morphologically and temporally distinct groups: the early, small *A. africanus* and the late, large *A. robustus*. The *A. robustus* specimens are probably late early Pleistocene in age; the *A. africanus* may be slightly or considerably older. A few very early Pleistocene australopithecine fragments have been collected in East Africa, but the well-known material from there is dated as about middle early Pleistocene and includes specimens that could be reasonably expected from an *A. robustus* population. With the East African *A. robustus* are other materials that might be considered either very advanced *Australopithecus* or a primitive *Homo*. This material, called *Homo habilis* by Leakey (but perhaps better classified as an early *H. erectus*), provides an intermediate between *A. africanus* and advanced *H. erectus* known from Africa, the East Indies, and southern Eurasia. If the South African *A. africanus* is as old as very early Pleistocene, the later stages of human evolution would be fully recorded in the *A. africanus—H. erectus—H. sapiens* sequence; *A. robustus* in any case represents a sterile side branch.

The ground-dwelling apes probably lived, as do the modern ground-dwelling primates, baboons and men, in social groups. In such groups, intelligent cooperation could offset physical disadvantage. Even the cleverest brain helps little when a baboon is caught alone in the open by a lion. Social organization reduces the chance of such an occurrence, but social organization also demands further mental adaptations. One suspects that man was a social animal before he could become a man.

Even so, what are baboons or ground apes alongside lions or water buffalo? Little except intelligent and therefore adaptable. But conditions

---

Fig. 21-9. THE PRIMATES. **(A)** *Notharctus;* early to middle Eocene. A primitive primate near the ancestry of the more advanced groups. **(B)** Skull of *Notharctus*—length slightly over 7 cm. **(C)** *Mesopithecus;* early Pliocene; length of skull about 7.5 cm. A monkey. **(D)** *Australopithecus;* Pleistocene; length approximately 17 cm. A very man-like ape or ape-like man. **(E)** *Homo erectus* Pleistocene. A primitive man. **(F)** Skeleton of *Homo sapiens neanderthalenis.* A primitive subspecies of our species. [**(A)** *and* **(B)** *after* Gregory, *Evolution Emerging.* Copyright, The American Museum of Natural History, New York. Used with permission. **(C)** *after* Gaudery; **(D)** *after* Broom *and* Robinson; **(E)** *after* Weidenreich.]

arise where adaptability is, itself, a superior adaptation. Those conditions arose at the close of the Pliocene. The primitive *Australopithecus* underwent an abortive radiation, one line leading to the short-lived *"robustus,"* the other to *Homo*. *Homo erectus* dispersed throughout Africa and southern Eurasia and evolved through the later Pleistocene toward the *Homo sapiens* level—attained in some local populations as much as three hundred thousand years ago. The simple hunting-collecting ecology of the Australopithecines was replaced by big-game hunting and then, after the last major glacial episode, by agriculture.

Here endeth the lesson.

REFERENCES

Brink, A. S. 1957. "Speculations on Some Advanced Mammalian Characteristics in the Higher Mammal-like Reptiles," *Palaeont. Africana*, vol. 4, pp. 77-96.

Caspari, E. 1963. "Selective Forces in the Evolution of Man," *American Nat.*, vol. 97, pp. 5-14.

Crompton, A. W. and F. A. Jenkins, Jr. 1967. "American Jurassic Symmetrodonts and Rhaetic 'Pantotheres'," *Science*, vol. 155, pp. 1006-1009.

Cys, J. M. 1967. "The Inability of Dinosaurs to Hibernate as a Possible Key Factor in Their Extinction," *Jour. of Paleontology*, vol. 41.

Hopson, J. A. 1964. "The Braincase of the Advanced Mammal-like Reptile *Bienotherium*," *Postilla, Peabody Mus. of Nat. Hist.*, no. 87, pp. 1-30.

Howell, F. C. and F. Bourliere (Eds.) 1964. *African Ecology and Human Evolution*. Chicago: Aldine Press.

Kermack, K. A. 1965. "The Origin of Mammals," *Science Jour.* (Sept., 1965), pp. 66-72.

Morris, W. J. 1966. "Fossil Mammals from Baja California: New Evidence on Early Tertiary Migrations," *Science*, vol. 153, pp. 1376-1378.

Pilbeam, D. R. and E. L. Simons. 1965. "Some Problems of Hominid Classification," *American Scientist*, vol. 53, pp. 237-259.

Radinsky, L. B. 1966. "The Adaptive Radiation of the Phenacodontid Condylarths and the Origin of the Perissodactyla," *Evolution*, vol. 20, pp. 408-417.

Robinson, J. T. 1962. "Australopithecines and the Origin of Man," *Ann. Rept., U.S. National Mus. for 1961*, pp. 479-500.

Russell, L. S. 1965. "Body Temperature of Dinosaurs and Its Relationship to Their Extinction," *Jour. of Paleontology*, vol. 39, pp. 497-501.

Sloan, R. E. and L. Van Valen. 1965. "Cretaceous Mammals from Montana," *Science*, vol. 148, pp. 220-227.

Van Valen, L. 1966. "*Deltatheridia*: a New Order of Mammals," *Bull. American Mus. of Nat. Hist.*, vol. 132, pp. 1-126. Includes a discussion of the origin of modern mammalian orders.

Watson, D. M. S. 1951. See ref. p. 455.

See also the references for Chapter 19.

Morphologic terms mentioned only in glossaries or illustrations are not indexed. References to illustrations are italicized.